Recent Advances in QSAR Studies

CHALLENGES AND ADVANCES IN COMPUTATIONAL CHEMISTRY AND PHYSICS

Volume 8

Series Editor:

JERZY LESZCZYNSKI

Department of Chemistry, Jackson State University, U.S.A.

For further volumes:
http://www.springer.com/series/6918

Recent Advances in QSAR Studies

Methods and Applications

Edited by

Tomasz Puzyn
University of Gdańsk, Gdańsk, Poland

Jerzy Leszczynski
Jackson State University, Jackson, MS, USA

Mark T.D. Cronin
Liverpool John Moores University, Liverpool, UK

 Springer

Editors
Dr. Tomasz Puzyn
Laboratory of Environmental
  Chemometrics
Faculty of Chemistry
University of Gdańsk
ul. Sobieskiego 18/19
80-952 Gdańsk
Poland
puzi@qsar.eu.org

Prof. Jerzy Leszczynski
Interdisciplinary Nanotoxicity
  Center
Department of Chemistry
Jackson State University
1325 Lynch St
Jackson, MS 39217-0510
USA
jerzy@icnanotox.org

Dr. Mark T. D. Cronin
School of Pharmacy and Chemistry
Liverpool John Moores University
Byrom Street
Liverpool L3 3AF
England
m.t.cronin@ljmu.ac.uk

ISBN 978-1-4020-9782-9   e-ISBN 978-1-4020-9783-6
DOI 10.1007/978-1-4020-9783-6
Springer Dordrecht Heidelberg London New York

Library of Congress Control Number: 2009930952

Springer is part of Springer Science+Business Media (www.springer.com)

PREFACE

Since the inception of this volume, the world's financial climate has radically changed. The emphasis has shifted from booming economies and economic growth to the reality of recession and diminishing outlook. With economic downturn comes opportunity, in all areas of chemistry from research and development through to product registration and risk assessment, replacements are being sought for costly time-consuming processes. Leading amongst the replacements are models with true predictive capability. Of these computational models are preferred.

This volume addresses a broad need within various areas of the "chemical industries", from pharmaceuticals and pesticides to personal products to provide computational methods to predict the effects, activities and properties of molecules. It addresses the use of models to design new molecules and assess their fate and effects both to the environment and to human health. There is an emphasis running throughout this volume to produce robust models suitable for purpose. The volume aims to allow the reader to find data and descriptors and develop, discover and utilise valid models.

Gdańsk, Poland
Jackson, MS, USA
Liverpool, UK
May 2009

Tomasz Puzyn
Jerzy Leszczynski
Mark T.D. Cronin

CONTENTS

13 The Role of QSAR Methodology in the Regulatory Assessment
    of Chemicals 367

Andrew Paul Worth

14 Nanomaterials – the Next Great Challenge for QSAR Modelers ... 383

*Tomasz Puzyn, Agnieszka Gajewicz, Danuta Leszczynska,
and Jerzy Leszczynski*

Part I
Theory of QSAR

CHAPTER 1

QUANTITATIVE STRUCTURE–ACTIVITY RELATIONSHIPS (QSARs) – APPLICATIONS AND METHODOLOGY

MARK T. D. CRONIN

School of Pharmacy and Chemistry, Liverpool John Moores University, Liverpool L3 3AF, England, e-mail: m.t.cronin@ljmu.ac.uk

Abstract: The aim of this introduction is to describe briefly the applications and methodologies involved in (Q)SAR and relate these to the various chapters in this volume. This chapter gives the reader an overview of how, why and where in silico methods, including (Q)SAR, have been utilized to predict endpoints as diverse as those from pharmacology and toxicology. It provides an illustration of how all the various topics in this book interweave to form a single coherent area of science.

Keywords: QSAR, In silico methods, Resources for QSAR

1.1. INTRODUCTION

If we can understand how a molecular structure brings about a particular effect in a biological system, we have a key to unlocking the relationship and using that information to our advantage. Formal development of these relationships on this premise has proved to be the foundation for the development of predictive models. If we take a series of chemicals and attempt to form a *quantitative relationship* between the biological effects (i.e. the *activity*) and the chemistry (i.e. the *structure*) of each of the chemicals, then we are able to form a *quantitative structure–activity relationship* or QSAR.

Less complex, or quantitative, understanding of the role of structure to govern effects, i.e. that a fragment or sub-structure could result in a certain activity, is often simply termed a *structure–activity relationship* or SAR. Together SARs and QSARs can be referred to as (Q)SARs and fall within a range of techniques known as in silico approaches. Generally, although there is no formal definition, in silico includes SARs and QSARs, as well as the use of existing data (e.g. searching within databases), category formation and read-across. It also borders into various other areas of chemoinformatics and bioinformatics.

3

T. Puzyn et al. (eds.), Recent Advances in QSAR Studies, 3–11.
DOI 10.1007/978-1-4020-9783-6_1, © Springer Science+Business Media B.V. 2010

A (Q)SAR comprises three parts: the (activity) data to be modelled and hence predicted, data with which to model and a method to formulate the model. These three components are described below and in greater detail in subsequent chapters.

1.2.   PURPOSE OF QSAR

QSAR should not be seen as an academic tool to allow for the post-rationalization of data. We wish to derive the relationships between molecular structure, chemistry and biology for good reason. From these relationships we can develop models, and with luck, good judgment and expertise these will be predictive. There are many practical purposes of a QSAR and these techniques are utilized widely in many situations. The purpose of in silico studies, therefore, includes the following:

- To predict biological activity and physico-chemical properties by rational means.
- To comprehend and rationalize the mechanisms of action within a series of chemicals.

Underlying these aims, the reasons for wishing to develop these models include

- Savings in the cost of product development (e.g. in the pharmaceutical, pesticide, personal products, etc. areas).
- Predictions could reduce the requirement for lengthy and expensive animal tests.
- Reduction (and even, in some cases, replacement) of animal tests, thus reducing animal use and obviously pain and discomfort to animals.
- Other areas of promoting green and greener chemistry to increase efficiency and eliminate waste by not following leads unlikely to be successful.

1.3.   APPLICATIONS OF QSAR

The ability to predict a biological activity is valuable in any number of industries. Whilst some QSARs appear to be little more than academic studies, there are a large number of applications of these models within industry, academia and governmental (regulatory) agencies. A small number of potential uses are listed below:

- The rational identification of new leads with pharmacological, biocidal or pesticidal activity.
- The optimization of pharmacological, biocidal or pesticidal activity.
- The rational design of numerous other products such as surface-active agents, perfumes, dyes, and fine chemicals.
- The identification of hazardous compounds at early stages of product development or the screening of inventories of existing compounds.
- The designing out of toxicity and side-effects in new compounds.
- The prediction of toxicity to humans through deliberate, occasional and occupational exposure.
- The prediction of toxicity to environmental species.
- The selection of compounds with optimal pharmacokinetic properties, whether it be stability or availability in biological systems.

- The prediction of a variety of physico-chemical properties of molecules (whether they be pharmaceuticals, pesticides, personal products, fine chemicals, etc.).
- The prediction of the fate of molecules which are released into the environment.
- The rationalization and prediction of the combined effects of molecules, whether it be in mixtures or formulations.

The key feature of the role of in silico technologies in all of these areas is that predictions can be made from molecular structure alone.

1.4.   METHODS

Predictive models of all types are reliant on the data on which they are based, the technique to develop the model and the overall quality of the information including the item to be modelled. In silico models for the prediction of the properties and effects of molecules are no different. In almost all cases two types of information are required for a model (the effect to be modelled and descriptors on the chemicals) and a technique(s) to formulate the relationship(s). These are denoted in Figure 1-1 in a typical spreadsheet for organizing the data. The data to be modelled are denoted as the X-matrix, the descriptors as the Y-matrix. From such a matrix various types of relationship may be obtained by statistical, or other, means. For instance, a structure–activity relationship will be formed for a categorical endpoint, e.g. active/non-active or toxic/non-toxic. In this case a molecular fragment or substructure is associated with an effect. A quantitative structure–activity relationship is based on a continuous endpoint, e.g. potency where activity (X) is a function of one or more descriptors (Y).

To develop a SAR, as few as a single compound might be required – should there be a very firm basis (such as a well-established mechanism of action) for developing the relationship. For instance, if a compound is known to elicit a particular effect, and the structural determinant is recognized, that structural fragment can be extracted. This may be in the form of a "structural alert" which can be coded easily into software. Obviously, the greater the number of compounds with the same structural determinant demonstrating the same effect, the greater the confidence that

Chemical Identifier	Activity (to be modelled)	Property/ Descriptor/ Fragment 1	Property/ Descriptor/ Fragment 2	Property/ Descriptor/ Fragment 3	...	Property/ Descriptor/ Fragment n
Molecule i	X_i	Y_{1i}	Y_{2i}	Y_{3i}	...	Y_{ni}
Molecule ii	X_{ii}	Y_{1ii}	Y_{2ii}	Y_{3ii}	...	Y_{nii}
Molecule iii	X_{iii}	Y_{1iii}	Y_{2iii}	Y_{3iii}	...	Y_{niii}
...	...	...	...	...	...	
Molecule n	X_n	Y_{1n}	Y_{2n}	Y_{3n}	...	Y_{nn}

Figure 1-1. Typical data matrix for a (Q)SAR study

can be demonstrated in the alert. The formation of SARs is usually appropriate for a qualitative (i.e. yes/no; active/inactive; presence of toxicity/absence of toxicity, etc.) endpoint.

To develop a QSAR, a more significant number of compounds is required to develop a meaningful relationship. An often asked question is "how many compounds are required to develop a QSAR?" There is no direct and simple response to this question – other than "as many as possible!" To provide some guide, it is widely accepted that between five and ten compounds are required for every descriptor in a QSAR [1, 2]. This does suggest that a one descriptor regression-based QSAR could be developed on five compounds. This is possible, but is very reliant on issues such as data distribution and range. Ideally "many more" compounds are required to obtain statistically robust QSARs, with some modelling techniques being considerably more data hungry than regression analysis.

In the history of developing in silico models, there have been many types of information integrated into (Q)SARs. These are summarized in Table 1-1. The biological effects are normally (though not exclusively) the property to be modelled; some aspect from the physical or structural chemistry of the molecules is related to the effects. Readers are welcome to extend this list according to their experience and requirements!

There has been a wide range of modelling approaches. A brief overview of these is given in Table 1-2. These can be very simplistic to extremely complex.

Table 1-1. Types of information included in in silico modelling approaches and reference to chapters for further reading

● Data to be modelled

    o Pharmacological effects (Chapter 9)
    o Toxicological effects (Chapters 7, 11, 12 and 14)
    o Physico-chemical properties (Chapters 12 and 14)
    o Pharmacokinetic properties governing bioavailability (Chapters 9 and 10)
    o Environmental fate (Chapter 12)

● Chemistry

    o Physico-chemical properties (Chapters 12 and 14)
    o Structural properties – 2-D and 3-D (Chapters 4, 5, 8 and 14)
    o Presence, absence and counts of atoms, fragments, sub-structures (Chapters 3 and7)
    o Quantum and computational chemistry (Chapters 2 and 14)

● Modelling

    o Formation of categories of "similar molecules" (Chapters 7, 13 and 14)
    o Statistical (Chapters 5, 6 and 12)
    o 3D/4D QSAR (Chapters 2, 4, 5, 9 and 14)

● Other issues

    o Data quality and reliability (Chapter 11)
    o Model and prediction reporting formats (Chapter 13)
    o Applicability domain (Chapters 12 and 13)
    o Robustness of model and validity of a prediction (Chapters 6 and 12)

Table 1-2. Summary of the main modelling approaches for the development of (Q)SARs and in silico techniques and where further details are available in this volume

(Q)SAR method	Chapters
Hansch analysis	9
Free-Wilson	9
Structural fragments and alerts	7, 12
Category formation and read-across	7
Linear regression analysis	5, 6, 12
Partial least squares	5, 6
Pattern recognition	6
Robust methods, outliers	6
Pharmacophores	4, 5, 9
3-D models	2, 4, 14
CoMFA	4

1.5. THE CORNERSTONES OF SUCCESSFUL PREDICTIVE MODELS

Predictive and intuitive models are widely used in all aspects of society and science. The user of a model accepts that it is a model and the results, or information it provides, should be used with circumspection. This is true whether one is accepting an actuarial prediction for one's pension planning or a weather forecast to determine whether to wear a raincoat. The same is true for a prediction, or any information (e.g. mechanism of action), that may be determined from a (Q)SAR. Therefore, the user must put the model in the context in which it exists and be aware of a number of possible problems and pitfalls.

Much has been written and said about the reality of using (Q)SARs. The concern is that scientists who are introduced to the field can place too much confidence in either a model or predictive system, only to see their expectations dashed. Alternatively, there is a point of view that (Q)SARs will not work and models are not to be trusted. A healthy dose of scepticism is important, but some form of balance is required to meet the hopes of the optimists and criticisms of the sceptics. In order to do that, some comment is required on the "successful use of (Q)SAR".

The requirements for a good model are quite straightforward. Some of the fundamentals are noted below and expanded upon in more detail in various chapters of this book.

(i) *The data to model.* The modeller, and user of a model, must consider the data to model. Data should, ideally, be of high quality, meaning they are reliable and consistent across the data set to be modelled. The definition of data quality is, at best, subjective and is likely to be different for any effect, endpoint or property. Therefore, the modeller or user should determine whether the data are performed in a standard manner, to a recognized protocol, and if they are taken from a single or multiple laboratories.

This author is of the belief that, within reason, poor-quality data can be used in models, but their limitations must be clearly understood, and the implications for the model appreciated. Therefore, to use a (Q)SAR successfully there should be complete access to the data used and a complete description of those data. Even then, producing public models from confidential business information may make this restrictive. To provide the source data is the responsibility of the modeller, and to assess the source data in terms of the model is the responsibility of the model user.

(ii) *Reasonable and honest use of statistics to describe a (Q)SAR*. Many (Q)SARs are accompanied by performance statistics of some kind. These statistics may assess statistical fit and predictivity for a QSAR or the predictive capability of a SAR. Generally statistics are helpful to the interpretation of a model. One would prefer to use a model with a good statistical fit between the effect to be modelled and the descriptors of the chemicals. However, it is important to ensure that the statistical fit of a model does not go beyond the experimental error of the data being modelled – should that happen it would suggest an overfit model. To develop a significant quantitative model, a significant range of effect values are ideally required. Also, one must be cautious of comparing the ubiquitous correlation coefficient between different data sets.

In the opinion of the author, whilst neither the model developer nor the user needs be a statistician, it is of great help to discuss the issues with a competent statistician. In addition, the developer or user must have confidence in the statistics they are applying and interpreting.

(iii) *The molecules for which predictions are being made must be within the applicability domain of the model*. The applicability domain of the model is the chemical, structural, molecular, biological and/or mechanistical space of the data set of the model. The definition of the applicability domain will vary for different types of model (e.g. SAR vs. QSAR), endpoints and effects. There are also a wide variety of methods to define it. The important fact is that the user of a model must assess whether a molecule is within the domain of a model, and thus how much confidence they can place in a predicted value.

(iv) *Ideally a (Q)SAR should be simple, transparent, interpretable and mechanistically relevant*. A simple model will have only one or a very small number of descriptors to form the relationship with the effect data. Transparency is usually dependent on the modelling approach itself; thus linear regression analysis can be thought of as being highly transparent, i.e. the algorithm is available, and predictions can be made easily. For the more multivariate and non-linear modelling techniques (e.g. a neural network), it is generally accepted that there is lower transparency.

The mechanistic relevance of a model is more difficult to define. Some data sets are based around a single, well-defined and understood mechanism of action. Other models comprise data where the mechanism may not be known or where there are many mechanisms. There is also a difference between biological mechanisms (e.g. receptor binding, concentration at an active site, accumulation in a membrane) and physico-chemical effects (e.g. the properties affecting solubility, ionization), which may be general across the chemical universe. There is no reason to exclude a model where the mechanism is not known or if there are multiple mechanisms. However, the advantages of a strong mechanistic basis to model are that it provides a clear capability to understand the model and should the descriptors be relevant to that mechanism it provides the user with extra confidence to use the model. Another advantage is that it can aid a priori descriptors selection.

In reality, (Q)SARs span the range from simple models to highly complex multivariate. It is important to remember that whilst a simple model is preferable in many circumstances if it provides comparable performance to something more complex, many multivariate models are routinely and successfully used. The requirements for model simplicity are highly dependent on the context and application of the model.

1.6.   A VALIDATED (Q)SAR OR A VALID PREDICTION?

Historically, much effort has been placed into performing some form of validation on a (Q)SAR. Often this has been in terms of a model's statistical fit; more recently the focus has turned to using an external test set, i.e. group of molecules not in the original data set on which the model has been developed. Confusion has arisen in some areas, due to the term "validated" which has a specific regulatory, and hence legal, connotation in replacing animal tests in toxicology.

As a result of the efforts to use (Q)SARs correctly, for the statistical validation of models it is more usual to refer to those algorithms that may be applied in drug discovery and lead optimization. Whilst statistical approaches may be applied to toxicological endpoints of regulatory significance, for the validation of a toxicological (Q)SAR to be used to assess hazard, for example for the purposes of registration of a product, a more formal validation process may be required.

In terms of toxicological predictive models, "Principles for the Validation of (Q)SARs" have been proposed by the Organization for Economic Co-operation and Development (OECD) and promoted widely. These principles are described in more detail in Chapter 13, and whilst they were originally derived with toxicity and fate endpoints in mind, they are generally applicable across all models to determine whether a (Q)SAR may be valid. The use of the OECD principles has brought to the forefront of whether a (Q)SAR can be "validated" in terms of being an acceptable alternative method. Probably of more importance is using these principles to evaluate and characterize a (Q)SAR and hence determine whether an individual prediction is valid.

1.7.   USING IN SILICO TECHNIQUES

This book will make it apparent that there are many models available for use in QSAR. Publication on paper is, of course, essential, but to make these models usable they must be presented in a user-friendly format. Thus, there have been many attempts to computerize these models. As computational power has increased, and hardware platforms became more sophisticated, the possibilities to produce useable algorithms have improved. Accessibility to software has also, of course, been made so much more convenient through the use of the Internet. As a result of the progress in these areas, many algorithms are now freely available. Sources of some, as well as other essential resources for (Q)SAR, are noted in Table 1-3.

Table 1-3. Invaluable resources for QSAR

Internet

There are obviously many Internet sites, wikis and blogs devoted to (Q)SAR, molecular modelling, drug design and predictive ADMET. Two of the most well established are

- The homepage of the International Chemoinformatics and QSAR Society: www.qsar.org – this is a good starting place for those in the field of QSAR; it also contains excellent listings of upcoming meetings and resources.
- The homepage of the Computational Chemistry List: www.ccl.net – this also contains excellent listings resources and freely downloadable software.

Journals

Papers relating to (Q)SAR are published in a very wide variety of journals from those in pure and applied chemistry to pharmacology, toxicology and risk assessment and as far as chemoinformatics and statistics. The following is a small number that is commonly used by the author; whilst the reader will hopefully find these suggestions useful, they are, by no means, an exhaustive list (see the resources section of www.qsar.org which lists over 250 journal titles).

- *Chemical Research in Toxicology*
- *Chemical Reviews*
- *Journal of Chemical Information and Modeling*
- *Journal of Enzyme Inhibition and Medicinal Chemistry*
- *Journal of Medicinal Chemistry*
- *Journal of Molecular Modelling*
- *"Molecular Informatics (formerly QSAR and Combinatorial Science)"*
- *SAR and QSAR in Environmental Research*

Books

There are many hundreds of books available in areas related to (Q)SAR. Again, the reader is referred to the resource section of www.qsar.org. A very short list is given below, clearly biased by the author's own interests and experience. Apologies are given for omission of other "favourite" or "essential" books that have not been listed.

- Cronin MTD, Livingstone DJ (eds) (2004) *Predicting Chemical Toxicity and Fate*, CRC Press, Boca Raton, FL.
- Helma C (ed) (2005) *Predictive Toxicology*, CRC Press, Boca Raton, FL.
- Livingstone DJ (1995) *Data Analysis for Chemists – Application to QSAR and Chemical Product Design*, Oxford University Press, Oxford.
- Todeschini R, Consonni V (2001) *Handbook of Molecular Descriptor*. Wiley, New York.
- Triggle DJ, Taylor JB (series eds) (2006) *Comprehensive Medicinal Chemistry II – Volumes 1–8*. Elsevier, Oxford.

Software

It is well beyond the scope or possibility of this section to note individual software for use in (Q)SAR. Experienced QSAR practitioners will no doubt be familiar with many of the freely available and commercial packages available. For the novice, in addition to the resources listed on www.qsar.org and www.ccl.net, there is information in the following chapters of this book in the three key areas to formulate a (Q)SAR:

- Activity to be modelled: Pharmacology (Chapters 4, 5, 9 and 10), ADMET (Chapters 4, 7, 10, 11, 12 and 14), physico-chemical properties (Chapters 8, 12 and 14)
- Descriptor calculation (Chapters 2, 3, 4, 5 and 14)
- Statistical analysis (Chapters 5, 6 and 12)

1.8.   NEW AREAS FOR IN SILICO MODELS

Understanding and forming the relationships between the effect of a molecule and its structure has a long history [3] – its nearly 50 years since Hansch, Fujita and co-workers first published in this area [4], over 150 years since the foundations of modern chemistry and millennia since man first determined the beneficial and harmful effects of plants. It is surprising therefore that there continues to be such continued interest in developing technologies for in silico models.

There are many reasons for the growth of in silico techniques. In particular, these can be in response to new problems. Areas where in silico approaches can play a particular role include

- integrating and harnessing new computational technologies and increasing speed and power of processing;
- ability to react to new disease states (e.g. HIV);
- ability to react to new toxicological problems (e.g. cardio-toxicity);
- modelling the new problems with regard to the impact of chemicals on the environment;
- new and emerging issues, problems and opportunities, e.g. nano-technology, properties of crystals, extension into other areas of chemistry, e.g. design of formulations;
- integration with the -omics technologies to improve all areas of molecular design.

1.9.   CONCLUSIONS

QSAR is a broadly used tool for developing relationships between the effects (e.g. activities and properties of interest) of a series of molecules with their structural properties. It is used in many areas of science. It is a dynamic area that integrates new technologies at a staggering rate. There have been many recent advances in the applications and methodologies of QSAR, which are summarized partially in Table 1-3 and more thoroughly described in this book.

REFERENCES

1. Topliss JG, Costello RJ (1972) Chance correlations in structure-activity studies using multiple regression analysis. J Med Chem 15:1066–1068.
2. Schultz TW, Netzeva TI, Cronin MTD (2003) Selection of data sets for QSARs: analyses of *Tetrahymena* toxicity from aromatic compounds. SAR QSAR Environ Res 14:59–81.
3. Selassie CD (2003) History of quantitative structure-activity relationships. In: Abraham DJ (ed) Burger's Medicinal Chemistry and Drug Discovery, 6th edn., Volume 1: Drug Discovery. John Wiley and Sons, Inc., New York.
4. Hansch C, Maloney PP, Fujita T et al. (1962) Correlation of biological activity of phenoxyacetic acids with Hammett substituent constants and partition coefficients. Nature 194:178–180.

CHAPTER 2

THE USE OF QUANTUM MECHANICS DERIVED DESCRIPTORS IN COMPUTATIONAL TOXICOLOGY

STEVEN J. ENOCH

*School of Pharmacy and Chemistry, Liverpool John Moores University, Liverpool L3 3AF, England,
e-mail: s.j.enoch@ljmu.ac.uk*

Abstract: The aim of this chapter is to outline the theoretical background and application of quantum mechanics (QM) derived descriptors in computational toxicology, specifically in (quantitative) structure–activity relationship models ((Q)SARs). The chapter includes a discussion of the mechanistic rationale for the need for such descriptors in terms of the underlying chemistry. Having established the mechanistic rationale for quantum mechanical descriptors, a brief discussion of the underlying mathematical theory to quantum mechanical methodologies is presented, the aim being to help the reader understand (in simple terms) the differences between the commonly used levels of theory that one finds when surveying the computational toxicological literature. Finally, the chapter highlights a number of (Q)SAR models in which QM descriptors have been utilised to model a range of toxicological effects

Keywords: Geometry optimisation, Semi-empirical methods, Density functional theory, Quantum mechanical descriptors

2.1. INTRODUCTION

Computational toxicology is concerned with rationalising the toxic effects of chemicals, with the hypothesis being that if the factors that are responsible for a given chemical's toxicity can be understood, then the toxicity of related chemicals can be predicted without the need for animal experiments. Unfortunately, there are many factors, some of them extremely complex, that govern whether even the simplest industrial chemical will be toxic. The majority of these factors (e.g. metabolism, bioavailability) are outside of the scope of this chapter. Instead the focus of this chapter is to highlight the importance of assessing the electronic state of a potentially toxic chemical, and how this information enables one to begin to rationalise and subsequently predict certain aspects of human health and environmental toxicology.

Knowledge of a chemical's mechanism of action is important if a chemical's potential toxic effects are to be understood. Broadly speaking, potential non-

13

T. Puzyn et al. (eds.), Recent Advances in QSAR Studies, 13–28.
DOI 10.1007/978-1-4020-9783-6_2, © Springer Science+Business Media B.V. 2010

receptor-mediated mechanisms of toxic action can be divided into non-covalent and covalent categories. One of the most important non-covalent mechanisms in aquatic systems involves the accumulation of a chemical within the cell membrane resulting in narcosis. Chemicals able to cause narcosis can be split into a number of mechanisms, the two most frequent being non-polar and polar narcosis. Non-polar narcotics are well modelled using hydrophobicity alone, whilst the modelling of the polar chemicals may require the inclusion of a parameter to account for the polarisation effect of an electronegative centre in the molecule. Such effects are well modelled using quantum mechanics derived descriptors such as E_{LUMO} and A_{max} (see Table 2-1 for definitions).

In contrast, covalent mechanisms of toxicity involve the formation of a chemical bond between proteins (or DNA) and the toxic chemical. Such mechanisms are irreversible and have little or no correlation with the chemical's hydrophobicity (assuming the hydrophobicity of the chemical is within a range that allows it to get to the reactive site). In order for a chemical to be toxic via a covalent mechanism, it must be electrophilic, that is to say some portion of it must be susceptible to attack (either directly or after either oxidative or metabolic conversion) from electron-rich amino acid (or nucleic acid) side chains. These covalent mechanisms have recently been rationalised in terms of simple electrophilic–nucleophilic organic chemistry reactions [1] (Figure 2-1).

The chemical reactions between toxicant and biomolecule can be rationalised in terms of hard–soft acid–base theory which states that for a chemical reaction to occur like should react with like, i.e. a soft electrophile (where an electrophile can be considered as an acid) prefers to react with a soft nucleophile (where the nucleophile can be considered as a base), whilst a hard electrophile preferentially reacts with a hard nucleophile [2]. This is related directly to the energies of the frontier molecular orbitals as a soft electrophile has a low E_{LUMO} which can readily interact with the energetically close high E_{HOMO} of the soft nucleophile. In contrast, a hard electrophile has a high E_{LUMO} that can readily interact with the energetically close low E_{HOMO} of a hard nucleophile.

In the simplest terms it is the relative differences between the nucleophile and electrophile orbitals that govern how reactive a given nucleophile–electrophile interaction will be (assuming factors such as entropy and steric hindrance at the reaction centre are equal). Clearly, in terms of covalent toxicity mechanisms, the more reactive a nucleophile–electrophile interaction is (in which the nucleophile is a protein or DNA and the electrophile is a chemical) the more toxic the chemical is likely to be. However, it is important to remember that toxicokinetics and toxicodynamics play an important role in a chemical's ability to produce a toxic effect, with the relative importance (compared to intrinsic reactivity) of such effects being mechanism dependent. Given the importance of the frontier molecular orbitals in hard–soft acid–base theory, it is clear that quantum mechanics methods that enable the molecular orbitals to be calculated play an important role in the rationalising and the subsequent modelling of such reactions.

Table 2-1 highlights some common descriptors used to model both covalent and non-covalent mechanisms. In addition, Schüürmann provides an excellent recent review of the theoretical background of such descriptors in more detail [3].

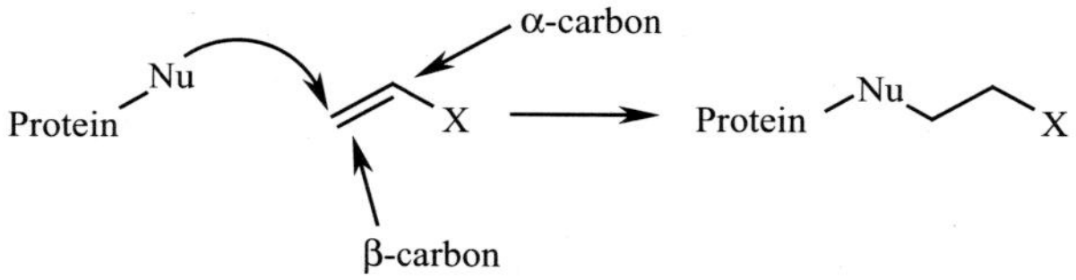

Michael addition: Characteristics: double or triple bond where X = electron withdrawing substituent (α and β alkene carbon atom as highlighted).

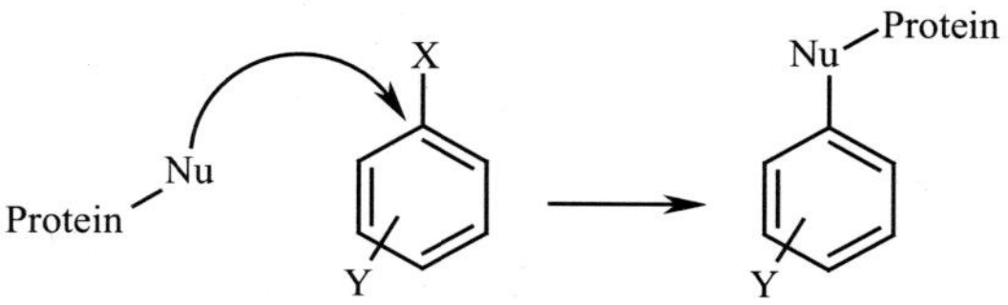

$S_N Ar$ electrophiles: Characteristics: X = halogen or pseudo-halogen. Y = (at least two) NO_2, CN, CHO, CF_3, halogen.

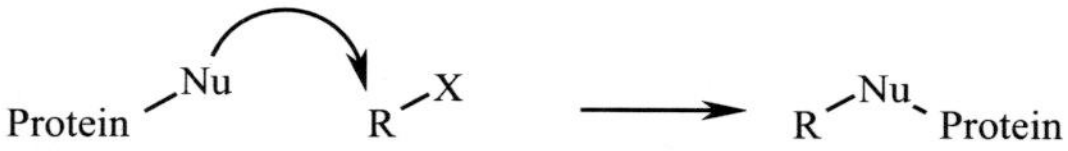

$S_N 2$ electrophiles: Characteristics: X = halogen or other electronegative leaving group.

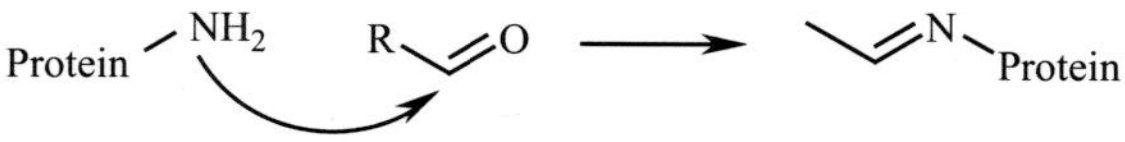

Schiff base formers: Characteristics: reactive carbonyl species such as aliphatic aldehyde or di-ketones.

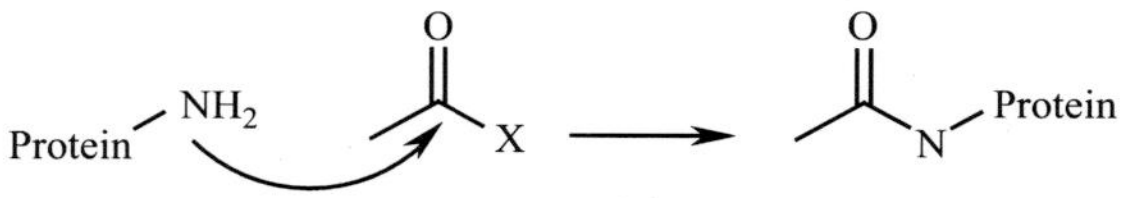

Acylating agents: Characteristics: X = halogen or electronegative leaving group

Figure 2-1. Electrophilic–nucleophilic reactions responsible for covalent mechanisms of toxic action

2.2.   THE SCHRÖDINGER EQUATION

Given the importance of the ability to calculate the electronic structure of a molecule in computational toxicology, it is important to outline, albeit briefly, the underlying theory that both the commonly used semi-empirical and density functional methods attempt to solve. The mathematics is complex and will be kept to an absolute minimum, the aim being to set the scene concerning the various components that must be dealt with if quantum mechanics is to be utilised to help understand the electronic structure of chemicals. The subsequent sections dealing with the commonly

Table 2-1. Common quantum mechanics derived molecular and atom-based descriptors

Name	Definition
E_{LUMO}	Energy of the lowest unoccupied molecular orbital
E_{HOMO}	Energy of the highest occupied molecular orbital
μ	Chemical potential (negative of electronegativity)
	$\mu = (E_{LUMO} + E_{HOMO})/2$
η	Chemical hardness
	$\eta = (E_{LUMO} - E_{HOMO})/2$
σ	Chemical softness
	$\sigma = 1 - \eta$
ω	Electrophilicity
	$\omega = \mu^2/2\eta$
AEI	Activation energy index
	$AEI = \Delta E_{HOMO-1} + \Delta E_{HOMO}$
	ΔE_{HOMO} and ΔE_{HOMO-1} are the changes in energy of the highest occupied molecular orbital and second highest occupied molecular orbital on going from the ground state to transition state in an $S_N Ar$ reaction
A_{max}	Maximum atomic acceptor superdelocalisability within a molecule, where acceptor superdelocalisability is a measure of an atom's ability to accept electron density
D_{max}	Maximum atomic donor superdelocalisability within a molecule, where donor superdelocalisability is a measure of an atom's ability to donate electron density
A_N	Atomic acceptor superdelocalisability for atom N
D_N	Atomic donor superdelocalisability for atom N
ω_m^+/ω_m^-	Atomic local philicity. Derived from Fukui functions [4], electrophilicity index ω and then applied to individual atoms using a charge scheme
Q_N	Atomic charge on atom N
B_{a-b}	Bond order between atom a and b

used semi-empirical and density functional approaches will highlight how each of these methods approximates these important mathematical components. The starting point of any discussion into quantum mechanics is always the time-independent Schrödinger equation (2-1):

$$\mathbf{H}\Psi = E\Psi \qquad (2\text{-}1)$$

where $\mathbf{H}$ is the *Hamiltonian operator*, E is the energy of the molecule and ψ is the wavefunction which is a function of the position of the electrons and nuclei within the molecule.

A number of solutions exist for Eq. (2-1), with each one representing a different electronic state of the molecule. Importantly the lowest energy solution represents the ground state. It is worth stating that the Schrödinger equation is an eigenvalue equation, which in mathematical terms means that the equation contains an operator acting upon a function that produces a multiple of the function itself as the result [Eq. (2-2)]:

$$\mathbf{Operator}^*\text{function} = \text{constant}^*\text{function} \qquad (2\text{-}2)$$

In Eq. (2-1) the wavefunction (ψ) can be approximated to the electronic state, this being the configuration of the electrons in a series of molecular orbitals. It is then possible to evaluate differing electronic configurations of the wavefunction in terms of their energies, with the lowest energy configuration being the ground state. It is the ground state energy that corresponds to the ground state geometry of a given molecule. For a given wavefunction the associated *Hamiltonian* is calculated according to Eq. (2-3):

$$\mathbf{H} = KE_{total} + PE_{total} \qquad (2\text{-}3)$$

where

> KE_{total} = total kinetic energy = $\sum$ (coulomb repulsion between each pair of charged entities)
>
> PE_{total} = total potential energy = $\sum$ (electron–nuclei attraction) + $\sum$ (electron– electron repulsion) + $\sum$ (nuclei–nuclei repulsion).

In order to evaluate the components of Eq. (2-3), a number of approximations are required that are complex and out of the scope of this chapter. A number of excellent texts exist that discuss these approximations in great detail [5, 6].

2.3.   HARTREE–FOCK THEORY

Having established the importance of the electronic wavefunction (ψ) in Eq. (2-1), it is now necessary to discuss the methods that enable the derivation of the electronic states for which Eq. (2-1) holds true. The following discussion is an outline of the fundamentals of Hartree–Fock theory from which both semi-empirical and density functional methods have been developed.

The first step towards obtaining an optimised electronic structure (i.e. the ground state) for a molecule is to consider the wavefunction as a series of molecular orbitals with differing electronic occupations. One of these sets of molecular orbitals will correspond to the ground state and hence have the lowest energy. Approximating the wavefunction to a series of molecular orbitals allows the substitution of the wavefunction in Eq. (2-1) with Eq. (2-4) resulting in Eq. (2-5) (both Eqs. (2-4) and (2-5) are simplified to illustrate the important conceptual idea that in Hartree–Fock theory the wavefunction is represented by a series of molecular orbitals).

$$\psi = \phi_1\phi_2\phi_3 \ldots \phi_n \qquad (2\text{-}4)$$

$$H(\phi_1\phi_2\phi_3 \ldots \phi_n) = E(\phi_1\phi_2\phi_3 \ldots \phi_n) \qquad (2\text{-}5)$$

where ϕ_i is the ith molecular orbital.

Having broken down the electronic wavefunction into a series of molecular orbitals, Hartree–Fock theory then makes use of so-called "basis functions". These functions are a series of one-electron mathematical representations that are localised on individual atoms, which can be thought of as representing the atomic orbitals.

The more basis functions included in the molecular orbital calculation, the more accurate the final representation. However as might be expected, this results in an increase in computational time. Both semi-empirical and density functional methods make use of basis functions to represent atomic orbitals (so-called basis sets). It is then possible to calculate the ground state electronic structure by making use of a mathematical procedure known as the variational principle.

The significant drawback within the Hartree–Fock formalisation is the incomplete treatment of so-called exchange–correlation effects when evaluating the energy of the wavefunction. These effects relate to the interactions between pairs of electrons with the same spin (exchange) and pairs of electrons with opposing spins (correlation). Thus, when evaluating the energy of the wavefunction within Hartree–Fock theory correlation effects are completely neglected, leading to an underestimation of the true energy of a given electronic state.

2.4.   SEMI-EMPIRICAL METHODS: AM1 AND RM1

Initial usage of Hartree–Fock theory was limited to very small systems for which the iterative process of locating the lowest energy wavefunction was amenable to early computers. Such limitations led to the development of so-called semi-empirical quantum mechanics methods, with the aim of allowing chemically meaningful systems to be investigated. As would be expected, one of the most time-consuming steps in the Hartree–Fock optimisation procedure is the manipulation of the mathematical representations of the molecular orbitals. In contrast, the semi-empirical Austin Method 1 (AM1) deals only with the valence electrons, thus significantly reducing the complexity and hence time of one of the most computationally expensive steps [7]. Additional computational savings are made in the use of parameterised functions for some of the terms in the *Hamiltonian*. These functions are developed using experimental data such as heats of formation, the aim being that the functions are optimised (often manually) until the resulting calculations can reproduce a series of experimental molecular properties. Such approximations obviously reduce the accuracy of the AM1 method (and semi-empirical methods in general), this being the major limitation. Semi-empirical methods generally perform well for calculations upon molecular systems for which the basis functions were optimised (for example, heats of formations are frequently well reproduced). However (and as might be expected) calculations into systems for which no experimental data existed (or was used) in the parameterisation procedure often perform poorly. The significant advantage of the computational efficiency resulting from the various approximations in the AM1 methodology is that it allows for a high number of chemicals to be investigated in a reasonable timeframe, and for calculations upon large molecular systems.

A recent re-parameterisation of the AM1 model has led to the development of the Recife Model 1 (RM1) semi-empirical method [8]. This methodology has been suggested to be a significant improvement over the original AM1 model as additional parameterisation data were included in its development. These data came from high-level density functional calculations allowing for a better definition of common geometrical variables poorly defined by existing experimental data. In addition, the description of the electron repulsion portion of the wavefunction was also improved.

2.5.   AB INITIO: DENSITY FUNCTIONAL THEORY

Density functional theory (DFT) is a closely related methodology to Hartree–Fock theory in that it attempts to provide a solution to the electronic state of a molecule directly from the electron density. One can view the methodologies as essentially analogous, for the purpose of this discussion, in terms of using basis functions for orbitals and in the use of the variational principle to locate the lowest energy wavefunction. However, the major difference is the inclusion of terms to account for both exchange and correlation when evaluating the energy of the wavefunction, resulting in a significantly improved description of the electronic structure. Differing functionals (for example, B3LYP) use differing mathematical approximations to describe the *Hamiltonian* and thus evaluate the energy of a given wavefunction. The discussion of how such functionals are calculated and thus their relative strengths and weaknesses is well outside the scope of this chapter. It is important only to realise that DFT (whatever the chosen functional) is a more complete description of the electronic structure than that offered from Hartree–Fock theory and is significantly more complete than semi-empirical methods. However, as would be expected by the inclusion of more complex mathematics, it is also the most time consuming. A more complete discussion of DFT and functionals can be found in several texts [6, 9].

2.6.   QSAR FOR NON-REACTIVE MECHANISMS OF ACUTE (AQUATIC) TOXICITY

The importance of quantum mechanics electronic parameters in toxicology becomes apparent when one examines the descriptors required to model the polar narcotic chemicals. Such relationships frequently involve the use of the energy of the lowest unoccupied molecular orbital (E_{LUMO}) to account for the increased electronegativity of these chemicals (compared to those that cause baseline narcosis). The most commonly used level of theory is the AM1 Hamiltonian. The descriptor E_{LUMO} in combination with the logarithm of the octanol–water partition coefficient (log P) (or other descriptor describing hydrophobicity) leads to excellent statistical relationships. Such two-parameter QSARs are commonly referred to as response-surface models [10]. Cronin and Schultz investigated the acute toxicity of 166 phenols to the ciliated protozoan *Tetrahymena pyriformis*, for which potential toxic mechanisms of action had previously been assigned [11]. Of the 166 chemicals in the training data, 120 were assigned as acting via polar narcosis, and response-surface analysis of the toxicity for these chemicals (IGC_{50}) produced the following relationship:

$$\text{Log } (IGC_{50})^{-1} = 0.67(0.02) \log P - 0.67(0.06)E_{LUMO} - 1.123(0.13)$$
$$n = 120, \ r^2 = 0.90, \ r_{cv}^2 = 0.89, \ s = 0.26, \ F = 523 \tag{2-6}$$

where

the figures in parentheses are the standard errors on the coefficients;
n is the number of compounds
r^2 is the square of the correlation coefficient
r^2_{cv} is the square of the leave-one-out cross-validated correlation coefficient
s is the standard error
F is Fisher's statistic

The importance of the AM1 Hamiltonian-derived parameter E_{LUMO} in Eq. (2-6) is reinforced by a QSAR model [Eq. (2-7)] in which E_{LUMO} was replaced by the experimentally determined Hammett constant σ, used to account for the polarising effect of substituent groups on the pK_a of the phenolic moiety [12]:

$$\text{Log } (IGC_{50})^{-1} = 0.64(0.04) \log P + 0.61(0.12)\sigma + 1.123(0.13)$$
$$n = 119, r^2 = 0.90, r^2_{cv} = 0.89, s = 0.265 \tag{2-7}$$

A number of related multiple linear regression models have been developed for chemicals acting via the polar narcosis mechanism; such studies on a range of fish species frequently make use of either E_{LUMO} or other equivalent electronic parameters derived using the AM1 Hamiltonian. Such models usually display excellent regression statistics indicating the mechanistic importance of hydrophobicity and an electronic descriptor related to electronegativity and/or polarisability [13, 14].

The benefits of electronic descriptors derived using AM1 theory and density functional theory (using the B3LYP functional coupled with a 6-31G(d,p) basis set) have been investigated [15]. The study utilised toxicity data to *Pimephales promelas* for 568 chemicals covering multiple mechanisms of action (covalent and non-covalent). A wide range of quantum mechanically derived descriptors were calculated at the two levels of theory, with two, three and four parameter models derived using multiple linear regression. The authors conclude that descriptors calculated at the AM1 level of theory resulted in QSAR models as statistically relevant as those constructed using the higher level of theory when modelling large multi-mechanism data sets. The simplest two-parameter equations for toxicity using AM1 and DFT, respectively, are:

AM1:$\text{Log } (LC_{50})^{-1} = 0.614(0.022)\log P - 0.240(0.026)E_{LUMO} - 0.392(0.062)$
$n = 568, r^2 = 0.663, r^2_{cv} = 0.658, s = 0.805, F = 555$

$$\tag{2-8}$$

DFT: $\text{Log } (LC_{50})^{-1} = 0.630(0.021)\log P - 0.242(0.025)E_{LUMO} - 0.603(0.057)$
$n = 568, r^2 = 0.667, r^2_{cv} = 0.663, s = 0.800, F = 565$

$$\tag{2-9}$$

Investigations into the prediction of physico-chemical properties have also demonstrated the comparable statistical performance between QSARs developed using RM1 and DFT derived descriptors [16].

2.7. QSARs FOR REACTIVE TOXICITY MECHANISMS

2.7.1. Aquatic Toxicity and Skin Sensitisation

Semi-empirical descriptors may be used to model the reactive covalent mechanisms of toxic action between electrophilic chemicals and nucleophilic centres in proteins. Karabunarliev et al. [17] investigated the ability of six mechanistically interpretable AM1 parameters (in addition to log P) to model the toxicity of 98 chemicals to *P. promelas*. The data were modelled within mechanistic domains resulting in 35 of the chemicals being assigned to the S_N2 mechanism, 18 to the Michael addition mechanism and 45 to the Schiff base mechanism (see Figure 2-1 for a summary of these mechanisms). Multiple linear regression analysis resulted in three QSAR models for the domains [Eqs. (2-10, 2-11 and 2-12)].

$$\textbf{S}_N\textbf{2:} \text{ Log } (1/LC_{50}) = -1.56(0.337)E_{LUMO} + 0.358(0.106) \log P + 4.43(0.283)$$
$$n = 35, r^2 = 0.69, r^2_{cv} = 0.41, s^2 = 0.67, F = 35.6$$

$$(2\text{-}10)$$

$$\textbf{Michael addition:} \text{ Log } (1/LC_{50}) = 28.6(0.525)A_R + 81.3(0.306)B_{\alpha-R}$$
$$+0.359(0.142)\log P - 89.1(0.290)$$
$$n = 18, r^2 = 0.78, r^2_{cv} = 0.43, s^2 = 0.33, F = 16.5$$

$$(2\text{-}11)$$

$$\textbf{Schiff base:} \text{ Log } (1/LC_{50}) = 0.466(0.059)\log P + 12.702(0.457)Q_O$$
$$+7.285(0.145)$$
$$n = 45, r^2 = 0.60, r^2_{cv} = 0.31, s^2 = 0.23, F = 31.3$$

$$(2\text{-}12)$$

where

A_R is the acceptor superdelocalisability for polarising atom
$B_{\alpha-R}$ is the bond order for the alpha carbon-polarising group bond
Q_O is the atom superdelocalisability for oxygen

As with the non-covalent mechanisms, a chemical's hydrophobicity is important in determining its overall toxicity. Equation 2-10 highlights the usefulness of E_{LUMO} in modelling a reactive mechanism. Mechanistically its inclusion is in keeping with the ideas presented previously detailing hard–soft acid–base theory, which for the S_N2 mechanism would involve the direct attack of the LUMO by the incoming nucleophile and then subsequent expulsion of the leaving group. In contrast, the QSARs for the Michael addition and Schiff base mechanisms utilise alternate quantum mechanics descriptors which are less interpretable in terms of the underlying reaction chemistry. The descriptors A_R and Q_O are derived from a family of descriptors known as superdelocalisability. These descriptors are atom specific and have been suggested to account for an atom's ability to either accept electron density (commonly denoted as A_N and A_{max}, where N = atom and A_{max} is the atom with the greatest ability to accept electron density within a molecule) or donate electron density (commonly denoted as D_N and D_{max}). A related study by the same authors into a series of benzene derivatives also using a similar set of AM1-derived descriptors produced similar results [18].

Aptula et al. [19, 20] recently introduced the activation energy index (AEI) based on the changes in energy of the frontier molecular orbitals for a series of chemicals acting via the Michael addition mechanism of action. The AEI was designed to model the alterations in orbital energies when an electrophile interacts with a nucleophile in the S_NAr mechanism. The analysis resulted in the AEI being calculated from the change in the highest occupied molecular orbital and second highest occupied molecular orbital energies upon formation of the ionic intermediate, both of which are optimised using the AM1 Hamiltonian equation (2-13):

$$AEI = \Delta E_{HOMO-1} + \Delta E_{HOMO} \qquad (2\text{-}13)$$

This parameter was first introduced to rationalise why two apparently related chemicals, 2-methylisothiazol-3-one and 5-chloro-2-methylisothiazol-3-one, both known to cause skin sensitisation, had been shown to react differently with the two nucleophiles producing different reaction products [19]. A follow-up study investigated the mechanistic rationale for the toxicity of 18 di- and tri-hydroxybenzenes to *T. pyriformis* [20]. The authors suggested that these chemicals exert their toxicity due to their ability to be oxidised to quinone-type species, which then react via subsequent electrophilic Michael addition. An initial quantitative relationship was developed for the 18 chemicals [Eq. (2-14)]:

$$Log\ (IGC_{50})^{-1} = -0.49(0.06)AEI + 6.85(0.69)$$
$$n = 18, r^2 = 0.810,\ s = 0.24,\ F = 73 \qquad (2\text{-}14)$$

Both of these studies highlight the ability of AM1 Hamiltonian-derived descriptors to model subtle electronic effects given a series of closely related chemicals in which the electronics of the system dominate the differences in the toxicity. The two studies also show how well-thought-out orbital analysis and subsequent calculations can aid significantly the mechanistic interpretation of a series of related chemicals.

A further descriptor that makes use of the frontier molecular orbitals has been developed, namely the electrophilicity index ω [21]. The electrophilicity index is based on two previously developed quantum mechanical properties, chemical potential (μ, which can be considered as the negative of chemical electronegativity) and chemical hardness (η) [22, 23]. Thus, ω can be calculated from E_{HOMO} and E_{LUMO} values as follows:

$$\omega = \mu^2/2\eta = (E_{HOMO} + E_{LUMO})^2/2(E_{LUMO} - E_{HOMO}) \qquad (2\text{-}15)$$

A number of recent articles have highlighted the ability of the electrophilicity index to model the site selectivity and reactivity in diene–dieneophile chemical reactions [24, 25]. In terms of toxicity prediction, these studies are relevant as the ability to rationalise the Michael acceptor reaction is of clear importance, given its involvement in reactive toxicity. In a more recent study, Domingo et al. [25] showed that the electrophilic index (calculated at the B3LYP/6-31Gd level of theory) was able to rank a series of Michael acceptors qualitatively. Importantly, the study also showed the quantitative relationships between experimentally determined rate constants and the electrophilicity index for several series of related chemicals. For example, the rate of the Michael reaction for piperidine reacting with a series of benzylidene-malononitriles was found to be reasonably correlated with ω ($r^2 = 0.75$). Inspection of the correlation revealed that the major deviation was due to lower than predicted reactivity of the para-NMe$_2$ species, with it being suggested that a significant solvation effect of the tertiary amide being responsible. Exclusion of this chemical improved the correlation significantly ($r^2 = 0.90$). In addition, the ω values for four chemicals from the same series were shown to be highly correlated with the available data for previously determined experimental measures of electrophilicity [26] ($r^2 = 0.98$). A similar correlation was also reported for five α-nitrostilbenes ($r^2 = 0.98$). It is important to note that these excellent correlations occurred after careful consideration of the reactivity applicability domain, that is to say within carefully considered chemical categories in which the electronic effects of the system were determined to be the major influence on the differing rates of reaction. No attempt was made to correlate ω in a global fashion with the reaction rates for all 39 chemicals in the study [25].

A local lymph node assay (LLNA) study into the skin sensitising potential of a series of Michael acceptor alkenes also highlighted the utility of the electrophilicity index [27]. The authors utilised ω (calculated using the B3LYP/6-31Gd level of theory) as a measure of electrophilic similarity within a well-defined alkene Michael acceptor category in order to perform quantitative mechanistic read-across. The methodology assumed that within the Michael reaction domain the skin-sensitising potential of a chemical is dominated by how electrophilic the chemical is and thus how readily it will react with skin proteins. This is in keeping with reactivity studies which have shown that within this domain reactivity is the driving force, with other factors such as toxicokinetics being of less importance [1, 28, 29]. Although not a statistically based QSAR study (in that no attempt was made to derive a linear model), the read-across methodology presented by the authors offered excellent predictions within the perceived experimental error of the local lymph node assay. In addition, the methodology allowed mechanistic outliers to be identified in terms of easily rationalised chemistry effects such as steric hindrance and ring strain release.

Other related studies have also demonstrated the utility of the electrophilicity index in modelling several reactive mechanisms that occur in the toxicity of industrial chemicals to *T. pyriformis*. In these studies the authors optimised a series of aliphatic and aromatic chemicals using Hartree–Fock theory and 6-31Gd basis set [30, 31]. In both studies chemicals were divided into a series of chemical categories and then modelled using a number of descriptors derived from ω, including so-called atom condensed philicity indices (ω_m^+ and ω_m^-) derived utilising Fukui functions and several charge schemes including Mulliken analysis [4, 32]. A range of QSAR models were developed, the best of which was for 18 amino alcohols [Eq. (2-16)]:

$$Log\ (IGC_{50}^{-1}) = -0.40\omega - 2.19\omega_m^- - 1.52$$
$$n = 18, r^2 = 0.93, s = 0.14 \tag{2-16}$$

Other density functional theory derived descriptors have been utilised to model the toxicity of 28 nitroaromatic chemicals to *P. promelas* [33]. Six mechanistically relevant descriptors were calculated using the B3LYP functional with a 6-31G(d,p) basis set. The resulting QSAR model obtained by stepwise regression is given by Eq. (2-17):

$$Log\ (LC_{50}) = -39.5\ E_{LUMO} + 16.9\ E_{HOMO} + 15.1\ Q_{NO2} + 4.17\ Q_c + 9.52$$
$$n = 28, r = 0.91, s = 0.36, F = 28.4$$
$$\tag{2-17}$$

where

Q_{NO2} is the charge on the nitro group
Q_c is the charge on the nitro carbon

As previously determined, this study demonstrated the importance of the frontier molecular orbitals in modelling hard–soft acid–base theory that plays an important role in the toxicity of these chemicals.

2.7.2.   QSARs for Mutagenicity

A chemical's ability to act as a genotoxic mutagen is considered to be related to its ability to form a covalent bond with nucleic acids [34–36]. The mechanistic basis for such interactions is similar to those discussed for excess aquatic toxicity and skin sensitisation, with the importance of the types of nucleophilic–electrophilic reactions (Figure 2-1) and hard–soft acid–base theory being applicable (nitrogen within nucleic acids acts as the nucleophile in genotoxic mutagenicity). Most mutagenicity studies have focused on the development of small local QSAR models based on a single chemical class. As has been discussed for aquatic toxicity and skin sensitisation, this type of approach leads to the most mechanistically interpretable model in which the inclusion of quantum mechanical descriptors can model a chemical's electrophilicity/reactivity.

Several studies of nitroaromatic chemicals in the TA98 and TA100 strains of *Salmonella typhimurium* noted that mutagenicity could be well modelled using log P and E_{LUMO} (calculated at the AM1 level of theory) [37–39], an example being Eq. (2-18):

$$log\ TA100 = 1.36(0.20)\ log\ P - 1.98(0.39)E_{LUMO} - 7.01(1.20)$$
$$n = 47, r = 0.91, s = 0.74, F = 99.9 \tag{2-18}$$

where TA100 is the number of revertants per millimole.

It was suggested that the inclusion of the electronic parameter E_{LUMO} accounted for the ability of the nitroaromatic chemicals to accept electrons, and thus be reduced to the mutagenic nitroso species.

A recent study into the mutagenic potential of a series of α, β-unsaturated aldehydes to the TA100 strain of *S. typhimurium* revealed a QSAR model (2-10) in which E_{LUMO} (calculated at the AM1 level of theory) also figured prominently [40]:

$$\text{Log TA100} = -4.58\ E_{LUMO} - 3.66\ MR + 72.46\ Q_{C-carb} + 2.55\ \log\ P$$
$$+13.09\ Q_{C-\beta} - 12.6 \tag{2-19}$$
$$n = 17, r^2 = 0.84$$

where

MR is the molar refractivity

$Q_{C\text{-carb}}$ is the partial charge on the carbonilic carbon atom

$Q_{C\text{-}\beta}$ is the partial charge on the β carbon atom

In addition, several binary classification models were also presented utilising hydrophobicity (log P) and electronic descriptors (E_{LUMO}). As previously discussed (in terms of aquatic toxicity and skin sensitisation), the inclusion of molecular orbital parameters to model the nucleophilic–electrophilic reaction thought to be responsible for the reactive toxicity of such chemicals shows the importance of quantum mechanical descriptors.

A related study investigated the important structural, quantum chemical and hydrophobic factors thought to be related to the mutagenic potential of 12 closely related heterocyclic amines to *S. typhimurium* TA98 [41]. The authors carried out a series of calculations using Hartree–Fock theory coupled with a 6-31Gd basis set to calculate a range of electronic descriptors. The study highlighted a number of quantum mechanical factors that were suggested to be important in the control of mutagenicity of the studied heterocyclic amines, these being low values for the dipole moment (p), calculated energy of the aromatic π system and chemical softness (measured as the gap between the HOMO and LUMO). A number of linear regression models were presented, such as Eq. (2-20):

$$\text{Log TA98} = -0.33\ p + 2.18\ \sigma - 1.85$$
$$n = 12, r^2_{adj} = 0.85, \text{RMSE} = 0.38, F = 39.1 \tag{2-20}$$

where

σ is chemical softness

RMSE is the root mean square error

2.8. FUTURE DIRECTIONS AND OUTLOOK

The increase in computational power has led to a parallel increase in the use of quantum mechanics derived molecular descriptors. This trend is likely to increase in the future as computational chemists/toxicologists seek to fully understand the underlying electronic effects of toxic mechanisms. This is especially true for the reactive mechanisms involving the formation of covalent bonds. One can envisage parameters such as the electrophilicity index being used to understand the electronic effects within a series of chemicals within a category (see Chapter 7 in this volume for a discussion of chemical categories). Such analysis will enable a theoretical understanding of the electronic effects to be added to weight of evidence approaches in regulatory chemical safety assessments. This information will compliment, not replace, other experimental investigations.

2.9.   CONCLUSIONS

This chapter has highlighted the mechanistic rationale for the use of quantum mechanics derived descriptors in the modelling of both non-covalent and covalent mechanisms of action. In addition, some of the underlying chemical and computational theory has been detailed to enable a qualitative understanding of the theoretical background to the calculation of such descriptors. Two aspects have been highlighted, the first being the inclusion of frontier molecular orbital descriptors such as E_{LUMO}, to aid the modelling of non-covalent mechanisms such as polar narcosis. The second and perhaps the more important being the relationship between the uses of such descriptors and hard–soft acid–base theory and how the two combine to help in the understanding of covalent mechanisms of toxicity involving nucleophilic–electrophilic chemistry.

For both non-covalent and covalent mechanisms, a number of examples have been presented to highlight the usage of quantum mechanics derived descriptors. The important conclusions from the examples presented are in the differing levels of computational theory required to model the two types of mechanism. It is clear that for non-covalent mechanisms the computationally efficient semi-empirical methods such as AM1 are sufficient for good predictions. In contrast, for covalent mechanisms higher levels of theory are required for successful modelling of these more chemically complex mechanisms. Finally, this chapter has demonstrated that given a well-defined mechanistic applicability domain, quantum mechanics derived molecular methods are extremely powerful tools that aid computational toxicologists in understanding the electronic structure of a chemical and how that structure influences both non-covalent and covalent toxic mechanisms.

ACKNOWLEDGEMENT

The funding of the European Union Sixth Framework CAESAR Specific Targeted Project (SSPI-022674-CAESAR) and the comments of Dr Judith Madden, Liverpool John Moores University, are gratefully acknowledged.

REFERENCES

1. Aptula AO, Roberts DW (2006) Mechanistic applicability domains for non-animal based prediction of toxicological end points: General principles and application to reactive toxicity. Chem Res Tox 19:1097–1105
2. Streitwieser A, Heathcock CH (1985) Introduction to organic chemistry. 3rd edn. Macmillan, New York

3. Schüürmann G (2004) Quantum chemical descriptors in structure–activity relationships – calculation, interpretation and comparison of methods. In: Cronin MTD, Livingstone DJ (eds) Predicting Chemistry Toxicity and Fate. Taylor and Francis, London
4. Fuentealba P, Perez P, Contreras R (2000) On the condensed Fukui function. J Chem Phys 113:2544–2551
5. Leach AR (2001) Molecular Modelling: Principles and Applications. Pearson Education Limited, Harlow
6. Foresman JB, Frisch A (1996) Exploring Chemistry with Electronic Structure Methods, 2nd edn. Gaussian Inc., Pittsburgh.
7. Dewar MJS, Zoebisch EG, Healy EF et al. (1985) AM1: A new general purpose quantum mechanical molecular model. J Am Chem Soc 107:3902–3909
8. Rocha GB, Freire RO, Simas AM et al. (2006) RM1: A reparameterisation of AM1 for H, C, N, O, P, S, F, Cl, Br, and I. J Comput Chem 27:1101–1111
9. Koch W, Holthausen MC (2000) A chemist's guide to density functional theory. Wiley-VCH, Weinheim.
10. Mekenyan OG, Veith GD (1993) Relationships between descriptors for hydrophobicity and soft electrophilicity in predicting toxicity. SAR QSAR Environ Res 1:335–344
11. Cronin MTD, Schultz TW (1996) Structure-toxicity relationships for phenols to *Tetrahymena pyriformis*. Chemosphere 32:1453–1468
12. Garg R, Kurup A, Hansch C (2001) Comparative QSAR: On the toxicology of the phenolic OH moiety. Crit Rev Toxicol 31:223–245
13. Cronin MTD, Schultz TW (1997) Validation of Vibrio fisheri acute toxicity data: mechanism of action-based QSARs for non-polar narcotics and polar narcotic phenols. Sci Total Environ 204: 75–88
14. Ramos EU, Vaes WHJ, Verhaar HJM et al. (1997) Polar narcosis: Designing a suitable training set for QSAR studies. Environ Sci Pollut Res 4:83–90
15. Netzeva TI, Aptula AO, Benfenati E et al. (2005) Description of the electronic structure of organic chemicals using semi-empirical and ab initio methods for development of toxicological QSARs. J Chem Inf Model 45:106–114
16. Puzyn T, Suzuki N, Haranczyk M et al. (2008) Calculation of quantum-mechanical descriptors for QSPR at the DFT level: Is it necessary? J Chem Inf Model 48:1174–1180
17. Karabunarliev S, Mekenyan OG, Karcher W et al. (1996) Quantum-chemical descriptors for estimating the acute toxicity of electrophiles to the fathead minnow (*Pimephales promelas*): An analysis based on molecular mechanisms. Quant Struc-Act Relat 15:302–310
18. Karabunarliev S, Mekenyan OG, Karcher W et al. (1996) Quantum-chemical descriptors for estimating the acute toxicity of substituted benzenes to the Guppy (*Poecilia reticulata*) and Fathead Minnow (*Pimephales promelas*). Quant Struc-Act Relat 15:311–320
19. Aptula AO, Roberts DW, Cronin MTD (2005) From experiment to theory: Molecular orbital parameters to interpret the skin sensitization potential of 5-chloro-2-methylisothiazol-3-one and 2-methylisothiazol-3-one. Chem Res Toxicol 18:324–329
20. Aptula AO, Roberts DW, Cronin MTD et al. (2005) Chemistry-toxicity relationships for the effects of di-and trihydroxybenzenes to *Tetrahymena pyriformis*. Chem Res Toxicol 18:844–854
21. Parr RG, Szentpaly LV, Liu S (1999) Electrophilicity index. J Am Chem Soc 121:1922–1924
22. Parr RG, Donnelly RA, Levy M et al. (1978) Electronegativity – density functional viewpoint. J Chem Phys 68:3801–3807
23. Parr RG, Pearson RG (1983) Absolute hardness – companion parameter to absolute electronegativity. J Am Chem Soc 105:7512–7516
24. Domingo LR, Aurell MJ, Perez P et al. (2002) Quantitative characterization of the local electrophilicity of organic molecules. Understanding the regioselectivity on Diels-Alder reactions. J Phys Chem A 106:6871–6875

25. Domingo LR, Perez P, Contreras R (2004) Reactivity of the carbon–carbon double bond towards nucleophilic additions. A DFT analysis. Tetrahedron 60:6585–6591

26. Lemek T, Mayr HJ (2003) Electrophilicity parameters for benzylidenemalononitriles. J Org Chem 68:6880–6886

27. Enoch SJ, Cronin MTD, Schultz TW et al. (2008) Quantitative and mechanistic read across for predicting the skin sensitization potential of alkenes acting via Michael addition. Chem Res Toxicol 21:513–520

28. Aptula AO, Patlewicz G, Roberts DW et al. (2006) Non-enzymatic glutathione reactivity and in vitro toxicity: A non-animal approach to skin sensitization. Toxicol In Vitro 20: 239–247

29. Schultz TW, Yarbrough JW, Hunter RS et al. (2007) Verification of the structural alerts for Michael acceptors. Chem Res Toxicol 20:1359–1363

30. Roy DR, Parthasarathi R, Maiti B et al. (2005) Electrophilicity as a possible descriptor for toxicity prediction. Bioorg Med Chem 13:3405–3412

31. Roy DR, Parthasarathi R, Subramanian V et al. (2006) An electrophilicity based analysis of toxicity of aromatic compounds towards *Tetrahymena pyriformis*. QSAR Combi Sci 25:114–122.

32. Contreras RR, Fuentealba P, Galvan M et al. (1999) A direct evaluation of regional Fukui functions in molecules. Chem Phys Lett 304:405–413

33. Yan X, Xiao HM, Ju XH et al. (2005) DFT study on the QSAR of nitroaromatic compound toxicity to the fathead minnow. Chin J Chem 23:947–952

34. Kazius J, McGuire R, Bursi R (2005) Derivation and validation of toxicophores for mutagenicity prediction. J Med Chem 48:312–320

35. Eder E, Scheckenbach S, Deininger C et al. (1993) The possible role of α,β-unsaturated carbonyl compounds in mutagenesis and carcinogenesis. Toxicol Lett 67:87–103

36. Passerini L (2003) QSARs for individual classes of chemical mutagens and carcinogens. In: Benigni R (ed) Quantitative Structure-Activity Relationships (QSAR) Models of Mutagens and Carcinogens. CRC Press LLC, Boca Raton

37. Lopez de Compadre RI, Debnath AK, Shusterman AJ et al. (1990) LUMO energies and hydrophobicity as determinants of mutagenicity by nitroaromatic compounds in *Salmonella typhimurium*. Environ Mol Mutagen 15:44–55

38. Debnath AK, Debnath G, Shusterman AJ et al. (1992) A QSAR investigation of the role of hydrophobicity in regulating mutagenicity in the AMES test. 1. Mutagenicity of aromatic and heteroaromatic amines in *Salmonella typhimurium* TA98 and TA100. Environ Mol Mutagen 19:37–52

39. Debnath AK, Decompadre RLL, Shusterman AJ et al. (1992) Quantitative structure-activity relationship investigation of the role of hydrophobicity in regulating mutagenicity in the AMES test. 2. Mutagenicity of aromatic and heteroaromatic nitro compounds in *Salmonella typhimurium* TA100. Environ Mol Mutagen 19:53–70

40. Benigni R, Conti L, Crebelli R et al. (2005) Simple and α,β-unsaturated aldehydes: Correct prediction of genotoxic activity through structure-activity relationship models. Environ Mol Mutagen 46:268–280

41. Knize MG, Hatch FT, Tanga MJ et al. (2006) A QSAR for the mutagenic potencies of twelve 2-amino-trimethylimidazopyridine isomers: Structural, quantum chemical, and hydropathic factors. Environ Mol Mutagen 47:132–146

CHAPTER 3

MOLECULAR DESCRIPTORS

VIVIANA CONSONNI, AND ROBERTO TODESCHINI

Milano Chemometrics & QSAR Research Group, Dept. of Environmental Sciences, University of Milano-Bicocca, Piazza della Scienza 1, 120126 Milan, Italy, e-mail: viviana.consonni@unimib.it

Abstract: In the last decades, several scientific researches have been focused on studying how to encompass and convert – by a theoretical pathway – the information encoded in the molecular structure into one or more numbers used to establish quantitative relationships between structures and properties, biological activities, or other experimental properties. Molecular descriptors are formally mathematical representations of a molecule obtained by a well-specified algorithm applied to a defined molecular representation or a well-specified experimental procedure. They play a fundamental role in chemistry, pharmaceutical sciences, environmental protection policy, toxicology, ecotoxicology, health research, and quality control. Evidence of the interest of the scientific community in the molecular descriptors is provided by the huge number of descriptors proposed up today: more than 5000 descriptors derived from different theories and approaches are defined in the literature and most of them can be calculated by means of dedicated software applications. Molecular descriptors are of outstanding importance in the research fields of quantitative structure–activity relationships (QSARs) and quantitative structure–property relationships (QSPRs), where they are the independent chemical information used to predict the properties of interest. Along with the definition of appropriate molecular descriptors, the molecular structure representation and the mathematical tools for deriving and assessing models are other fundamental components of the QSAR/QSPR approach. The remarkable progress during the last few years in chemometrics and chemoinformatics has led to new strategies for finding mathematical meaningful relationships between the molecular structure and biological activities, physico-chemical, toxicological, and environmental properties of chemicals. Different approaches for deriving molecular descriptors here reviewed and some of the most relevant descriptors are presented in detail with numerical examples.

Keywords: Molecular representation, Topological indexes, Autocorrelation descriptors, Geometrical descriptors

3.1.   INTRODUCTION

3.1.1.   Definitions

In the last decades, much scientific research has focused on how to capture and convert – by a theoretical pathway – the information encoded in a molecular

29

T. Puzyn et al. (eds.), Recent Advances in QSAR Studies, 29–102.
DOI 10.1007/978-1-4020-9783-6_3, © Springer Science+Business Media B.V. 2010

structure into one or more numbers used to establish quantitative relationships between structures and properties, biological activities, or other experimental properties. Molecular descriptors are formal mathematical representations of a molecule, obtained by a well-specified algorithm, and applied to a defined molecular representation or a well-specified experimental procedure: *the molecular descriptor is the final result of a logic and mathematical procedure which transforms chemical information encoded within a symbolic representation of a molecule into a useful number or the result of some standardized experiment* [1].

Molecular descriptors play a fundamental role in developing models for chemistry, pharmaceutical sciences, environmental protection policy, toxicology, ecotoxicology, health research, and quality control. Evidence of the interest of the scientific community in molecular descriptors is provided by the huge number of descriptors that have been proposed: more than 5000 descriptors [1] derived from different theories and approaches are defined and computable by using dedicated software of chemical structure.

There are three main parts to the QSAR/QSPR approach in scientific research: the concept of molecular structure, the definition of molecular descriptors, and the chemoinformatic tools. The concept of molecular structure, its representation by theoretical molecular descriptors, and its relationship with experimental properties of molecules is an inter-disciplinary network. Many theories, knowledge, and methodologies and their inter-relationships are required. These have led to a new scientific research field resulting in several practical applications. Molecular descriptors are numerical indexes encoding some information related to the molecular structure. They can be both experimental physico-chemical properties of molecules and theoretical indexes calculated by mathematical formulas or computational algorithms. Thus, molecular descriptors, which are closely connected to the concept of molecular structure, play a fundamental role in scientific research, being the theoretical core of a complex network of knowledge, as it is shown in Figure 3-1.

Molecular descriptors are derived by applying principles from several different theories, such as quantum-chemistry, information theory, organic chemistry, graph theory. They are used to model several different properties of chemicals in scientific fields such as toxicology, analytical chemistry, physical chemistry, medicinal, pharmaceutical, and environmental chemistry. Moreover, in order to obtain reliable estimates of molecular properties, identify the structural features responsible for biological activity, and select candidate structures for new drugs, molecular descriptors have been processed by a number of statistical, chemometrics, and chemoinformatics methods. In particular, for about 30 years chemometrics has been developing classification and regression methods able to provide – although not always – reliable models, for both reproducing known experimental data and predicting unknown values. The interest in predictive models able to provide effective reliable estimates has been growing in the last few years as they are more and more considered to be useful and safer tools for predicting data for chemicals. In recent years, *"The use of information technology and management has become a critical part of the drug discovery process. Chemoinformatics is the*

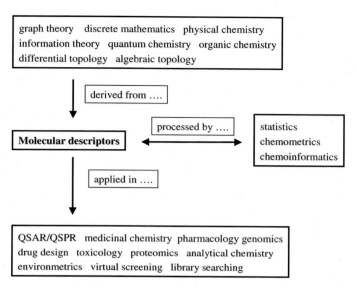

graph theory   discrete mathematics   physical chemistry
information theory   quantum chemistry   organic chemistry
differential topology   algebraic topology

derived from ....

Molecular descriptors

processed by ....

statistics
chemometrics
chemoinformatics

applied in ....

QSAR/QSPR   medicinal chemistry   pharmacology genomics
drug design   toxicology   proteomics   analytical chemistry
environmetrics   virtual screening   library searching

Figure 3.1. General scheme of the relationships among molecular structure, molecular descriptors, chemoinformatics, and QSAR/QSPR modeling

mixing of those information resources to transform data into information and information into knowledge for the intended purpose of making better decisions faster in the area of drug lead identification and organization" [2]. In fact, chemoinformatics encompasses the design, creation, organization, management, retrieval, analysis, dissemination, visualization, and use of chemical information [3,4]; molecular descriptors play a fundamental role in all these processes being the basic tool to transform chemical information into a numerical code suitable for applying informatic procedures.

3.1.2.   History

The history of molecular descriptors is closely related to the history of what can be considered one of the most important scientific concepts of the last part of the nineteenth century and the whole twentieth century, that is, the concept of molecular structure. The years between 1860 and 1880 were characterized by a strong dispute about the concept of molecular structure, arising from the studies on substances showing optical isomerism and the studies of Kekulé (1861–1867) on the structure of benzene. The concept of the molecule thought of as a three-dimensional body was first proposed by Butlerov (1861–1865), Wislicenus (1869–1873), Van't Hoff (1874–1875), and Le Bel (1874). The publication in French of the revised edition of "La chimie dans l'éspace" by Van't Hoff in 1875 is considered a milestone of the three-dimensional understanding of the chemical structures.

Molecular descriptors can be considered as the most important realization of the theory of Crum-Brown. His M.D. Thesis at the University of Edinburgh (1861), entitled "On the Theory of Chemical Combination", shows that he was a pioneer of mathematical chemistry science. In it, he developed a system of graphical representation of compounds which is basically identical to that used today. His formulae were the first that showed clearly both valency and linking of atoms in organic compounds. Toward the conclusion of his M.D. thesis he wrote:

> It does not seem to me improbable that we may be able to form a mathematical theory of chemistry, applicable to all cases of composition and recomposition.

In 1864, he published an important paper on the "Theory of Isomeric Compounds" in which, using his graphical formulae, he discussed various types of isomerism [5], guessing the link between mathematics and chemistry. Later, Crum-Brown and Fraser [6,7] proposed the existence of a correlation between biological activity of different alkaloids and their molecular constitution. More specifically, the physiological action of a substance in a certain biological system (Φ) was defined as a function (f) of its chemical constitution (C) [Eq. (3-1)]:

$$\Phi = f(C) \tag{3-1}$$

Thus, an alteration in chemical constitution, ΔC, would be reflected by an effect on biological activity, $\Delta \Phi$. This equation can be considered the first general formulation of a quantitative structure–activity relationship.

Another hypothesis on the existence of correlations between molecular structure and physico-chemical properties was reported in the work of Körner [9], which dealt with the synthesis of di-substituted benzenes and the discovery of *ortho*, *meta*, and *para* derivatives: the different colors of di-substituted benzenes were thought of to be related to differences in molecular structure and the indicator variables for *ortho*, *meta*, and *para* substitution can be considered as the first three molecular descriptors [8,9].

From the Hammett equation [10,11], the seminal work of Hammett gave rise to the "σ–ρ" culture in the delineation of substituent effects for organic reactions. The aim of this work was the search for linear free energy relationships (LFER) [12]: steric, electronic, and hydrophobic constants were derived for several substituents and used in an additive model to estimate the biological activity of congeneric series of compounds. The first theoretical QSAR/QSPR approaches, that related biological activities and physico-chemical properties to theoretical numerical indexes derived from the molecular structure, date back to the end of 1940s. The *Wiener index* [13] and the *Platt number* [14], proposed in 1947 to model the boiling point of hydrocarbons, were the first theoretical molecular descriptors based on the graph theory. In the 1950s, the fundamental work of Taft in physical organic chemistry was the foundation of relationships between physico-chemical properties and solute–solvent interaction energies (linear solvation energy relationships, LSER), based on steric, polar, and resonance parameters for substituent groups in congeneric compounds [15,16].

In the mid-1960s, led by the pioneering work of Hansch [17–19], the QSAR/QSPR approach began to assume its modern look. The definition of Hansch models led to an explosion in the development of QSAR analysis and related approaches [20]. This approach, known by the name of *Hansch analysis*, became and still is a basic tool for QSAR modeling. At the same time, Free and Wilson developed a model of additive substituent contributions to biological activities [21], giving a further push to the development of QSAR strategies. In the 1960s, several other molecular descriptors were proposed, which signaled the beginning of systematic studies on molecular descriptors, mainly based on the graph theory [22–26].

The use of quantum-chemical descriptors in QSAR/QSPR modeling dates back to early 1970s [27], although they actually were conceived several years before to encode information about relevant properties of molecules in the framework of quantum-chemistry. The fundamental work of Balaban [28], Randić [29], Kier and Hall [30] led to further significant developments of the QSAR approaches based on topological indexes. As a natural extension of the topological representation of a molecule, the geometrical aspects of molecular structures have been taken into account since the mid-1980s, leading to the development of the 3D-QSAR, which exploits information on molecular geometry. Geometrical descriptors were derived from the 3D spatial coordinates of a molecule and, among them, there were shadow indexes [31], charged partial surface area descriptors [32], WHIM descriptors [33], gravitational indexes [34], EVA descriptors [35], 3D-MoRSE descriptors [36], and GETAWAY descriptors [37].

In the late 1980s, a new strategy for describing molecule characteristics was proposed, based on molecular interaction fields, which are comprised of interaction energies between a molecule and probes, at specified spatial points in 3D space. Different probes (such as a water molecule, methyl group, hydrogen) were used for evaluating the interaction energies in thousands of grid points where the molecule was embedded. As the final result of this approach, a scalar field (a lattice) of interaction energy values characterizing the molecule was obtained. The formulation of a lattice model to compare molecules by aligning them in 3D space and extracting chemical information from molecular interaction fields was first proposed by Goodford [38] in the GRID method and then by Cramer, Patterson, Bunce [39] in the comparative molecular field analysis (CoMFA).

Finally, an increasing interest of the scientific community has been shown in recent years for virtual screening and design of chemical libraries, for which several similarity/diversity approaches, cell-based methods, and scoring functions have been proposed based mainly on *substructural descriptors* such as molecular fingerprints [3,40].

3.1.3. Theoretical vs. Experimental Descriptors

Molecular descriptors are divided into two main classes: *experimental measurements*, such as log P, molar refractivity, dipole moment, polarizability, and, in general, physico-chemical properties, and *theoretical molecular descriptors*, which are derived from a symbolic representation of the molecule and can be further classified according to the different types of *molecular representation*. The fundamental difference between theoretical descriptors and experimentally measured ones is that theoretical descriptors contain no statistical error due to experimental noise, as opposed to experimental measurements. However, the assumptions needed to facilitate calculation and numerical approximation are themselves associated with an inherent error, although in most cases the direction, but not the magnitude, of the error is known. Moreover, within a series of related compounds the error term is usually considered to be approximately constant. All kinds of error are absent only for the most simple theoretical descriptors such as counts of structural features or for descriptors directly derived from exact mathematical theories.

Theoretical descriptors derived from physical and physico-chemical theories show some natural overlap with experimental measurements. Several quantum-chemical descriptors, surface areas, and volume descriptors are examples of such descriptors also having an experimental counterpart. With respect to experimental measurements, the greatest recognized advantages of the theoretical descriptors are usually (but not always) in terms of cost, time, and availability.

Each molecular descriptor takes into account a small part of the whole chemical information contained into the real molecule and, as a consequence, the number of descriptors is continuously increasing with the increasing request of deeper investigations on chemical and biological systems. Different descriptors have independent methods or perspectives to view a molecule, taking into account the various features of chemical structure. Molecular descriptors have now become some of the most important variables used in molecular modeling, and, consequently, managed by statistics, chemometrics, and chemoinformatics.

The availability of molecular descriptors has not only provided a new opportunity to search for new relationships, but has been stimulated a great change of the research paradigm in this field: in effect, the use of the molecular descriptors – calculated theoretically – has permitted for the first time a link between the experimental knowledge and theoretical information arising from molecular structure. Whereas, until the 1960s–1970s molecular modeling mainly consisted of the search for mathematical relationships between experimentally measured quantities, nowadays it is mainly performed to search for relationships between a measured property and molecular descriptors able to capture structural chemical information.

A general consideration about the use of molecular descriptors in modeling problems concerns their information content. This depends on the type of molecular representation used and the defined algorithm for their calculation. There are simple molecular descriptors derived by counting some atom types or structural fragments in the molecule, as well as physico-chemical and bulk properties such as, for example, molecular weight, number of hydrogen bond donors/acceptors, number of OH-groups, and so on. Other molecular descriptors are derived from algorithms applied to a topological representation. These are usually termed topological, or 2D-descriptors. Other molecular descriptors are derived from the spatial (x, y, z) coordinates of the molecule, usually called geometrical, or 3D-descriptors; another class of molecular descriptors, called 4D-descriptors, is derived from the interaction energies between the molecule, imbedded into a grid, and some probe.

It is true that geometrical 3D- or 4D-descriptors have higher information content than other simpler descriptors, such as counts of atoms/fragments or topological descriptors which often show significant levels of degeneracy. Thus, there is a point of view that it is better to use the most informative descriptors in all modeling processes. This thinking is not correct because the "best descriptors" are those whose information content is comparable with the information content of the response for which the model is sought. In effect, too much information in the independent variables (the descriptors) with respect to the response is often seen as noise in the model, thus giving instable or unpredictive models. For example, a property whose values are equal or similar for isomeric structures is better modelled by a simple descriptor with appropriate values for isomeric structures. In this case, descriptors able to discriminate among the isomeric structures have redundant information which cannot be integrated in the model. In conclusion, it can be stated that the best descriptor(s) valid for all the problems does(do) not exist.

In general, molecular descriptors, besides trivial invariance properties, should satisfy some basic requirements. A list of desirable requirements of chemical descriptors suggested by Randić [41] is shown in Table 3-1.

Table 3-1. List of desirable attributes of molecular descriptors for use in (Q)SAR studies

#	*Descriptors should be associated with the following desirable features*
1	Structural interpretation
2	Show good correlation with at least one property
3	Preferably allow for the discrimination of isomers
4	Applicable to local structure
5	Generalizable to "higher" descriptors
6	Independence
7	Simplicity
8	Not to be based on properties
9	Not to be trivially related to other descriptors
10	Allow for efficient construction
11	Use familiar structural concepts
12	Show the correct size dependence
13	Show gradual change with gradual change in structures

3.2. MOLECULAR REPRESENTATION

Molecular representation is the manner in which a molecule, i.e., a phenomenological real body, is symbolically represented by a specific formal procedure and conventional rules. The quantity of chemical information which is transferred to the molecule symbolic representation depends on the kind of representation [42,43]. The simplest molecular representation is the *chemical formula* (or *molecular formula*), which is the list of the different atom types, each accompanied by a subscript representing the number of occurrences of the atoms in the molecule. For example, the chemical formula of 4-chlorotoluene is C_7H_7Cl, indicating the presence in the molecule of $A = 8$ (number of atoms, hydrogen excluded), $N_C = 7$, $N_H = 7$, and $N_{Cl} = 1$ (the subscript "1" is usually omitted in the chemical formula). This representation is independent of any knowledge concerning the molecular structure and, hence, molecular descriptors obtained from the chemical formula can be referred to as *0D-molecular descriptors*. Examples are the number of atoms A, the molecular weight MW, the number N_X of atoms of type X, and, in general, any function of atomic properties. Atomic properties are usually the weighting schemes used to characterize the atoms in a molecule and express chemical information regarding a molecular structure. The most common atomic properties for molecular descriptor calculation are atomic masses, atomic charges, van der Waals radii, atomic polarizabilities, and electronegativities. Atoms can also be characterized by the local vertex invariants (LOVIs) derived from graph theory.

The *substructure list representation* can be considered as a one-dimensional representation of a molecule and consists of a list of structural fragments of a molecule. The list is as simple as a partial list of fragments, functional groups, or substituents of interest present in the molecule. Thus, it does not require a complete knowledge of molecular structure. The descriptors derived from this representation are holographic vectors or bit-strings, usually referred to as *1D-molecular descriptors*. These are typically used in substructural analysis, similarity/diversity analysis of molecules, and virtual screening and design of molecule libraries. 0D and 1D descriptors can be always easily calculated, are naturally interpreted, do not require optimization of the molecular structure, and are independent of any conformational problem. They usually show a very high degeneration, i.e., many molecules have the same values, for example, isomers. Their information content is low, but nevertheless they can play an important role in modeling several physico-chemical properties or can be included in more complex models.

The two-dimensional representation of a molecule considers how the atoms are connected, that is, it defines the connectivity of atoms in the molecule in terms of the presence and, ultimately, nature of chemical bonds. The representation of a molecule in terms of the *molecular graph* is commonly known as the *topological representation*. The molecular graph depicts the connectivity of atoms in a molecule irrespective of the metric parameters such as equilibrium interatomic distances between nuclei, bond angles, and torsion angles. In Figure 3-2, examples of H-depleted molecular graphs are given for 2-methyl-3-butenoic acid, 1-ethyl-2-methyl-cyclobutan, and 5-methyl-1,3,4-oxathiazol-2-one.

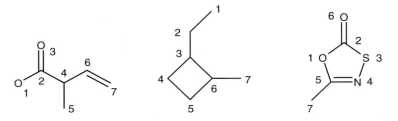

Figure 3-2. Some molecular graph representations of molecules

Molecular descriptors derived from algorithms applied to a topological representation are referred to as *2D-molecular descriptors*; they include the so-called *topological indexes*.

Linear notation systems are alternative two-dimensional representations to the molecular graph. These include, for instance, Wiswesser line notation (WLN) system [44], SMILES [45,46], and SMARTS [47]. Three-dimensional molecular representation considers a molecule as a rigid geometrical object in space and allows a representation not only of the nature and connectivity of the atoms, but also the overall spatial configuration of molecule atoms. This representation of a molecule is called *geometrical representation* and defines a molecule in terms of atom types constituting the molecule and the set of (x, y, z)-coordinates associated to each atom. Figure 3-3 shows a geometrical representation of lactic acid. Molecular descriptors derived from this representation are referred to as *3D-molecular descriptors* or *geometrical descriptors*; several of them were proposed to measure the steric and size properties of molecules.

Several molecular descriptors derive from multiple molecular representations and can only be classified with difficulty. For example, graph invariants derived from a molecular graph weighted by properties obtained by methods of computational chemistry are both 2D and 3D descriptors. The bulk representation of a molecule describes the molecule in terms of a physical object with 3D attributes such as bulk and steric properties, surface area, and volume. The stereoelectronic representation (or lattice representation) of a molecule is a molecular description related to those molecular properties arising from electron distribution, interaction of the molecule with probes characterizing the space surrounding them (e.g., molecular interaction fields, see Chapter 4). This representation is typical of the grid-based QSAR techniques. Descriptors at this level can be considered *4D-molecular descriptors*, being characterized by a scalar field, that is, a lattice of scalar numbers, associated with the 3D molecular geometry (Figure 3-4).

GRID [38] and CoMFA [39] approaches were the first methods based on the calculation of the interaction energy between molecule and a probe. The focus of these approaches is to identify and characterize quantitatively the interactions between the

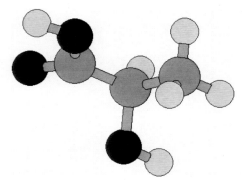

Figure 3-3. The 3D-structure representation of a molecule

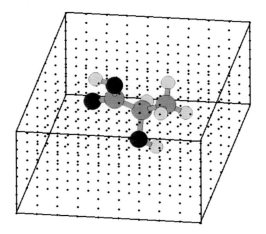

Figure 3-4. A lattice of grid points with an embedded molecule

molecule and the receptor's active site. They place molecules in a 3D lattice constituted by several thousands of evenly spaced grid points and use a probe (a steric, electrostatic, hydrophilic, etc., atom, ion, or fragment) to map the surface of the molecule on the basis of the molecule interaction with the probe. It is noteworthy that the use of interaction energies at the grid points for molecular modeling requires careful alignment of the data set molecules in such a way that each of the thousands of grid points represents, for all the molecules, the same kind of information and not spurious information due to the lack of invariance to rotation of molecules. However, an advantage of these approaches is that the scalar fields can be efficiently visualized and used to display information visually for new drug candidates, thus resulting in a very helpful tool in the drug discovery process [48,49].

Several other methods that are also based on molecular interaction fields have been successively developed. Among them are comparative molecular similarity indexes analysis (CoMSIA) [50], compass method [51], G-WHIM descriptors [52], Voronoi field analysis [53], SOMFA [54], VolSurf descriptors [55], and GRIND [56]. Finally, the *stereodynamic representation* of a molecule is a time-dependent representation which adds structural properties to the 3D representations, such as flexibility, conformational behavior, transport properties. Dynamic QSAR [57], 4D molecular similarity analysis [58], and 4D-QSAR analysis [59] are examples of a multi-conformational approach.

3.3.   TOPOLOGICAL INDEXES

3.3.1.   Molecular Graphs

A molecular graph is a topological representation of a chemical; it is a connected graph where vertexes and edges are chemically interpreted as atoms and covalent bonds [60]. The molecular graph is usually denoted as G =(V,E), where V is a set of vertexes which correspond to the molecule atoms and E is a set of elements representing the binary relationship between pairs of vertexes; unordered vertex pairs are called edges, which correspond to bonds between atoms. A molecular graph obtained excluding all the hydrogen atoms is called H-depleted molecular graph, while a molecular graph where hydrogens are included is called a H-filled molecular graph (or, simply, molecular graph). A *walk* in G is a sequence of pairwise adjacent edges leading from one vertex to another vertex in the graph; any vertex or edge can be traversed several times. The walk length is the number of edges traversed by the walk. A *path* (or *self-avoiding walk*) is a walk without any repeated vertexes. The *path length* is the number of edges associated with the path. The *topological distance* between two vertexes is the length of the shortest path between them.

   Graph invariants are mathematical quantities derived from a graph representation of the molecule and represent graph–theoretical properties that are preserved by isomorphism, i.e., properties with identical values for isomorphic graphs. A graph invariant may be a characteristic polynomial, a sequence of numbers, or a single numerical index obtained by the application of algebraic operators to graph–theoretical matrixes and whose values are independent of vertex numbering or labeling [61–69].

3.3.2.   Definition and Calculation of Topological Indexes (TIs)

Single indexes derived from a molecular graph are called *topological indexes* (TIs). These are numerical quantifiers of molecular topology that are mathematically derived in a direct and unambiguous manner from the structural graph of a molecule, usually an H-depleted molecular graph. They can be sensitive to one or more structural features of the molecule such as size, shape, symmetry, branching, and cyclicity and can also encode chemical information concerning atom type and bond multiplicity. In fact, it has been proposed to divide topological indexes into two categories: *topostructural* and *topochemical indexes* [70]. Topostructural indexes only encode information about the adjacency and distances between atoms in the molecular structure; topochemical indexes quantify information about topology, but also specific chemical properties of atoms such as their chemical identity and hybridization state.

   Topological indexes are based mainly on distances between atoms calculated by the number of intervening bonds and are thus considered *through-bond* indexes. They differ from geometrical descriptors which are, instead, considered *through-space* indexes because they are based on interatomic geometric distances [71,72]. In general, TIs do not uniquely characterize molecular topology, but different structures may have some of the same TIs. A consequence of topological index non-uniqueness is that TIs do not, in general, allow for the re-construction of a molecule. Therefore, suitably defined ordered sequences of TIs can be used to characterize molecules with higher discrimination.

   There are several ways to calculate topological indexes. Simple TIs consist of the counts of some specific graph elements; examples are the *Hosoya Z index* [73], *path counts* [74], *self-returning walk counts* [26], *Kier shape descriptors* [75], *path/walk shape indexes* [76]. However, the most common TIs are derived by applying some graph operators (e.g., the *Wiener operator*, that is, the half-sum of the matrix elements) to graph–theoretical matrixes. Among them there are the *Wiener index* [77], *spectral indexes* [69], and *Harary indexes* [78]. In the last few years, several efforts have been made to formalize the several formulae and algorithms dealing with molecular graph information: " *a graph operator applies a mathematical equation to compute a whole class of related molecular graph descriptors, using different molecular matrixes and various weighting schemes... In this way, molecular graph operators introduce a systematization of topological indexes and graph invariants by assembling together all descriptors computed with the same mathematical formula or algorithm, but with different parameters or molecular matrixes.*" [79].

The most common functions to derive topological indexes from graph–theoretical matrixes are listed in Table 3-2. Note that, in functions D_1 and D_2, the most common parameter values are $\alpha = 1/2$ and $\lambda = 1$. Moreover, it should be noted that function D_2, is restricted to pairs of adjacent vertexes, a_{ij} being the elements of the adjacency matrix. which are equal to one for pairs of adjacent vertexes, and zero otherwise. Function D_3 is used to generate descriptors derived from the matrix determinant and function D_4 descriptors that are linear combinations of the coefficients of the characteristic polynomial of a graph–theoretical matrix, such as the *Hosoya-type indexes* [80]. Function D_5 is based on the eigenvalues calculated from graph–theoretical matrixes and the related molecular descriptors are the so-called *spectral indexes*. Function D_6 makes use of the matrix row sums VS_i as the local vertex invariants (LOVIs) and, then, adds up the contributions from different graph fragments (e.g., edges), each weighted by the product of the local invariants of all the vertexes contained in the fragment; *connectivity-like indexes* [81] and *Balaban-like indexes* [81] are calculated according to this function. Function D_7 for $\alpha = 1/2$ and $\lambda = 2$ generates the *hyper-Wiener-type indexes* [81].

Table 3-2. Classical functions to derive molecular descriptors from graph–theoretical matrixes $\mathbf{M}$; n is the matrix dimension, $c(Ch(\mathbf{M}; x))_i$ is the ith coefficient of the characteristic polynomial of $\mathbf{M}$, $\Lambda(\mathbf{M})$ indicates the graph spectrum (i.e., the set of eigenvalues of $\mathbf{M}$), α and λ are real parameters. In function D_6, $VS_i(\mathbf{M})$ is the ith matrix row sum, K is the total number of selected graph fragments, and n_k the number of vertexes in kth fragment, a_{ij} indicates the elements of the adjacency matrix which are equal to one for pairs of adjacent vertexes, and zero otherwise

	Function		Function
1	$D_1(\mathbf{M};\alpha,\lambda) = \alpha \cdot \sum\limits_{i=1}^{n} \sum\limits_{j=1}^{n} [\mathbf{M}]_{ij}^{\lambda}$	5	$D_5(\mathbf{M}) = f(\Lambda(\mathbf{M}))$
2	$D_2(\mathbf{M};\alpha,\lambda) = \alpha \cdot \sum\limits_{i=1}^{n} \sum\limits_{j=1}^{n} a_{ij} \cdot [\mathbf{M}]_{ij}^{\lambda}$	6	$D_6(\mathbf{M};\alpha,\lambda) = \alpha \cdot \sum\limits_{k=1}^{K} \left(\prod\limits_{i=1}^{n_k} VS_i(\mathbf{M}) \right)_k^{\lambda}$
3	$D_3(\mathbf{M};\alpha) = \alpha \cdot \det(\mathbf{M})$	7	$D_7(\mathbf{M};\alpha,\lambda) = \alpha \cdot \sum\limits_{i=1}^{n} \sum\limits_{j=1}^{n} \left([\mathbf{M}]_{ij}^{\lambda} + [\mathbf{M}]_{ij} \right)$
4	$D_4(\mathbf{M};\alpha,\lambda) = \alpha \cdot \sum\limits_{i=0}^{n} \left| c(Ch(\mathbf{M};x))_i^{\lambda} \right|$	8	$D_8(\mathbf{M};\alpha,\lambda) = \alpha \cdot \max_{ij} \left([\mathbf{M}]_{ij}^{\lambda} \right)$

Other topological indexes can be obtained by using suitable functions applied to local vertex invariants (LOVIs); the most common functions are atom and/or bond additive, resulting into descriptors which correlate well physico-chemical properties that are atom and/or bond additive themselves. For example, *Zagreb indexes* [82], *Randić connectivity index* [83], related *higher-order connectivity indexes* [84], and the *Balaban distance connectivity index* [85] are derived according to this approach. *Local vertex invariants* (LOVIs) are numerical quantities associated with graph vertexes independent of any arbitrary vertex numbering used to characterize local properties in a molecule. They can be either purely topological if heteroatoms are not distinguished from carbon atoms, or chemical if the heteroatoms are assigned distinct values from carbon atoms, even when these are topologically equivalent [86].

Some functions to derive molecular descriptors D from local vertex invariants, denoted by L, are presented in Table 3-3. It should be noted that function D_4, that is the well-known Randić-type formula for $\alpha = 1$ and $\lambda = -1/2$, is restricted to pairs of adjacent vertexes, a_{ij} being the elements of the adjacency matrix, which are equal to one for pairs of adjacent vertexes, and zero otherwise. Function D_6 is an extension of function D_4 to any type of graph fragments as in the Kier–Hall connectivity indexes [84]. Function D_7 gives autocorrelation descriptors, while function D_8 gives maximum auto-cross-correlation descriptors. Moreover, similar functions can be applied to local edge invariants L_{ij} in place of local vertex invariants L_i so that other sets of molecular descriptors can be generated.

Another way to derive topological indexes is by generalizing the existing indexes or graph–theoretical matrixes. Moreover, *topological information indexes* are indexes based on information theory and calculated as the information content

Table 3-3. Classical functions to derive molecular descriptors from local vertex invariants. L_i and L_j are local invariants associated to the vertexes v_i and v_j, respectively; A is the number of graph vertexes, V denotes the set of graph vertexes, and $\delta(d_{ij};k)$ is a Dirac delta function equal to one for pairs of vertexes at topological distance d_{ij} equal to k, and zero otherwise. In function D_4, a_{ij} indicates the elements of the adjacency matrix which are equal to one for pairs of adjacent vertexes, and zero otherwise. In function D_6, the summation goes over fragments of a given type, K is the total number of selected graph fragments, and n_k the number of vertexes in the kth fragment

Function		Function	
1	$D_1 (L;\alpha,\lambda) = \alpha \cdot \sum_{i=1}^{A} L_i^{\lambda}$	5	$D_5 (L;\alpha,\lambda) = \alpha \cdot \sum_{i=1}^{A} \sum_{j=1}^{A} (L_i \cdot L_j)^{\lambda} \quad j \neq i$
2	$D_2 (L;\alpha,\lambda) = \alpha \cdot \left(\prod_{i=1}^{A} L_i \right)^{\lambda}$	6	$D_6 (L;\alpha,\lambda) = \alpha \cdot \sum_{k=1}^{K} \left(\prod_{i=1}^{n_k} L_i \right)_k^{\lambda}$
3	$D_3 (L;\alpha) = \alpha \cdot \max_{i \in V} (L_i)$	7	$D_7 (L;\alpha,\lambda,k) = \alpha \cdot \sum_{i=1}^{A} \sum_{j=1}^{A} (L_i \cdot L_j)^{\lambda} \cdot \delta (d_{ij}; k)$
4	$D_4 (L;\alpha,\lambda) = \alpha \cdot \sum_{i=1}^{A} \sum_{j=1}^{A} a_{ij} \cdot (L_i \cdot L_j)^{\lambda}$	8	$D_8 (L;\alpha,\lambda,k) = \alpha \cdot \max_{i,j \in V} \left[(L_i \cdot L_j)^{\lambda} \cdot \delta (d_{ij}; k) \right]$

of specified equivalence relationships on the molecular graph [87]. Several *fragment topological indexes* can be derived by any topological index calculated for molecular subgraphs [88].

Particular topological indexes are derived from weighted molecular graphs where vertexes and/or edges are weighted by quantities representing some 3D features of the molecule, such as those obtained by methods of computational chemistry. The topological indexes obtained in this way encode both information on molecular topology and molecular geometry. *BCUT descriptors* [89] are an example of these topological descriptors. *Triplet topological indexes* were proposed based on a general matrix-vector multiplication approach [90].

Several graph invariants can also be derived by the *vector-matrix-vector multiplication approach* (or *VMV approach*) proposed by Estrada [91]. This approach allows the generation of graph invariants D according to the following equation (3-2):

$$D\left(\mathbf{M}, \mathbf{v}_1, v_2; \alpha, \lambda\right) = \alpha \cdot \left(v_1^{\mathrm{T}} \cdot \mathbf{M}^{\lambda} \cdot \mathbf{v}_2\right) \qquad (3\text{-}2)$$

where $\mathbf{v}_1$ and $\mathbf{v}_2$ are two-column vectors collecting atomic properties or local vertex invariants, and $\mathbf{M}$ is a graph–theoretical matrix; α and λ are two real parameters.

Topological indexes have been successfully applied in characterizing the structural similarity/dissimilarity of molecules and in QSAR/QSPR modeling. Due to the large proliferation of graph invariants and as the result of many authors following the procedures outlined above and other general schemes, some rules are needed to critically analyze such invariants, paying particular attention to their effective role in correlating physico-chemical properties, biological responses, and other experimental responses as well as their chemical meaning. In this respect, a list of desirable attributes for topological indexes (Table 3-1) was suggested by Randić [41].

3.3.3.   Graph-Theoretical Matrixes

Molecular matrixes are the most common mathematical tool to encode structural information of molecules. Very popular molecular matrixes are the graph–theoretical matrixes, a huge number of which were proposed in the last decades in order to derive topological indexes and describe molecules from a topological point of view. Graph–theoretical matrixes are matrixes derived from a molecular graph G (often from an H-depleted molecular graph). A comprehensive collection of graph–theoretical matrixes is reported in [92]. Graph–theoretical matrixes can be either *vertex matrixes*, if both rows and columns refer to graph vertexes (atoms) and matrix elements encode some property of pairs of vertexes, or *edge matrixes*, if both rows and columns refer to graph edges (bonds) and matrix elements encode some property of pairs of edges. Vertex matrixes are square matrixes of dimension $A \times A$, A being the number of graph vertexes, while edge matrixes are square matrixes of dimension $B \times B$, B being the number of graph edges. Together with vertex matrixes and edge matrixes, *incidence matrixes* are another class of important graph–theoretical matrixes used to characterize a molecular graph. These are matrixes whose rows can represent either vertexes or edges and columns some subgraphs, such as edges, paths, or cycles.

Vertex matrixes are undoubtedly the graph–theoretical matrixes most frequently used for characterizing a molecular graph. The matrix entries encode different information about pairs of vertexes such as their connectivities, topological distances, sums of the weights of the atoms along the connecting paths; the diagonal entries can encode chemical information about the vertexes. A huge number of topological indexes were proposed from vertex matrixes.

The most important vertex matrixes are the *adjacency matrix* **A** which encodes information about vertex connectivities and the *distance matrix* **D** which also encodes information about relative locations of graph vertexes. The *adjacency matrix* **A** is symmetric with dimension $A \times A$, where A is the number of vertexes, and it is usually derived from the H-depleted molecular graph; the entries a_{ij} of the matrix equal one if vertexes v_i and v_j are adjacent (i.e., the atoms i and j are bonded), and zero otherwise [Eq. (3-3)]:

$$[\mathbf{A}]_{ij} = \begin{cases} 1 & \text{if } (i, j) \in E\,(G) \\ 0 & \text{otherwise} \end{cases} \tag{3-3}$$

where E(G) is the set of the graph edges.

The ith row sum of the adjacency matrix is called *vertex degree*, denoted by δ_i, and defined as follows [Eq. (3-4)]:

$$\delta_i = \sum_{j=1}^{A} a_{ij} \tag{3-4}$$

An example of calculation of the adjacency matrix **A** and vertex degrees δ_i is shown for the H-depleted molecular graph of 2-methylpentane.

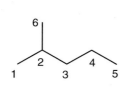

$$\mathbf{A} =$$

Atom	1	2	3	4	5	6	δ_i
1	0	1	0	0	0	0	1
2	1	0	1	0	0	1	3
3	0	1	0	1	0	0	2
4	0	0	1	0	1	0	2
5	0	0	0	1	0	0	1
6	0	1	0	0	0	0	1

The *total adjacency index*, denoted as A_V, is a measure of the graph connectedness and is calculated as the sum of all the entries of the adjacency matrix of a molecular graph, which is twice the number B of graph edges [Eq. (3-5)] [26]:

$$A_V = \sum_{i=1}^{A} \sum_{j=1}^{A} a_{ij} = \sum_{i=1}^{A} \delta_i = 2 \cdot B \qquad (3\text{-}5)$$

For example, the total adjacency index of 2-methylpentane is $A_V = 1 + 3 + 2 + 2 + 1 + 1 = 10$, which is twice the number of edges equal to five in the H-depleted molecular graph of this molecule. The total adjacency index is sometimes calculated as the half-sum of the adjacency matrix elements.

The *distance matrix* **D** is a symmetric $A \times A$ matrix whose elements are the topological distances between all the pairs of graph vertexes; the topological distance d_{ij} is the number of edges along the shortest path $^{min}P_{ij}$ between the vertexes v_i and v_j [Eq. (3-6)]:

$$[D]_{ij} = \begin{cases} d_{ij} = |^{min}P_{ij}| & \text{if } i \neq j \\ 0 & \text{if } i = j \end{cases} \qquad (3\text{-}6)$$

The off-diagonal entries of the distance matrix equal one if vertexes v_i and v_j are adjacent (that is, the atoms i and j are bonded and $d_{ij} = a_{ij} = 1$, where a_{ij} are elements of the adjacency matrix **A**) and are greater than one otherwise. For vertex- and edge-weighted graphs, the distance matrix entry i–j could be defined as the minimum sum of edge weights along the path between the vertexes v_i and v_j, which is not necessarily the shortest possible path between them, or otherwise as the sum of the weights of the edges along the shortest path between the considered vertexes. Diagonal entries usually are the vertex weights. Different weighting schemes for vertex and/or edges were proposed from which a number of weighted distance matrixes were derived [93].

The *distance degree* (or *distance sum*), denoted as σ_i, is defined as the distance matrix row sum [Eq. (3-7)]:

$$\sigma_i = \sum_{j=1}^{A} d_{ij} \qquad (3\text{-}7)$$

The maximum value entry in the ith row of the distance matrix is called *atom eccentricity* (or *vertex eccentricity*) and denoted as η_i [Eq. (3-8)]:

$$\eta_i = \max_j \left(d_{ij} \right) \qquad (3\text{-}8)$$

The atom eccentricity is a local vertex invariant representing the maximum distance from a vertex to any other vertex in the graph. An example of calculation of the distance matrix **D**, vertex distance degrees σ_i, and atom eccentricities η_i is here reported for 2-methylpentane.

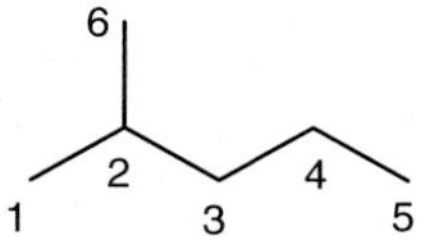

D =

Atom	1	2	3	4	5	6	σ_i	η_i
1	0	1	2	3	4	2	12	4
2	1	0	1	2	3	1	8	3
3	2	1	0	1	2	2	8	2
4	3	2	1	0	1	3	10	3
5	4	3	2	1	0	4	14	4
6	2	1	2	3	4	0	12	4

Vertex distance degrees are local vertex invariants: high values are observed for terminal vertexes (e.g., in 2-methylpentane, $\sigma = 12$ for terminal vertexes 1 and 6, and $\sigma = 14$ for terminal vertex 5), while low values for central vertexes. Moreover, among the terminal vertexes, vertex distance degrees are small if the vertex is next to a branching site (e.g., in 2-methylpentane, vertexes 1 and 6 are directly bonded to vertex 2 which represents a branching site) and larger if the terminal vertex is far away (e.g., in 2-methylpentane, terminal vertex 5 is three bonds far away from the branching site 2).

The half-sum of all the elements d_{ij} of the distance matrix [73], which is equal to the half-sum of the distance degrees σ_i of all the vertexes [94], is the well known *Wiener index W*, which is one of the most popular topological indexes used in QSAR modeling [Eq. (3-9)] [77]:

$$W = \frac{1}{2} \cdot \sum_{i=1}^{A} \sum_{j=1}^{A} d_{ij} = \frac{1}{2} \cdot \sum_{i=1}^{A} \sigma_i \tag{3-9}$$

where A is the number of graph vertexes.

The total sum of the entries of the distance matrix is another topological index called *Rouvray index* and denoted as I_{ROUV}, which is twice the Wiener index W [Eq. (3-10)]:

$$I_{\mathrm{ROUV}} = \sum_{i=1}^{A} \sum_{j=1}^{A} d_{ij} = \sum_{i=1}^{A} \sigma_i = 2\,W \tag{3-10}$$

For example, in 2-methylpentane, the Rouvray index derived from distance values is $I_{\mathrm{ROUV}} = 10\times1 + 10\times2 + 6\times3 + 4\times4 = 64$ or, alternatively, derived from distance degrees is $I_{\mathrm{ROUV}} = 12 + 8 + 8 + 10 + 14 + 12 = 64$.

From the vertex eccentricity definition, a graph can be immediately characterized by two molecular descriptors known as *topological radius R* and *topological diameter D*. The topological radius of a molecule is defined as the minimum vertex eccentricity and the topological diameter is defined as the maximum vertex eccentricity, according to the following equation (3-11) [26]:

$$R = \min_i (\eta_i) \quad \text{and} \quad D = \max_i (\eta_i) \tag{3-11}$$

From the topological radius and the topological diameter, a graph–theoretical shape index, called *Petitjean index*, is defined as follows [Eq. (3-12)] [95]:

$$I_2 = \frac{D - R}{R} \qquad 0 \le I_2 \le 1 \tag{3-12}$$

For strictly cyclic graphs, $D = R$ and $I_2 = 0$. For example, the radius of 2-methylpentane is 2, while the diameter is 4; the Petitjean index is 1.

The *detour matrix* $\mathbf{\Delta}$ of a graph G (or *maximum path matrix*) is a square symmetric $A \times A$ matrix, A being the number of graph vertexes, whose entry i–j is the length of the longest path from vertex v_i to vertex v_j ($^{max}P_{ij}$) [Eq. (3-13)] [26]:

$$[\mathbf{\Delta}]_{ij} = \begin{cases} \mathbf{\Delta}_{ij} = |^{max}P_{ij}| & \text{if } i \ne j \\ 0 & \text{if } i = j \end{cases} \tag{3-13}$$

This definition is the exact "opposite" of the definition of the distance matrix whose off-diagonal elements are the lengths of the shortest paths between the considered vertexes. However, the distance and detour matrixes coincide for acyclic graphs, there being only one path connecting any pair of vertexes.

The *maximum path sum* of the ith vertex, denoted by $MPVS_i$, is a local vertex invariant defined as the sum of the lengths of the longest paths between vertex v_i and any other vertex in the molecular graph, i.e., Eq. (3-14):

$$MPVS_i = \sum_{j=1}^{A} [\mathbf{\Delta}]_{ij} \tag{3-14}$$

A Wiener-type index, originally called the *MPS topological index* [96] but usually known as the *detour index* and denoted by w [97], was proposed as the half-sum of the detour distances between any two vertexes in the molecular graph [Eq. (3-15)]:

$$w = \frac{1}{2} \cdot \sum_{i=1}^{A} \sum_{j=1}^{A} [\mathbf{\Delta}]_{ij} = \frac{1}{2} \cdot \sum_{i=1}^{A} MPVS_i \tag{3-15}$$

where $MPVS_i$ is the maximum path sum of the ith vertex. Calculation of detour matrix $\mathbf{\Delta}$, maximum path sums $MPVS_i$, and detour index w is illustrated with that for ethylbenzene.

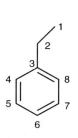

$$\Delta = \begin{array}{c|cccccccc|c} Atom & 1 & 2 & 3 & 4 & 5 & 6 & 7 & 8 & MSVP_i \\ \hline 1 & 0 & 1 & 2 & 7 & 6 & 5 & 6 & 7 & 34 \\ 2 & 1 & 0 & 1 & 6 & 5 & 4 & 5 & 6 & 28 \\ 3 & 2 & 1 & 0 & 5 & 4 & 3 & 4 & 5 & 24 \\ 4 & 7 & 6 & 5 & 0 & 5 & 4 & 3 & 4 & 34 \\ 5 & 6 & 5 & 4 & 5 & 0 & 5 & 4 & 3 & 32 \\ 6 & 5 & 4 & 3 & 4 & 5 & 0 & 5 & 4 & 30 \\ 7 & 6 & 5 & 4 & 3 & 4 & 5 & 0 & 5 & 32 \\ 8 & 7 & 6 & 5 & 4 & 3 & 4 & 5 & 0 & 34 \end{array}$$

$$w = \frac{1}{2} \times (34 + 28 + 24 + 34 + 32 + 30 + 32 + 34) = 124$$

The *Laplacian matrix* **L** is a square $A \times A$ symmetric matrix, A being the number of graph vertexes, defined as the difference between the vertex degree matrix **V** and the adjacency matrix **A** [Eq. (3-16)] [98,99]:

$$\mathbf{L} = \mathbf{V} - \mathbf{A} \qquad (3\text{-}16)$$

where **V** is a diagonal matrix of dimension $A \times A$, whose diagonal entries are the vertex degrees δ_i. The entries of the Laplacian matrix formally are given by Eq. (3-17):

$$[\mathbf{L}]_{ij} = \begin{cases} \delta_i & \text{if } i = j \\ -1 & \text{if } (i,j) \in \mathrm{E(G)} \\ 0 & \text{if } (i,j) \notin \mathrm{E(G)} \end{cases} \qquad (3\text{-}17)$$

where E(G) is the set of edges of the molecular graph G. Important molecular descriptors are derived from the eigenvalues of the Laplacian matrix (see spectral indexes). An example of calculation of the Laplacian matrix **L** is shown for 2-methylpentane.

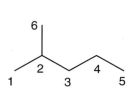

$$\mathbf{L} = \begin{array}{c|cccccc} Atom & 1 & 2 & 3 & 4 & 5 & 6 \\ \hline 1 & 1 & -1 & 0 & 0 & 0 & 0 \\ 2 & -1 & 3 & -1 & 0 & 0 & -1 \\ 3 & 0 & -1 & 3 & -1 & 0 & 0 \\ 4 & 0 & 0 & -1 & 2 & -1 & 0 \\ 5 & 0 & 0 & 0 & -1 & 1 & 0 \\ 6 & 0 & -1 & 0 & 0 & 0 & 1 \end{array}$$

3.3.4.   Connectivity Indexes

Connectivity indexes are among the most popular topological indexes and are calculated from the vertex degrees δ_i of the atoms in the H-depleted molecular graph.
The *Randić connectivity index* was the first connectivity index proposed [100]; it is
defined as [Eq. (3-18)]

$$\chi_R \equiv {}^1\chi = \sum_{i=1}^{A-1} \sum_{j=i+1}^{A} a_{ij} \cdot (\delta_i \cdot \delta_j)^{-1/2} \tag{3-18}$$

where the summation goes over all the pairs of vertexes v_i and v_j in the molecular
graph, but only contributions from pairs of adjacent vertexes are accounted for, a_{ij}
being the elements of the adjacency matrix **A**.

The term $(\delta_i \cdot \delta_j)^{-1/2}$ for each pair of adjacent vertexes is called *edge connectivity*
and can be used to characterize edges as a primitive bond order accounting for bond
accessibility, i.e., the accessibility of a bond to encounter another bond in intermolecular interactions, as the reciprocal of the vertex degree δ is the fraction of the
total number of non-hydrogen sigma electrons contributing to each bond formed
with a particular atom [101]. This interpretation places emphasis on the possibility
of bimolecular encounters among molecules, reflecting the collective influence of
the accessibilities of the bond in each molecule to other molecules in its immediate
environment.

Kier and Hall defined [84,102] a general scheme based on the Randić index to
also calculate zero-order and higher-order descriptors; these are called *molecular
connectivity indexes* (MCIs), also known as *Kier–Hall connectivity indexes*. They
are calculated by the following equations (3-19):

$$ {}^0\chi = \sum_{i=1}^{A} \delta_i^{-1/2} \quad {}^1\chi = \sum_{b=1}^{B} (\delta_i \cdot \delta_j)_b^{-1/2} \quad {}^2\chi = \sum_{k=1}^{{}^2P} (\delta_i \cdot \delta_l \cdot \delta_j)_k^{-1/2} \quad {}^m\chi_t = \sum_{k=1}^{K} \left(\prod_{i=1}^{n} \delta_i \right)_k^{-1/2} \tag{3-19} $$

where k runs over all of the mth order subgraphs constituted by n atoms ($n = m+1$
for acyclic subgraphs); K is the total number of mth order subgraphs present in
the molecular graph. The product is over the simple vertex degrees δ of all the
vertexes involved in each subgraph. The subscript "t" refers to the type of molecular
subgraph and is ch for chain or ring, pc for path–cluster, c for cluster, and p for path.
Obviously, the first-order Kier–Hall connectivity index is the Randić connectivity
index.

Calculation of 0-, 1-, and 2-order connectivity indexes is illustrated for 2-
methylpentane:

Atoms	1	2	3	4	5	6
δ_i	1	3	2	2	1	1

$$^0\chi = \delta_1^{-1/2} + \delta_2^{-1/2} + \delta_3^{-1/2} + \delta_4^{-1/2} + \delta_5^{-1/2} + \delta_6^{-1/2} =$$
$$= 1^{-1/2} + 3^{-1/2} + 2^{-1/2} + 2^{-1/2} + 1^{-1/2} + 1^{-1/2} = 4.992$$

$$^1\chi = \left(\delta_1 \times \delta_2\right)^{-1/2} + \left(\delta_2 \times \delta_3\right)^{-1/2} + \left(\delta_3 \times \delta_4\right)^{-1/2} + \left(\delta_4 \times \delta_5\right)^{-1/2} + \left(\delta_2 \times \delta_6\right)^{-1/2} =$$
$$= \left(1 \times 3\right)^{-1/2} + \left(3 \times 2\right)^{-1/2} + \left(2 \times 2\right)^{-1/2} + \left(2 \times 1\right)^{-1/2} + \left(3 \times 1\right)^{-1/2} = 2.770$$

$$^2\chi = \left(\delta_1 \times \delta_2 \times \delta_3\right)^{-1/2} + \left(\delta_2 \times \delta_3 \times \delta_4\right)^{-1/2} + \left(\delta_3 \times \delta_4 \times \delta_5\right)^{-1/2} + \left(\delta_1 \times \delta_2 \times \delta_6\right)^{-1/2} + \left(\delta_3 \times \delta_2 \times \delta_6\right)^{-1/2} =$$
$$= \left(1 \times 3 \times 2\right)^{-1/2} + \left(3 \times 2 \times 2\right)^{-1/2} + \left(2 \times 2 \times 1\right)^{-1/2} + \left(1 \times 3 \times 1\right)^{-1/2} + \left(2 \times 3 \times 1\right)^{-1/2} = 2.183$$

Connectivity-like indexes are molecular descriptors calculated applying the same mathematical formula of the connectivity indexes, but substituting the vertex degree δ with any local vertex invariant (LOVI) [Eq. (3-20)]:

$$^m Chi_t(\text{L}) = \sum_{k=1}^{K} \left(\prod_{i=1}^{n} \text{L} \right)_k^{-1/2} \tag{3-20}$$

where L_i is the general symbol for local vertex invariants, the summation goes over all the subgraphs of type t constituted by n atoms and m edges; K is the total number of such mth order subgraphs present in the molecular graph, and each subgraph is weighted by the product of the local invariants associated to the vertexes contained in the subgraph. Connectivity-like indexes may also be calculated by replacing local vertex invariants L_i with physico-chemical atomic properties P_i.

The general formula for the calculation of connectivity-like indexes, which uses the row sums VS_i of a graph–theoretical matrix as the local vertex invariants, was called by Ivanciuc *Chi* operator [81]. Specifically, for any square symmetric $(A \times A)$ matrix $\mathbf{M}(w)$ representing a molecular graph with A vertexes and a weighting scheme w, the *Chi* operator is defined as follows [Eq. (3-21)]:

$$^m Chi\,(\mathbf{M};w) = \sum_{k=1}^{K} \left(\prod_{i=1}^{n} VS_i\,(\mathbf{M},w) \right)_k^{-1/2} \tag{3-21}$$

where VS_i indicates the matrix row sums.

Moreover, *generalized connectivity indexes* are a generalization of the Kier–Hall connectivity indexes in terms of a variable exponent λ as given by Eq. (3-22):

$$^m\chi_t = \sum_{k=1}^{K} \left(\prod_{i=1}^{n} \delta_i \right)_k^{\lambda} \tag{3-22}$$

where λ is any real exponent. If $\lambda = 1$ and $m = 1$, the *second Zagreb index* M_2 [82] is obtained:

$$M_2 = \sum_{b=1}^{B} \left(\delta_{b(1)} \cdot \delta_{b(2)} \right)_b \tag{3-23}$$

where the summation goes over all the edges in the molecular graph and B is the total number of edges in the graph; the subscripts $b(1)$ and $b(2)$ represent the two vertexes connected by the edge b.

Values of $\lambda = -1$ and $\lambda = 1/2$ were considered by Altenburg [103] and values of $\lambda = -1/3$ and $\lambda = -1/4$ were also investigated [104].

Related to Randić-like indexes are the *Balaban-like indexes*, which only differ by the normalization factor [Eq. (3-24)]:

$$J(\mathbf{M};w) = \frac{B}{C+1} \cdot \sum_{i=1}^{A-1} \sum_{j=i+1}^{A} a_{ij} \cdot \left(VS_i(\mathbf{M};w) \cdot VS_j(\mathbf{M};w) \right)^{-1/2} \tag{3-24}$$

where $\mathbf{M}$ is a graph–theoretical matrix, a_{ij} the elements of the adjacency matrix $\mathbf{A}$ equal to one for pairs of adjacent vertexes and zero otherwise, and w the weighting scheme applied to represent molecules containing heteroatoms and/or multiple bonds; VS is the vertex sum operator applied to the matrix $\mathbf{M}$. A is the number of graph vertexes, B the number of graph edges, and C the cyclomatic number, i.e., the number of rings. The denominator $C + 1$ is a normalization factor against the number of rings in the molecule.

This formula for the calculation of the Balaban-like indexes was called the *Ivanciuc-Balaban operator* by Ivanciuc [81,105]. It is a generalization of the *Balaban distance connectivity index* denoted by J and defined as [Eq. (3-25)] [85]

$$J = \frac{B}{C+1} \cdot \sum_{i=1}^{A-1} \sum_{j=i+1}^{A} a_{ij} \cdot (\sigma_i \cdot \sigma_j)^{-1/2} \tag{3-25}$$

where σ_i and σ_j are the vertex distance degrees of the vertexes v_i and v_j, which are the row sums of the distance matrix $\mathbf{D}$. The Balaban index is a very discriminating molecular descriptor and its values do not increase substantially with molecule size or number of rings.

3.3.5.   Characteristic Polynomial

The characteristic polynomial of the molecular graph is the characteristic polynomial of a graph–theoretical matrix $\mathbf{M}$ derived from the graph [Eq. (3-26)] [105–107]:

$$Ch(\mathbf{M};w;x) \det(x\mathbf{I} - \mathbf{M}(w)) = \sum_{i=0}^{n}(-1)^i c_i x^{n-i} = x^n - c_1 x^{n-1} + c_2 x^{n-2}$$

$$+ \cdots + (-1)^{n-1} c_{n-1} x + (-1)^n c_n \qquad (3\text{-}26)$$

where "det" denotes the matrix determinant, $\mathbf{I}$ is the identity matrix of dimension $n \times n$, x is a scalar variable, and c_i are the $n + 1$ polynomial coefficients; $\mathbf{M}(w)$ is any square $n \times n$ matrix computed on weighted or unweighted molecular graphs; w is the weighting scheme applied to the molecular graph in order to encode chemical information. Note that $w = 1$ denotes unweighted graphs. If $\mathbf{M}$ is a vertex matrix, then n is equal to A, the number of graph vertexes, while, if $\mathbf{M}$ is an edge matrix, then n is equal to B, the number of graph edges. Polynomial coefficients are graph invariants and are thus related to the structure of a molecular graph.

A large number of graph polynomials have been proposed in the literature. They differ from each other according to the molecular matrix $\mathbf{M}$ they are derived from and the weighting scheme w used to characterize heteroatoms and bond multiplicity of molecules. The most known polynomial is the characteristic polynomial of the adjacency matrix ($\mathbf{M} = \mathbf{A}$), which is usually referred to as the *graph characteristic polynomial* [Eq. (3-27)] [26]:

$$Ch(\mathbf{A};1; x) = \det(x\mathbf{I} - \mathbf{A}) \qquad (3\text{-}27)$$

For any acyclic graph, the absolute values of $Ch(\mathbf{A}; 1; x)$ coefficients are equal to the coefficients of the Z-counting polynomial $Q(G; x)$, which are the non-adjacent numbers $a(G, k)$ of order k, i.e., the numbers of k mutually non-incident edges [108]. An H-depleted molecular graph and adjacency matrix $\mathbf{A}$ are shown for 2-methylpentane:

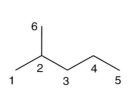

Atom	1	2	3	4	5	6
1	0	1	0	0	0	0
2	1	0	1	0	0	1
3	0	1	0	1	0	0
4	0	0	1	0	1	0
5	0	0	0	1	0	0
6	0	1	0	0	0	0

$\mathbf{A} =$

The characteristic polynomial of the adjacency matrix of 2-methylpentane is given by Eq. (3-28):

$$Ch(\mathbf{A};x) = x^6 - 5 \cdot x^4 + 5 \cdot x^2 \qquad (3\text{-}28)$$

where coefficients c_1, c_3, c_5, and c_6 are zero. Absolute values of non-zero coefficients are: $|c_0| = 1$, which corresponds to the non-adjacent number of zero order, $a(G, 0) = 1$ (by definition); $|c_2| = 5$, which corresponds to the non-adjacent number of first order, $a(G, 1) = 5$ (the number of graph edges); $|c_4| = 5$, which corresponds to the non-adjacent number of second order, $a(G, 2) = 5$ (the number of ways two edges may be selected so that they are non-adjacent).

Depending on the elements of the matrix **M**, a characteristic polynomial can have very large coefficients, and spanning the x axis, often, asymptotic curves are obtained, whose characteristic points are not very representative as graph descriptors. To deal with this problem, the characteristic polynomial can be transformed according to some Hermite-like wave functions for graphs, as given by Eq. (3-29) [109]:

$$\Psi \equiv Ch\,(\mathbf{M};w;x) \cdot \exp\left(-\frac{x^2}{2}\right) \tag{3-29}$$

where $Ch(\mathbf{M};w;x)$ is the characteristic polynomial of a graph. The most significant difference is that the area under the curve becomes finite in this approach, thus allowing the definition of more reliable graph invariants, such as the *area under the curve* (AUC), the *maximum* Ψ *value* (Ψ^{max}) and the *maximum amplitude* (MA) of the obtained sinusoidal curve.

By analogy with the *Hosoya Z index* [73] which, for acyclic graphs, can be calculated as the sum of the absolute values of the coefficients of the characteristic polynomial of the adjacency matrix, the *stability index* (or *modified Z index*) is a molecular descriptor calculated for any graph as the sum of the absolute values of the coefficients c_{2i} appearing alternatively in the characteristic polynomial of the adjacency matrix [Eq. (3-30)] [110]:

$$\tilde{Z} = \sum_{i=0}^{[A/2]} |c_{2i}| \tag{3-30}$$

where the square brackets indicate the greatest integer not exceeding $A/2$ and A is the number of graph vertexes. The same approach applied to the distance polynomial led to the definition of the *Hosoya Z′ index* (or *Z′ index*) [Eq. (3-31)] [111]:

$$Z' = \sum_{i=0}^{A} |c_i| \tag{3-31}$$

where c_i are the coefficients of the distance polynomial of the molecular graph.

An extension of the Z' index are the *Hosoya-type indexes*, which are defined as the sum of the absolute values of the coefficients of the characteristic polynomial of any square graph–theoretical matrix **M** [Eq. (3-32)] [80,105]:

$$\text{Ho}\,(\mathbf{M};w) = \sum_{i=0}^{n} |c_i| \qquad (3\text{-}32)$$

where n is the matrix dimension and w the weighting scheme applied to compute the matrix $\mathbf{M}$. The formula for the calculation of Hosoya-type indexes was called by Ivanciuc the *Hosoya operator*. For any graph, when $\mathbf{M}$ is the distance matrix of a simple graph, $\text{Ho}(\mathbf{D};1) = Z'$, when $\mathbf{M}$ is the adjacency matrix of a simple graph, $\text{Ho}(\mathbf{A};1) = \tilde{Z}$; moreover, for acyclic graphs, when $\mathbf{M}$ is the adjacency matrix of a simple graph, $\text{Ho}(\mathbf{A};1) = \tilde{Z} = Z$ (Hosoya Z index).

3.3.6. Spectral Indexes

Spectral indexes are molecular descriptors defined in terms of the eigenvalues of a square graph–theoretical matrix $\mathbf{M}$ of size $(n \times n)$. The eigenvalues are the roots of the characteristic polynomial of the matrix $\mathbf{M}$ and the set of the eigenvalues is the matrix spectrum $\mathbf{\Lambda}(\mathbf{M}) = \{\lambda_1, \lambda_2, \ldots, \lambda_n\}$; the eigenvalues are conventionally labeled so that $\lambda_1 \geq \lambda_2 \geq \cdots \geq \lambda_n$. The most common eigenvalue functions used to derive spectral indexes are given below in a general form which can be applied to any molecular matrix $\mathbf{M}(w)$, calculated with the weighting scheme w [Eq. (3-33)] [69,112]:

$$SpSum^k\,(\mathbf{M},w) = \sum_{i=1}^{n} |\lambda_i|^k \qquad SpSum_+^k\,(\mathbf{M},w) = \sum_{i=1}^{n^+} \left(\lambda_i^+\right)^k \qquad SpSum_-^k\,(\mathbf{M},w) = \sum_{i=1}^{n^-} \left|\lambda_i^-\right|^k$$

$$SpAD\,(\mathbf{M},w) = \sum_{i=1}^{n} |\lambda_i - \bar{\lambda}| \qquad SpMAD\,(\mathbf{M},w) = \sum_{i=1}^{n} |\lambda_i - \bar{\lambda}|/n$$

$$SpMin\,(\mathbf{M},w) = \min_i\{\lambda_i\} \qquad SpMax\,(\mathbf{M},w) = \max_i\{\lambda_i\}$$

$$SpAMax\,(\mathbf{M},w) = \max_i\{|\lambda_i|\} \qquad SpDiam\,(\mathbf{M},w) = SpMax - SpMin$$

$$(3\text{-}33)$$

where k is a real exponent, usually taken to be equal to one; for negative values of k, eigenvalues equal to zero must not be considered. For $k = 1$, *SpSum* is the sum of the n absolute values of the spectrum eigenvalues; this quantity calculated on the adjacency matrix of simple graphs *SpSum*($\mathbf{A}$) was called *graph energy* and denoted by E [113,114]; the same quantity derived from the Laplacian matrix was called *Laplacian graph energy* [115]; *SpSum$_+$* is the sum of the n^+ positive eigenvalues; *SpSum$_-$* is the sum of the absolute values of the n^- negative eigenvalues; *SpAD* is the sum of the absolute deviations of the eigenvalues from their mean and is called *generalized graph energy* [112]; *SpMAD* is the mean absolute deviation and is called *generalized average graph energy* [112]; *SpMin* is the minimum eigenvalue; *SpMax* is the maximum eigenvalue, called *leading eigenvalue* or *spectral radius*; *MaxSpA* is the maximum absolute value of the spectrum; and *SpDiam* is the *spectral diameter* of the molecular matrix, defined as the difference between *SpMax* and *SpMin*. These kinds of functions were called by Ivanciuc *matrix spectrum operators* [116]. It has been demonstrated that the leading eigenvalue of a symmetric matrix $\mathbf{M}$ is bounded from above and from below by its largest and smallest row sum [Eq. (3-34)]:

$$\min_i \left[VS_i \left(\mathbf{M} \right) \right] \le MaxSp \left(\mathbf{M} \right) \equiv \lambda_1 \left(\mathbf{M} \right) \le \max_i \left[VS_i \left(\mathbf{M} \right) \right] \tag{3-34}$$

where *VS* indicates the matrix row sums.

Spectral moments of the matrix $\mathbf{M}(w)$ are molecular descriptors defined in terms of the *k*th power of the eigenvalues [Eq. (3-35)]:

$$\mu^k \left(\mathbf{M};w \right) = \sum_{i=1}^{n} \lambda_i^k \tag{3-35}$$

where $k = 1,\ldots,n$ is the order of the spectral moment. It is noteworthy that for even *k* values, spectral moments μ^k coincide with spectral indexes $SpSum^k$.

Spectral indexes and spectral moments were tested in QSAR/QSPR modeling, calculated from a number of graph–theoretical matrixes [117]. Important spectral indexes are defined in terms of the eigenvalues of the adjacency matrix $\mathbf{A}$; these eigenvalues take both positive and negative values, their sum being equal to zero. The largest eigenvalue of adjacency matrix $\mathbf{A}$ is among the most popular graph invariants and is known as the *Lovasz-Pelikan index* λ_1^{LP} [118]: $\lambda_1^{LP} \equiv SpMax \left(\mathbf{A} \right)$ This eigenvalue has been suggested as an index for molecular branching, the smallest values corresponding to chain graphs and the highest to the most branched graphs. It is not a very discriminatory index because in many cases the same value is obtained for two or more non-isomorphic graphs. The eigenvalues of the Laplacian matrix $\mathbf{L}$ have some relevant properties [119,120]; among these, three important ones are

(a) the Laplacian eigenvalues are non-negative numbers;
(b) the last eigenvalue λ_A is always equal to zero;
(c) the eigenvalue λ_{A-1} is greater than zero if, and only if, the graph G is connected; therefore, for a molecular graph all the Laplacian eigenvalues except the last are positive numbers.

Moreover, the sum of the positive eigenvalues is equal to twice the number *B* of graph edges, i.e., Eq. [3-36]

$$\sum_{i=1}^{A-1} \lambda_i = 2 \cdot B \tag{3-36}$$

The sum of the reciprocal positive eigenvalues was proposed as a molecular descriptor [121,122] and called the *quasi-Wiener index* W^* [123]; it is defined as [Eq. (3-37)]

$$W^* = A \cdot \sum_{i=1}^{A-1} \frac{1}{\lambda_i} \tag{3-37}$$

For acyclic graphs, the quasi-Wiener index W^* coincides with the Wiener index *W*, whereas for cycle-containing graphs the two descriptors differ. The product of the positive $A - 1$ eigenvalues of the Laplacian matrix gives the *spanning tree number* T^* of the molecular graph G as [Eq. (3-38)]

$$T^* = \frac{1}{A} \cdot \prod_{i=1}^{A-1} \lambda_i = \frac{|a|}{A} \tag{3-38}$$

where the spanning tree is a connected acyclic subgraph containing all the vertexes of G [120]. The term a in the second equality is the coefficient of the linear term in the *Laplacian polynomial* [124]. The number of spanning trees of a graph is used as a measure of molecular complexity for polycyclic graphs; it increases with the complexity of the molecular structure. Moreover, the *spanning-tree density* (STD) and the *reciprocal spanning-tree density* (RSTD) were defined as [Eq. (3-39)] [125]

$$\text{STD} = \frac{T^*}{{}^e N} \quad \text{STD} \leq 1 \quad \text{RSTD} = \frac{{}^e N}{T^*} \quad \text{RSTD} \geq 1 \tag{3-39}$$

where ${}^e N$ is the number of ways of choosing any $A - 1$ edges belonging to the set E(G) of graph edges. RSTD was proposed as a measure of *intricacy* of a graph, that is, the larger RSTD the more intricate G. Also derived from the Laplacian matrix are the *Mohar indexes* TI_1 and TI_2, defined as Eqs. (3-40) and (3-41):

$$\text{TI}_1 = 2 \cdot A \cdot \log\left(\frac{B}{A}\right) \cdot \sum_{i=1}^{A-1} \frac{1}{\lambda_i} = 2 \cdot \log\left(\frac{B}{A}\right) \cdot W^* \tag{3-40}$$

$$\text{TI}_2 = \frac{4}{A \cdot \lambda_{A-1}} \tag{3-41}$$

where λ_{A-1} is the first non-zero eigenvalue and W^* the quasi-Wiener index [120]. Being $W^* = W$ for acyclic graphs, it also derives that the first Mohar index TI_1 is closely related to the Wiener index W for acyclic graphs.

3.4. AUTOCORRELATION DESCRIPTORS

3.4.1. Introduction

Spatial autocorrelation coefficients are frequently used in molecular modeling and QSAR to account for spatial distribution of molecular properties. The simplest descriptor P for a molecular property is obtained by summing the (squared) atomic property values. Mathematically it is defined as

$$P = \sum_{i=1}^{A} p_i^2 \tag{3-42}$$

where A is the number of atoms in a molecule and P the global property which depends on the kind of molecule atoms and not on the molecular structure; p_i is the property of the ith atom. An extension of this global property descriptor that combines chemical information given by property values in specified molecule regions and structural information are the spatial autocorrelation descriptors. These are based on a conceptual dissection of the molecular structure and the application of an autocorrelation function to molecular properties measured in different molecular regions. Autocorrelation functions AC_k for ordered discrete sequence of n values $f(x_i)$ are based on summation of the products of the ith value and the $(i + k)$th value as.

$$AC_k = \frac{1}{(n-k) \cdot \sigma^2} \cdot \sum_{i=1}^{n-k} \left[(f(x_i) - \mu) \cdot (f(x_{i+k}) - \mu) \right] \qquad (3\text{-}43)$$

where $f(x)$ is any function of the variable x and k is the *lag* representing an interval of x, σ^2 is the variance of the function values, and μ their mean. The lag assumes values between 1 and K, where the maximum value K can be $n - 1$; however, in several applications, K is chosen equal to a small number (K < 8). A lag value of zero corresponds to the sum of the square-centered values of the function. The function $f(x)$ is usually a time-dependent function such as a time-dependent electrical signal, or a spatial-dependent function such as the population density in space. Then, autocorrelation measures the strength of a relationship between observations as a function of the time or space separation between them [126]. Autocorrelation descriptors of chemical compounds are calculated by using various molecular properties that can be represented at the atomic level or molecular surface level or else.

Based on the same principles as the autocorrelation descriptors, but calculated contemporarily on two different properties $f(x)$ and $g(x)$, *cross-correlation descriptors* are calculated to measure the strength of relationships between the two considered properties. For any two-ordered sequences comprised of a number of discrete values, the cross-correlation is calculated by summing the products of the ith value of the first sequence and the $(i + k)$th value of the second sequence, as

$$CC_k = \frac{1}{(n-k) \cdot \sigma_{f(x)} \cdot \sigma_{g(x)}} \cdot \sum_{i=1}^{n-k} \left[(f(x_i) - \mu_{f(x)}) \cdot (g(x_{i+k}) - \mu_{g(x)}) \right] \qquad (3\text{-}44)$$

where n is the lowest cardinality of the two sets.

The most common spatial autocorrelation molecular descriptors are obtained taking the molecule atoms as the set of discrete points in space and an atomic property as the function evaluated at those points. Common weighting schemes w used to describe atoms in the molecule are atomic masses, van der Waals volumes, atomic electronegativities, atomic polarizabilities, covalent radii, etc. Alternatively, the weighting scheme for atoms can be based on quantities, which are local vertex invariants derived from the molecular graph, such as the topological vertex degrees (i.e., the number of adjacent vertexes) and the Kier–Hall intrinsic states or E-state indexes [127]. For spatial autocorrelation molecular descriptors calculated on a molecular graph, the lag k coincides with the topological distance between any pair of vertexes (i.e., the number of edges along the shortest path between two vertexes).

Autocorrelation descriptors can also be calculated from 3D-spatial molecular geometry. In this case, the distribution of a molecular property can be evaluated by a mathematical function $f(x,y,z)$, x, y, and z being the spatial coordinates, either defined for each point of molecular space or molecular surface (i.e., a continuous property such as electronic density or molecular interaction energy) or only for points occupied by atoms (i.e., atomic properties) [128–130].

The plot of an ordered sequence of autocorrelation descriptors from lag 0 to lag K is called *autocorrelogram*; this is a vectorial descriptor usually used to describe a chemical compound in similarity/diversity analysis. Autocorrelation descriptors have been demonstrated to be useful in QSAR studies as they are unique for a given geometry, are sensitive to changes in conformation, and do not require any molecule alignment being invariant to roto-translation. A typical disadvantage of all the autocorrelation descriptors might be that the original information on the molecular structure or surface cannot be reconstructed.

3.4.2. Moreau–Broto Autocorrelation Descriptors

Moreau and Broto were the researchers who applied first an autocorrelation function to the molecular graph to measure the distribution of atomic properties on the molecule topology [131–133]. They termed the final vectorial descriptor comprised of autocorrelation functions *autocorrelation of a topological structure (ATS)*. This was calculated as follows [Eq. (3-45)]:

$$\mathrm{ATS}_k = \frac{1}{2} \cdot \sum_{i=1}^{A} \sum_{j=1}^{A} w_i \cdot w_j \cdot \delta(d_{ij};k) \qquad (3\text{-}45)$$

where w is any atomic property, A is the number of atoms in a molecule, k is the lag, and d_{ij} is the topological distance between ith and jth atoms; $\delta(d_{ij};k)$ is a Kronecker delta function equal to 1 if $d_{ij} = k$, zero otherwise. The autocorrelation ATS_0 defined for path of length zero is calculated as [Eq. (3-46)]:

$$\mathrm{ATS}_0 = \sum_{i=1}^{A} w_i^2 \qquad (3\text{-}46)$$

that is, the sum of the squares of the atomic properties. Typical atomic properties are atomic masses, polarizabilities, charges, electronegativities. Atomic properties w should be centered by subtracting the average property value in the molecule in order to obtain proper autocorrelation values. Hollas demonstrated that, only if properties are centered are all autocorrelation descriptors uncorrelated, thus resulting more suitable for subsequent statistical analysis [134]. For each atomic property

w, the set of autocorrelation terms defined for all existing topological distances in the graph is the ATS descriptor defined as in Eq. (3-47):

$$\{ATS_0, ATS_1, ATS_2, \ldots, ATS_D\}_w \tag{3-47}$$

where D is the topological diameter, that is, the maximum distance in the graph.

Average spatial autocorrelation descriptors are obtained by dividing each term by the corresponding number of contributions, thus avoiding any dependence on molecular size [Eq. (3-48)]:

$$\overline{ATS}_k = \frac{1}{2\Delta_k} \cdot \sum_{i=1}^{A} \sum_{j=1}^{A} w_i \cdot w_j \cdot \delta(d_{ij};k) \tag{3-48}$$

where Δ_k is the sum of the Kronecker delta, i.e., the total number of vertex pairs at distance equal to k [130]. An example of calculation of Moreau–Broto autocorrelation descriptors is reported for 4-hydroxy-2-butanone. The H-depleted molecular graph is given below: .

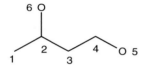

Atomic masses are used as the weighting scheme for molecule atoms: $w_1 = w_2 = w_3 = w_4 = 12$; $w_5 = w_6 = 16$. Then, autocorrelation terms for lag k from 0 to 4 are

$$
\begin{aligned}
ATS_0 &= w_1^2 + w_2^2 + w_3^2 + w_4^2 + w_5^2 + w_6^2 \\
&= 12^2 + 12^2 + 12^2 + 12^2 + 16^2 + 16^2 = 1088 \\
ATS_1 &= w_1 \cdot w_2 + w_2 \cdot w_3 + w_3 \cdot w_4 + w_4 \cdot w_5 + w_2 \cdot w_6 \\
&= 12 \cdot 12 + 12 \cdot 12 + 12 \cdot 12 + 12 \cdot 16 + 12 \cdot 16 = 816 \\
ATS_2 &= w_1 \cdot w_3 + w_1 \cdot w_6 + w_2 \cdot w_4 + w_3 \cdot w_5 + w_3 \cdot w_6 \\
&= 12 \cdot 12 + 12 \cdot 16 + 12 \cdot 12 + 12 \cdot 16 + 12 \cdot 16 = 864 \\
ATS_3 &= w_1 \cdot w_4 + w_2 \cdot w_5 + w_4 \cdot w_6 \\
&= 12 \cdot 12 + 12 \cdot 16 + 12 \cdot 16 = 528 \\
ATS_4 &= w_1 \cdot w_5 + w_6 \cdot w_5 \\
&= 12 \cdot 16 + 16 \cdot 16 = 448
\end{aligned}
$$

Three-dimensional topological distance-based descriptors (*3D-TDB descriptors*) are a variant of the average Moreau–Broto autocorrelations also encoding information about the 3D spatial separation between two atoms [135]. TDB-steric descriptors, denoted by S, are defined for each lag k as [Eq. (3-49)]

$$S_k = \frac{1}{\Delta_k} \cdot \sum_{i=1}^{A-1} \sum_{j=i+1}^{A} \left(R_i^{\text{cov}} \cdot r_{ij} \cdot R_j^{\text{cov}} \right) \cdot \delta \left(d_{ij};k \right) \tag{3-49}$$

where Δ_k is the number of atom pairs located at a topological distance k, r_{ij} is the geometric distance between the ith and jth atoms, and R^{cov} is the atomic covalent radius accounting for steric properties of atoms. In a similar way, TDB-electronic descriptors, denoted by X, are defined as [Eq. (3-50)]

$$X_k = \frac{1}{\Delta_k} \cdot \sum_{i=1}^{A-1} \sum_{j=i+1}^{A} \left(\chi_i \cdot r_{ij} \cdot \chi_j \right) \cdot \delta \left(d_{ij};k \right) \tag{3-50}$$

where χ is the sigma orbital electronegativity accounting for electronic properties of atoms.

Together with steric and electronic descriptors, TDB atom-type descriptors, denoted by I, are defined as [Eq. (3-51)]

$$I_k (u,u) = \frac{1}{2} \cdot \sum_{i=1}^{A} \sum_{j=1}^{A} \delta_{ij} (u,u) \cdot \delta \left(d_{ij};k \right) \tag{3-51}$$

where u denotes an atom type and $\delta_{ij}(u,u)$ is a Kronecker delta equal to one if both atoms i and j are of type u. These atom-type autocorrelations are calculated only for pairs of atoms of the same type. Moreover, unlike the previous two TDB descriptors (S_k and X_k), this autocorrelation descriptor does not account for 3D information.

3.4.3. Moran and Geary Coefficients

Moran and Geary coefficients are autocorrelation functions applied mainly in ecological studies to measure spatial distribution of environmental properties. They are applied to molecular structure in the same way as the Moreau–Broto function; however, unlike the Moreau–Broto function, Moran and Geary functions give real autocorrelation accounting explicitly for the mean and standard deviation of each property. The Moran coefficient, applied to a molecular graph, is calculated as follows [Eq. (3-52)] [136]:

$$I_k = \frac{\frac{1}{\Delta_k} \cdot \sum_{i=1}^{A} \sum_{j=1}^{A} (w_i - \bar{w}) \cdot (w_j - \bar{w}) \cdot \delta(d_{ij};k)}{\frac{1}{A} \cdot \sum_{i=1}^{A} (w_i - \bar{w})^2} \tag{3-52}$$

where w_i is any atomic property, $\bar{w}$ is its average value on the molecule, A is the number of atoms, k is the considered lag, and d_{ij} is the topological distance between ith and jth atoms; $\delta(d_{ij};k)$ is the Kronecker delta equal to 1 if $d_{ij} = k$, zero otherwise.

Δ_k is the number of vertex pairs at distance equal to k. Moran coefficient usually takes value in the interval $[-1,+1]$. Positive autocorrelation corresponds to positive values of the coefficient, whereas negative autocorrelation produces negative values. The Geary coefficient, denoted by c_k, is defined as [Eq. (3-53)] [137].

$$c_k = \frac{\frac{1}{2\Delta_k} \cdot \sum_{i=1}^{A} \sum_{j=1}^{A} (w_i - w_j)^2 \cdot \delta(d_{ij};k)}{\frac{1}{(A-1)} \cdot \sum_{i=1}^{A} w_i - \bar{w}^2} \tag{3-53}$$

where w_i is any atomic property, $\bar{w}$ is its average value on the molecule, A is the number of atoms, k is the considered lag, and d_{ij} is the topological distance between ith and jth atoms; $\delta(d_{ij};k)$ is the Kronecker delta equal to 1 if $d_{ij} = k$, zero otherwise.

The Geary coefficient is a distance-type function varying from zero to infinity. Strong autocorrelation produces low values of this index; moreover, positive autocorrelation translates to values between 0 and 1, whereas negative autocorrelation produces values larger than 1; therefore, the reference "no correlation" is $c_k = 1$.

In Table 3-4, Moran and Geary coefficients are listed together with Moreau–Broto autocorrelation values for 22 N,N-dimethyl-α-bromo-phenethylamines, whose parent structure is shown in Figure 3-5. Carbon-scaled atomic masses were used as the weighting scheme for molecule atoms for the calculation of all the autocorrelation functions.

3.4.4.   Auto-cross-covariance Transforms

Auto-cross-covariance (ACC) transforms are autocovariances and cross-covariances calculated from sequential data with the aim of transforming them into uniform-length descriptors suitable for QSAR modeling. ACC transforms were originally proposed to describe peptide sequences [138,139]. In order to calculate ACC transforms, each amino acid position in the peptide sequence was defined in terms of a number of amino acid properties; in particular, three orthogonal z-scores, derived from a principal component analysis (PCA) of 29 physico-chemical properties of the 20 coded amino acids, were originally used to describe each amino acid. Then, for each peptide sequence, auto- and cross-covariances with lags $k = 1, 2, \ldots, K$ were calculated as [Eq. (3-54)]

$$\text{ACC}_k (j, j) = \sum_{i=1}^{n-k} \frac{z_i (j) \cdot z_{i+k} (j)}{n - k} \qquad \text{ACC}_k (j, m) = \sum_{i=1}^{n-k} \frac{z_i (j) \cdot z_{i+k} (m)}{n - k} \tag{3-54}$$

where j and m indicate two different amino acid properties, n is the number of amino acids in the sequence, and index i refers to amino acid position in the sequence. Z-score values, being derived from PCA, are used directly because they are already mean centered.

Table 3-4. Some autocorrelation descriptors for a set of phenethylamines (Figure 3-5) *ATS*: Moreau–Broto autocorrelations; *I*: Moran coefficient; *c*: Geary coefficient. Calculations are based on the carbon-scaled atomic mass as the weighting scheme for atoms

Mol	X	Y	ATS_1	ATS_2	ATS_3	ATS_4	I_1	I_2	I_3	I_4	c_1	c_2	c_3	c_4
1	H	H	2.952	3.313	3.473	3.554	−0.006	−0.056	−0.139	−0.319	0.504	0.804	1.364	2.193
2	H	F	3.032	3.422	3.567	3.598	−0.006	−0.055	−0.134	−0.322	0.512	0.782	1.299	2.198
3	H	Cl	3.096	3.508	3.641	3.635	−0.006	−0.077	−0.177	−0.368	0.531	0.811	1.306	2.114
4	H	Br	3.251	3.708	3.819	3.728	−0.005	−0.098	−0.203	−0.320	0.549	0.838	1.174	1.485
5	H	I	3.392	3.884	3.977	3.818	−0.004	−0.090	−0.174	−0.223	0.542	0.828	1.053	1.042
6	H	Me	3.003	3.383	3.533	3.582	−0.005	−0.045	−0.112	−0.290	0.503	0.768	1.280	2.176
7	F	H	3.032	3.422	3.567	3.640	−0.006	−0.055	−0.134	−0.299	0.512	0.782	1.299	2.034
8	Cl	H	3.096	3.508	3.641	3.710	−0.006	−0.077	−0.177	−0.366	0.531	0.811	1.306	2.008
9	Br	H	3.251	3.708	3.819	3.876	−0.005	−0.098	−0.203	−0.374	0.549	0.838	1.174	1.645
10	I	H	3.392	3.884	3.977	4.027	−0.004	−0.090	−0.174	−0.300	0.542	0.828	1.053	1.363
11	Me	H	3.003	3.383	3.533	3.609	−0.005	−0.045	−0.112	−0.261	0.503	0.768	1.280	2.009
12	Cl	F	3.165	3.598	3.828	3.748	−0.006	−0.073	−0.159	−0.365	0.539	0.793	1.208	2.025
13	Br	F	3.310	3.783	4.081	3.909	−0.005	−0.089	−0.194	−0.364	0.554	0.816	1.235	1.655
14	Me	F	3.079	3.485	3.663	3.651	−0.005	−0.045	−0.106	−0.266	0.511	0.752	1.168	2.025
15	Cl	Cl	3.221	3.671	3.966	3.780	−0.007	−0.095	−0.159	−0.403	0.561	0.825	1.201	1.969
16	Br	Cl	3.359	3.843	4.264	3.936	−0.006	−0.109	−0.141	−0.398	0.574	0.845	1.180	1.655
17	Me	Cl	3.140	3.566	3.763	3.686	−0.006	−0.063	−0.157	−0.302	0.529	0.778	1.210	1.942
18	Cl	Br	3.359	3.843	4.264	3.861	−0.006	−0.109	−0.141	−0.333	0.574	0.845	1.180	1.417
19	Br	Br	3.480	3.991	4.636	4.006	−0.005	−0.130	−0.029	−0.372	0.594	0.875	1.069	1.386
20	Me	Br	3.289	3.756	3.993	3.775	−0.005	−0.080	−0.214	−0.257	0.545	0.802	1.257	1.361
21	Me	Me	3.052	3.449	3.617	3.636	−0.005	−0.037	−0.083	−0.241	0.503	0.740	1.149	2.008
22	Br	Me	3.289	3.756	3.993	3.897	−0.005	−0.080	−0.214	−0.342	0.545	0.802	1.257	1.633

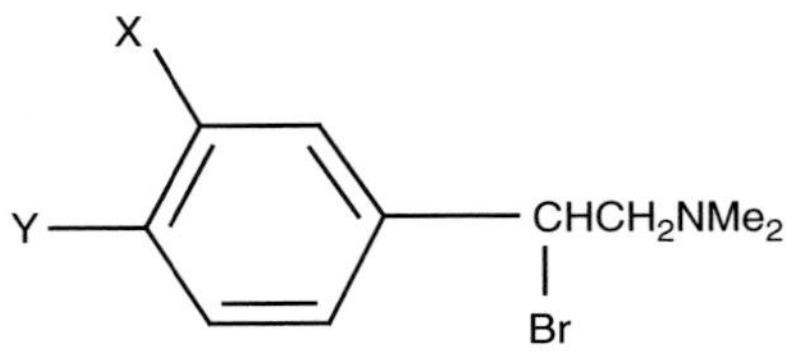

Figure 3-5. Parent structure of *N*, *N*-dimethyl-α-bromo-phenethylamines

ACC transforms were also used to encode information contained in molecular interaction fields typical of CoMFA analysis using as the lag the distance between grid points along each coordinate axis, along the diagonal, or along any intermediate direction [140]. The cross-correlation terms were calculated by the products of the interaction energy values for steric and electrostatic fields in grid points at distances equal to the lag. Different kinds of interactions, namely positive–positive, negative–negative, and positive–negative, were kept separated, thus resulting in 10 ACC terms for each lag. The major drawback of these ACC transforms is that their values depend on molecule orientation along the axes.

Topological maximum auto-cross-correlation (*TMACC*) descriptors are a variant of the ACC transforms for molecular graphs [141]. These are cross-covariances calculated taking into account the topological distance d_{ij} between the atoms i and j and four basic atomic properties: (1) Gasteiger-Marsili partial charges, accounting for electrostatic properties [142]; (2) Wildman–Crippen molar refractivity parameters, accounting for steric properties and polarizabilities [143]; (3) Wildman–Crippen $\log P$ values, accounting for hydrophobicity [143]; and (4) $\log S$ values, accounting for solubility and solvation phenomena [144].

The general formula for the calculation of *TMACC* descriptors is given by Eq. (3-55):

$$TMACC\,(x,y;k) = \frac{1}{\Delta_k} \cdot \sum_{i=1}^{A} \sum_{j=1}^{A} x_i \cdot y_j \cdot \delta\left(d_{ij};k\right) \qquad (3\text{-}55)$$

where x and y are two atomic properties, A is the number of atoms in the molecule, k is the lag, and d_{ij} is the topological distance between the ith and jth atoms; Δ_k is the number of atom pairs located at topological distance k and $\delta(d_{ij};k)$ is the Kronecker delta equal to 1, if the topological distance equals the lag, zero otherwise. If only one property is considered, i.e., $x = y$, autocovariances are obtained. Because all the selected properties, except for molar refractivity, can assume both positive and negative values, these are treated as different properties and cross-covariance terms are also calculated between positive and negative values of each property. Therefore, 7 autocovariance terms and 12 cross-covariance terms constitute the final *TMACC* vector.

3.4.5. Autocorrelation of Molecular Surface Properties

The autocorrelation of molecular surface properties is a general approach for the description of property measures on the molecular surface by using uniform-length descriptors, which are comprised of the same number of elements regardless of the size of the molecule [130,145]. This approach is an extension of Moreau–Broto autocorrelation function to 3D molecular geometry. Since geometrical distances r_{ij} can have any real positive value, some ordered distance intervals need to be specified, each defined by a lower and upper value of r_{ij}. All distances falling in the same interval are considered identical.

To generate 3D autocorrelation descriptors of molecular surface properties, first, a number of points are randomly distributed on the molecular surface with a user-defined density and in an orderly manner to ensure a continuous surface. Then, the *surface autocorrelation vector* (SAV) is derived by calculating for each lag k the sum of the products of the property values at two surface points located at a distance falling into the kth distance interval. This value is then normalized by the number Δ_k of the geometrical distances r_{ij} in the interval [Eq. (3-56)]:

$$A\,(k) = \frac{1}{\Delta_k} \cdot \sum_{i=1}^{N-1} \sum_{j=i+1}^{N} w_i \cdot w_j \cdot \delta\left(r_{ij};k\right) \qquad (3\text{-}56)$$

where N is the number of surface points and k represents a distance interval defined by a lower and upper bound.

It was demonstrated that to obtain the best surface autocorrelation vectors for QSAR modeling, the van der Waals surface is better than other molecular surfaces. Then, the surface should have no fewer than five grid points per Å^2 and a distance interval no greater than 1 Å should be used in the distance binning scheme. Autocorrelation values calculated for a number of distance intervals constitute a unique fingerprint of the molecule, thus resulting suitable for similarity/diversity analysis of molecules. Figure 3-6 shows the autocorrelation vector of estradiol calculated by using molecular electrostatic potential (MEP) as the surface property.

3.4.6. Atom Pairs

A special case of autocorrelation descriptors is the *atom-type autocorrelation* (*ATAC*), which is calculated by summing property values only of atoms of given types. The simplest atom-type autocorrelation is given by Eq. (3-57):

$$ATAC_k\,(u,v) = \sum_{i=1}^{A} \sum_{j=1}^{A} \delta\,(i;u) \cdot \delta\,(j;v) \cdot \delta\left(d_{ij};k\right) \qquad (3\text{-}57)$$

where u and v denote two different atom types; A is the number of molecule atoms; and $\delta(i;u)$ is a Kronecker delta function equal to 1 if the atom i is of type u, and zero

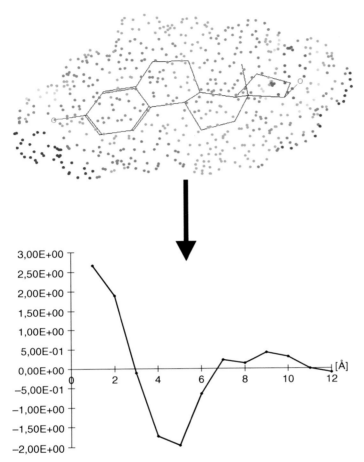

Figure 3-6. Surface autocorrelation vector of estradiol calculated by using molecular electrostatic potential (MEP) as the surface property

otherwise; analogously, $\delta(j;v)$ is a Kronecker delta function equal to 1 if the atom j is of type v, and zero otherwise; $\delta(d_{ij}; k)$ is a Kronecker delta function equal to one if the interatomic distance d_{ij} is equal to the lag k, and zero otherwise.

This descriptor is defined for each pair of atom types and simply encodes the occurrence numbers of the given atom type pair at different distance values. It can be normalized by using two procedures: the first one consists in dividing each $ATAC_k$ value by the total number of atom pairs at distance k independently of their types; the second one consists in dividing each $ATAC_k$ value by a constant, which can be equal to the total number of atoms in the molecule or, alternatively, to the total number of (u,v) atom type pairs in the molecule. Note that if atom types u and v coincide, i.e., $u = v$, then the atom-type autocorrelation is calculated as [Eq. (3-58)]

$$ATAC_k(u,u) = \frac{1}{2} \cdot \sum_{i=1}^{A} \sum_{j=1}^{A} \delta(i;u) \cdot \delta(j;u) \cdot \delta(d_{ij};k) \qquad (3\text{-}58)$$

in order to avoid to count twice a given pairs of atom types.

Atom types can be defined in different ways; they can be defined in terms of the simple chemical elements or account also for atom connectivity, hybridization states, and pharmacophoric features. Atom-type autocorrelations have been used to derive some vectors of substructure descriptors such as atom pairs [146] and CATS descriptors [147]. Substructure descriptors are counts of the occurrences of predefined structural features (functional groups, augmented atoms, pharmacophore point pairs, atom pairs and triangles, surface triangles, etc.) in molecules or binary variables specifying their presence/absence. These constitute string representations of chemical structures usually designed to enhance the efficiency of chemical database screening and analysis. Each bin or set of bins of the string is associated with a structural feature or pattern. The string length can vary depending on the amount of structural information to be encoded.

Atom pairs are substructure descriptors defined in terms of any pair of atoms and bond types connecting them. An atom pair (AP) is composed of two non-hydrogen atoms and an interatomic separation [Eq. (3-59)] [146]:

$$AP = \{[i\text{th atom description}][\text{separation}][j\text{th atom description}]\} \qquad (3\text{-}59)$$

The two considered atoms need not be directly connected and the *separation* can be the topological distance between them; these descriptors are properly called topological atom pairs being based on the topological representation of the molecules. Atom type is defined by the element itself, number of heavy-atom connections and number of π electron pairs on each atom.

Atom pairs are sensitive to long-range correlations between atoms in molecules and therefore to small changes even in one part of large molecules. Atom pair descriptors usually are Boolean variables encoding the presence or absence of a particular atom pair in each molecule.

Distance-counting descriptors (or *SE-vectors*) are a particular implementation of topological atom pairs proposed by Clerc and Terkovics in 1990 [148]. These are holographic vectors encoding information on the occurrence frequency of any combination of two atom types and a distance relationship between them. All the paths and not only the shortest one between any pair of atom types are considered in the original proposal. Based on the shortest path, revised SE-vectors were proposed by Baumann in 2002 and called *SESP-Top vectors* and *SESP-Geo vectors* [149].

CATS descriptors are a particular implementation of atom pairs descriptors based on pharmacophore point types [147,150]. CATS descriptors are holographic vectors where each bin encodes the number of times a potential pharmacophore point-pair (PPP-pair) occurs in the molecule. The five defined potential pharmacophore points (PPPs) are hydrogen-bond donor (D), hydrogen-bond acceptor (A), positively charged or ionizable (P), negatively charged or ionizable (N), and lipophilic (L).

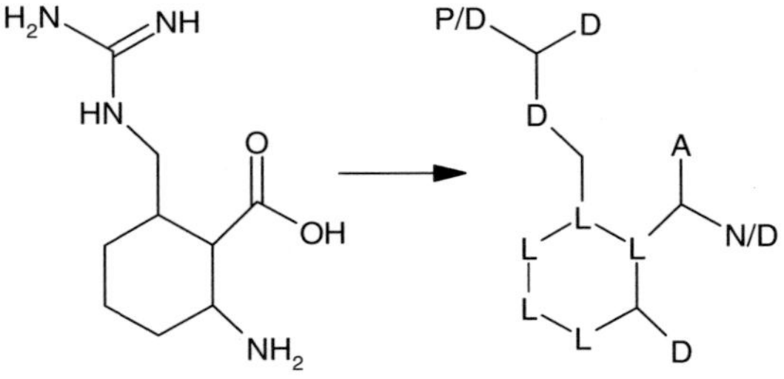

Figure 3-7. Conversion of a two-dimensional molecular representation into the molecular graph, in which pharmacophore point types are assigned as implemented in CATS

If an atom does not belong to any of the five PPP types it is not considered. Moreover, an atom is allowed to be assigned to one or two PPP types (Figure 3-7). For each molecule, the number of occurrences of all 15 possible pharmacophore point-pairs (DD, DA, DP, DN, DL, AA, AP, AN, AL, PP, PN, PL, NN, NL, LL) is determined and then associated with the number of intervening bonds between the two considered points, whereby the shortest path length is used. Topological distances of 0–9 bonds are considered leading to a 150-dimensional autocorrelation vector. Finally, PPP-pair counts are scaled by the total frequency in the molecule.

CATS3D descriptors [151] are based on geometrical distances between PPPs rather than topological distances; hydrogens are also considered. Pairs of PPPs are considered to fall into one of 20 equal-spaced bins from 0 to 20 Å. Multiple potential pharmacophore point assignments of one atom are not allowed. Moreover, an additional type is defined to account for atoms assigned to none of the five PPP types: a total of 21 possible PPP-pairs is thus obtained and to each of them, 20 distance bins are assigned, resulting into a 420-dimensional vector. *SURFCATS descriptors* [152] are based on the spatial distance between PPPs on the Connolly surface area. Surface points are calculated with a spacing of 2 Å and assigned to the pharmacophore type of the nearest atoms. *CATS-charge descriptors* [150]. map the partial atom charges of a molecule to predefined spatial distance bins. The geometrical distances of all atom pair combinations in one molecule are calculated. Distances within a certain range (0.1 Å) are allocated to the same bin. The charges of the two atoms that form a pair are multiplied to yield a single charge value per pair. Charge values that are assigned to the same bin are summed up. Distances from 0 to 10 Å are considered at increments of 0.1 Å. All distances greater than 10 Å are associated with the last bin. The output is a 100-dimensional vector, which characterizes the molecule by means of its atom partial charge distribution.

3.4.7.   Estrada Generalized Topological Index

Variable molecular descriptors are local and graph invariants containing adjustable parameters whose values are optimized in order to improve the statistical quality of a given regression model. Sometimes also called *flexible descriptors* or *optimal*

descriptors, their flexibility in modeling is useful to obtain good models; however, due to the increased number of parameters needing to be optimized, they require more intensive validation procedures to generate predictive models. These molecular descriptors are called "variable" because their values are not fixed for a molecule but change depending on the training set and the property to be modelled.

The *Estrada generalized topological index (GTI)* is a general strategy to search for optimized quantitative structure–property relationship models based on variable topological indexes [153,154]. The main objective of this approach is to obtain the best optimized molecular descriptors for each property under study. The family of GTI descriptors is comprised of autocorrelation functions defined by the following general form [Eq. (3-60)]:

$$GTI = \sum_{k=1}^{D} C_k(x_0, p_0) \cdot \eta^{(k)} \qquad (3\text{-}60)$$

where the summation goes over the different topological distances in the graph, D being the topological diameter, that is, the maximum topological distance in the graph and accounts for the contributions $\eta^{(k)}$ of pairs of vertexes located at the same topological distance k. Each contribution $\eta^{(k)}$ is scaled by two real parameters x_0 and p_0 through the $C_k(x_0, p_0)$ coefficient defined as [Eq. (3-61)].

$$C_k(x_0, p_0) = k^{p_0} \cdot x_0^{p_0(k-1)} \qquad (3\text{-}61)$$

By definition, the C_k coefficient is equal to one for any pair of adjacent vertexes ($k = 1$), independently of the parameter values. Note that the coefficients C_k are the elements of the so-called *generalized molecular-graph matrix* Γ, which is a square symmetric $A \times A$ matrix, defined as [Eq. (3-62)] [155]

$$[\Gamma(x_0, p_0)]_{ij} = \begin{cases} 1 & \text{if } d_{ij} = 1 \\ \left[d_{ij} \cdot x_0^{(d_{ij}-1)} \right]^{p_0} & \text{if } i \neq j \quad \wedge \quad d_{ij} > 1 \\ 0 & \text{if } i = j \end{cases} \qquad (3\text{-}62)$$

where d_{ij} is the topological distance between vertexes v_i and v_j. This matrix was also defined in terms of interatomic geometric distances. The term $\eta^{(k)}$ defines the contribution of all those interactions due to the pairs of vertexes at distance k in the graph as given by Eq. (3-63):

$$\eta^{(k)} = \frac{1}{2} \cdot \sum_{i=1}^{A} \sum_{j=1}^{A} \langle i, j \rangle \cdot \delta(d_{ij}; k) \qquad (3\text{-}63)$$

where A is the number of vertexes in the graph, $\delta(d_{ij}; k)$ is a Kronecker delta function equal to one if the topological distance d_{ij} is equal to k, and zero otherwise.

The term$\langle i, j \rangle$ is the "geodesic-bracket" term encoding information about the molecular shape on the basis of a connectivity-like formula as [Eq. (3-64)].

$$\langle i, j \rangle = \frac{1}{2} \cdot \left(u_i \cdot v_j + v_i \cdot u_j \right) \tag{3-64}$$

where u and v are two functions of the variable parameters x and p and can be considered as generalized vertex degrees defined as Eqs. [3-65] and [3-66]:

$$u_i (x_1, p_1, \mathbf{w}) = \left[w_i + \delta_i + \sum_{k=2}^{D} k \cdot x_1^{k-1} \cdot {}^k f_i \right]^{p_1} \tag{3-65}$$

$$v_i (x_2, p_2, \mathbf{s}) = \left[s_i + \delta_i + \sum_{k=2}^{D} k \cdot x_2^{k-1} \cdot {}^k f_i \right]^{p_2} \tag{3-66}$$

where δ_i is the simple vertex degree of the ith vertex, i.e., the number of adjacent vertexes and ${}^k f_i$ is its vertex distance count, i.e., the number of vertexes at distance k from the ith vertex. The scalars $x_0, x_1, x_2, p_0, p_1, p_2, \mathbf{w}$ and $\mathbf{s}$ define a $(2A + 6)$-dimensional real space of parameters; $\mathbf{w}$ and $\mathbf{s}$ are two A-dimensional vectors collecting atomic properties. The first six parameters x_0, x_1, x_2, p_0, p_1, and p_2 are free parameters to be optimized, whereas the parameters $\mathbf{w}$ and $\mathbf{s}$ are predefined quantities used to distinguish among the different atom types. For each combination of the possible values of these parameters a different topological index is obtained for a molecule. It has to be noted that several of the well-known topological indexes can be calculated by the GTI formula by settling specific combinations of the parameters; for instance, for $\mathbf{w} = (0, 0, ..., 0)$ and $\mathbf{s} = (0, 0, ..., 0)$, the index *GTI* reduces to the *Wiener index* when $x_0 = 1, x_1 = $ any, $x_2 = $ any, $p_0 = 1, p_1 = 0, p_2 = 0$, while *GTI* coincides with the *Randić connectivity index* when $x_0 = 0, x_1 = 0, x_2 = 0, p_0 = 1, p_1 = -1/2, p_2 = -1/2$.

3.5.   GEOMETRICAL DESCRIPTORS

3.5.1.   Introduction

Geometrical molecular descriptors, also called 3D-molecular descriptors, are derived from a geometrical representation of the molecule, more specifically from the x, y, z Cartesian coordinates of the molecule atoms. These are molecular descriptors defined in several different ways but always derived from the three-dimensional structure of the molecule [156,157]. Generally, geometrical descriptors are calculated either on some optimized molecular geometry obtained by the methods of computational chemistry or from crystallographic coordinates. *Topographic indexes* constitute a special subset of geometrical descriptors, being calculated on the graph representation of molecules but using the geometric distances between atoms instead of the topological distances [71,158,159].

Since a geometrical representation involves the knowledge of the relative positions of the atoms in 3D space, geometrical descriptors usually provide more information and discrimination power for similar molecular structures and molecule conformations than topological descriptors. Despite their high information content, geometrical descriptors usually show some drawbacks. They require geometry optimization and therefore the cost to calculate them. Moreover, for flexible molecules, several molecule conformations can be available; on one hand, new information is available and can be exploited, but, on the other hand, the problem complexity can significantly increase.

For these reasons, topological descriptors, fingerprints based on fragment counts and other simple descriptors are usually preferred for the screening of large databases of molecules. On the other hand, searching for relationships between molecular structures and complex properties, such as biological activities, can often be performed efficiently by using geometrical descriptors, exploiting their large information content. Moreover, it is important to remember that the biologically active conformation of the studied chemicals is seldom known. Some authors overcome this problem by using a multi-conformation dynamic approach [57].

Most of the geometrical descriptors are calculated directly from the x,y,z coordinates of the molecule atoms and other quantities derived from the coordinates such as interatomic distances or distances from a specified origin (e.g., the molecule barycenter). Many of these are derived from the molecular *geometry matrix* defined by all the geometrical distances r_{ij} between atom pairs. In order to account for more chemical information, the atoms in the molecule can be represented by their atomic masses and molecular descriptors can be derived from the molecule inertia matrix, from atom distances with respect to the centers of mass, and by weighting interatomic distances with functions of atomic masses.

3.5.2.   Indexes from the Geometry Matrix

The *geometry matrix*, denoted by **G**, is a simple molecular representation where atoms are viewed as single points in the 3D molecule space. It is a square symmetric matrix $A \times A$, A being the number of molecule atoms, where each entry r_{ij} is the Euclidean distance between the atoms i and j [Eq. (3-67)]:

$$\mathbf{G} \equiv \begin{vmatrix} 0 & r_{12} & \cdots & r_{1A} \\ r_{21} & 0 & \cdots & r_{2A} \\ \cdots & \cdots & \cdots & \cdots \\ r_{A1} & r_{A2} & \cdots & 0 \end{vmatrix} \tag{3-67}$$

Diagonal entries are always zero, by definition. Geometric distances are intramolecular interatomic distances.

The geometry matrix contains information about molecular configurations and conformations; however, the geometry matrix does not contain information about atom connectivity. Thus, for several applications, it is accompanied by a connectivity table where, for each atom, the identification number of the bonded atoms is listed. The geometry matrix can also be calculated on geometry-based standardized bond lengths and bond angles and derived by embedding a graph on a regular two-dimensional or three-dimensional grid; in these cases, the geometry matrix is often referred to as the *topographic matrix* **T** and the interatomic distance to as the *topographic distance* [72]. Depending on the kind of grid used for graph embedding, different topographic matrixes can be obtained. The *bond length-weighted adjacency matrix*, or *3D-adjacency matrix*, is obtained from the geometry matrix **G** as [Eq. (3-68)] [160]

$$^b\mathbf{A} = \mathbf{G} \otimes \mathbf{A} \qquad (3\text{-}68)$$

where $\otimes$ indicates the Hadamard matrix product and **A** is the adjacency matrix, whose elements are equal to one for pairs of bonded atoms, and zero otherwise. Thus, the elements of the 3D-adjacency matrix are the bond lengths for pairs of bonded atoms, and zero otherwise. The ith row sum of the geometry matrix is called *geometric distance degree* (or *Euclidean degree* [161]) and denoted by $^G\sigma_i$; it is defined as [Eq. (3-69)]

$$^G\sigma_i = \sum_{j=1}^{A} r_{ij} \qquad (3\text{-}69)$$

In general, the row sum of this matrix represents a measure of the centrality of an atom; atoms that are close to the center of the molecule have smaller atomic sums, while those far from the center have large atomic sums. The smallest and the largest row sums give the extreme values of the first eigenvalue of the geometry matrix; therefore when all the atoms are equivalent, i.e., the distance degrees are all the same, the geometric distance degree yields exactly the first eigenvalue. The average sum of all geometric distance degrees is a molecular descriptor called *average geometric distance degree*, i.e., Eq. (3-70)

$$^G\bar{\sigma} = \frac{1}{A} \cdot \sum_{i=1}^{A} {}^G\sigma_i = \frac{1}{A} \cdot \sum_{i=1}^{A} \sum_{j=1}^{A} r_{ij} \qquad (3\text{-}70)$$

while the half-sum of all geometric distance degrees is another molecular descriptor called *3D-Wiener index* by analogy with the Wiener index calculated from the topological distance matrix. The 3D-Wiener index is calculated as [Eq. (3-71)] [162]

$$^{3D}W_H = \frac{1}{2} \cdot \sum_{i=1}^{A} \sum_{j=1}^{A} r_{ij} \qquad (3\text{-}71)$$

where r_{ij} is the interatomic distance between the ith and jth atom. This index is obviously more discriminatory than the 2D-Wiener index as it accounts for spatial molecular geometry; it shows different values for different molecular conformations, the largest values corresponding to the most extended conformations, the smallest to the most compact conformations. Therefore, it is considered a measure of molecular shape since it decreases with increasing sphericity of a structure [163]. The 3D-Wiener index can be calculated both considering ($^{3D}W_H$) and not considering (^{3D}W) hydrogen atoms [164]. Moreover, a strictly related molecular descriptor is the *bond length-weighted Wiener index* calculated by using as the distance between two atoms the sum of the bond lengths along the shortest path [165].

The *3D-connectivity indexes* denoted by $\chi\chi$ are defined as connectivity-like indexes derived using the geometric distance degree $^{G}\sigma$ in place of the topological vertex degree δ [Eq. (3-72)] [166]:

$$^{m}\chi\chi_t = \sum_{k=1}^{K} \left(\prod_{i=1}^{n} {}^{G}\sigma_i \right)_k^{-1/2} \qquad (3\text{-}72)$$

where k runs over all of the mth order subgraphs constituted by n vertexes; K is the total number of mth order subgraphs and each subgraph is weighted by the product of the local invariants associated to the vertexes contained in the subgraph. The subscript "t" refers to the type of molecular subgraph and is ch for chain or ring, pc for path–cluster, c for cluster, and p for path.

The *Euclidean connectivity index* is another geometrical descriptor defined by using a Randić-like formula applied to geometric distance degrees $^{G}\sigma$ as given by Eq. (3-73) [161]:

$$\chi^E = \sum_{i=1}^{A-1} \sum_{j=i+1}^{A} \left({}^{G}\sigma_i \cdot {}^{G}\sigma_j \right)^{-1/2} \qquad (3\text{-}73)$$

This index discriminates geometrical isomers and can be considered as a measure of the compactness of a molecule in the 3D space. Note that all possible atom pairs are considered instead of the pairs of bonded atoms because in 3D space there exists a Euclidean distance between every pair of atoms. The *3D-Balaban index*, denoted as ^{3D}J, is a Balaban-like index derived from the geometry matrix as [Eq. (3-74)] [160]

$$^{3D}J = \frac{B}{C+1} \cdot \sum_{i=1}^{A-1} \sum_{j=i+1}^{A} a_{ij} \cdot \left({}^{G}\sigma_i \cdot {}^{G}\sigma_j \right)^{-1/2} \qquad (3\text{-}74)$$

where $^G\sigma_i$ and $^G\sigma_j$ are the geometric distance degrees of the atoms i and j and a_{ij} are the elements of the adjacency matrix used to account only for contributions from pairs of bonded atoms. A variant of the geometric distance degree was proposed as [Eq. (3-75)] [167,168].

$$3DW_i = \sum_{j=1}^{A} (1 - a_{ij}) \cdot \exp\left(r_{ij}^{-2}\right) \qquad j \neq i \qquad (3\text{-}75)$$

where the summation accounts only for contributions from non-bonded atoms, a_{ij} being the elements of the adjacency matrix equal to one for pairs of bonded atoms, and zero otherwise. The exponential form of the distance was chosen from a series of terms approximating the attracting interatomic potentials. From this 3D local invariant, Zagreb-like ($3DM_1$ and $3DM_2$), connectivity-like ($3D^0\chi$ and $3D^1\chi$), and Wiener-type indexes ($3DWi$) were derived, as given by Eqs. (3-76, 3-77, 3-78, 3-79, 3-80):

$$3DM_1 = \sum_{i=1}^{A} 3DW_i \qquad (3\text{-}76)$$

$$3DM_2 = \sum_{i=1}^{A-1} \sum_{j=i+1}^{A} a_{ij} \cdot \left(3DW_i \cdot 3DW_j\right) \qquad (3\text{-}77)$$

$$3D^0\chi = \sum_{i=1}^{A} \left(3DW_i\right)^{-1/2} \qquad (3\text{-}78)$$

$$3D^1\chi = \sum_{i=1}^{A-1} \sum_{j=i+1}^{A} a_{ij} \cdot \left(3DW_i \cdot 3DW_j\right)^{-1/2} \qquad (3\text{-}79)$$

$$3DWi = \sum_{i=1}^{A-1} \sum_{j=i+1}^{A} \exp\left(r_{ij}\right) \qquad (3\text{-}80)$$

where A is the number of molecule atoms.

Molecular descriptors based on this kind of local vertex invariant were termed *method of ideal symmetry (MIS) indexes*, based on a partial optimization procedure of the molecular geometry, where bond lengths and bond angles are kept fixed and only free rotations around C–C bonds are varied [169]. The maximum value entry in the ith row of the geometry matrix is a local descriptor called *geometric eccentricity* $^{G}\eta_i$ representing the longest geometric distance from the ith atom to any other atom in the molecule [Eq. (3-81)]:

$$^{G}\eta_i = \max_j \left(r_{ij} \right) \tag{3-81}$$

From the eccentricity definition, the *geometric radius* ^{G}R and *geometric diameter* ^{G}D can immediately characterize a molecule. The radius of a molecule is defined as the minimum geometric eccentricity and the diameter is defined as the maximum geometric eccentricity in the molecule, according to the following equation (3-82):

$$^{G}D = \max_i \left(^{G}\eta_i \right) \ and \ ^{G}D = \max_i \left(^{G}\eta_i \right) \tag{3-82}$$

These parameters are measures of molecule size which also depend on molecular shape. The *geometrical shape coefficient* I_3 is defined in a similar way to the Petitjean index I_2 [Eq. (3.12)] as a function of the geometric radius and diameter as given by Eq. (3-83) [170]:

$$I_3 = \frac{^{G}D - ^{G}R}{^{G}R} \tag{3-83}$$

An example of calculation of some geometrical indexes is reported for 2-methylpentane. The H-depleted molecular graph and geometry matrix **G** of 2-methylpentane are

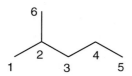

Atom	1	2	3	4	5	6	$^{G}\sigma_i$	$^{G}\eta_i$
1	0	1.519	2.504	3.856	5.014	2.498	15.391	5.014
2	1.519	0	1.530	2.521	3.864	1.521	10.955	3.864
3	2.504	1.530	0	1.521	2.509	2.507	10.571	2.509
4	3.856	2.521	1.521	0	1.511	3.038	12.447	3.856
5	5.014	3.864	1.511	1.511	0	4.348	17.246	5.014
6	2.498	1.521	3.038	3.038	4.348	0	13.912	4.348

with **G =** at left of the table.

where $^G\sigma_i$ indicates the geometric distance degrees and $^G\eta_i$ the atomic eccentricities. Then, geometric radius ^GR, diameter ^GD, average geometric distance degrees $^G\bar\sigma$, and 3D-Wiener index ^{3D}W are, respectively:

$$
\begin{aligned}
^GR &= \min_i \left(^G\eta_i\right) = 2.509 \quad ^G\bar\sigma = \tfrac{1}{6} \cdot (15.391 + 10.955 + 10.571 \\
&\quad + 12.447 + 17.246 + 13.912) = 13.420 \\
^GD &= \max_i \left(^G\eta_i\right) = 5.014 \quad ^{3D}W = \tfrac{1}{6} \cdot (15.391 + 10.955 + 10.571 \\
&\quad + 12.447 + 17.246 + 13.912) = 40.261
\end{aligned}
$$

The *3D-Schultz index*, denoted as ^{3D}MTI, is derived from both geometry matrix **G** and 3D-adjacency matrix ^{b}**A** as given by Eq. (3-84) [160]:

$$
^{3D}\text{MTI} = \sum_{i=1}^{A} \left[\left(^b\mathbf{A} + \mathbf{G}\right) \cdot \mathbf{v} \right]_i \tag{3-84}
$$

where **v** is an A-dimensional column vector collecting the vertex degrees (i.e., number of adjacent vertexes) of the A vertexes in the H-depleted molecular graph. Moreover, the *3D-MTI' index* is defined as [Eq. (3-85)].

$$
^{3D}\text{MTI}' = \sum_{i=1}^{A} \sum_{j=1}^{A} [\mathbf{A} \cdot \mathbf{G}]_{ij} = \sum_{i=1}^{A} \left[\mathbf{v}^{\mathrm{T}} \cdot \mathbf{G}\right]_i \tag{3-85}
$$

where **A** is the topological adjacency matrix and $\mathbf{v}^{\mathrm{T}}$ is the transpose of the vector **v** defined above.

When the information carried by the atom masses is added to the interatomic distances, several other molecular descriptors can be defined. Among these, the *gravitational indexes* are geometrical descriptors reflecting the mass distribution in a molecule, defined as follows [Eqs. (3-86) and (3-87)] [34]:

$$
G_1 = \sum_{i=1}^{A-1} \sum_{j=i+1}^{A} \frac{m_i \cdot m_j}{r_{ij}^2} \tag{3-86}
$$

$$
G_2 = \sum_{b=1}^{B} \left(\frac{m_i \cdot m_j}{r_{ij}^2} \right)_b \tag{3-87}
$$

where m_i and m_j are the atomic masses of the considered atoms, r_{ij} the corresponding interatomic distances, A and B the number of atoms and bonds of the molecule, respectively. The G_1 index takes into account all atom pairs in the molecule, while the G_2 index is restricted to pairs of bonded atoms. These indexes are related to the bulk cohesiveness of the molecules accounting, simultaneously, for both atomic masses (volumes) and their distribution within the molecular space. For modeling purposes the square root and cubic root of the gravitational indexes were also proposed. Both indexes can be extended to any other atomic property different from atomic mass, such as atomic polarizability, van der Waals volume, and electronegativity.

Triangular descriptors (or *triplet descriptors*) can be easily calculated from the geometry matrix. They describe the relative positions of three atoms or group centroids in the molecule. Each possible triplet of non-hydrogen atoms is taken as a triangle, and different triangle measures have been proposed such as individual triangle side lengths (i.e., geometric interatomic distances), triangular perimeter, and area; these measures are integerized and transformed into single-bit integers of defined length by different procedures, and their distribution is used to describe the molecule. They are used both to characterize molecular shape and for 3D pharmacophore database searching [171–176]. Similar to the triangular descriptors are potential pharmacophore point-pairs (*PPP pairs*) and potential pharmacophore point triangles *(PPP triangles)*, which are 3D fingerprints encoding, respectively, the distance information between all possible combinations of two and three potential pharmacophore points [177]. Potential pharmacophore points usually considered are hydrogen bond donors and acceptors, sites of potential negative and positive charge, and hydrophobic atoms. Moreover, a set of molecular descriptors can be obtained by summing the geometric distances between all possible combinations of predefined heteroatom-type pairs, such as N...N, N...O, O...O, N...S, N...P, N... Cl, O...P.

Derived from the geometry matrix, the *neighborhood geometry matrix*, (or *neighborhood Euclidean matrix*), denoted as $^{N}\mathbf{G}$, was proposed, as given in Eq. (3-88) [178]:

$$\left[^{N}\mathbf{G}\right]_{ij} = \begin{cases} r_{ij} & \text{if } r_{ij} \leq R_t \\ 0 & \text{if } r_{ij} > R_t \end{cases} \tag{3-88}$$

where R_t is a user-defined distance threshold. This matrix was originally used to calculate numerical indexes characterizing proteomics maps by the additional constraint that the matrix element i–j is set at zero also for non-connected protein spots.

The *reciprocal geometry matrix*, denoted as $\mathbf{G}^{-1}$, is obtained by inverting the interatomic distances collected in the geometry matrix as the following equation (3-89):

$$[\mathbf{G}^{-1}]_{ij} = \begin{cases} r_{ij}^{-1} & i \neq j \\ 0 & i = j \end{cases} \tag{3-89}$$

Other important derived matrixes are the generalized geometry matrixes obtained by raising to different powers the elements of the geometry matrix. These matrixes were used to calculate *molecular profiles*, which are molecular descriptors denoted by ^{k}D and defined as the average row sum of all interatomic distances raised to the kth power, normalized by the factor k! [Eq. (3-90)] [179]:

$$^kD = \frac{1}{k!} \frac{\sum_{i=1}^{A} \sum_{j=1}^{A} r_{ij}^k}{A} \tag{3-90}$$

where r_{ij}^k is the kth power of the i–j entry of the geometry matrix and A is the number of atoms.

Using several increasing k values, a sequence of molecular invariants called *molecular profile* is obtained, as given by Eq. (3-91):

$$\{^1D, {}^2D, {}^3D, {}^4D, {}^5D, {}^6D, \dots\} \tag{3-91}$$

As the exponent k increases, the contributions of the most distant pairs of atoms become the most important. Moreover, *distance/distance matrixes*, denoted as **D/D**, were defined as quotient matrixes in terms of the ratio of geometric r_{ij} or topographic distances t_{ij} over topological distances d_{ij} in order to unify 2D and 3D information about the structure of molecules [180,181]. The row sums of these matrixes contain information on the molecular folding; in effect, in highly folded structures, they tend to be relatively small as the interatomic distances are small while the topological distances increase as the size of the structure increases. Therefore, the average row sum is a molecular descriptor called *average distance/distance degree*, i.e., Eq. (3-92)

$$\text{ADDD} = \frac{1}{A} \cdot \sum_{i=1}^{A} \sum_{j=1}^{A} \frac{r_{ij}}{d_{ij}} \quad j \neq i \tag{3-92}$$

while the half-sum of all distance/distance matrix entries is another molecular descriptor called *D/D index*, i.e., Eq. (3-93).

$$D/D = \frac{1}{2} \cdot \sum_{i=1}^{A} \sum_{j=1}^{A} \frac{r_{ij}}{d_{ij}} \quad j \neq i \tag{3-93}$$

From the largest eigenvalue of the distance/distance matrix, a *folding degree index* ϕ was also defined [180,181]; this is the largest eigenvalue λ_1^{DD} obtained by the diagonalization of the distance/distance matrix **D/D**, then normalized dividing it by the number of atoms A [Eq. (3-94)]:

$$\phi = \frac{\lambda_1^{DD}}{A} \quad 0 < \phi < 1 \tag{3-94}$$

This quantity tends to one for linear molecules (of infinite length) and decreases in correspondence with the folding of the molecule. For example, ϕ values for transoid-molecules are always greater than the values for the corresponding cisoid-molecules. Thus, ϕ can be thought of as a measure of the degree of folding of the molecule because it indicates the degree of departure of a molecule from strict linearity. The folding degree index is a measure of the conformational variability of the molecule, i.e., the capability of a flexible molecule (often macromolecules, proteins) to assume conformations close over upon itself. This index allows a quantitative measure of similarity between chains of the same length but with different geometries, it is sensitive to conformational changes. The *folding profile* of a molecule is proposed, as given by Eq. (3-95):

$$\{^1\phi, ^2\phi, ^3\phi, \ldots, ^k\phi, \ldots\} \tag{3-95}$$

where $^k\phi$ is the normalized first eigenvalue of the kth order distance/distance matrix, whose elements are derived raising to the kth power the elements of the matrix **D/D**. Obviously, $^1\phi$ is the folding degree index. These vectorial descriptors were used to study the folding of peptide sequences [182].

3.5.3.   WHIM Descriptors

Weighted holistic invariant molecular (WHIM) descriptors are geometrical descriptors based on statistical indexes calculated on the projections of the atoms along principal axes [33,183]. WHIM descriptors are generated in such a way as to capture relevant molecular 3D information regarding molecular size, shape, symmetry, and atom distribution with respect to invariant reference frames. Within the WHIM approach, a molecule is seen as a configuration of points (the atoms) in the three-dimensional space defined by the Cartesian axes (x,y,z). In order to obtain a unique reference frame, principal axes of the molecule are calculated. Then, projections of the atoms along each of the principal axes are performed and their dispersion and distribution around the geometric center are evaluated. More specifically, the algorithm consists of calculating the eigenvalues and eigenvectors of a weighted covariance matrix of the centered Cartesian coordinates of the atoms of a molecule, obtained from different *weighting schemes* for the atoms [Eq. (3-96)]:

$$s_{jk} = \frac{\sum_{i=1}^{A} w_i \left(q_{ij} - \bar{q}_j\right)\left(q_{ik} - \bar{q}_k\right)}{\sum_{i=1}^{A} w_i} \tag{3-96}$$

where s_{jk} is the weighted covariance between the jth and kth atomic coordinates, A is the number of atoms, w_i is the weight of the ith atom, q_{ij} and q_{ik} represent the jth and kth coordinate $(j, k = x, y, z)$ of the ith atom, respectively, and $\bar{q}$ the corresponding average value.

Six different weighting schemes, providing *WHIM weighted covariance matrixes* (WWC matrixes), were proposed: (1) the unweighted case u ($w_i = 1$ $i = 1, A$, where A is the number of atoms for each compound), (2) atomic mass m, (3) the van der Waals volume v, (4) the Sanderson atomic electronegativity e, (5) the atomic polarizability p, and (6) the electrotopological state indexes of Kier and Hall, S

[127]. All the weights are scaled with respect to the carbon atom. Depending on the kind of weighting scheme, different covariance matrixes and, therefore, different principal axes are obtained. For example, using atomic masses as the weighting scheme, the directions of the three principal axes are the directions of the principal inertia axes. Thus, the WHIM approach can be viewed as a generalization of searching for the principal axes with respect to a defined atomic property (the weighting scheme). Based on the same principles of the WHIMs, COMMA descriptors were later proposed [184]. They consist of 11 descriptors given by moment expansions for which the zero-order moment of a property field is non-vanishing. WHIM descriptors are divided into two main classes:*directional WHIM descriptors* and *global WHIM descriptors*. A summary of WHIMs is shown in Table 3-5.

Table 3-5. WHIM descriptors. λ refers to eigenvalues of the weighted covariance matrix; t refers to atomic coordinates with respect to the principal axes; A is the number of molecule atoms; n_s is the number of symmetric atoms along a principal axis; and n_a is the number of asymmetric atoms

Equation	Formula	Name	Molecular feature
(3-97)	$\lambda_m \quad m = 1,2,3$	d-WSIZ indexes	Axial dimension
(3-98)	$T = \lambda_1 + \lambda_2 + \lambda_3$	WSIZ index	Global dimension
(3-99)	$A = \lambda_1\lambda_2 + \lambda_1\lambda_3 + \lambda_2\lambda_3$	WSIZ index	Global dimension
(3-100)	$V = \prod\limits_{m=1}^{3}(1 + \lambda_m) - 1 = T + A + \lambda_1\lambda_2\lambda_3$	WSIZ index	Global dimension
(3-101)	$\vartheta_m = \dfrac{\lambda_m}{\sum_m \lambda_m} \quad m = 1,2,3$	d-WSHA indexes	Axial shape
(3-102)	$K = \dfrac{3}{4} \cdot \sum\limits_{m=1}^{3} \left| \dfrac{\lambda_m}{\sum_m \lambda_m} - \dfrac{1}{3} \right|$	WSHA index	Global shape
(3-103)	$\eta_m = \dfrac{\lambda_m^2 \cdot A}{\sum_i t_i^4} \quad m = 1,2,3$	d-WDEN indexes	Axial density
(3-104)	$D = \eta_1 + \eta_2 + \eta_3$	WDEN index	Global density
(3-105)	$\gamma_m = \left\{ 1 - \left[\dfrac{n_s}{A} \cdot \log_2 \dfrac{n_s}{A} + n_a \cdot \left(\dfrac{1}{A} \cdot \log_2 \dfrac{1}{A} \right) \right] \right\}^{-1}$ $m = 1,2,3$	d-WSYM indexes	Axial symmetry
(3-106)	$G = (\gamma_1 \cdot \gamma_2 \cdot \gamma_3)^{1/3}$	WSYM index	Global symmetry

Directional WHIM descriptors are calculated as some univariate statistical indexes on the projections of the atoms along each individual principal axis, while the global WHIMs are directly calculated as a combination of the former, thus simultaneously accounting for the variation of molecular properties along the three principal directions in the molecule. In this case, any information related individually to each principal axis disappears and the description is related only to a global view of the molecule.

WHIM descriptors are invariant to translation due to the centering of the atomic coordinates and invariant to rotation due to the uniqueness of the principal axes, thus resulting free from prior alignment of molecules. To make the WHIM approach clearer, it is illustrated with a simple example. Considering chlorobenzene to be the molecule for analysis, it can be thought of as the configuration of points shown in Figure 3-8, the atomic Cartesian coordinates being those shown in Table 3-6.

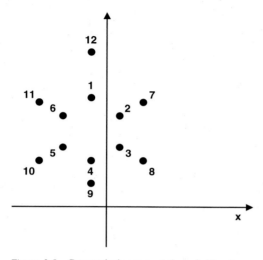

Figure 3-8. Geometrical representation of chlorobenzene based on the Cartesian coordinates (Table 3-6). The chlorine atom (12) shows the highest distance from the aromatic ring

Table 3-6. Cartesian atomic coordinates of an optimized geometry of chlorobenzene (see Figure 3-8)

ID	Atom	x	y	z
1	C	−0.662	4.186	0
2	C	0.549	3.489	0
3	C	0.547	2.093	0
4	C	−0.662	1.395	0
5	C	−1.871	2.093	0
6	C	−1.873	3.489	0
7	H	1.511	4.030	0
8	H	1.502	1.540	0
9	H	−0.662	0.291	0
10	H	−2.826	1.540	0
11	H	−2.835	4.030	0
12	Cl	−0.662	5.911	0

This reference frame is obviously not unique as it depends on how the molecule was drawn and its conformation was optimized. Thus, to calculate unique molecular descriptors independent of the reference frame the principal axes have to be searched for. Figure 3-9 shows the principal axes of chlorobenzene computed by considering each point weighted by the corresponding atomic mass. Note that the first principal axis is along the direction of the heteroatom and the origin of the reference frame coincides with the geometrical center of the molecule. The atomic coordinates with respect to the new reference frame together with scaled atomic masses are shown in Table 3-7. In general, the effect of weighting atoms with atomic properties consists of redirecting the principal axes toward molecular regions with large property values.

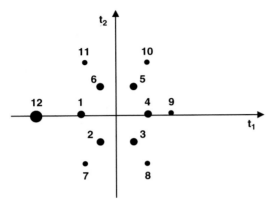

Figure 3-9. Geometrical representation of chlorobenzene in the space of principal axes. Point size is proportional to atomic mass. The chlorine atom shows (12) the highest distance from the aromatic ring and the largest mass resulting for the most important atom in determining the first axis direction (t_1)

Table 3-7. Scaled atomic masses and coordinates with respect to the principal axes of chlorobenzene

ID	Atom	m	t_1	t_2	t_3
1	C	1	−1.346	0	0
2	C	1	−0.649	−1.211	0
3	C	1	0.748	−1.209	0
4	C	1	1.446	0	0
5	C	1	0.748	1.209	0
6	C	1	−0.649	1.211	0
7	H	0.084	−1.189	−2.173	0
8	H	0.084	1.300	−2.164	0
9	H	0.084	2.549	0	0
10	H	0.084	1.300	2.164	0
11	H	0.084	−1.189	2.173	0
12	Cl	2.952	−3.071	0	0

A fundamental role in the WHIM descriptor calculation process is played by the eigenvalues λ_1, λ_2, and λ_3 of the weighted covariance matrix of the molecule atomic coordinates. Each eigenvalue represents a dispersion measure (i.e., the weighted variance) of the projected atoms along the considered principal axis, thus accounting for the molecular size along that principal direction [Eqs. (3-97,3-98,3-99,3-100)].

In the case of chlorobenzene, the mass-weighted eigenvalues are 3.709, 0.794, and 0, highlighting that the molecule is much more elongated along the first axis with respect to the second one due to the relatively large mass of the chlorine atom. Note that if the three unweighted eigenvalues 2.333. 2.055, and 0 were considered, then this difference would be less significant, as expected from a purely geometrical point of view. Moreover, the third eigenvalue is zero as expected for planar molecules, there being no variance out of the molecular plane.

Relationships among the eigenvalues are used to describe the molecular shape [Eqs. (3-101 and (3-102)]. For example, for an ideal straight molecule both λ_2 and λ_3 are equal to zero and the global shape K is equal to 1 (maximum value); for an ideal spherical molecule all three eigenvalues are all equal to 1/3 and $K = 0$. In the case of chlorobenzene, the mass-weighted global shape Km is equal to 0.736, highlighting once again the role of the chlorine mass in amplifying the molecule linearity with respect to the unweighted case Ku = 0.500.

Exploiting the new coordinates t_m of the atoms along the principal axes, the atom distribution and density around the molecule center can be evaluated by an inverse function [Eq. (3-103)] of the kurtosis κ ($\eta = 1/\kappa$). Low values of the kurtosis are obtained when the data points (i.e., the atom projections) assume opposite values with respect to the center. When an increasing number of data values are within the extreme values along a principal axis, the kurtosis value increases (i.e., kurtosis equal to 1.8 for a uniform distribution of points and equal to 3.0 for a normal distribution). When the kurtosis value tends to infinity, the corresponding η value tends to zero.

In an analogous way, from the analysis of the new coordinates t_m of the atoms, molecular symmetry is evaluated on the basis of the number n_s of symmetric atoms with respect to the molecule center, i.e., atoms with opposite coordinates along the considered axis, and the number n_a of asymmetric atoms [Eq. (3-105)].

In conclusion, for each weighting scheme w, 11 molecular directional WHIM descriptors (ϑ_3 is excluded) were proposed, thus resulting in a total of 66 directional WHIM descriptors. For planar compounds, λ_3, λ_3, and η_3 are always equal to zero. The global WHIMs are five for each of the six proposed weighting schemes w, plus the symmetry indexes Gu, Gm, and Gs, giving a total number of 33 descriptors. An example of calculation of WHIMs is given in Table 3-8.

3.5.4.   GETAWAY Descriptors

Geometry, topology, and atom-weights assembly (GETAWAY) descriptors have been proposed as chemical structure descriptors derived from the molecular influence matrix. This is a matrix representation of molecules denoted by **H** and defined as the following equation (3-107) [37,185]:

$$\mathbf{H} = \mathbf{M} \times (\mathbf{M}^{\mathbf{T}} \times \mathbf{M})^{-1} \times \mathbf{M}^{\mathbf{T}} \tag{3-107}$$

where **M** is the molecular matrix comprising the centered Cartesian coordinates x, y, z of the molecule atoms (hydrogens included) in a chosen conformation. Atomic coordinates are assumed to be calculated with respect to the geometrical center of

Table 3-8. Some WHIM descriptors for 22 *N,N*-dimethyl-α-bromo-phenethylamines (Figure 3-5). The scaled atomic mass (m) is the weighting scheme for atoms; u refers to the unitary weighting scheme

Mol	X	Y	$\lambda_1 u$	$\lambda_2 u$	$\lambda_3 u$	$\lambda_1 m$	$\lambda_2 m$	$\lambda_3 m$	Tu	Tm	Ku	Km	Du	Dm
1	H	H	8.038	1.465	0.909	4.678	2.735	0.472	10.411	7.885	0.658	0.410	0.501	0.602
2	H	F	8.085	1.716	0.665	6.501	2.408	0.572	10.466	9.482	0.659	0.528	0.426	0.804
3	H	Cl	8.225	1.719	0.677	8.447	2.264	0.543	10.621	11.255	0.662	0.626	0.420	0.788
4	H	Br	8.301	1.725	0.674	12.468	1.934	0.471	10.699	14.873	0.664	0.757	0.417	0.846
5	H	I	8.308	1.700	0.719	16.081	1.716	0.403	10.728	18.200	0.662	0.825	0.416	0.988
6	H	Me	10.551	1.611	0.691	6.391	2.350	0.587	12.853	9.327	0.731	0.528	0.425	0.521
7	F	H	8.040	1.751	0.657	5.843	2.402	0.789	10.448	9.034	0.654	0.470	0.426	0.758
8	Cl	H	8.104	1.794	0.677	7.170	2.264	0.915	10.575	10.349	0.650	0.539	0.423	0.704
9	Br	H	8.129	1.812	0.687	10.116	1.946	0.935	10.628	12.997	0.647	0.667	0.421	0.708
10	I	H	8.155	1.836	0.697	12.912	1.701	0.882	10.688	15.494	0.645	0.750	0.417	0.824
11	Me	H	9.335	2.053	0.718	5.789	2.472	0.776	12.106	9.038	0.657	0.461	0.426	0.547
12	Cl	F	8.138	1.820	0.676	7.885	2.199	1.129	10.634	11.214	0.648	0.555	0.411	0.887
13	Br	F	8.163	1.856	0.671	9.935	1.991	1.301	10.690	13.226	0.645	0.627	0.405	0.865
14	Me	F	8.989	2.232	0.684	6.670	2.171	0.877	11.905	9.719	0.633	0.529	0.436	0.591
15	Cl	Cl	8.278	1.819	0.693	9.447	2.121	1.098	10.790	12.667	0.651	0.619	0.405	0.887
16	Br	Cl	8.320	1.871	0.672	11.006	2.022	1.334	10.863	14.362	0.649	0.650	0.396	0.863
17	Me	Cl	9.136	2.260	0.669	8.222	2.057	0.840	12.064	11.119	0.636	0.609	0.431	0.625
18	Cl	Br	8.357	1.836	0.679	12.877	1.844	1.008	10.873	15.728	0.653	0.728	0.399	0.916
19	Br	Br	8.381	1.880	0.669	13.632	1.943	1.269	10.930	16.844	0.650	0.714	0.392	0.948
20	Me	Br	9.195	2.220	0.709	11.412	1.763	0.721	12.124	13.895	0.638	0.732	0.434	0.657
21	Me	Me	10.698	2.108	0.741	6.843	2.164	0.877	13.547	9.884	0.685	0.539	0.437	0.415
22	Br	Me	10.553	1.704	0.744	8.461	2.377	1.226	13.001	12.065	0.718	0.552	0.410	0.684

the molecule in order to obtain translational invariance. The molecular influence matrix is a symmetric $A \times A$ matrix, where A represents the number of atoms, and shows rotational invariance with respect to the molecule coordinates, thus resulting independent of any alignment of data set molecules.

The diagonal elements h_{ii} of the molecular influence matrix, called *leverages* being the elements of the *leverage matrix* defined in statistics, range from 0 to 1 and encode atomic information related to the *"influence"* of each molecule atom in determining the whole shape of the molecule; in effect, mantle atoms always have higher h_{ii} values than atoms near the molecule center. Moreover, the magnitude of the maximum leverage in a molecule depends on the size and shape of the molecule. As derived from the geometry of the molecule, leverage values are effectively sensitive to significant conformational changes and to the bond lengths that account for atom types and bond multiplicity.

Each off-diagonal element h_{ij} represents the degree of accessibility of the *j*th atom to interaction with the *i*th atom, or, in other words, the attitude of the two considered atoms to interact with each other. A negative sign for the off-diagonal elements means that the two atoms occupy opposite molecular regions with respect to the center, hence the degree of their mutual accessibility should be low. Table 3-9 shows the molecular influence matrix of chlorobenzene, whose three-dimensional structure has been optimized by minimizing the conformational energy. Atom numbering of chlorobenzene is shown in Figure 3-10.

It can be noted that the outer atoms of chlorobenzene (Cl and hydrogens) have larger leverage values (0.337, 0.242, 0.250, 0.232) than the carbon atoms of the aromatic ring (0.065, 0.075, 0.079). Then, among the outer atoms, the chlorine atom has the largest value (0.337), its bond length being larger than the bond distances of hydrogens. It must also be noted that equal leverage values are obtained for symmetric atoms, such as (C_2, C_6), (C_3, C_5), (H_7, H_{11}), and (H_8, H_{10}). Moreover, the off-diagonal terms give, to some extent, information on the relative spatial position of pairs of atoms. For instance, atoms C_1, C_2, C_6, H_7, and H_{11} have positive off-diagonal values with respect to the chlorine atom and, among these, C_1 has the largest value being the nearest one.

Combining the elements of the molecular influence matrix **H** with those of the geometry matrix **G**, which encodes spatial relationships between pairs of atoms, another symmetric $A \times A$ molecular matrix, called *influence/distance matrix* and denoted by **R**, was derived as the following equation (3-108):

$$[\mathbf{R}]_{ij} \equiv \left[\frac{\sqrt{h_i \cdot h_j}}{r_{ij}} \right]_{ij} \qquad i \neq j \qquad (3\text{-}108)$$

where h_i and h_j are the leverages of the atoms i and j, and r_{ij} is their geometric distance. The diagonal elements of the matrix **R** are zero, while each off-diagonal element i–j is calculated by the ratio of the geometric mean of the corresponding *i*th and *j*th diagonal elements of the matrix **H** over the interatomic distance r_{ij} provided by the geometry matrix **G**. The square root product of the leverages of two atoms is

Table 3-9. Molecular influence matrix **H** of chlorobenzene. Atom numbering is shown in Figure 3-9

	C_1	C_2	C_3	C_4	C_5	C_6	H_7	H_8	H_9	H_{10}	H_{11}	Cl_{12}
C_1	0.065	0.031	-0.036	-0.070	-0.036	0.031	0.057	-0.063	-0.123	-0.063	0.057	0.148
C_2	0.031	0.075	0.042	-0.034	-0.077	-0.044	0.134	0.076	-0.059	-0.136	-0.079	0.071
C_3	-0.036	0.042	0.079	0.039	-0.039	-0.077	0.075	0.141	0.068	-0.071	-0.138	-0.082
C_4	-0.070	-0.034	0.039	0.075	0.039	-0.034	-0.061	0.067	0.132	0.067	-0.061	-0.159
C_5	-0.036	-0.077	-0.039	0.039	0.079	0.042	-0.138	-0.071	0.068	0.141	0.075	-0.082
C_6	0.031	-0.044	-0.077	-0.034	0.042	0.075	-0.079	-0.136	-0.059	0.076	0.134	0.071
H_7	0.057	0.134	0.075	-0.061	-0.138	-0.079	0.242	0.135	-0.108	-0.246	-0.141	0.130
H_8	-0.063	0.076	0.141	0.067	-0.071	-0.136	0.135	0.250	0.118	-0.129	-0.246	-0.143
H_9	-0.123	-0.059	0.068	0.132	0.068	-0.059	-0.108	0.118	0.232	0.118	-0.108	-0.280
H_{10}	-0.063	-0.136	-0.071	0.067	0.141	0.076	-0.246	-0.129	0.118	0.250	0.135	-0.143
H_{11}	0.057	-0.079	-0.138	-0.061	0.075	0.134	-0.141	-0.246	-0.108	0.135	0.242	0.130
Cl_{12}	0.148	0.071	-0.082	-0.159	-0.082	0.071	0.130	-0.143	-0.280	-0.143	0.130	0.337

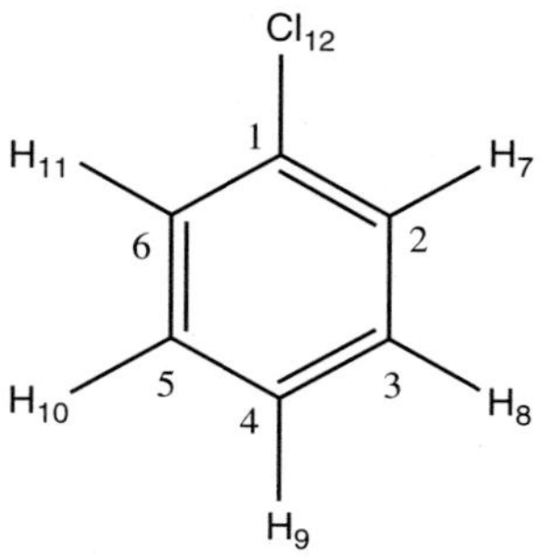

Figure 3-10. H-filled molecular graph of chlorobenzene

divided by their interatomic distance in order to make less significant contributions from pairs of atoms far apart, according to the basic idea that interactions between atoms in the molecule decrease as their distance increases. Obviously, the largest values of the matrix elements are derived from the most external atoms (i.e., those with high leverages) and simultaneously next to each other in the molecular space (i.e., those having small interatomic distances).

A set of the GETAWAY descriptors (H_{GM}, I_{TH}, I_{SH}, *HIC*, *RARS*, *RCON*, *REIG*) was derived by applying some traditional matrix operators and concepts of information theory both to the molecular influence matrix **H** and the influence/distance matrix **R**. Most of these descriptors are simply calculated only by the leverages used as the atomic weightings. Their formulae and definitions are described in Table 3-10.

The index H_{GM} (Eq 3-109) has been proposed as the geometric mean of the leverage values in order to encompass information related to molecular shape. It has, in effect, been found that in an isomeric series of hydrocarbons, the H_{GM} index is sensitive to the molecular shape increasing from linear to more branched molecules; it is also inversely related to molecular size, decreasing as the number of atoms in the molecule increases.

The *total* and *standardized information content on the leverage equality* [Eqs. (3-110) and (3-111)] mainly encode information on molecular symmetry; if all the atoms have different leverage values, i.e., the molecule does not show any element of symmetry, $I_{TH} = A_0 \log A_0$ and $I_{SH} = 1$; otherwise, if all the atoms have equal leverage values (a perfectly symmetric theoretical case), $I_{TH} = 0$ and $I_{SH} = 0$. The total information content on the leverage equality I_{TH} is more discriminating than I_{SH}, because of its dependence on molecular size, and thus it could be thought of as a measure of molecular complexity. These indexes were demonstrated to be useful in modeling physico-chemical properties related to entropy and symmetry [185].

The *HIC* descriptor [Eq. (3-112)] seems to encompass more information related to molecular complexity than the total and standardized information content on the leverage equality. Unlike I_{TH} and I_{SH}, *HIC* can, for example, recognize the different substituents in a series of monosubstituted benzenes. It is also sensitive to the presence of multiple bonds.

Table 3-10. GETAWAY descriptors based on matrix operators and information indexes. A is the number of molecule; atoms (hydrogen included); A_0 is the number of non-hydrogen atoms; N_g is the number of atoms with the same leverage value and G the number of equivalence classes; $M = 1$, 2, or 3 (1 for linear, 2 for planar, and 3 for non-planar molecules); B is the number of molecule bonds; $VS_i(\mathbf{R})$ indicates the row sums of the influence/distance matrix $\mathbf{R}$; a_{ij} is equal to one for pairs of bonded atoms, and zero otherwise

Equation	Formula	Name
(3-109)	$H_{GM} = 100 \cdot \left(\prod_{i=1}^{A} h_{ii} \right)^{1/A}$	Geometric mean of the leverage magnitude
(3-110)	$I_{TH} = A_0 \cdot \log_2 A_0 - \sum_{g=1}^{G} N_g \cdot \log_2 N_g$	Total information content on the leverage equality
(3-111)	$I_{SH} = \dfrac{I_{TH}}{A_0 \cdot \log_2 A_0} = 1 - \dfrac{\sum_{g=1}^{G} N_g \cdot \log_2 N_g}{A_0 \cdot \log_2 A_0}$	Standardized information content on the leverage equality
(3-112)	$HIC = - \sum_{i=1}^{A} \dfrac{h_{ii}}{M} \cdot \log_2 \dfrac{h_{ii}}{M}$	Mean information content on the leverage magnitude
(3-113)	$RARS = \dfrac{1}{A} \cdot \sum_{i=1}^{A} \sum_{j=1}^{A} \dfrac{\sqrt{h_{ii} \cdot h_{jj}}}{r_{ij}} = \dfrac{1}{A} \cdot \sum_{i=1}^{A} VS_i(\mathbf{R})$	Average row sum of the influence/distance matrix
(3-114)	$RCON = \sum_{i=1}^{A} \sum_{j=1}^{A} a_{ij} \cdot \left(VS_i(\mathbf{R}) \cdot VS_j(\mathbf{R}) \right)^{1/2}$	R-connectivity index
(3-115)	$REIG = S_p Max(\mathbf{R})$	R-matrix leading eigenvalue

Both *RARS* [Eq. (3-113)] and *RCON* [Eq. (3-114)] are based on the row sums of the influence/distance matrix since these encode some useful information that could be related to the presence of significant substituents or fragments in the molecule. It was, in effect, observed that larger row sums correspond to terminal atoms that are located very next to other terminal atoms such as those in substituents on a parent structure. Moreover, the *RCON* index is very sensitive to the molecular size as well as to conformational changes and cyclicity. The *REIG* descriptor [Eq. (3-115)] has been defined by analogy with the Lovasz-Pelikan index [118], that is an index of molecular branching calculated as the first eigenvalue of the adjacency matrix. *RARS* and *REIG* indexes are closely related; their values decrease as the molecular size increases and seem to be a little more sensitive to molecular branching than to cyclicity and conformational changes.

The calculation of some GETAWAY descriptors is illustrated for acrylic acid. The hydrogen-filled molecular graph, molecular influence matrix **H**, and influence/distance matrix **R** for acrylic acid are shown in Figure 3-11, Tables 3-11 and 3-12, respectively. The matrixes were calculated from the x, y, z coordinates of the atoms in the minimum energy conformation optimized by the AM1 semi-empirical method (see Chapter 2).

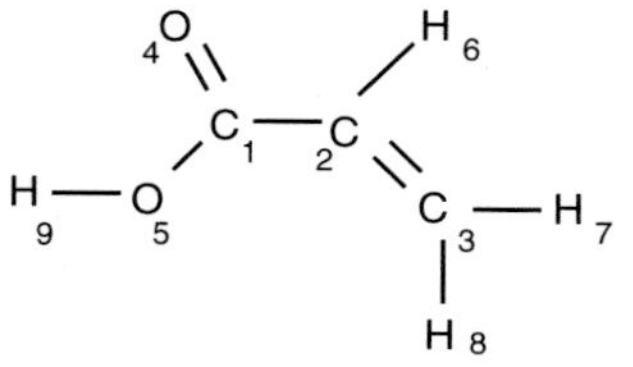

Figure 3-11. H-filled molecular graph of acrylic acid

Table 3-11. Molecular influence matrix **H** of acrylic acid. Atom numbering is shown in Figure 3-11.

	C_1	C_2	C_3	O_4	O_5	H_6	H_7	H_8	H_9
C_1	0.056	0.004	−0.076	0.130	0.017	0.037	−0.114	−0.110	0.056
C_2	0.004	0.054	0.009	0.040	−0.096	0.134	0.049	−0.071	−0.122
C_3	−0.076	0.009	0.109	−0.171	−0.048	−0.018	0.170	0.135	−0.109
O_4	0.130	0.040	−0.171	0.321	−0.017	0.163	−0.233	−0.293	0.059
O_5	0.017	−0.096	−0.048	−0.017	0.179	−0.225	−0.136	0.082	0.243
H_6	0.037	0.134	−0.018	0.163	−0.225	0.347	0.061	−0.230	−0.270
H_7	−0.114	0.049	0.170	−0.233	−0.136	0.061	0.291	0.157	−0.247
H_8	−0.110	−0.071	0.135	−0.293	0.082	−0.230	0.157	0.292	0.038
H_9	0.056	−0.122	−0.109	0.059	0.243	−0.270	−0.247	0.038	0.351

Table 3-12. Influence/distance matrix **R** of acrylic acid. Atom numbering is shown in Figure 3-11. VS_i indicates the matrix row sums

	C_1	C_2	C_3	O_4	O_5	H_6	H_7	H_8	H_9	VS_i
C_1	0	0.037	0.031	0.108	0.073	0.064	0.037	0.046	0.073	0.469
C_2	0.037	0	0.058	0.054	0.041	0.124	0.059	0.059	0.043	0.475
C_3	0.031	0.058	0	0.052	0.050	0.091	0.162	0.162	0.052	0.658
O_4	0.108	0.054	0.052	0	0.109	0.125	0.067	0.077	0.150	0.742
O_5	0.073	0.041	0.050	0.109	0	0.074	0.059	0.091	0.258	0.755
H_6	0.064	0.124	0.091	0.125	0.074	0	0.126	0.102	0.086	0.792
H_7	0.037	0.059	0.162	0.067	0.059	0.126	0	0.157	0.066	0.733
H_8	0.046	0.059	0.162	0.077	0.091	0.102	0.157	0	0.092	0.786
H_9	0.073	0.043	0.052	0.150	0.258	0.086	0.066	0.092	0	0.820

$$H_{GM} = 100 \times \left(\prod_{i=1}^{9} h_i \right)^{1/9} = 100 \times (0.059 \times 0.054 \times 0.109 \times 0.321 \times 0.179$$
$$\times 0.347 \times 0.291 \times 0.292 \times 0.351)^{1/9} = 179.8$$

$$I_{TH} = 5 \times \log_2 5 - \sum_{g=1}^{5} N_g$$
$$\times \log_2 N_g = 11.61 - 5 \times (1 \times \log_2 1) = 11.61$$
$$I_{SH} = \frac{I_{TH}}{5 \times \log_2 5} = \frac{11.61}{11.61} = 1$$

$$HIC = -\sum_{i=1}^{9} \frac{h_i}{2} \times \log_2 \frac{h_i}{2} = -\frac{0.056}{2} \times \log_2 \frac{0.056}{2} - \frac{0.054}{2} \times \log_2 \frac{0.054}{2} - \frac{0.109}{2}$$

$$\times \log_2 \frac{0.109}{2} - \frac{0.321}{2} \times \log_2 \frac{0.321}{2} - \frac{0.179}{2} \times \log_2 \frac{0.179}{2}$$

$$-\frac{0.347}{2} \times \log_2 \frac{0.347}{2} - \frac{0.291}{2} \times \log_2 \frac{0.291}{2} - \frac{0.292}{2} \times \log_2 \frac{0.292}{2} - \frac{0.351}{2}$$

$$\times \log_2 \frac{0.351}{2} = 2.938$$

$$RARS = \frac{1}{9} \times (0.469 + 0.475 + 0.658 + 0.742 + 0.755 + 0.792 + 0.733$$
$$+0.786 + 0.820) = 0.692$$

$$RCON = (0.469 \times 0.475 + 0.469 \times 0.742 + 0.469 \times 0.755 + 0.475$$
$$\times 0.658 + 0.475 \times 0.792 + 0.658 \times 0.733 + 0.658 \times 0.786 + 0.755$$
$$\times 0.820)^{1/2} = 5.028$$

The set of eigenvalues of the influence/distance matrix $\mathbf{R}$ is: 0.713, 0.159, 0.022, −0.037, −0.103, −0.149, −0.166, −0.177, −0.263. Therefore, $REIG = 0.713$.

The other set of GETAWAY descriptors, shown in Table 3-13, is comprised of autocorrelation vectors obtained by double-weighting the molecule atoms in such a way as to account for atomic mass, polarizability, van der Waals volume, and electronegativity together with 3D information encoded by the elements of the molecular influence matrix $\mathbf{H}$ and influence/distance matrix $\mathbf{R}$.

Table 3-13. GETAWAY descriptors based on autocorrelation functions A is the number of molecule atoms (hydrogen included); D is the topological diameter; d_{ij} is the topological distance between atoms i and j; w_i is a physico-chemical property of the ith atom

Equation	Formula		Name
(3-116)	$HATS_k (w) = \sum_{i=1}^{A-1} \sum_{j>i} (w_i \cdot h_{ii}) \cdot (w_j \cdot h_{jj}) \cdot \delta (d_{ij};k)$	$k = 0,1,2,\ldots,D$	HATS indexes
(3-117)	$HATS (w) = HATS_0 (w) + 2 \cdot \sum_{k=1}^{D} HATS_k (w)$		HATS total index
(3-118)	$H_k (w) = \sum_{i=1}^{A-1} \sum_{j>i} h_{ij} \cdot w_i \cdot w_j \cdot \delta (d_{ij};h_{ij};k)$	$k = 0,1,2,\ldots,D$	H indexes
(3-119)	$HT (w) = H_0 (w) + 2 \cdot \sum_{k=1}^{D} H_k (w)$		H total index
(3-120)	$R_k (w) = \sum_{i=1}^{A-1} \sum_{j>i} \frac{\sqrt{h_{ii} \cdot h_{jj}}}{r_{ij}} \cdot w_i \cdot w_j \cdot \delta (d_{ij};k)$	$k = 1,2,\ldots,D$	R indexes
(3-121)	$RT (w) = 2 \cdot \sum_{k=1}^{D} R_k (w)$		R total index
(3-122)	$R_k^+ (w) = \max_{ij} \left(\frac{\sqrt{h_i \cdot h_j}}{r_{ij}} \cdot w_i \cdot w_j \cdot \delta (d_{ij};k) \right)$	$i \neq j \quad k = 1,2,\ldots,D$	Maximal R indexes
(3-123)	$RT^+ (w) = \max_k \left(R_k^+ (w) \right)$		Maximal R total index

HATS indexes [Eqs. (3-116) and (3-117)] are defined by analogy with the Moreau–Broto autocorrelation descriptors *ATS* [Eq. (3-45)], but weighting each atom of the molecule by physico-chemical properties combined with the diagonal elements of the molecular influence matrix **H**, thus they also account for the 3D features of the molecules. The calculation of *HATS*(m) indexes is illustrated for acrylic acid (Figure 3-11). Atomic masses scaled on the carbon atom were used as the weighting scheme for molecule atoms: $m(C) = 1$, $m(H) = 0.084$, $m(O) = 1.332$. The molecular influence matrix **H** of acrylic acid is reported in Table 3-11. Because the topological diameter D is equal to 5, six *HATS* indexes ($k = 0, 5$) can be derived. Examples of calculation for $k = 0$ and $k = 3$ are reported. For $k = 0$, the summation goes over the single atoms, then

$$HATS_0(m) = \sum_{i=1}^{9} (m_i \cdot h_i)^2 = 0.003 + 0.003 + 0.012 + 0.183 + 0.057 + 0.001$$
$$+0.001 + 0.001 + 0.001 = 0.262$$

For $k = 3$, the summation goes over all of the atom pairs at topological distance three:

$$HATS_3(m) = (m_1 \cdot h_1) \cdot (m_7 \cdot h_7) + (m_1 \cdot h_1) \cdot (m_8 \cdot h_8)$$
$$+(m_2 \cdot h_2) \cdot (m_9 \cdot h_9) + (m_3 \cdot h_3) \cdot (m_4 \cdot h_4)$$
$$+(m_3 \cdot h_3) \cdot (m_5 \cdot h_5) + (m_4 \cdot h_4) \cdot (m_9 \cdot h_9)$$
$$+(m_4 \cdot h_4) \cdot (m_6 \cdot h_6) + (m_5 \cdot h_5) \cdot (m_6 \cdot h_6) + (m_6 \cdot h_6) \cdot (m_7 \cdot h_7)$$
$$+(m_6 \cdot h_6) \cdot (m_8 \cdot h_8) = 0.001 + 0.001 + 0.002 + 0.047 + 0.026$$
$$+0.013 + 0.012 + 0.007 + 0.001 + 0.001 = 0.110$$

H indexes [Eqs. (3-118 and 3-119)] are filtered autocorrelation descriptors. The function $\delta(k;d_{ij};h_{ij})$ used for the calculation of these indexes is a Dirac-delta function defined as follows [Eq. (3-124)]:

$$\delta\left(k;d_{ij};h_{ij}\right) = \begin{cases} 1 & \text{if } d_{ij} = k \quad \text{and} \quad h_{ij} > 0 \\ 0 & \text{if } d_{ij} \neq k \quad \text{or} \quad h_{ij} \leq 0 \end{cases} \tag{3-124}$$

While the *HATS* indexes [Eq. (3-116)] make use of the diagonal elements of the matrix **H**, the *H* indexes [Eq. (3-118)] exploit the off-diagonal elements, which can be either positive or negative. In order to emphasize interactions between spatially near atoms, only off-diagonal positive h values are used. In effect, for a given *lag* (i.e., topological distance) the product of the atom properties is multiplied by the corresponding h_{ij} value and only those contributions with a positive h_{ij} value are considered. This means that, for a given atom i, only those atoms j at topological distance d_{ij} with a positive h_{ij} value are considered, because they may have the chance to interact with the ith atom. The *maximal R indexes* [Eq. (3-122)] were proposed in order to take into account local aspects of the molecule and allow reversible decoding; only the maximum property product between atom pairs at a given topological distance (*lag*) is retained.

An example for the calculation of $H(m)$, $R(m)$, and $R^+(m)$ indexes for acrylic acid (Figure 3-11) is reported. The molecular influence matrix and influence/distance matrix of acrylic acid are given in Tables 3-11 and 3-12, respectively. Calculation is based on the atomic mass weighting scheme scaled on the carbon atom: $m(C) = 1$, $m(H) = 0.084$, $m(O) = 1.332$. Only indexes for $k = 3$ are reported

$$H_3(m) = m_4 \times m_9 \times h_{49} + m_4 \times m_6 \times h_{46} + m_6 \times m_7 \times h_{67}$$
$$= 0.0182 + 0.0066 + 0.0004 = 0.025$$

$$R_3(m) = [\mathbf{R}]_{1,7} \cdot m_1 \cdot m_7 + [\mathbf{R}]_{1,8} \cdot m_1 \cdot m_8 + [\mathbf{R}]_{2,9} \cdot m_2 \cdot m_9 + [\mathbf{R}]_{3,4} \cdot m_3 \cdot m_4$$
$$+ [\mathbf{R}]_{3,5} \cdot m_3 \cdot m_5 + [\mathbf{R}]_{4,9} \cdot m_4 \cdot m_9 + [\mathbf{R}]_{4,6} \cdot m_4 \cdot m_6 + [\mathbf{R}]_{5,6}$$
$$\cdot m_5 \cdot m_6 + [\mathbf{R}]_{6,7} \cdot m_6 \cdot m_7 + [\mathbf{R}]_{6,8} \cdot m_6 \cdot m_8 = 0.003 + 0.004 + 0.004$$
$$+ 0.069 + 0.067 + 0.017 + 0.014 + 0.008 + 0.001 + 0.001 = 0.188$$

$$R_3^+(m) = \max (0.003; 0.004; 0.004; 0.069; 0.067; 0.017; 0.014; 0.008; 0.001; 0.001)$$
$$= 0.069$$

Note that the $R_3^+(m)$ index identifies the structural fragment $C_3 = C_2 - C_1 = O_4$.

The atomic weighting schemes applied for GETAWAY descriptor calculation are those proposed for the WHIM descriptors, that is, atomic mass (m), atomic polarizability (p), Sanderson atomic electronegativity (e), atomic van der Waals volume (v), plus the unit weighting scheme (u).

HATS, *H*, *R*, and maximal *R* indexes are vectorial descriptors for structure–property correlations, but they can also be used as molecular profiles suitable for similarity/diversity analysis studies. These descriptors, as based on spatial autocorrelation, encode information on structural fragments and therefore seem to be particularly suitable for describing differences in congeneric series of molecules. Unlike the Moreau–Broto autocorrelations, GETAWAYs are geometrical descriptors encoding information on the effective position of substituents and fragments in the molecular space. Moreover, they are independent of molecule alignment and, to some extent, account also for information on molecular size and shape as well as for specific atomic properties.

A joint use of GETAWAY and WHIM descriptors is advised, exploiting both local information of the former and holistic information of the latter set of descriptors. The GETAWAY descriptors have been used for modeling several data sets of pharmacological and environmental interest [185–188].

3.5.5.   Molecular Transforms

Molecular transforms are vectorial descriptors based on the concept of obtaining information from the 3D atomic coordinates by the transform used in electron diffraction studies for preparing theoretical scattering curves [189,190].

A generalized scattering function can be used as the functional basis for deriving, from a known molecular structure, the specific analytic relationship of both X-ray and electron diffraction. The general molecular transform is given by Eq. (3-125):

$$G(\mathbf{s}) = \sum_{i=1}^{A} f_i \cdot \exp\left(2\pi i \cdot \mathbf{r}_i \cdot \mathbf{s}\right) \tag{3-125}$$

where $\mathbf{s}$ represents the scattering in various directions by a collection of A atoms located at points $\mathbf{r}_i$; f_i is a form factor taking into account the direction dependence of scattering from a spherical body of finite size. The scattering parameter s has the dimension of a reciprocal distance and depends on the scattering angle, as given by Eq. (3-126):

$$s = \frac{4\pi}{\lambda} \cdot \sin\left(\vartheta/2\right) \tag{3-126}$$

where ϑ is the scattering angle and λ is the wavelength of the electron beam.

Usually, the above equation is used in a modified form as suggested in 1931 by Wierl [191]. On substituting the form factor by an atomic property w_i, considering the molecule to be rigid and setting the instrumental constant equal to one, the following function, usually called *radial distribution function*, is used to calculate molecular transforms [Eq. (3-127)]:

$$I(s) = \sum_{i=1}^{A-1} \sum_{j=i+1}^{A} w_i \cdot w_j \cdot \frac{\sin\left(s \cdot r_{ij}\right)}{s \cdot r_{ij}} \tag{3-127}$$

where $I(s)$ is the scattered electron intensity, w is an atomic property, chosen as the atomic number Z by Soltzberg and Wilkins [189], r_{ij} is the geometric distance between the ith and jth atom, and A is the number of atoms in the molecule. The sum is performed over all the pairs of atoms in the molecule.

Soltzberg and Wilkins introduced a number of simplifications in order to obtain a binary code. Only the zero crossing of the $I(s)$ curve, i.e., the s values at which $I(s) = 0$, in the range 1–31 Å^{-1} were considered. The s range was then divided into 100 equal-sized bins, each described by a binary variable equal to 1 if the bin contains a zero crossing, and zero otherwise. Thus, a vectorial descriptor consisting of 100 bins was finally calculated for each molecule.

Raevsky and co-workers applied the molecular transform to study ligand–receptor interactions by using hydrogen-bond abilities, hydrophobicity, and charge of the atoms, instead of the atomic number Z [192]. For each atomic property, a spectrum of interatomic distances, called *interatomic interaction spectrum*, was derived to represent the 3D structure of molecules and the scattered intensities in selected regions of the spectrum were used as the molecular descriptors [193]. These *descriptors* are based on local characteristics of different pairs of centers

in the molecule. For a selected distance R, the following function was evaluated
[Eq. (3-128)] [193,194]:

$$I(R) = \sum_{r^{\min}}^{r^{\max}} \sum_{i=1}^{A} \sum_{j=1}^{A} \frac{w_i \cdot w_j}{1 + \sqrt{\frac{(R-r_{ij})^2}{0.1}}} \qquad i \neq j \qquad (3\text{-}128)$$

where A is the number of atoms in the molecule, w_i and w_j are atomic properties of
the ith and jth atom, respectively, r_{ij} is the geometric interatomic distance; $r^{\min}$ and
$r^{\max}$ define a distance range around R, which accounts for vibrations of atoms and
allows to obtain a band instead of a line in the final spectrum for each pair of centers
defined by R. Distances R are varied from 1.1 to 20 Å with step 0.1 Å, resulting in a
total of 190 signals per spectrum.

Superimposition of all the bands for all the possible pairs of centers forms the
final interatomic interaction spectrum. Seven types of spectrum are calculated for
each molecule by using different atomic properties w: steric interaction spectrum,
spectrum of interactions between positively charged atoms, spectrum of interactions between negatively charged atoms, spectrum of interactions of positively
charged atoms with negatively charged atoms, spectrum of interactions between
hydrogen-bond donors, spectrum of interactions between hydrogen-bond acceptors, and spectrum of interactions of hydrogen-bond donors with hydrogen-bond
acceptors.

The *integrated molecular transform* (FT_m) is a molecular descriptor calculated
from the square of the molecular transform by integrating the squared molecular
transform in a selected interval of the scattering parameter s to obtain the area under
the curve and finally taking the square root of the area [195,196].

To calculate *3D-molecule representation of structures based on electron diffraction descriptors* (or *3D-MoRSE*), Gasteiger et al. [197,198] returned to the initial
$I(s)$ curve and maintained the explicit form of the curve. As for the atomic weighting scheme w, various physico-chemical properties such as atomic mass, partial
atomic charges, atomic polarizability were considered. In order to obtain uniform-
length descriptors, the intensity distribution $I(s)$ was made discrete, calculating its
value at a sequence of evenly distributed values of, e.g., 32 or 64 values in the range
of $1\text{--}31$ Å^{-1}. Clearly, the more values are chosen, the finer the resolution in the
representation of the molecule.

Radial distribution function (*RDF*) *descriptors* were proposed based on a radial
distribution function different from that commonly used to calculate molecular
transforms $I(s)$ [199,200]. The radial distribution function here selected is that quite
often used for the interpretation of the diffraction patterns obtained in powder X-ray
diffraction experiments. Formally, the radial distribution function of an ensemble
of A atoms can be interpreted as the probability distribution to find an atom in a
spherical volume of radius R. The general form of the radial distribution function is
represented by Eq. (3-129):

$$g(R) = f \cdot \sum_{i=1}^{A-1} \sum_{j=i+1}^{A} w_i \cdot w_j \cdot e^{-\beta \cdot (R-r_{ij})^2} \tag{3-129}$$

where f is a scaling factor, w are characteristic atomic properties of the atoms i and j, r_{ij} is the interatomic distance between the ith and jth atom, and A is the number of atoms. The exponential term contains the distance r_{ij} between the atoms i and j and the smoothing parameter β, that defines the probability distribution of the individual interatomic distances; β can be interpreted as a temperature factor that defines the movement of atoms. $g(R)$ is generally calculated at a number of discrete points with defined intervals. A RDF vector of 128 values was proposed, using a step size for R about 0.1–0.2 Å, while the β parameter is fixed in the range between 100 and 200 Å^{-2}. By including characteristic atomic properties w of the atoms i and j, RDF descriptors can be used in different tasks to fit the requirements of the information to be represented. These atomic properties enable the discrimination of the atoms of a molecule for almost any property that can be attributed to an atom.

The radial distribution function in this form meets all the requirements for a 3D structure descriptor: it is independent of the number of atoms, i.e., the size of a molecule, it is unique regarding the three-dimensional arrangement of the atoms, and invariant against translation and rotation of the entire molecule. Additionally, the RDF descriptors can be restricted to specific atom types or distance ranges to represent specific information in a certain three-dimensional structure space, e.g., to describe steric hindrance or structure/activity properties of a molecule. Moreover, the RDF vectorial descriptor is interpretable by using simple rules and, thus, it provides a possibility of reversible decoding. Besides information about distribution of interatomic distances in the entire molecule, the RDF vector provides further valuable information, e.g., about bond distances, ring types, planar and non-planar systems, and atom types. This fact is a most valuable consideration for a computer-assisted code elucidation.

3.6.   CONCLUSIONS

The scientific community is showing an increasing interest in the field of QSAR. Several chemoinformatics methods were specifically conceived trying to solve QSAR problems, answering the demand to know in a deeper manner the chemical systems and their relationships with biological systems. Nowadays, the need to deal with biological systems described by peptide/protein or DNA sequences, to describe proteomics maps, or to give effective answers to ecological and health problems, pushes new borders further where mathematics, statistics, chemistry, and biology and their inter-relationships may produce new effective useful knowledge. Several molecular descriptors have been proposed in the last few years, illustrating the great interest the scientific community has shown in the theoretical approach to capture information about chemical compounds and the need for more sophisticated molecular descriptors useful for the development of predictive QSAR/QSPR models [92].

REFERENCES

1. Todeschini R, Consonni V (2000) Handbook of molecular descriptors. Wiley-VCH, Weinheim
2. Brown FK (1998) Chemoinformatics: What is it and how does it impact drug discovery. Annu Rep Med Chem 33:375–384
3. Gasteiger J (ed) (2003) Handbook of chemoinformatics. From data to knowledge in 4 volumes. Wiley-VCH, Weinheim
4. Oprea TI (2003) Chemoinformatics and the quest for leads in drug discovery. In: Gasteiger J. (ed) Handbook of chemoinformatics. Wiley-VCH, Weinheim
5. Crum-Brown A (1864) On the theory of isomeric compounds. Trans Roy Soc Edinburgh 23:707–719
6. Crum-Brown A (1867) On an application of mathematics to chemistry. Proc Roy Soc (Edinburgh) 73:89–90
7. Crum-Brown A, Fraser TR (1868) On the connection between chemical constitution and physiological action. Part 1. On the physiological action of salts of the ammonium bases, derived from strychnia, brucia, thebia, codeia, morphia and nicotia. Trans Roy Soc Edinburgh 25:151–203
8. Körner W (1869) Fatti per servire alla determinazione del luogo chimico nelle sostanze aromatiche. Giornale di Scienze Naturali ed Economiche 5:212–256
9. Körner W (1874) Studi sulla Isomeria delle Così Dette Sostanze Aromatiche a Sei Atomi di Carbonio. Gazz Chim It 4:242
10. Hammett LP (1935) Reaction rates and indicator acidities. Chem Rev 17:67–79
11. Hammett LP (1937) The effect of structure upon the reactions of organic compounds. Benzene derivatives. J Am Chem Soc 59:96–103
12. Hammett LP (1938) Linear free energy relationships in rate and equilibrium phenomena. Trans Faraday Soc 34:156–165
13. Wiener H (1947) Influence of interatomic forces on paraffin properties. J Chem Phys 15:766
14. Platt JR (1947) Influence of neighbor bonds on additive bond properties in paraffins. J Chem Phys 15:419–420
15. Taft RW (1952) Polar and steric substituent constants for aliphatic and o-benzoate groups from rates of esterification and hydrolysis of esters. J Am Chem Soc 74:3120–3128
16. Taft RW (1953) Linear steric energy relationships. J Am Chem Soc 75:4538–4539
17. Hansch C, Maloney PP, Fujita T et al. (1962) Correlation of biological activity of phenoxyacetic acids with Hammett substituent constants and partition coefficients. Nature 194:178–180
18. Hansch C, Muir RM, Fujita T et al. (1963) The correlation of biological activity of plant growth regulators and chloromycetin derivatives with Hammett constants and partition coefficients. J Am Chem Soc 85:2817–2824
19. Fujita T, Iwasa J, Hansch C (1964) A new substituent constant, π, derived from partition coefficients. J Am Chem Soc 86:5175–5180
20. Hansch C, Leo A (1995) Exploring QSAR. Fundamentals and applications in chemistry and biology. American Chemical Society, Washington DC
21. Free SM, Wilson JW (1964) A mathematical contribution to structure–activity studies. J Med Chem 7:395–399
22. Gordon M, Scantlebury GR (1964) Non-random polycondensation: Statistical theory of the substitution effect. Trans Faraday Soc 60:604–621
23. Smolenskii EA (1964) Application of the theory of graphs to calculations of the additive structural properties of hydrocarbons. Russ J Phys Chem 38:700–702
24. Spialter L (1964) The atom connectivity matrix (ACM) and its characteristic polynomial (ACMCP). J Chem Doc 4:261–269

25. Balaban AT, Harary F (1968) Chemical graphs. V. Enumeration and proposed nomenclature of benzenoid catacondensed polycyclic aromatic hydrocarbons. Tetrahedron 24:2505–2516
26. Harary F (1969) Graph theory. Addison-Wesley, Reading MA
27. Kier LB (1971) Molecular orbital theory in drug research. Academic Press, New York
28. Balaban AT, Harary F (1971) The characteristic polynomial does not uniquely determine the topology of a molecule. J Chem Doc 11:258–259
29. Randić M (1974) On the recognition of identical graphs representing molecular topology. J Chem Phys 60:3920–3928
30. Kier LB, Hall LH, Murray WJ et al. (1975) Molecular connectivity I: Relationship to nonspecific local anesthesia. J Pharm Sci 64:1971–1974
31. Rohrbaugh RH, Jurs PC (1987) Descriptions of molecular shape applied in studies of structure/activity and structure/property relationships. Anal Chim Acta 199:99–109
32. Stanton DT, Jurs PC (1990) Development and use of charged partial surface area structural descriptors in computer-assisted quantitative structure–property relationship studies. Anal Chem 62:2323–2329
33. Todeschini R, Lasagni M, Marengo E (1994) New molecular descriptors for 2D- and 3D-structures. Theory. J Chemom 8:263–273
34. Katritzky AR, Mu L, Lobanov VS et al. (1996) Correlation of boiling points with molecular structure. 1. A training set of 298 diverse organics and a test set of 9 simple inorganics. J Phys Chem 100:10400–10407
35. Ferguson AM, Heritage TW, Jonathon P et al. (1997) EVA: A new theoretically based molecular descriptor for use in QSAR\QSPR analysis. J Comput Aid Mol Des 11:143–152
36. Schuur J, Selzer P, Gasteiger J (1996) The coding of the three-dimensional structure of molecules by molecular transforms and its application to structure-spectra correlations and studies of biological activity. J Chem Inf Comput Sci 36:334–344
37. Consonni V, Todeschini R, Pavan M (2002) Structure/response correlations and similarity/diversity analysis by GETAWAY descriptors. Part 1. Theory of the novel 3D molecular descriptors. J Chem Inf Comput Sci 42:682–692
38. Goodford PJ (1985) A computational procedure for determining energetically favorable binding sites on biologically important macromolecules. J Med Chem 28:849–857
39. Cramer III RD, Patterson DE, Bunce JD (1988) Comparative molecular field analysis (CoMFA). 1. Effect of shape on binding of steroids to carrier proteins. J Am Chem Soc 110:5959–5967
40. Oprea TI (2004) 3D QSAR modeling in drug design. In: Bultinck P, De Winter H, Langenaeker W et al. (eds) Computational medicinal chemistry for drug discovery. Marcel Dekker, New York
41. Randić M (1991) Generalized molecular descriptors. J Math Chem 7:155–168
42. Testa B, Kier LB (1991) The concept of molecular structure in structure–activity relationship studies and drug design. Med Res Rev 11:35–48
43. Jurs PC, Dixon JS, Egolf LM (1995) Representations of molecules. In: van de Waterbeemd H (ed) Chemometrics methods in molecular design. VCH Publishers, New York
44. Smith EG, Baker PA (1975) The Wiswesser line-formula chemical notation (WLN). Chemical Information Management, Cherry Hill, NJ
45. Weininger D (1988) SMILES, a chemical language and information system. 1. Introduction to methodology and encoding rules. J Chem Inf Comput Sci 28:31–36
46. Weininger D. (2003) SMILES – A language for molecules and reactions. In: van de Waterbeemd H (ed) Handbook of chemoinformatics. Wiley-VCH, Weinheim
47. SMARTS Tutorial. Daylight chemical information systems. Santa Fe, New Mexico
48. Buolamwini JK, Assefa H (2002) CoMFA and CoMSIA 3D QSAR and docking studies on conformationally-restrained cinnamoyl HIV-1 integrase inhibitors: Exploration of a binding mode at the active site. J Med Chem 45:841–852

49. Xu M, Zhang A, Han S et al. (2002) Studies of 3D-quantitative structure–activity relationships on a set of nitroaromatic compounds: CoMFA, advanced CoMFA and CoMSIA. Chemosphere 48:707–715

50. Klebe G, Abraham U, Mietzner T (1994) Molecular similarity indices in a comparative analysis (CoMSIA) of drug molecules to correlate and predict their biological activity. J Med Chem 37:4130–4146

51. Jain AN, Koile K, Chapman D (1994) Compass: Predicting biological activities from molecular surface properties. Performance comparisons on a steroid benchmark. J Med Chem 37: 2315–2327

52. Todeschini R, Moro G, Boggia R et al. (1997) Modeling and prediction of molecular properties. Theory of grid-weighted holistic invariant molecular (G-WHIM) descriptors. Chemom Intell Lab Syst 36:65–73

53. Chuman H, Karasawa M, Fujita T (1998) A novel 3-dimensional QSAR procedure – voronoi field analysis. Quant Struct – Act Relat 17:313–326

54. Robinson DD, Winn PJ, Lyne PD et al. (1999) Self-organizing molecular field analysis: A tool for structure–activity studies. J Med Chem 42:573–583

55. Cruciani G, Pastor M, Guba W (2000) VolSurf: a new tool for the pharmaceutic optimization of lead compounds. Eur J Pharm Sci 11(Suppl.):S29–S39

56. Pastor M, Cruciani G, McLay IM et al. (2000) GRid-INdependent descriptors (GRIND): A novel class of alignment-independent three-dimensional molecular descriptors. J Med Chem 43:3233–3243

57. Mekenyan O, Ivanov J, Veith GD et al. (1994) Dynamic QSAR: A new search for active conformations and significant stereoelectronic indices. Quant Struct – Act Relat 13:302–307

58. Duca JS, Hopfinger AJ (2001) Estimation of molecular similarity based on 4D-QSAR analysis: Formalism and validation. J Chem Inf Comput Sci 41:1367–1387

59. Hopfinger AJ, Wang S, Tokarski JS et al. (1997) Construction of 3D-QSAR models using the 4D-QSAR analysis formalism. J Am Chem Soc 119:10509–10524

60. Harary F (1969) Proof techniques in graph theory. Academic Press, San Diego CA

61. Kier LB, Hall LH (1976) Molecular connectivity in chemistry and drug research. Academic Press, New York

62. Bonchev D, Trinajstić N (1977) Information theory, distance matrix, and molecular branching. J Chem Phys 67:4517–4533

63. Balaban AT, Motoc I, Bonchev D et al. (1983) Topological indices for structure–activity correlations. In: Charton M, Motoc I (eds) Steric effects in drug design (Topics in Current Chemistry, Vol. 114). Springer-Verlag, Berlin

64. Rouvray DH (1983) Should we have designs on topological indices? In: King RB (ed) Chemical applications of topology and graph theory. Studies in physical and theoretical chemistry. Elsevier, Amsterdam

65. Basak SC, Magnuson VR, Veith GD (1987) Topological indices: Their nature, mutual relatedness, and applications. In: Charton M, Motoc I (eds) Mathematical modelling in science and technology. Pergamon Press, Oxford

66. Trinajstić N (1992) Chemical graph theory. CRC Press, Boca Raton, FL

67. Randić M. (1993) Comparative regression analysis. Regressions based on a single descriptor. Croat Chem Acta 66:289–312

68. Diudea MV, Gutman I (1998) Wiener-type topological indices. Croat Chem Acta 71:21–51

69. Ivanciuc O, Balaban AT (1999) The graph description of chemical structures. In: Devillers J, Balaban AT (eds) Topological indices and related descriptors in QSAR and QSPR. Gordon & Breach Science Publishers, Amsterdam

70. Basak SC, Gute BD, Grunwald GD (1997) Use of topostructural, topochemical, and geometric parameters in the prediction of vapor pressure: A hierarchical QSAR approach. J Chem Inf Comput Sci 37:651–655

71. Diudea MV, Horvath D, Graovac A (1995) Molecular topology. 15. 3D distance matrices and related topological indices. J Chem Inf Comput Sci 35:129–135

72. Balaban AT (1997) From chemical graphs to 3D molecular modeling. In: Balaban AT (ed) From chemical topology to three-dimensional geometry. Plenum Press, New York

73. Hosoya H (1971) Topological index. A newly proposed quantity characterizing the topological nature of structural isomers of saturated hydrocarbons. Bull Chem Soc Jap 44:2332–2339

74. Randić M, Wilkins CL (1979) Graph theoretical ordering of structures as a basis for systematic searches for regularities in molecular data. J Phys Chem 83:1525–1540

75. Kier LB (1985) A shape index from molecular graphs. Quant Struct – Act Relat 4:109–116

76. Randić M (2001) Novel shape descriptors for molecular graphs. J Chem Inf Comput Sci 41:607–613

77. Wiener H (1947) Structural determination of paraffin boiling points. J Am Chem Soc 69:17–20

78. Ivanciuc O, Balaban T-S, Balaban AT (1993) Design of topological indices. Part 4. Reciprocal distance matrix, related local vertex invariants and topological indices. J Math Chem 12:309–318

79. Ivanciuc O (2000) QSAR comparative study of Wiener descriptors for weighted molecular graphs. J Chem Inf Comput Sci 40:1412–1422

80. Ivanciuc O (1999) Design of topological indices. Part 11. Distance-valency matrices and derived molecular graph descriptors. Rev Roum Chim 44:519–528

81. Ivanciuc O, Ivanciuc T, Diudea MV (1997) Molecular graph matrices and derived structural descriptors. SAR & QSAR Environ Res 7:63–87

82. Gutman I, Trinajstić N (1972) Graph theory and molecular orbitals. Total π-electron energy of alternant hydrocarbons. Chem Phys Lett 17:535–538

83. Randić M (1975) Graph theoretical approach to local and overall aromaticity of benzenoid hydrocarbons. Tetrahedron 31:1477–1481

84. Kier LB, Hall LH (1977) The nature of structure–activity relationships and their relation to molecular connectivity. Eur J Med Chem 12:307–312

85. Balaban AT (1982) Highly discriminating distance-based topological index. Chem Phys Lett 89:399–404

86. Balaban AT (1987) Numerical modelling of chemical structures: Local graph invariants and topological indices. In: King RB, Rouvray DH (eds) Graph theory and topology in chemistry. Elsevier, Amsterdam

87. Bonchev D (1983) Information theoretic indices for characterization of chemical structures. Research Studies Press, Chichester

88. Mekenyan O, Bonchev D, Balaban AT (1988) Topological indices for molecular fragments and new graph invariants. J Math Chem 2:347–375

89. Burden FR (1997) A chemically intuitive molecular index based on the eigenvalues of a modified adjacency matrix. Quant Struct – Act Relat 16:309–314

90. Filip PA, Balaban T-S, Balaban AT (1987) A new approach for devising local graph invariants: Derived topological indices with low degeneracy and good correlation ability. J Math Chem 1:61–83

91. Estrada E, Rodriguez L, Gutièrrez A (1997) Matrix algebraic manipulation of molecular graphs. 1. Distance and vertex-adjacency matrices. MATCH Commun Math Comput Chem 35:145–156

92. Todeschini R., Consonni V (2009) Molecular Descriptors for Chemoinformatics (in 2 volumes). Wiley-VCH, Weinheim

93. Ivanciuc O, Ivanciuc T, Cabrol-Bass D et al. (2000) Comparison of weighting schemes for molecular graph descriptors: application in quantitative structure–retention relationship models for alkylphenols in gas–liquid chromatography. J Chem Inf Comput Sci 40:732–743

94. Harary F (1959) Status and contrastatus. Sociometry 22:23–43

95. Petitjean M (1992) Applications of the radius-diameter diagram to the classification of topological and geometrical shapes of chemical compounds. J Chem Inf Comput Sci 32:331–337

96. Ivanciuc O, Balaban AT (1994) Design of topological indices. Part 8. Path matrices and derived molecular graph invariants. MATCH Commun Math Comput Chem 30:141–152

97. Amić D, Trinajstić N (1995) On the detour matrix. Croat Chem Acta 68:53–62

98. Mohar B (1989) Laplacian matrices of graphs. Stud Phys Theor Chem 63:1–8

99. Mohar B (1989) Laplacian matrices of graphs. In: Graovac A (ed) MATH/CHEM/COMP 1988. Elsevier, Amsterdam

100. Randić M (1975) On characterization of molecular branching. J Am Chem Soc 97:6609–6615

101. Kier LB, Hall LH (2000) Intermolecular accessibility: The meaning of molecular connectivity. J Chem Inf Comput Sci 40:792–795

102. Kier LB, Hall LH (1986) Molecular connectivity in structure–activity analysis. Research Studies Press – Wiley, Chichester

103. Altenburg K (1980) Eine Bemerkung zu dem Randicschen "Molekularen Bindungs-Index". Z. Phys Chemie 261:389–393

104. Randić M, Hansen PJ, Jurs PC (1988) Search for useful graph theoretical invariants of molecular structure. J Chem Inf Comput Sci 28:60–68

105. Ivanciuc O (2001) Design of topological indices. Part 19. Computation of vertex and molecular graph structural descriptors with operators. Rev Roum Chim 46:243–253

106. Graham RL, Lovasz L (1978) Distance matrix polynomials of trees. Adv Math 29:60–88

107. Diudea MV, Gutman I, Jäntschi L (2001) Molecular topology. Nova Science Publishers, Huntington NY

108. Nikolić S, Plavšić D, Trinajstić N (1992) On the Z-counting polynomial for edge-weighted graphs. J Math Chem 9:381–387

109. Gálvez J, García-Domenech R, De Julián-Ortiz V (2006) Assigning wave functions to graphs: A way to introduce novel topological indices. MATCH Commun Math Comput Chem 56:509–518

110. Hosoya H, Hosoi K, Gutman I (1975) A topological index for the total π-electron energy. Proof of a generalized hückel rule for an arbitrary network. Theor Chim Acta 38:37–47

111. Hosoya H, Murakami M, Gotoh M (1973) Distance polynomial and characterization of a graph. Natl Sci Rept Ochanomizu Univ 24:27–34

112. Consonni V, Todeschini R (2008) New spectral indices for molecule description. MATCH Commun Math Comput Chem 60:3–14

113. Gutman I (1978) The energy of a graph. Ber Math Statist Sekt Forschungszentrum 103:1–22

114. Gutman I (2005) Topology and stability of conjugated hydrocarbons. The dependence of total π-electron energy on molecular topology. J Serb Chem Soc 70:441–456

115. Gutman I, Zhou B (2006) Laplacian energy of a graph. Lin Alg Appl 414:29–37

116. Ivanciuc O (2003) Topological indices. In: Gasteiger J (ed) Handbook of chemoinformatics. Wiley-VCH, Weinheim

117. Ivanciuc O (2001) Design of topological indies. Part 26. Structural descriptors computed from the Laplacian matrix of weighted molecular graphs: Modeling the aqueous solubility of aliphatic alcohols. Rev Roum Chim 46:1331–1347

118. Lovasz L, Pelikan J (1973) On the eigenvalue of trees. Period Math Hung 3:175–182

119. Mohar B (1991) The Laplacian spectrum of graphs. In: Alavi Y, Chartrand C, Ollermann OR (eds) Graph theory, combinatorics, and applications. Wiley, New York

120. Trinajstić N, Babic D, Nikolić S et al. (1994) The Laplacian matrix in chemistry. J Chem Inf Comput Sci 34:368–376

121. Mohar B, Babic D, Trinajstić N (1993) A novel definition of the Wiener index for trees. J Chem Inf Comput Sci 33:153–154

122. Gutman I, Yeh Y-N, Lee S-L et al. (1993) Some recent results in the theory of the Wiener number. Indian J Chem 32A:651–661

123. Marković S, Gutman I, Bancevic Z (1995) Correlation between Wiener and quasi-Wiener indices in benzenoid hydrocarbons. J Serb Chem Soc 60:633–636

124. Nikolić S, Trinajstić N, Jurić A et al. (1996) Complexity of some interesting (chemical) graphs. Croat Chem Acta 69:883–897

125. Mallion RB, Trinajstić N (2003) Reciprocal spanning-tree density: A new index for characterizing the intricacy of a (poly)cyclic molecular graph. MATCH Commun Math Comput Chem 48:97–116

126. Moreau G, Turpin C (1996) Use of similarity analysis to reduce large molecular libraries to smaller sets of representative molecules. Analusis 24:M17–M21

127. Kier LB, Hall LH (1990) An electrotopological-state index for atoms in molecules. Pharm Res 7:801–807

128. Broto P, Moreau G, Vandycke C (1984) Molecular structures: perception, autocorrelation descriptor and SAR studies. Use of the autocorrelation descriptors in the QSAR study of two non-narcotic analgesic series. Eur J Med Chem 19:79–84

129. Broto P, Devillers J (1990) Autocorrelation of properties distributed on molecular graphs. In: Karcher W, Devillers J (eds) Practical applications of quantitative structure–activity relationships (QSAR) in environmental chemistry and toxicology. Kluwer, Dordrecht

130. Wagener M, Sadowski J, Gasteiger J (1995) Autocorrelation of molecular surface properties for modeling *Corticosteroid Binding Globulin* and cytosolic *Ah* receptor activity by neural networks. J Am Chem Soc 117:7769–7775

131. Moreau G, Broto P (1980) The autocorrelation of a topological structure: A new molecular descriptor. Nouv J Chim 4:359–360

132. Moreau G, Broto P (1980) Autocorrelation of molecular structures, application to SAR studies. Nouv J Chim 4:757–764

133. Broto P, Moreau G, Vandycke C (1984) Molecular structures: Perception, autocorrelation descriptor and SAR studies. Autocorrelation descriptor. Eur J Med Chem 19:66–70

134. Hollas B (2002) Correlation properties of the autocorrelation descriptor for molecules. MATCH Commun Math Comput Chem 45:27–33

135. Klein CT, Kaiser D, Ecker G (2004) Topological distance based 3D descriptors for use in QSAR and diversity analysis. J Chem Inf Comput Sci 44:200–209

136. Moran PAP (1950) Notes on continuous stochastic phenomena. Biometrika 37:17–23

137. Geary RC (1954) The contiguity ratio and statistical mapping. Incorp Statist 5:115–145

138. Wold S, Jonsson J, Sjöström M et al. (1993) DNA and peptide sequences and chemical processes multivariately modelled by principal component analysis and partial least-squares projections to latent structures. Anal Chim Acta 277:239–253

139. Sjöström M, Rännar S, Wieslander Å (1995) Polypeptide sequence property relationships in *Escherichia coli* based on auto cross covariances. Chemom Intell Lab Syst 29:295–305

140. Clementi S, Cruciani G, Riganelli D et al. (1993) Autocorrelation as a tool for a congruent description of molecules in 3D QSAR studies. Pharm Pharmacol Lett 3:5–8

141. Melville JL, Hirts JD (2007) TMACC: Interpretable correlation descriptors for quantitative structure–activity relationships. J Chem Inf Model 47:626–634

142. Gasteiger J, Marsili M (1980) Iterative partial equalization of orbital electronegativity: A rapid access to atomic charges. Tetrahedron 36:3219–3228

143. Wildman SA, Crippen GM (1999) Prediction of physicochemical parameters by atomic contributions. J Chem Inf Comput Sci 39:868–873

144. Hou T-J, Xia K, Zhang W et al. (2004) ADME evaluation in drug discovery. 4. Prediction of aqueous solubility based on atom contribution approach. J Chem Inf Comput Sci 44:266–275

145. Sadowski J, Wagener M, Gasteiger J (1995) Assessing similarity and diversity of combinatorial libraries by spatial autocorrelation functions and neural networks. Angew Chem Int Ed Engl 34:2674–2677

146. Carhart RE, Smith DH, Venkataraghavan R (1985) Atom pairs as molecular features in structure–activity studies: Definition and applications. J Chem Inf Comput Sci 25:64–73

147. Schneider G, Neidhart W, Giller T et al. (1999) "Scaffold-Hopping" by topological pharmacophore search: A contribution to virtual screening. Angew Chem Int Ed Engl 38:2894–2895

148. Clerc JT, Terkovics AL (1990) Versatile topological structure descriptor for quantitative structure/property studies. Anal Chim Acta 235:93–102

149. Baumann K (2002) An alignment-independent versatile structure descriptor for QSAR and QSPR based on the distribution of molecular features. J Chem Inf Comput Sci 42:26–35

150. Fechner U, Franke L, Renner S et al. (2003) Comparison of correlation vector methods for ligand-based similarity searching. J Comput Aid Mol Des 17:687–698

151. Renner S, Noeske T, Parsons CG et al. (2005) New allosteric modulators of metabotropic glutamate receptor 5 (mGluR5) found by ligand-based virtual screening. ChemBioChem 6:620–625

152. Renner S, Schneider G (2006) Scaffold-hopping potential of ligand-based similarity concepts. ChemMedChem 1:181–185

153. Estrada E (2001) Generalization of topological indices. Chem Phys Lett 336:248–252

154. Estrada E, Matamala AR (2007) Generalized topological indices. Modeling gas-phase rate coefficients of atmospheric relevance. J Chem Inf Model 47:794–804

155. Estrada E (2003) Generalized graph matrix, graph geometry, quantum chemistry, and optimal description of physicochemical properties. J Phys Chem A 107:7482–7489

156. Ivanciuc O (2001) 3D QSAR models. In: Diudea MV (ed) QSPR/QSAR studies by molecular descriptors. Nova Science, Huntington NY

157. Todeschini R, Consonni V (2003) Descriptors from molecular geometry. In: Gasteiger J (ed) Handbook of chemoinformatics. Wiley-VCH, Weinheim

158. Randić M, Wilkins CL (1979) Graph theoretical study of structural similarity in benzomorphans. Int J Quantum Chem Quant Biol Symp 6:55–71

159. Balaban AT (1997) From chemical topology to 3D geometry. J Chem Inf Comput Sci 37:645–650

160. Mihalić Z, Nikolić S, Trinajstić N (1992) Comparative study of molecular descriptors derived from the distance matrix. J Chem Inf Comput Sci 32:28–37

161. Balasubramanian K (1995) Geometry-dependent connectivity indices for the characterization of molecular structures. Chem Phys Lett 235:580–586

162. Mekenyan O, Peitchev D, Bonchev D et al. (1986) Modelling the interaction of small organic molecules with biomacromolecules. I. Interaction of substituted pyridines with anti-3-azopyridine antibody. Arzneim Forsch 36:176–183

163. Nikolić S, Trinajstić N, Mihalić Z et al. (1991) On the geometric-distance matrix and the corresponding structural invariants of molecular systems. Chem Phys Lett 179:21–28

164. Basak SC, Gute BD, Ghatak S (1999) Prediction of complement-inhibitory activity of benzamidines using topological and geometric parameters. J Chem Inf Comput Sci 39:255–260

165. Castro EA, Gutman I, Marino D et al. (2002) Upgrading the Wiener index. J Serb Chem Soc 67:647–651

166. Randić M (1988) Molecular topographic descriptors. Stud Phys Theor Chem 54:101–108

167. Toropov AA, Toropova AP, Ismailov T et al. (1998) 3D weighting of molecular descriptors for QSAR/QSPR by the method of ideal symmetry (MIS). 1. Application to boiling points of alkanes. J Mol Struct (Theochem) 424:237–247

168. Krenkel G, Castro EA, Toropov A A (2002) 3D and 4D molecular models derived from the ideal symmetry method: prediction of alkanes normal boiling points. Chem Phys Lett 355:517–528

169. Toropov AA, Toropova AP, Muftahov RA et al. (1994) Simulation of molecular systems by the ideal symmetry method for revealing quantitative structure–property relations. Russ J Phys Chem 68:577–579

170. Bath PA, Poirrette AR, Willett P et al. (1995) The extent of the relationship between the graph–theoretical and the geometrical shape coefficients of chemical compounds. J Chem Inf Comput Sci 35:714–716

171. Pepperrell CA, Willett P (1991) Techniques for the calculation of three-dimensional structural similarity using inter-atomic distances. J Comput Aid Mol Des 5:455–474

172. Bemis GW, Kuntz ID (1992) A fast and efficient method for 2D and 3D molecular shape description. J Comput Aid Mol Des 6:607–628

173. Nilakantan R, Bauman N, Venkataraghavan R (1993) New method for rapid characterization of molecular shapes: Applications in drug design. J Chem Inf Comput Sci 33:79–85

174. Bath PA, Poirrette AR, Willett P et al. (1994) Similarity searching in files of three-dimensional chemical structures: Comparison of fragment-based measures of shape similarity. J Chem Inf Comput Sci 34:141–147

175. Good AC, Kuntz ID (1995) Investigating the extension of pairwise distance pharmacophore measures to triplet-based descriptors. J Comput Aid Mol Des 9:373–379

176. Good AC, Ewing TJA, Gschwend DA et al. (1995) New molecular shape descriptors: Application in database screening. J Comput Aid Mol Des 9:1–12

177. Brown RD, Martin YC (1996) Use of structure-activity data to compare structure-based clustering methods and descriptors for use in compound selection. J Chem Inf Comput Sci 36:572–584

178. Bajzer Ž, Randić M, Plavšić D et al. (2003) Novel map descriptors for characterization of toxic effects in proteomics maps. J Mol Graph Model 22:1–9

179. Randić M (1995) Molecular profiles. Novel geometry-dependent molecular descriptors. New J Chem 19:781–791

180. Randić M, Kleiner AF, De Alba LM (1994) Distance/distance matrices. J Chem Inf Comput Sci 34:277–286

181. Randić M, Krilov G (1999) On a characterization of the folding of proteins. Int J Quant Chem 75:1017–1026

182. Randić M (1997) On characterization of chemical structure. J Chem Inf Comput Sci 37:672–687

183. Todeschini R, Gramatica P (1997) 3D-modelling and prediction by WHIM descriptors. Part 5. Theory development and chemical meaning of WHIM descriptors. Quant Struct – Act Relat 16:113–119

184. Silverman BD, Platt DE (1996) Comparative molecular moment analysis (CoMMA): 3D-QSAR without molecular superposition. J Med Chem 39:2129–2140

185. Consonni V, Todeschini R, Pavan M et al. (2002) Structure/response correlations and similarity/diversity analysis by GETAWAY descriptors. Part 2. Application of the novel 3D molecular descriptors to QSAR/QSPR studies. J Chem Inf Comput Sci 42:693–705

186. Fedorowicz A, Singh H, Soderholm S et al. (2005) Structure-activity models for contact sensitization. Chem Res Toxicol 18:954–969

187. Pérez González M, Terán C, Teijeira M et al. (2005) GETAWAY descriptors to predicting A_{2A} adenosine receptors agonists. Eur J Med Chem 40:1080–1086

188. Saiz-Urra L, Pérez González M, Fall Y et al. (2007) Quantitative structure–activity relationship studies of HIV-1 integrase inhibition. 1. GETAWAY descriptors. Eur J Med Chem 42:64–70

189. Soltzberg LJ, Wilkins CL (1976) Computer recognition of activity class from molecular transforms. J Am Chem Soc 98:4006

190. Soltzberg LJ, Wilkins CL (1977) Molecular transforms: A potential tool for structure–activity studies. J Am Chem Soc 99:439–443

191. Wierl K (1931) Elektronenbeugung und Molekulbau. Ann Phys (Leipzig) 8:521–564

192. Novikov VP, Raevsky OA (1982) Representation of molecular structure as a spectrum of inter-atomic distances for the study of structure-biological activity relations. Khimico-Farmaceucheskii Zhurnal 16:574–581

193. Raevsky OA, Dolmatova L, Grigor'ev VJ et al. (1995) Molecular recognition descriptors in QSAR. In: Sanz F, Giraldo J, Manaut F (eds) QSAR and Molecular Modelling: Concepts, Computational Tools and Biological Applications. Prous Science, Barcelona

194. Raevsky OA, Trepalin SV, Razdol'skii AN (2000) New QSAR descriptors calculated from interatomic interaction spectra. Pharm Chem J 34:646–649

195. King JW, Kassel RJ, King BB (1990) The integrated molecular transform as a correlation parameter. Int J Quant Chem Quant Biol Symp 17:27–34

196. Famini GR, Kassel RJ, King JW et al. (1991) Using theoretical descriptors in quantitative structure–activity relationships: Comparison with the molecular transform. Quant Struct -Act Relat 10:344–349

197. Schuur J, Gasteiger J (1996) 3D-MoRSE code – A new method for coding the 3D structure of molecules. In: Gasteiger J (ed) Software Development in Chemistry. Fachgruppe Chemie-Information-Computer (CIC). Frankfurt am Main, Germany

198. Schuur J, Gasteiger J (1997) Infrared spectra simulation of substituted benzene derivatives on the basis of a 3D structure representation. Anal Chem 69:2398–2405

199. Hemmer MC, Steinhauer V, Gasteiger J (1999) Deriving the 3D structure of organic molecules from their infrared spectra. Vibrat Spect 19:151–164

200. Selzer P, Gasteiger J, Thomas H et al. (2000) Rapid access to infrared reference spectra of arbitrary organic compounds: scope and limitations of an approach to the simulation of infrared spectra by neural networks. Chem Eur J 6:920–927

CHAPTER 4

3D-QSAR – APPLICATIONS, RECENT ADVANCES, AND LIMITATIONS

WOLFGANG SIPPL

Department of Pharmaceutical Chemistry, Martin-Luther-Universität Halle-Wittenberg, 06120 Halle (Saale), Germany, e-mail: wolfgang.sippl@pharmazie.uni-halle.de

Abstract: Three-dimensional quantitative structure–activity relationship (3D-QSAR) techniques are the most prominent computational means to support chemistry within drug design projects where no three-dimensional structure of the macromolecular target is available. The primary aim of these techniques is to establish a correlation of biological activities of a series of structurally and biologically characterized compounds with the spatial fingerprints of numerous field properties of each molecule, such as steric demand, lipophilicity, and electrostatic interactions. The number of 3D-QSAR studies has exponentially increased over the last decade, since a variety of methods are commercially available in user-friendly, graphically guided software. In this chapter, we will review recent advances, known limitations, and the application of receptor-based 3D-QSAR

Keywords: 3D-QSAR, CoMFA, CoMSIA, GRID/GLOPE, AFMoC, Receptor-based QSAR

4.1. INTRODUCTION

An important goal in computer-aided design is to find a correlation between the structural features of ligands and their biological activity, that is, their ability to bind to specific target proteins. In some cases, simple mathematical models may provide a means for identifying the property related to biological activity; in other cases a multitude of parameters are necessary to describe the complex behavior of a compound in a biological system. In general, the necessary parameters can be derived by forming a relationship between those properties that describe the structural variation within the group of molecules under investigation and those that describe their biological activities. This relationship is termed a quantitative structure–activity relationship (QSAR) [1–3]. Historically, the primary objective of QSAR was to understand which properties are important to control a specific biological activity of a series of compounds. However, the main objective of these techniques nowadays is the prediction of novel compounds on the basis of previously synthesized molecules.

T. Puzyn et al. (eds.), Recent Advances in QSAR Studies, 103–125.
DOI 10.1007/978-1-4020-9783-6_4, © Springer Science+Business Media B.V. 2010

Usually, a QSAR model is derived using a training set of already characterized ligands. Using statistical methods, one considers the molecular descriptors and the effects of substituents on biological activities. Molecular descriptors are measured or calculated physico-chemical properties, such as log P, pK_a, boiling point, molecular refraction, molecular surface areas, or molecular interaction fields. If prepared correctly, this strategy identifies which structural variations are relevant and influence changes in biological activities. Usually, the mathematical models obtained by regression analysis are validated, in terms of their predictive power, by assessing their capability to predict correctly the biological data of compounds belonging to a so-called test set, that is, a set of molecules with determined biological activity that was not used to generate the initial QSAR model.

4.2.   WHY IS 3D-QSAR SO ATTRACTIVE?

The era of quantitative analysis for the correlation of molecular structures with biological activities started in the 1960s from the classical equation for 2D-QSAR analysis proposed by Hansch [4]. Since then a variety of QSAR approaches have been reported [5–8]. The first applicable 3D-QSAR method was proposed by Cramer et al. in 1988 [6]. His program, CoMFA, was a major breakthrough in the field of 3D-QSAR. The primary aim of 3D-QSAR methods is to establish a correlation of biological activities of a series of structurally and biologically characterized compounds with the spatial fingerprints of numerous field properties of each molecule, such as steric demand, lipophilicity, and electrostatic interactions. Typically, a 3D-QSAR analysis allows the identification of the pharmacophoric arrangement of molecular features in space and provides guidelines for the design of next-generation compounds with enhanced bioactivity or selectivity.

No 3D-QSAR method would be applied to a data set unless one expects that the analysis will reveal insights into useful 3D structure–activity relationships. Since chemists and biologists know that 3D properties of molecules govern biological activity, it is especially informative to see a 3D picture of how structural changes influence biological activities. Approaches that do not provide such a graphical representation are often less attractive to the scientific community. An advantage of 3D-QSAR – over the traditional 2D-QSAR – method is that it takes into account the 3D structures of ligands and additionally is applicable to sets of structurally diverse compounds.

The number of 3D-QSAR studies has increased exponentially over the last decade, since a variety of methods have been made commercially available in user-friendly software [6, 9]. As of the end of 2007, the number of papers dealing with 3D-QSAR is greater than 2500 when the CAS (Chemical Abstracts Service) service is searched using the keywords "3D-QSAR" or "CoMFA" (Figure 4-1). However, it seems that the initial "QSAR hype" is over, as indicated by the constant number of new 3D-QSAR applications in the last few years. The major drawback of 3D-QSAR is that it is not applicable to huge data sets containing more than several thousand compounds, which are usually considered in high-throughput screening. For these kinds of studies, novel faster and simpler methods have been developed, which use

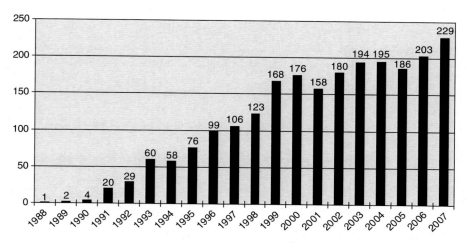

Figure 4-1. Number of 3D-QSAR articles published between 1988 and 2007

the original 3D descriptors (i.e., molecular interaction fields or surface descriptors) as inputs for the generation of alignment-independent models. Examples for this kind of programs recently developed are Volsurf and Almond (for review see [10]).

The most frequently applied methods include comparative molecular field analysis (CoMFA), comparative molecular similarity indices analysis (CoMSIA) [9], and the GRID/GOLPE program (generating optimal linear PLS estimations) [11]. Several reviews have been published in the last few years dealing with the basic theory, the pitfalls, and the application of 3D-QSAR approaches [9–15]. Apart from the commercial distribution, a major factor for the ongoing enthusiasm for CoMFA-related approaches comes from the proven ability of several of these methods to correctly estimate the biological activity of novel compounds. However, very often the predictive ability of QSAR models is only tested in retrospective studies rather than taking the ability to design and develop novel bioactive molecules. Despite the known limitations of 3D-QSAR, the possibility to predict biological data is gaining respect as scientists realize that we are far away from the hoped-for fast and accurate forecast of affinity from (the structure of a) protein–ligand complexes by free-energy perturbation or empirical scoring methods [17–19].

4.3.   LIGAND ALIGNMENT

Establishing the molecular alignment of 3D structures of the investigated ligands is an important prerequisite for several methodologies in drug design, e.g., 3D similarity analysis, prediction of biological activities, and even the estimation of ADME parameters [1–3]. Various procedures and pharmacophore strategies for the superposition of small ligands have thus been proposed in the past. An alignment generation procedure usually considers two steps: superimposing the molecules and scoring of the resulting alignments. Superposition techniques may either utilize information obtained from a binding site of a target protein (direct target-based methods) or be based solely on information obtained from the ligands themselves (indirect ligand-based methods). Some common assumptions, especially for the ligand-based methods, are that the aligned molecules interact with the same amino acids within a binding pocket and exhibit a unique binding mode. Additionally, the generated alignment ideally contains the molecules studied in their bioactive conformation. Superposition methods differ in how they treat flexibility and how the molecules are represented. Ligands can be considered as flexible or rigid; alternatively, flexibility can also be modelled via a limited set of rigid ligand conformers. The molecules to be aligned can be represented by their atoms, shape, or molecular interaction fields [1, 2].

The prediction of biological activity of novel ligands with improved activity/selectivity based on their structure is one of the major challenges in today's drug design. A prerequisite for most approaches is the correct alignment of the ligands under study. Similar to the alignment procedures, the prediction methods can be classified into two major groups: indirect ligand-based and direct structure-based approaches. Ligand-based methods, including traditional quantitative structure–activity relationships (QSARs) [4] and modern 3D-QSAR techniques [5], are based entirely on experimental structure–activity relationships for receptor ligands. 3D-QSAR methods are currently used as standard tools in drug design, since they are computationally feasible and afford fast generation of models from which the biological activity of newly synthesized molecules can be predicted. The basic assumption is that a suitable sampling of the molecular interaction fields around a set of aligned molecules might provide all the information necessary for an understanding of their biological activities [6]. A suitable sampling is achieved by calculating interaction energies between each molecule and an appropriate probe placed at regularly spaced grid points surrounding the molecules. The resulting energies derived from simple potentials can then be contoured in order to give a quantitative spatial description of molecular properties. If correlated with biological data, 3D-fields can be generated which describe the contribution of a region of interest surrounding the ligands to the target properties. However, there is a significant difficulty in the application of 3D-QSAR methods: in order to obtain a correct model, a spatial arrangement of the ligands toward one another has to be found which is representative for the relative differences in the binding geometry at the protein-binding site. The success of a molecular field analysis is therefore determined by the choice of the ligand superposition [7–9]. In most cases, the first step in a 3D-QSAR study is the generation of a reliable pharmacophore. Many alignment strategies have been reported and compared that accomplish. Depending on the molecular flexibility and the structural diversity of the investigated molecules, the task of generating a reliable pharmacophore can become less feasible. Despite the difficulties concerning ligand alignment, many successful 3D-QSAR case studies applying different programs have been reported in the last few years. Most CoMFA applications in drug design have been comprehensively listed and discussed in some reviews [10–13] and books [7, 10, 16].

Target-based methods, on the other hand, incorporate information from the target protein and are able to calculate fairly accurately the position and orientation of a potential ligand in a protein-binding site [20–27]. Over the last decade, a broad spectrum of competitive methods for scoring protein–ligand interactions has emerged [19–27]. Established approaches have been further improved, e.g., in the area of the regression-based scoring functions or methods based on first principles. In addition, well-known techniques have been applied to protein–ligand scoring by using atom–atom contact potentials to develop knowledge-based scoring functions. The major problem of modern docking programs is the inability to evaluate the free energy of binding required to correctly score different ligand–receptor complexes. The main problem in affinity prediction is that the underlying molecular interactions are highly complex and that the experimental data (structural as well as biological data) are far away from being perfect for computational approaches. Numerous terms have to be taken into account when trying to quantify the free energy of binding correctly [27–29]. Elaborate methods, such as the free-energy perturbation or the thermodynamic integration methods, have been shown to be able – at least to some extent – to predict the binding free energy correctly. However, these approaches have the drawback of being computationally very expensive.

In order to exploit the strengths of both approaches, i.e., incorporation of protein information by docking programs and generation of predictive models for related molecules by 3D-QSAR methods, it was suggested to use a combination of both methods resulting in an automated unbiased procedure named receptor-based 3D-QSAR [30–37]. In this context, the three-dimensional structure of a target protein is used within a docking protocol to guide the alignment for a comparative molecular field analysis. This approach allows the generation of a kind of target-specific scoring method considering all the structure–activity data known for a related ligand data set. The comprehensive utility of this approach is exemplified by a variety of successful case studies published in the last few years.

4.4. CoMFA AND RELATED METHODS

4.4.1. CoMFA

For many years, 3D-QSAR has been used as a synonym for CoMFA [6], which was the first method that implemented the concept into a QSAR method, i.e., that the biological activity of a ligand can be predicted from its three-dimensional structure. Until now, CoMFA is probably the most commonly applied 3D-QSAR method [6, 12]. A CoMFA study normally starts with traditional pharmacophore modeling in order to suggest a bioactive conformation of each molecule and ways to superimpose the molecules under study. The underlying idea of CoMFA is that differences in a target property, e.g., biological activity, are often closely related to equivalent changes in shapes and strengths of non-covalent interaction fields surrounding the molecules. Or stated in a different way, the steric and electrostatic fields provide all information necessary for understanding the biological properties of a set of compounds. Hence, the molecules are placed in a cubic grid and the interaction energies between the molecule and a defined probe are calculated for each grid point. Normally, only two potentials, namely a steric potential in the form of a Lennard-Jones function and an electrostatic potential in the form of a simple Coulomb function, are used within a CoMFA study. It is obvious that the description of molecular similarity is not a trivial task nor is the description of the interaction process of ligands with corresponding biological targets. In the standard application of CoMFA, only enthalpic contributions of the free energy of binding are provided by the potentials used. However, many binding effects are governed by hydrophobic and entropic contributions. Therefore, one has to characterize in advance the expected main contributions of forces and whether under these conditions CoMFA will actually be able to find realistic results.

In the original CoMFA report, field values were systematically calculated for ligands at each grid point of a regularly sampled 3D grid box that extended 4 Å beyond the dimension of all molecules in the data set, using a sp^3 carbon atom with +1 charge as probe [6]. The grid resolution should be in a range to produce the field information that is necessary to describe variations in biological activity. On the other hand, introduction of too much irrelevant data to statistical analysis may result in a decrease of predictivity of the model. Typically, a resolution of 2 Å is utilized. Often, superior results are derived using a grid spacing of 2 Å as opposed to the more accurate 1 Å spacing [7]. In addition, the CoMFA program provides a variety of other parameters (probe atoms, charges, energy scaling, energy cut-offs, etc.) which can be adjusted by the user. This flexibility in parameter settings enables the user to fit the whole procedure as closely as possible to his problem. However, it enhances the possibility of chance correlations. Interestingly, nearly all of the successful CoMFA analyses have been achieved with default parameters.

4.4.2.   CoMSIA

Due to the problems associated with the functional form of the Lennard-Jones potential used in most CoMFA methods [12], Klebe et al. [9] have developed a similarity indices-based CoMFA method named CoMSIA (comparative molecular similarity indices analysis). Instead of grid-based fields, CoMSIA is based on similarity indices that are obtained using a functional form that is adapted from the SEAL algorithm. Three different indices related to steric, electrostatic, and hydrophobic potentials were used in their study of the classical steroid benchmark data set. Models of comparable statistical quality with respect to cross-validation of the training set, as well as predictivities of a test set, were derived using CoMSIA. The advantage of this method lies in the functions used to describe the molecules studied, as well as the resulting contour maps. The contour maps obtained from CoMSIA are generally easier to interpret, compared to the ones obtained by the CoMFA approach. CoMSIA also avoids cut-off values used in CoMFA to restrict potential functions by assuming unacceptably large values. For a detailed description of the method as well as its application, the reader is referred to the literature [9, 10].

4.4.3. GRID/GOLPE

GRID [38] has been used by a number of authors as an alternative to the original CoMFA method for calculating interaction fields. An advantage of the GRID approach, apart from the large number of chemical probes available, is the use of a 6-4 potential function, which is smoother than the 6-12 form of the Lennard-Jones type, for calculating the interaction energies at the grid lattice points. Good statistical results have been obtained; for example, in an analysis of glycogen phosphorylase b inhibitors by Cruciani et al. [39]. They used GRID interaction fields in combination with the GOLPE program [39], which accomplishes the necessary chemometrical analysis. The particularly interesting aspect of this data set is that the crystal structures of the protein–ligand complexes have been solved. This allowed the authors to test the predictive abilities of the applied 3D-QSAR techniques.

A further refinement of the original CoMFA technique has been realized by introducing the concept of variable selection and reduction [39, 40]. The large number of variables in the descriptor matrix (i.e., the interaction energies) represents a statistical problem in the CoMFA approach. These variables make it increasingly difficult for multivariate projection methods, such as PLS, to distinguish the useful information contained in the descriptor matrix from that of less quality or noise. Thus, approaches for separating the useful variables from the less useful ones were needed. The GOLPE approach was developed in order to identify which variables are meaningful for the prediction of the biological activity and to remove those with no predictivity. Within this approach, fractional factorial design (FFD) is applied initially to test multiple combinations of variables. For each combination, a PLS model is generated and only variables which significantly increase the predictivity are considered. Variables are then classified considering their contribution to predictivity. A further advance in GOLPE is the implementation of the smart region definition (SRD) procedure that aims to select the cluster of variables mainly responsible for activity rather than a single variable. The SRD technique was found to be less prone to change correlation than any single variable selection, and improves the interpretability of the models.

4.4.4. 4D-QSAR and 5D-QSAR

Recently developed QSAR methods include the so-called 4D-QSAR approach, where an ensemble of conformations for each ligand represents the fourth dimension [41], and 5D-QSAR, which considers in addition hypotheses for changes that might occur in a conformation of a receptor due to ligand binding (induced fit) as a fifth dimension [42, 43]; whether these novel QSAR approaches show increased quality regarding the predictive ability and interpretation of the results must be demonstrated by future case studies.

4.4.5. AFMoC

A novel method which might overcome the problem of neglecting the protein information in a 3D-QSAR analysis has been recently developed by Klebe et al. [44].

In this approach named AFMoC (adaptation of fields for molecular comparison) or "inverted CoMFA," potential fields derived from a scoring function (Drug Score) are generated in the binding pocket of a target protein. Methodologically, the program is related to CoMFA and CoMSIA but with the advantage of including the protein environment to the 3D-QSAR analysis. Instead of only using the Coulomb or Lennard-Jones potential, AFMoC starts with a grid of pre-assigned values. The numbers at the individual grid points consider the Drug Score potential values. By use of ligands with known binding mode and biological data, the deliberately placed ligand atoms introduce an activity-based weighting of the individual Drug Score potential values. The resulting interaction fields are then evaluated by classical PLS. It has been shown that AFMoC-derived QSAR models achieve much better correlations between experimentally derived and computed activities compared with the original scoring function Drug Score [45, 46].

4.5.   RELIABILITY OF 3D-QSAR MODELS

The quality and reliability of any 3D-QSAR model is strongly dependent on the careful examination of each step within a 3D-QSAR analysis. As with any QSAR method, an important point is the question of whether the biological activities of all compounds studied are of comparable quality. Preferably, the biological activity should be obtained in the same laboratory under the same conditions. All compounds being tested in a system must have the same mechanism (binding mode) and all inactive compounds must be shown to be truly inactive. Only in vitro data should be considered, since only in vitro experiments are able to reach a true equilibrium. All other test systems undergo time-dependent changes by multiple coupling to parallel biochemical processes (for example, transport processes). Another critical point is the existence of transport phenomena and diffusion gradients underlying all biological data. One has to bear in mind that all 3D-QSAR approaches were developed to describe only one interaction step in the lifetime of ligands. In all cases, where non-linear phenomena result from drug transport and distribution, any 3D-QSAR technique should be applied with caution. The biological activities of the molecules used in a CoMFA study should ideally span a range of at least three orders of magnitude. For all molecules under study, the exact 3D structure has to be reported. If no information on the exact stereochemistry of the tested compounds is given (mixtures of enantiomers or diastereomers), these compounds should be excluded from a CoMFA study.

The search for the bioactive conformation and a molecular alignment constitutes a serious problem within all 3D-QSAR studies. It is one of the most important sources of incorrect conclusions and errors in all 3D-QSAR analysis. The risk of deriving irrelevant geometries can be reduced by considering rigid analogs. Even then, the alignment poses problems, because there are some cases of different binding modes of seemingly closely related compounds [8]. Even if the binding modes are comparable, choice of wrong ligand conformations may dramatically influence the result of a 3D-QSAR analysis. Problems in the generation of conformations and the correct alignment could be avoided by deriving them from the 3D structures of ligand–protein complexes which are known from X-ray crystallography, NMR, or homology modeling [34].

The final stage of a 3D-QSAR analysis consists of statistical validation in order to assess the significance of the model and hence its ability to predict biological activities of other (novel) compounds. In most 3D-QSAR case studies published in the literature, the leave-one-out (LOO) cross-validation procedure has been used for this purpose. The output of this procedure is the cross-validated q^2 which is commonly regarded as an ultimate criterion of both robustness and predictive ability of a model. The simplest cross-validation method is LOO, where one object at a time is removed from the data set and predicted by the model generated. A more robust and reliable method is the leave-several-out cross-validation. For example, in the leave-20%-out cross-validation, five groups of approximately the same size are generated. Thus, 80% of the compounds are randomly selected for the generation of a model, which is then used to predict the remaining compounds. This operation must be repeated numerous times in order to obtain reliable statistical results. The leave-20%-out or also the more demanding leave-50%-out cross-validation results are much better indicators for the robustness and the predictive ability of a 3D-QSAR model than the usually used LOO procedure [47, 48].

Despite the known limitations of the LOO procedure, it is still uncommon to test 3D-QSAR models for their ability to correctly predict the biological activities of compounds not included in the training set. Regardless, many authors claim that their models, showing high LOO q^2 values, have high predictive ability in the absence of external validation (for a detailed discussion on this problem, see [48–52]). Contrary to such expectations, it has been shown by several studies that a correlation between the LOO cross-validated q^2 value for the training set and the correlation coefficient r^2 between the predicted and observed activities for the test set does not exist [49, 51].

In an attempt to get an idea of the predictive nature of 3D-QSAR models, Doweyko has analyzed 61 models from 37 papers published in the last decade [52]. These papers were selected in a near random manner, focusing on those models for which LOO and externalized test set data were listed. The average 3D-QSAR model of the study contained 48 training set ligands, which showed good internal consistency ($r^2 = 0.93$) and appeared to be reasonably predictive ($q^2 = 0.67$). The average test set consisted of a smaller number of 17 compounds with an r^2_{pred} equal to 0.46. Doweyko then analyzed the correlation between q^2 of the training set and the r^2_{pred} for the test set and tried to answer the question whether a 3D-QSAR model with high q^2 value is predictive.

The author found that there is no obvious correlation between q^2 and r^2_{pred}, an observation, which has been already reported in the literature as the "Kubinyi paradox" [53]. Poor q^2 models as well as good q^2 models were found to be well spread between high and low test set r^2_{pred}, indicating that there is no relationship between q^2 and the model's ability to predict an external test set. The author stated that this may be due to several reasons: (1) a low q^2 value for a small training set may simply reflect the importance of each member of the training set to the model and have nothing to do with predictivity and (2) a high q^2 may reflect redundancy in the training set, and once again have nothing to do with predictivity. It is also reported in the literature that the predictive ability of a 3D-QSAR model is strongly dependent on the structural similarity between the training and test set molecules. Therefore, instances where a high q^2 is associated with a high r^2_{pred} may be attributed to the care taken to choose a test set that covers the same descriptor space as utilized by the training set.

4.6.   RECEPTOR-BASED 3D-QSAR

The combination of ligand-based and receptor-based approaches has been shown to provide an interesting strategy for ligands for which the binding site is known, but the exact binding mode has not been determined experimentally. This has been demonstrated by a variety of applications published within the last 10 years. One of the earliest approaches in this field was published by Marshall et al. [54]. The VALIDATE program uses 12 physico-chemical and energetic parameters, including the electrostatic and steric interaction energy between a receptor protein and ligands computed with the AMBER force field to correlate these descriptors with biological activities. The method has been validated on 51 diverse protein–ligand X-ray structures. The ligands ranged in size from 24 to 1512 atoms and spanned a pK_i range from 2.47 to 14.0. The best-fit equation, using PLS analysis, yielded an $r^2 = 0.85$ with a standard error of 1.0 log units and a cross-validated $r^2 = 0.78$. This QSAR was found to be predictive for at least two of three test sets of enzyme inhibitor complexes: 14 structurally diverse crystalline complexes (predictive $r^2 = 0.81$), 13 HIV protease inhibitors (predictive $r^2 = 0.57$), and 11 thermolysin inhibitors (predictive $r^2 = 0.72$). VALIDATE has also been successfully applied to the design of non-peptidic HIV-1 protease inhibitors [55].

Another approach which utilizes the molecular interaction energy between the receptor and ligand is the COMBINE approach developed by Wade et al. [56]. It employs a unique method that partitions the interaction energy between receptor and ligand fragments and subjects them to a statistical analysis. This is proposed to enhance contributions from mechanistically important interaction terms and to tune out noise due to inaccuracies in the potential energy functions and molecular models. For a set of 26 phospholipase A2 inhibitors, the direct correlation between interaction energies computed using the CFF91 DISCOVER force field and percent enzyme inhibition was very low, $r = 0.21$. However, with the COMBINE approach, employing PLS fitting and the GOLPE variable selection procedure, good correlations with the percent inhibition rate were observed ($q_{LOO}^2 = 0.82$). Predictive models were also obtained for a variety of other biological targets and their ligands: acetylcholinesterase (AChE) inhibitors ($n = 35$, $q_{LOO}^2 = 0.76$) [57], factor Xa inhibitors ($n = 133$, $q_{LOO}^2 = 0.61$) [58], periplasmic oligopeptide binding component (OppA) ligands ($n = 28$, $q_{LOO}^2 = 0.73$) [59], neuraminidase inhibitors ($n = 39$, $q_{LOO}^2 = 0.78$) [60], cyclooxygenase-2 inhibitors ($n = 58$, $q_{LOO}^2 = 0.64$) [61], and cytochrome P450 1A2 ligands ($n = 12$, $q_{LOO}^2 = 0.74$) [37]. A comprehensive review on a variety of COMBINE applications has been recently published [62].

Recent approaches that primarily employ the combination of structure-based alignment strategies and comparative molecular field analysis to predict ligand affinity have included studies of ligand binding to enzymes and receptor X-ray structures, as well as protein homology models. Garland Marshall was one of the first who applied this technique. He studied the binding of 59 HIV-1 protease inhibitors from different structural classes [63]. The availability of X-ray crystallographic data for at least one representative from each class bound to HIV-1 protease provided information regarding not only the active conformation of each inhibitor, but also via superposition of protease backbones, the relative positions of each ligand with respect to one another in the active site of the enzyme. The molecules were aligned and served as templates on which additional ligands were field-fit minimized. The predictive ability of the derived models was subsequently evaluated using external test set molecules, for which X-ray structural information was available.

Tropsha et al. used the X-ray structures of the three AChE inhibitors bound to the enzyme as a template onto which other structurally analogous AChE inhibitors were superimposed. In order to obtain quantitative relationship between the structure and biological activities of the inhibitors, CoMFA in combination with a variable-selection method (cross-validated r^2 guided region selection (q^2-GRS) routine [64]) was applied. Using the resulting alignment of 60 AChE inhibitors and CoMFA/q^2-GRS yielded a highly predictive QSAR model with a q^2 of 0.73. Whereas in the latter two studies, manually derived protein-based alignments were used as input for a 3D-QSAR analysis, several case studies have been recently reported where an automated docking procedure was applied for structure-based alignment generation.

Mügge et al. have generated a series of CoMFA models from docking-based and atom-based alignments for biphenyl carboxylic acid matrix-metalloproteinase-2 (MMP-3) inhibitors [65]. The underlying statistics of these approaches was assessed in order to determine whether a docking approach can be employed as an automated alignment tool for the development of 3D-QSAR models. The docking-based alignment provided by a DOCK/PMF scoring protocol yielded statistically significant, cross-validated CoMFA models. Field-fit minimization was successfully applied to refine the docking-based alignments. The statistically best CoMFA model has been created by the ligand-based alignment that has been found, however, to be inconsistent with the stromelysin crystal structure. The refined docking-based alignment has resulted in a final alignment that is consistent with the crystal structure and only slightly statistically inferior to the ligand-based aligned CoMFA model.

Pelliciari et al. used the combination of ligand docking and 3D-QSAR analysis to build a predictive model for 46 poly (ADP-ribose) polymerase (PARP) inhibitors [66]. The PARP inhibitors were docked into the crystallographic structure of the catalytic domain of PARP by using the AutoDock 2.4 software. The docking study provided an alignment scheme that was crucial for superimposing all the remaining inhibitors. Based on this alignment, a 3D-QSAR model was established [67]. The resulting statistical analysis yielded a predictive model which was able to explain much of the variance of the 46-compound data set ($q^2 = 0.74$).

Matter et al. examined a series of 138 inhibitors of the blood coagulation enzyme factor Xa using CoMFA and CoMSIA [68]. To rationalize biological affinity and to provide guidelines for further design, all compounds were docked into the factor Xa binding site. Those docking studies were based on X-ray structures of factor Xa in complex with literature-known inhibitors. The docking results were validated by four X-ray crystal structures of representative ligands in factor Xa. The 3D-QSAR models based on a superposition rule derived from these docking studies were validated using conventional and cross-validated q^2 values. This led to consistent and highly predictive 3D-QSAR models with which were found to correspond to experimentally determined factor Xa binding site topology in terms of steric, electrostatic, and hydrophobic complementarity ($q^2 = 0.75$). The same strategy was successfully applied to a data set of 90 MMP-8 matrix-metalloproteinase inhibitors ($q^2 = 0.57$) [69].

Poso et al. examined the binding of 92 catechol-O-methyltransferase inhibitors (COMT) [70]. They used a combination of FlexX molecular docking method with a GRID/GOLPE 3D-QSAR to analyze possible interactions between COMT and its inhibitors and to encourage the design of new inhibitors. The GRID/GOLPE models were made using bioactive conformations from docking experiments, which yielded a q^2 value of 0.64. The docking results, the COMT X-ray structure, and the 3D-QSAR models were found to be in good agreement with each other. Interest was also focused on how well the calculated FlexX total energy scores correlated with the experimental biological activity. FlexX total energy scores for the 92 compounds were correlated with the corresponding pIC_{50} values resulting in an r^2 value of 0.30, indicating the problem of the used scoring function.

In a study from the same group, receptor-based alignment techniques for 3D-QSAR have been analyzed and compared with traditional atom-based approaches. A set of 113 HIV-1 protease inhibitors was used to generate CoMFA and CoMSIA models [71]. Inhibitors that were docked automatically with GOLD were in agreement with information obtained from existing X-ray structures. The protein- as well as the ligand-based alignment strategy produced statistically significant CoMFA and CoMSIA models (best q^2 value of 0.65 and best predictive r^2 value of 0.75), whereas the GOLD-based alignment gave more robust models for predicting the activities of the molecules of the external test set.

Several groups have applied the docking-based alignment strategy to develop 3D-QSAR models for nuclear hormone receptor ligands. During the last decade several X-ray structures of nuclear hormone receptors in complex with hormones, agonist, and antagonists have been resolved and used for structure-based drug design [72]. In general, automated docking programs were shown to be successful in docking ligands to this receptor class [34, 73, 74]. Therefore, it was quiet appealing to use structure-based 3D-QSAR approaches also for this class of targets. Predictive and robust receptor-based 3D-QSAR models have been reported for estrogen receptor agonists ($n = 30$, $q_{LOO}^2 = 0.90$, $q_{L50\%O}^2 = 0.82$) [34] and ($n = 36$, $q_{LOO}^2 = 0.63$) [75], as well as for androgen receptor ligand ($n = 67$, $q_{LOO}^2 = 0.66$) [76] and ($n = 25$, $q_{LOO}^2 = 0.78$) [77].

Moro et al. used a homology model of the A3 adenosine receptor to generate a target-based alignment [78]. Docking-based structure superimposition was used to perform a 3D-QSAR analysis using the CoMFA program. A correlation coefficient q^2 of 0.84 was obtained for a set of 106 A3 receptor ligands. Both steric and electrostatic contour plots, obtained from the CoMFA analysis, were found to be in agreement with the hypothetical binding site achieved by molecular docking. Following the reported computational approach, 17 new ligands were designed, synthesized, and tested. The predicted K_i values were consistently very close to the experimental values.

The near exponential growth of the Protein Data Bank in the last few years has resulted in a huge number of 3D structures of interesting target proteins which can be analyzed by means of structure-based drug design methods. It has also been shown on numerous high-resolution protein–ligand structures that docking methods are now able to predict the position of ligands in the corresponding binding sites with reasonable accuracy. Therefore, it is not surprising that an increasing number of receptor-based 3D-QSAR models are now published. Combination of docking and comparative molecular field analysis has been successfully applied to enzyme inhibitors of the following pharmaceutically relevant targets: non-nucleoside HIV-1 reverse transcriptase inhibitors ($n = 29$, $q_{LOO}^2 = 0.72$) [79], Raf-1 kinase inhibitors ($n = 91$, $q_{LOO}^2 = 0.53$) [80], aldose reductase inhibitors ($n = 45$, $q_{LOO}^2 = 0.56$) [81], cyclooxygenase-2 inhibitors ($n = 88$, $q_{LOO}^2 = 0.84$) [82], HIV-1 reverse transcriptase inhibitors ($n = 70$, $q_{LOO}^2 = 0.84$) [83], EGFR kinase inhibitors ($n = 96$, $q_{LOO}^2 = 0.64$) [84], Yersinia protein tyrosine phosphatase YopH inhibitors ($n = 34$, $q_{LOO}^2 = 0.83$) [85], HIV-1 integrase inhibitors ($n = 66$, $q_{LOO}^2 = 0.72$) [86], HIV-1 reverse transcriptase inhibitors ($n = 50$, $q_{LOO}^2 = 0.78$) [87], dihydrofolate reductase inhibitors ($n = 240$, $q_{L10\%O}^2 = 0.65$) [88], and type-B monoamine-oxydase inhibitors ($n = 130$, $q_{L10\%O}^2 = 0.73$) [89].

Sippl et al. applied the combination of receptor-based 3D-QSAR to several drug design projects [34, 35, 90–95]. The ultimate goal was a prediction of biological activities and a prioritization of synthesis of proposed compounds a priori. The receptor-based 3D-QSAR approach was applied for the design of novel AChE inhibitors [94]. AChE has been the focus of many drug discovery projects aimed at maintaining the acetylcholine level in Alzheimer patients via mild or reversible inhibition [96]. They started with a series of morpholine derivatives including minaprine which were shown to be weak AchE inhibitors. Starting with the lead structure minaprine and the available X-ray structures of AChE in complex with inhibitors [97], a variety of minaprine derivatives were synthesized [98]. In order to obtain ideas for the synthesis of modified, more potent, inhibitors, a combination of automated docking and 3D-QSAR was applied. AutoDock in combination with a force-field refinement yielded good results when docking the AChE inhibitors [94]. The docked minaprine derivatives showed an interaction with both the catalytic and the peripheral anionic site and showed mainly hydrophobic and π–π interactions with the residues of the binding pocket (Figure 4-2).

The docking positions were subsequently extracted from the protein environment and were taken as an input for a GRID/GOLPE analysis. Applying the variable-selection strategy incorporated within GOLPE, a significant model was obtained. The significance was tested applying a variety of validation procedures, such as leave-20%-out cross-validation ($q^2 = 0.91$), leave-50%-out cross-validation

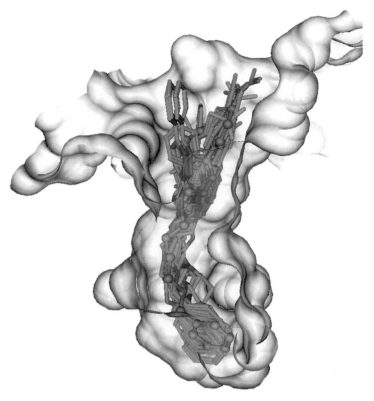

Figure 4-2. Receptor-based alignment of all investigated AChE inhibitors as obtained by the docking analysis. The Conolly surface of the binding pocket is displayed. The most potent inhibitor **4j** is colored *magenta*

($q^2 = 0.90$), or scrambling tests. The statistical results gave confidence that the derived model could also be useful to guide the further optimization process.

To get an impression of which parts of the AChE inhibitors are correlated with variation in activity, the PLS coefficient plots (obtained by using the water and the methyl probe) were analyzed and compared with the amino acid residues of the binding pocket. The plots indicate those lattice points where a particular property significantly contributes and thus explains the variation in biological activity data (Figure 4-3). The plot obtained with the methyl probe indicated that, close to the arylpyridazine moiety, a region with positive coefficients exists (region A in Figure 4-3). The interaction energies in region A are positive; therefore the decrease in activity is due to a steric overlap within this region. Thus, it should be possible to get active inhibitors by reducing the ring size compared to compound **4j** (which is shown in Figure 4-3 together with the PLS coefficient maps). For that reason, several novel ligands containing hydrophobic groups were proposed (Table 4-1). A second interesting field was observed located above the arylpyridazine moiety in the model obtained using the water probe. Here a region exists where

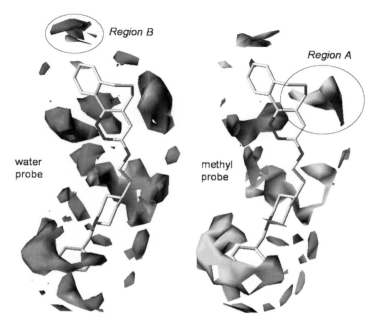

Figure 4-3. PLS coefficient maps obtained using the water probe (*left side*) and the methyl probe (*right side*). *Green* and *cyan* fields are contoured at −0.003, *yellow* and *orange* fields are contoured at +0.003 (compound **4j** is shown for comparison)

polar interactions increase activity (region B in Figure 4-3). After analysis of the entrance of the binding pocket (the interaction site for the arylpyridazine system), we rationalized the design of compounds bearing polar groups. In the calculated AChE-aminopyridazine complexes, we observed two polar amino acid residues (Asn280 and Asp285) located at the entrance of the gorge, which could serve as an additional binding site for the substituted arylpyridazine system. To test this hypothesis, several inhibitors possessing polar groups with hydrogen bond donor and acceptor properties were synthesized and tested. The designed inhibitors were docked into the binding pocket applying the developed procedure and their biological activities were predicted using the PLS models. In Table 4-1 the predicted and experimentally determined inhibitor activities are listed for the novel compounds. In general, good agreement between predicted and experimentally determined values was observed, indicated by the low $SDEP_{ext}$ values of 0.40. The reduction of the size of the aminopyridazine ring system resulted in highly potent inhibitors **4g–4i**. The molecules of the second series of designed inhibitors containing polar groups were also predicted accurately. The gain in activity compared to the non-substituted compound is moderate, indicating that the potential interaction with the two polar residues at the entrance does not play an important role. Since the two residues are located at the entrance of the binding pocket, it is suggested that these residues have a stronger interaction with water molecules than with the protein side chains.

Further support for the reported docking study came from the crystal structure of AChE in complex with donepezil [99]. Similar to the most potent inhibitors of

Table 4-1. Predicted and experimentally determined activities of novel AChE inhibitors

Compound	Structure	pIC$_{50}$ observed[a]	pIC$_{50}$ predicted[b]	pIC$_{50}$ predicted[c]
4g		8.00	7.00	7.20
4h		7.41	7.62	7.66
4i		7.66	7.48	7.56
6g		7.24	6.90	6.77
6h		7.24	7.05	7.11
6i		7.27	7.25	7.2
6j		7.14	6.88	6.92

[a] Inhibitory activity measured on the AChE of *Torpedo californica*.

[b] Predicted activity using the GRID/GOLPE model (water probe).

[c] Predicted activity using the GRID/GOLPE model (C3 probe).

the series, donepezil also contains a benzylpiperidine moiety. The comparison of docking positions and crystal structures revealed that both kinds of inhibitors adopt a comparable conformation in the narrow binding pocket [94].

4.7.   CONCLUSION

In this contribution several studies have been reviewed where a combination of 3D-QSAR and receptor-based alignments has led to predictive and meaningful models. Apart from good predictive ability, the derived models are also able to indicate which interaction sites in the binding pocket might be responsible for the variance in biological activities. Therefore, it is not surprising that more and more studies are published where receptor-based 3D-QSAR is applied [100–117]. In the last decade, structure-based methods have become major tools in drug design, including lead finding and optimization. It has also been shown that structure-based methods are now able to predict, with a reasonable degree of accuracy, the position of a ligand in its binding site. Apart from the accurate prediction of experimental data, modern docking methods have become more and more efficient. Meanwhile, docking programs have been developed which accomplish the docking of highly flexible ligands in a few seconds/minutes on modern PCs. The major problem is still the prediction of the binding affinity, probably limited by the approximation used in today's scoring and force field methods [18]. The application of 3D-QSAR methods may facilitate the prediction of binding affinities if one has a series of compounds, which bind in a similar way to a target protein. Up until now, the imprecise nature of docking and scoring makes blind virtual screening of a large number of compounds, without any information about true actives or known experimental complex structures, a risky exercise. It has been recently shown by Norinder et al. that limited experimental information and proper multivariate statistical treatment of the scoring data dramatically increases the value of these kinds of computations [118]. They generated scoring matrices for known actives and potential inactives for four different targets, using docking followed by scoring with seven different scoring functions. Based on these matrices multivariate classifiers were generated and evaluated with external test sets, and compared to classical consensus scoring and single scoring functions. It was found that proper multivariate analysis of scoring data is very rewarding in terms of recall of known actives and enrichment of true actives in the set of predicted actives. It is suggested that the combination of different approaches as described, e.g., by Norinder [118] or Klebe [44] might represent a way out of the known limitations of 3D-QSAR (such as the "Kubinyi paradox" [53]).

REFERENCES

1. Miller MD, Sheridan RP, Kearsley SK (1999) A program for rapidly producing pharmacophorically relevent molecular superpositions. J Med Chem 42:1505–1514
2. Lemmen C, Lengauer T (2000) Computational methods for the structural alignment of molecules. J Comput-Aided Mol Des 14:215–232

3. Jain AN (2004) Ligand-based structural hypotheses for virtual screening. J Med Chem 47:947–961
4. Hansch C, Leo A (1995) Exploring QSAR: Fundamentals and applications in chemistry and biology. American Chemical Society, Washington, DC
5. Kubinyi H (1997) QSAR and 3D QSAR in drug design. Drug Discov Today 2:457–467
6. Cramer RD III, Patterson DE, Bunce JD (1988) Comparative molecular field analysis (CoMFA) 1. Effect of shape on binding of steroids to carrier proteins. J Am Chem Soc 110:5959–5967
7. Folkers G, Merz A, Rognan D (1993) CoMFA: Scope and limitations. In: Kubinyi H (ed) 3D QSAR in drug design. Theory, methods and applications. ESCOM Science Publishers BV, Leiden
8. Klebe G, Abraham U (1993) On the prediction of binding properties of drug molecules by comparative molecular field analysis. J Med Chem 36:70–80
9. Klebe G, Abraham U, Mietzner T (1994) Molecular similarity indices in a comparative analysis (CoMSIA) of drug molecules to correlate and predict their biological activity. J Med Chem 37:4130–4146
10. Cruciani G (ed) (2006) Methods and principles in medicinal chemistry – molecular interaction fields. VCH Publisher, New York
11. Norinder U (1998) Recent progress in CoMFA methodology and related techniques. Perspect Drug Discov Des 12:3–23
12. Kim KH, Greco G, Novellino EA (1998) Critical review of recent CoMFA applications. Perspect Drug Discov Des 12:257–315
13. Podlogar BL, Ferguson DM (2000) QSAR and CoMFA: A perspective on the practical application to drug discovery. Drug Des Discov 1:4–12
14. Akamatsu M (2002) Current state and perspectives of 3D QSAR. Curr Top Med Chem 2:1381–1394
15. Kubinyi H, Folkers G, Martin YC (eds) (1998) 3D QSAR in drug design. Ligand–protein interactions and molecular similarity. Kluwer/ESCOM, Dodrecht
16. Kubinyi H, Folkers G, Martin YC (eds) (1998) 3D QSAR in drug design. Recent advances. Kluwer/ESCOM, Dodrecht
17. Masukawa KM, Kollman PA, Kuntz ID (2003) Investigation of neuraminidase-substrate recognition using molecular dynamics and free energy calculations. J Med Chem 46:5628–5637
18. Tame JR (2005) Scoring functions: The first 100 years. J Comput Aided Mol Des 19:441–451
19. Kitchen DB, Decornez H, Furr JR et al. (2004) Docking and scoring in virtual screening for drug discovery: Methods and applications. Nat Rev Drug Discov 11:935–949
20. Morris GM, Goodsell DS, Huey R et al. (1994) Distributed automatic docking of flexible ligands to proteins. J Comput Aided Mol Des 8:243–256
21. Verdonk ML, Cole JC, Hartshorn M et al. (2003) Improved protein–ligand docking using GOLD. Proteins 52:609–623
22. Meng E, Shoichet BK, Kuntz ID (1992) Automated docking with grid-based energy evaluation. J Comp Chem 13:505–524
23. Kontoyianni M, McClellan LM, Sokol GS (2004) Evaluation of docking performance: Comparative data on docking algorithms. J Med Chem 47:558–565
24. Böhm HJ (1998) Prediction of binding constants of protein–ligands: A fast method for the prioritisation of hits obtained from de-novo design or 3D database search programs. J Comput Aided Mol Des 12:309–323
25. Tame JRH (1999) Scoring functions: A view from the bench. J Comput Aided Mol Des 13:99–108
26. Wang R, Lu Y, Fang X et al. (2004) An extensive test of 14 scoring functions using the PDBbind refined set of 800 protein–ligand complexes. J Chem Inf Comput Sci 44:2114–2125
27. Perola E, Walters WP, Charifson PS (2004) A detailed comparison of current docking and scoring methods on systems of pharmaceutical relevance. Proteins 56:235–249

28. Gohlke H, Klebe G (2002) Approaches to the description and prediction of the binding affinity of small molecule ligands to macromolecular receptors. Angew Chem Int Ed 41:2644–2676

29. Huang D, Caflisch A (2004) Efficient evaluation of binding free energy using continuum electrostatics solvation. J Med Chem 47:5791–5797

30. Waller CL, Oprea TI, Giolitti A et al. (1993) Three-dimensional QSAR of human immunodeficiency virus (I) protease inhibitors. 1. A CoMFA study employing experimentally-determined alignment rules. J Med Chem 36:4152–4160

31. Cho SJ, Garsia ML, Bier J et al. (1996) Structure-based alignment and comparative molecular field analysis of acetylcholinesterase inhibitors. J Med Chem 39:5064–5071

32. Vaz RJ, McLEan LR, Pelton JT (1998) Evaluation of proposed modes of binding of (2S)-2-[4-[[(3S)-1-acetimidoyl-3-pyrrolidinyl]oxyl]phenyl]-3-(7-amidino-2-naphtyl)-propanoic acid hydrochloride and some analogs to factor Xa using a comparative molecular field analysis. J Comput Aided Mol Des 12:99–110

33. Sippl W, Contreras JM, Rival YM et al. (1998) Comparative molecular field analysis of aminopyridazine acetylcholinesterase inhibitors. In: Gundertofte K (ed) Proceedings of the 12th European symposium on QSAR – molecular modelling and predicting of bioactivity. Plenum Press, Copenhagen

34. Sippl W (2000) Receptor-based 3D quantitative structure–activity relationships of estrogen receptor ligands. J Comput Aided Mol Des 14:559–572

35. Sippl W (2002) Binding affinity prediction of novel estrogen receptor ligands using receptor-based 3D QSAR methods. Bioorg Med Chem 10:3741–3755

36. Ortiz AR, Pisabarro MT, Gago F et al. (1995) Prediction of drug binding affinities by comparative binding energy analysis. J Med Chem 38:2681–2691

37. Lozano JJ, Pastor M, Cruciani G et al. (2000) 3D QSAR methods on the basis of ligand-receptor complexes. Application of combine and GRID/GOLPE methodologies to a series of CYP1A2 inhibitors. J Comput Aided Mol Des 13:341–353

38. Reynolds CA, Wade RC, Goodford PJ (1989) Identifying targets for bioreductive agents: Using GRID to predict selective binding regions of proteins. J Mol Graph 7:103–108

39. Cruciani G, Watson K (1994) Comparative molecular field analysis using GRID force field and GOLPE variable selection methods in a study of inhibitors of glycogen phosphorylase b. J Med Chem 37:2589–2601

40. Cruciani G, Crivori P, Carrupt P-A et al. (2000) Molecular fields in quantitative structure–permeation relationships. J Mol Struct 503:17–30

41. Duca S, Hopfinger AJ (2001) Estimation of molecular similarity based on 4D-QSAR analysis: Formalism and validation. J Chem Inf Comput Sci 41:1367–1387

42. Vedani A, Dobler M (2002) 5D-QSAR: The key for simulating induced fit? J Med Chem 45:2139–2149

43 Vedani A, Dobler M, Lill MA (2005) Combining protein modeling and 6D-QSAR. Simulating the binding of structurally diverse ligands to the estrogen receptor. J Med Chem 48:3700–3703

44. Gohlke H, Klebe G (2002) Drug score meets CoMFA: Adaptation of fields for molecular comparison (AFMoC) or how to tailor knowledge-based pair-potentials to a particular protein. J Med Chem 45:4153–4170

45. Silber K, Heidler P, Kurz T et al. (2005) AFMoC enhances predictivity of 3D QSAR: A case study with DOXP-reductoisomerase. J Med Chem 48:3547–3563.

46. Hillebrecht A, Supuran CT, Klebe G (2006) Integrated approach using protein and ligand information to analyze selectivity- and affinity-determining features of carbonic anhydrase isozymes. ChemMedChem 1:839–853

47. Oprea TI, Garcia AE (1996) Three-dimensional quantitative structure–activity relationships of steroid aromatase inhibitors. J Comput Aided Mol Des 10:186–200

48. Golbraikh A, Trophsa A (2002) Beware of q2! J Mol Graph Model 20:269–276
49. Kubinyi H, Hamprecht FA, Mietzner T (1998) Three-dimensional quantitative similarity–activity relationships (3D QSiAR) from SEAL similarity matrices. J Med Chem 41:2553–2564
50. Golbraikh A, Shen M, Xiao Z et al. (2003) Rational selection of training and test sets for the development of validated QSAR models. J Comput Aided Mol Des 17:241–253
51. Norinder U (1996) Single and domain made variable selection in 3D QSAR applications. J Chemomet 10:95–105
52. Doweyko AM (2004) 3D QSAR illusions. J Comput Aided Mol Des 18:587–596
53. van Drie JH (2004) Pharmacophore discovery: A critical review. In: Bultinck P, De Winter H, Langenaeker, W et al., (eds) Computational medicinal chemistry for drug discovery. Marcel Dekker, New York
54. Head RD, Smythe ML, Oprea TI et al. (1996) VALIDATE: A new method for the receptor-based prediction of binding affinities of novel ligands. J Am Chem Soc 118:3959–3969
55. Di Santo R, Costi R, Artico M et al. (2002) Design, synthesis and QSAR studies on *N*-aryl heteroarylisopropanolamines, a new class of non-peptidic HIV-1 protease inhibitors. Bioorg Med Chem 10:2511–2526
56. Ortiz AR, Pisabarro MT, Gago F et al. (1995) Prediction of drug binding affinities by comparative binding energy analysis. J Med Chem 38:2681–2691
57. Martin-Santamaria S, Munoz-Muriedas J, Luque FJ et al. (2004) Modulation of binding strength in several classes of active site inhibitors of acetylcholinesterase studied by comparative binding energy analysis. J Med Chem 47:4471–4482
58. Murcia M, Ortiz AR (2004) Virtual screening with flexible docking and COMBINE-based models. Application to a series of factor Xa inhibitors. J Med Chem 47:805–820
59. Wang T, Wade RC (2002) Comparative binding energy (COMBINE) analysis of OppA-peptide complexes to relate structure to binding thermodynamics. J Med Chem 45:4828–4837
60. Wang T, Wade RC (2001) Comparative binding energy (COMBINE) analysis of influenza neuraminidase-inhibitor complexes. J Med Chem 44:961–997
61. Kim HJ, Chae CH, Yi KY et al. (2004) Computational studies of COX-2 inhibitors: 3D QSAR and docking. Bioorg Med Chem 12:1629–1641
62. Lushington GH, Guo JX, Wang JL (2007) Whither combine? New opportunities for receptor-based QSAR. Curr Med Chem 14:1863–1877
63. Waller CL, Oprea TI, Giolitti A et al. (1993) Three-dimensional QSAR of human immunodeficiency virus (I) protease inhibitors. 1. A CoMFA study employing experimentally-determined alignment rules. J Med Chem 36:4152–4160
64. Cho SJ, Garsia ML, Bier J et al. (1996) Structure-based alignment and comparative molecular field analysis of acetylcholinesterase inhibitors. J Med Chem 39:5064–5071
65. Muegge I, Podlogary BL (2001) 3D-quantitative structure–activity relationships of biphenyl carboxylic acid MMP-3 inhibitors: Exploring automated docking as alignment method. Quant Struct-Act Relat 20:215–223
66. Costantino G, Macchiarulo A, Camaioni E et al. (2001) Modeling of poly(ADP-ribose)polymerase (PARP) inhibitors. Docking of ligands and quantitative structure–activity relationship analysis. J Med Chem 44:3786–3794
67. Matter H, Defossa E, Heinelt U et al. (2002) Design and quantitative structure–activity relationship of 3-amidinobenzyl-1*H*-indole-2-carboxamides as potent, nonchiral, and selective inhibitors of blood coagulation factor Xa. J Med Chem 45:2749–2769
68. Matter H, Schudok M, Schwab W et al. (2002) Tetrahydroisoquinoline-3-carboxylate based matrix-metalloproteinase inhibitors: Design, synthesis and structure–activity relationship. Bioorg Med Chem 10:3529–3544

69. Tervo AJ, Nyroenen TH, Ronkko T et al. (2003) A structure–activity relationship study of catechol-*O*-methyltransferase inhibitors combining molecular docking and 3D QSAR methods. J Comput Aided Mol Des 17:797–810

70. Tervo AJ, Nyroenen TH, Ronkko T et al. (2004) Comparing the quality and predictiveness between 3D QSAR models obtained from manual and automated alignment. J Chem Inf Comput Sci 44:807–816

71. Egea PF, Klahoz BP, Moras D (2000) Ligand–protein interactions in nuclear receptors of hormones. FEBS Lett 476:62–67

72. Bissantz C, Folkers G, Rognan D (2000) Protein-based virtual screening of chemical databases: 1. Evaluation of different docking/scoring combinations. J Med Chem 43:4759–4767

73. Chen YZ, Zhi DG (2001) Ligand–protein inverse docking and its potential use in the computer search of protein targets of a small molecule. Proteins 43:217–226

74. Wolohan P, Reichert DE (2003) CoMFA and docking study of novel estrogen receptor subtype selective ligands. J Comput Aided Mol Des 17:313–328

75. Soderholm AA, Lehtovuori PT, Nyronen TH (2005) Three-dimensional structure–activity relationships of nonsteroidal ligands in complex with androgen receptor ligand-binding domain. J Med Chem 48:917–925

76. Ai N, DeLisle RK, Yu SJ et al. (2003) Computational models for predicting the binding affinities of ligands for the wild-type androgen receptor and a mutated variant associated with human prostate cancer. Chem Res Toxicol 16:1652–1660

77. Moro S, Braiuca P, Deflorian F et al. (2005) Combined target-based and ligand-based drug design approach as a tool to define a novel 3D-pharmacophore model of human A3 adenosine receptor antagonists: Pyrazolo[4,3-*e*]1,2,4-triazolo[1,5-*c*]pyrimidine derivatives as a key study. J Med Chem 48:152–162

78. Medina-Franco JL, Rodrýguez-Morales S, Juarez-Gordiano CA et al. (2004) Docking-based CoMFA and CoMSIA studies of non-nucleoside reverse transcriptase inhibitors of the pyridinone derivative type. J Comput Aided Mol Des 18:345–360

79. Thaimattam R, Daga P, Rajjak SA et al. (2004) 3D QSAR CoMFA, CoMSIA studies on substituted areas as Raf-1 kinase inhibitors and its confirmation with structure-based studies. Bioorg Med Chem 12:6415–6425

80. Sun WS, Park YS, Yoo J et al. (2003) Rational design of an indolebutanoic acid derivative as a novel aldose reductase inhibitor based on docking and 3D QSAR studies of phenethylamine derivatives. J Med Chem 46:5619–5627

81. Kim HJ, Chae CH, Yi KY et al. (2004) Computational studies of COX-2 inhibitors: 3D-QSAR and docking. Bioorg Med Chem 12:1629–1641

82. Ragno R, Artico M, De Martino G et al. (2005) Docking and 3D QSAR studies on indolyl aryl sulfones. Binding mode exploration at the HIV-1 reverse transcriptase non-nucleoside binding site and design of highly active *N*-(2-hydroxyethyl)carboxamide and *N*-(2-hydroxyethyl)carbohydrazide derivatives. J Med Chem 48:213–223

83. Assefa H, Kamath S, Buolamwini JK (2003) 3D-QSAR and docking studies on 4-anilinoquinazoline and 4-anilinoquinoline epidermal growth factor receptor (EGFR) tyrosine kinase inhibitors. J Comput Aided Mol Des 17:475–493

84. Hu X, Stebbins CE (2005) Molecular docking and 3D QSAR studies of yersinia protein tyrosine phosphatase YopH inhibitors. Bioorg Med Chem 13:1101–1109

85. Kuo CL, Assefa H, Kamath S et al. (2004) Application of CoMFA and CoMSIA 3D QSAR and docking studies in optimization of mercaptobenzenesulfonamides as HIV-1 integrase inhibitors. J Med Chem 47:385–399

86. Zhou Z, Madura JD (2004) 3D QSAR analysis of HIV-1 RT non-nucleoside inhibitors, TIBO derivatives based on docking conformation and alignment. J Chem Inf Comput Sci 44:2167–2178

87. Sutherland J, Weaver DF (2004) Three-dimensional quantitative structure–activity and structure–selectivity relationships of dihydrofolate reductase inhibitors. J Comput Aided Mol Des 18:309–331

88. Carrieri A, Carotti A, Barreca ML et al. (2002) Binding models of reversible inhibitors to type-B monoamine oxidase. J Comput Aided Mol Des 16:769–778

89. Sippl W, Höltje H-D (2000) Structure-based 3D-QSAR – merging the accuracy of structure-based alignments with the computational efficiency of ligand-based methods. J Mol Struct (Theochem) 503:31–50

90. Cinone N, Höltje H-D, Carotti A (2000) Development of a unique 3D interaction model of endogenous and synthetic peripheral benzodiazepine receptor ligands. J Comput Aided Mol Des 14:753–768

91. Hammer S, Spika L, Sippl W et al. (2003) Glucocorticoid receptor interactions with glucocorticoids: Evaluation by molecular modeling and functional analysis of glucocorticoid receptor mutants. Steroids 68:329–339

92. Classen-Houben D, Sippl W, Höltje HD (2002) Molecular modeling on ligand-receptor complexes of protein-tyrosine-phosphatase 1B. In: Ford M, Livingstone D, Dearden J et al. (eds). EuroQSAR 2002 designing drugs and crop protectants: Processes, problems and solutions. Blackwell Publishing, Bournemouth

93. Broer BM, Gurrath M, Höltje H-D (2003) Molecular modelling studies on the ORL1-receptor and ORL1-agonists. J Comput Aided Mol Des 17:739–754

94. Sippl W, Contreras JM, Parrot I et al. (2001) Structure-based 3D QSAR and design of novel acetylcholinesterase inhibitors. J Comput Aided Mol Des 15:395–410

95. Wichapong K, Lindner M, Pianwanita S et al. (2008) Receptor-based 3D-QSAR studies of checkpoint wee1 kinase inhibitors. Eur J Med Chem 44:1383–1395

96. Barner EL, Gray SL (1998) Donepezil in alzheimer disease. Ann Pharmacother 32:70–77

97. Raves ML, Harel M, Pang YP et al. (1997) Structure of acetylcholinesterase complexed with the nootropic alkaloid huperzine A. Nat Struct Biol 4:57–63

98. Contreras JM, Rival Y, Chayer S et al. (1999) Aminopyridazines as acetylcholinesterase inhibitors. J Med Chem 42:730–741

99. Kryger G, Silman I, Sussman JL (1999) Structure of acetylcholinesterase complexed with E2020 (aricept): Implications for the design of new anti-alzheimer drugs. Structure Fold Des 15:297–307

100. San Juan AA (2008) 3D-QSAR models on clinically relevant K103N mutant HIV-1 reverse transcriptase obtained from two strategic considerations. Bioorg Med Chem Lett 18:1181–1194

101. Pately PD, Pately MR, Kaushik-Basu N et al. (2008) 3D QSAR and molecular docking studies of benzimidazole derivatives as hepatitis C virus NS5B polymerase inhibitors. J Chem Inf Model 48:42–51

102. Murumkar PR, Giridhar R, Yadav MR (2008) 3D-quantitative structure–activity relationship studies on benzothiadiazepine hydroxamates as inhibitors of tumor necrosis factor-α converting enzyme. Chem Biol Drug Des 71:363–373

103. Muddassar M, Pasha FA., Chung HW et al. (2008) Receptor guided 3D-QSAR: A useful approach for designing of IGF-1R inhibitors. J Biomed Biotech 2008:837653

104. San Juan AA (2008) Towards predictive inhibitor design for the EGFR autophosphorylation activity. Eur J Med Chem 43:781–791

105. Fischer B, Fukuzawa K, Wenzel W (2008) Receptor-specific scoring functions derived from quantum chemical models improve affinity estimates for in silico drug discovery. Proteins 70:1264–1273

106. Roncaglioni A, Benfenati E (2008) In silico-aided prediction of biological properties of chemicals: Oestrogen receptor-mediated effects. Chem Soc Rev 37:441–450

107. Sadeghian H, Seyedi SM, Saberi MR et al. (2008) Design and synthesis of eugenol derivatives, as potent 15-lipoxygenase inhibitors. Bioorg Med Chem 16:890–901

108. Holder S, Lilly M, Brown ML (2007) Comparative molecular field analysis of flavonoid inhibitors of the PIM-1 kinase. Bioorg Med Chem 15:6463–6473

109. Abu Hammad AM, Afifi FU, Taha MO (2007) Combining docking, scoring and molecular field analyses to probe influenza neuraminidase-ligand interactions. J Mol Graph Model 26:443–456

110. Dezi C, Brea J, Alvarado M et al. (2007) Multistructure 3D-QSAR studies on a series of conformationally constrained butyrophenones docked into a new homology model of the 5-HT2A receptor. J Med Chem 50:3242–3255

111. Korhonen S-P, Tuppurainen K, Asikainen A et al. (2007) SOMFA on large diverse xenoestrogen dataset: The effect of superposition algorithms and external regression tools. QSAR Comb Sci 26:809–819

112. Du L, Li M, You Q et al. (2007) A novel structure-based virtual screening model for the hERG channel blockers. Biochem Biophys Res Commun 355:889–894

113. Tuccinardi T, Ortore G, Rossello A et al. (2007) Homology modeling and receptor-based 3D-QSAR study of carbonic anhydrase IX. J Chem Inf Model 47:2253–2262

114. Pissurlenkar RRS, Shaikh MS, Coutinho EC (2007) 3D-QSAR studies of dipeptidyl peptidase IV inhibitors using a docking based alignment. J Mol Mod 13:1047–1071

115. Tuccinardi T, Nuti E, Ortore G. et al. (2007) Analysis of human carbonic anhydrase II: Docking reliability and receptor-based 3D-QSAR study. J Chem Inf Model 47:515–525

116. Huang H, Pan X, Tan N et al. (2007) 3D-QSAR study of sulfonamide inhibitors of human carbonic anhydrase II. Eur J Med Chem 42:365–372

117. Jójárt B, Márki A (2007) Receptor-based QSAR studies of non-peptide human oxytocin receptor antagonists. J Mol Graph Model 25:711–720

118. Jacobsson M, Liden P, Stjernschantz E et al. (2003) Improving structure-based virtual screening by multivariate analysis of scoring data. J Med Chem 46:5781–5789

CHAPTER 5

VIRTUAL SCREENING AND MOLECULAR DESIGN BASED ON HIERARCHICAL QSAR TECHNOLOGY

VICTOR E. KUZ'MIN, A.G. ARTEMENKO, EUGENE N. MURATOV,
P.G. POLISCHUK, L.N. OGNICHENKO, A.V. LIAHOVSKY,
A.I. HROMOV, AND E.V. VARLAMOVA

*A.V. Bogatsky Physical-Chemical Institute NAS of Ukraine, Lustdorfskaya Doroga 86,
Odessa 65080, Ukraine, e-mail: victor@ccmsi.us*

Abstract: This chapter is devoted to the hierarchical QSAR technology (HiT QSAR) based on simplex representation of molecular structure (SiRMS) and its application to different QSAR/QSPR tasks. The essence of this technology is a sequential solution (with the use of the information obtained on the previous steps) of the QSAR paradigm by a series of enhanced models based on molecular structure description (in a specific order from 1D to 4D). Actually, it's a system of permanently improved solutions. Different approaches for domain applicability estimation are implemented in HiT QSAR. In the SiRMS approach every molecule is represented as a system of different simplexes (tetratomic fragments with fixed composition, structure, chirality, and symmetry). The level of simplex descriptors detailed increases consecutively from the 1D to 4D representation of the molecular structure. The advantages of the approach presented are an ability to solve QSAR/QSPR tasks for mixtures of compounds, the absence of the "molecular alignment" problem, consideration of different physical–chemical properties of atoms (e.g., charge, lipophilicity), and the high adequacy and good interpretability of obtained models and clear ways for molecular design. The efficiency of HiT QSAR was demonstrated by its comparison with the most popular modern QSAR approaches on two representative examination sets. The examples of successful application of the HiT QSAR for various QSAR/QSPR investigations on the different levels (1D–4D) of the molecular structure description are also highlighted. The reliability of developed QSAR models as the predictive virtual screening tools and their ability to serve as the basis of directed drug design was validated by subsequent synthetic, biological, etc. experiments. The HiT QSAR is realized as the suite of computer programs termed the "HiT QSAR" software that so includes powerful statistical capabilities and a number of useful utilities.

Keywords: HiT QSAR, Simplex representation, SiRMS

Abbreviations

A/I/EVS	Automatic/Interactive/Evolutionary Variables Selection
ACE	Angiotensin Converting Enzyme

127

T. Puzyn et al. (eds.), Recent Advances in QSAR Studies, 127–176.
DOI 10.1007/978-1-4020-9783-6_5, © Springer Science+Business Media B.V. 2010

AchE	Acetylcholinesterase
CoMFA	Comparative Molecular Fields Analysis QSAR approach
CoMSIA	Comparative Molecular Similarity Indexes Analysis QSAR approach
DA	Applicability Domain
DSTP	dispirotripiperazine
EVA	Eigenvalue Analysis QSAR approach
GA	Genetic Algorithm
HiT QSAR	Hierarchical QSAR Technology
HQSAR	Hologram QSAR approach
HRV	Human Rhinovirus
HSV	Herpes Simplex Virus
MLR	Multiple Linear Regression statistical method
PLS	Partial Least Squares or Projection on Latent Structures statistical method
Q^2	cross-validation determination coefficient
QSAR/QSPR	Quantitative Structure-Activity/Property Relationship
R^2	determination coefficient for training set
R^2_{test}	determination coefficient for test set
SD	Simplex Descriptor
SI	Selectivity Index
SiRMS	Simplex Representation of Molecular Structure QSAR approach
TV	Trend-Vector statistical method

5.1.   INTRODUCTION

Nowadays the creation of a new medicine costs more than one billion dollars and the price of this process is growing steadily day by day [1]. During recent decades different theoretical approaches have been used to facilitate and accelerate the process of new drugs creation that is not only very expensive, but also is a multistep and long-term activity [2]. The choice of approaches depends on a presence or absence of information regarding a biological target and the substances interacting with it. A situation, when we have a set of biologically active compounds (ligands) and have no information about a biological target (e.g., receptor) is the most common. Different quantitative structure–activity relationship (QSAR) approaches are used in this case. For many years, QSAR has been used successfully for the analysis of huge variety of endpoints, e.g., antiviral and anticancer activity, toxicity [3–14]. Its staying power may be attributed to the strength of its initial postulate that activity is a function of structure and the rapid and extensive development of the methodology and computational techniques. The overall goals of QSAR retain their original essence and remain focused on the predictive ability of the approach and its receptiveness to mechanistic interpretation [15].

Many different QSAR methods [16–20] have been developed since the second half of the last century and new techniques and improvements are still being created

[21]. These approaches differ mainly by the principles and levels of representation and description of molecular structure. The degree of adequacy of molecular structure models varies from 1D to 4D level:

- 1D models consider only the gross formula of a molecule (for example, glycine – $C_2H_4NO_2$). Actually, such models reflect only a composition of a molecule. Obviously, it is quite impossible to solve adequately the "structure–activity" tasks using such approaches. So, usually these models have an auxiliary role only, but sometimes they can be used as independent virtual screening tools [22].
- 2D models contain information regarding the structure of a compound and are based on its structural formula [20]. Such models reflect only the topology of the molecule. These models are very popular [3, 23]. The capacity of such approaches is due to the fact that the topological models of molecular structure, in an implicit form, contain information about possible conformations of the compound. Our operational experience shows that 2D level of representation of the molecular structure is enough for the solution of more than 90% of existing QSAR/QSPR tasks.
- 3D-QSAR models [16, 17, 19, 20] give full structural information taking into account composition, topology, and spatial shape of molecule for one conformer only. These models are widespread. However, the choice of the conformer of the molecule analyzed is mostly accidental.

The description of the molecular structure is realized more adequately by 4D-QSAR models [10, 24]. These models are similar to 3D models, but compared to them the structural information is considered for a set of conformers (conditionally the fourth dimension), instead of one fixed conformation (also see Chapter 3).

The description of compounds from 1D to 4D models reflects the hierarchy of molecular structure representation. However, it's only one of the principles of HiT QSAR. In this work the hierarchic strategy related to all the aspects of the QSAR models development has been considered.

The developed strategy has been realized as a complex of computer programs known as the "HiT QSAR" software. Innovative aspect and main advantages of HiT QSAR involve

- Simplex representation of molecular structure that provides universality, diversity, and flexibility of the description of compounds related to different structural types.
- HiT QSAR that, depending on the concrete aims of research, allows for the construction of the optimal strategy for QSAR model generation, avoiding at the same time superfluous complications that do not result in an increase in the adequacy of the model.
- HiT QSAR does not have the restrictions of such well-known and widely used approaches as CoMFA, CoMSIA, and HASL. Usage of such methods is limited by the requirement for a structurally homogeneous set of molecules and the use of only one conformer.

- HiT QSAR does not have the HQSAR restrictions that are related to the ambiguity of descriptor system formation.
- At every stage of HiT QSAR use, we can determine the molecular structural features that are important for the studied activity and exclude the rest. It shows unambiguously the limits of QSAR models' complication and ensures that resources are not wasted on needless calculations.

The efficiency of the HiT QSAR has been demonstrated through the example of various QSAR tasks, e.g., given in [3, 10–12, 22, 25–37].

5.2. MULTI-HIERARCHICAL STRATEGY OF QSAR INVESTIGATION

5.2.1. HiT QSAR Concept

In this chapter, the hierarchic QSAR technology (HiT QSAR) [31, 32, 36, 37] based on the simplex representation of molecular structure (SiRMS) has been considered. This method has proved efficient in numerous studies for solving different "structure–activity/property" problems [3, 10–12, 22, 25–37]. The essence of the strategy presented is based on the solution of QSAR problems via the sequence of the permanently improved molecular structure models (from 1D to 4D) (Figure 5-1). Thus, at each stage of the hierarchical system, the QSAR task is not solved ab ovo, but with the use of the information received from a previous stage. In fact, it is proposed to deal with a system of permanently improved solutions. It leads to more effective interpretation of the obtained QSAR models because the approach reveals molecular fragments/models for which the detailed development of structure is important.

The main feature of the strategy presented consists of the multiple-aspect hierarchy (Figure 5-1), related to

- models describing molecular structure (1D → 2D → 3D → 4D);
- scales of activity estimation (binomial → nominal → ordinal → continual);
- mathematical methods used to establish structure–activity relationships [pattern recognition → rank correlation → multivariate regression → partial least squares (PLS)];
- final aims of the solution of the QSAR task (prediction → interpretation → structure optimization → molecular design).

The set of different QSAR models that supplement each other results from the HiT QSAR application. These models altogether, in combination, solve the problems of virtual screening, evaluation of the influence of structural factors on activity, modification of known molecular structures, and the design of new high-potency potential antiviral agents or other compounds with desired properties.

The scheme for HiT QSAR is shown in Figure 5-1. The information from the lowest level QSAR models has been transferred (curved arrow) to the highest

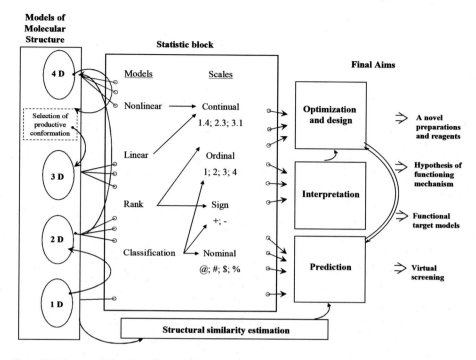

Figure 5-1. Scheme of the hierarchical QSAR technology

level models following corresponding statistical processing ("Statistic block" in Figure 5-1), during which the most significant structural parameters have been chosen. It is necessary to note that after the 2D modeling, the QSAR task is solved at the 4D level, because there is no a priori information available about a "productive" conformation (the conformer that interacts with a biological target most effectively) for 3D-QSAR models. This information comes only after the development of 4D-QSAR models and activity calculation for all conformers considered. Then the information about the "productive" (the most active) conformation is transferred to the 3D-QSAR level. This is the main difference between HiT QSAR and ordinary 3D-QSAR approaches, where the investigated conformers have been chosen through a less vigorous process. When an investigated activity is mainly determined by the interaction of the exact "productive" conformation (not by the set of conformers) with a biological target, it is possible to construct the most adequate "structure–property" models at this stage. In all cases (1D–4D), different statistical methods can be used to obtain the QSAR models (the "Statistic block" in Figure 5-1).

The principal feature of the HiT QSAR is its multi-hierarchy, i.e., not only the hierarchy of different models but also that the hierarchy of the aims has been taken into account (Figure 5-1, unit –"Final Aims"). Evidently, it is very difficult to obtain a model that can solve all the problems related to the influence of the structure of the studied molecules to the property examined. Thus, to solve every definitive task, it is necessary to develop a set of different QSAR models, where some of them are more suitable for the prediction of the studied property, the others for the interpretation of the obtained relationships, and the third for molecular design. These models altogether, in combination, solve the problem of the creation of the new compounds and issue relating to the desired set of properties. The important feature of such an approach is that the general results obtained from a few different independent models always are more relevant. It's also necessary to note that these resulting QSAR models have been chosen in accordance with the QECD principles for the validation of (Q)SARs [38], i.e., they have a defined endpoint, an unambiguous algorithm, a defined domain of applicability (DA), mechanistic interpretation, have good statistical fit, and are robust and predictive. Thus, we assume that the proposed strategy provides a solution to solve all problems dealing with virtual screening, modeling of functional (biological) targets, advancement of hypotheses regarding mechanisms of action, and, finally, the design of the new compounds with desired properties.

5.2.2.   Hierarchy of Molecular Models

5.2.2.1.   *Simplex Representation of Molecular Structure (SiRMS)*

In the framework of SiRMS, any molecule can be represented as a system of different simplexes (tetratomic fragments of fixed composition, structure, chirality, and symmetry) [29, 31, 32, 39] (Figure 5-2).

1D models. At the 1D level, a simplex is a combination of four atoms contained in the molecule (Figure 5-2). The simplex descriptor (SD) at this level is the number of quadruples of atoms of the definite composition. For the compound $(A_aB_bC_cD_dE_eF_f...)$, the value of SD $(A_iB_jC_lD_m)$ is $K = f(i) \cdot f(j) \cdot f(l) \cdot f(m)$, where, for example Eq. (5-1),

$$f(i)\frac{a!}{(a-i)! \cdot i!} \qquad (5\text{-}1)$$

The values of $f(j)$, $f(l)$, $f(m)$ have been calculated analogically. It is possible to define the number of smaller fragments ("pairs," "triples") by the same scheme. In this case some of i, j, l, m parameters are equal to zero.

2D models. At the 2D level, the connectivity of atoms in simplex, atom type, and bond nature (single, double, triple, aromatic) has been considered. Atoms in simplex can be differentiated on the basis of different characteristics, especially

- atom individuality (nature or more detailed type of atom);
- partial atom charge [40] (see Figure 5-2) (reflects electrostatic properties);
- lipophilicity of atom [41] (reflects hydrophobic properties);
- atomic refraction [42] (partially reflects the ability of the atom to dispersion interactions);

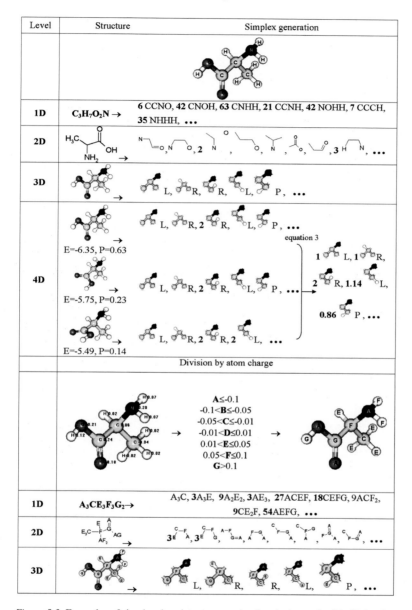

Level	Structure	Simplex generation

Figure 5-2. Examples of simplex descriptors generation for alanine at the 1D–4D levels

- a mark that characterizes the atom as a possible a Hydrogen donor or acceptor (A – Hydrogen acceptor in H-bond, D – Hydrogen donor in H-bond, I – no bond).

For atomic characteristics, which have real values (charge, lipophilicity, refraction, etc.) the division of values range into definite discrete groups is carried out at the preliminary stage. The number of groups (G) is a tuning parameter and can be varied (as a rule $G = 3$–7).

The usage of sundry variants of simplex vertexes (atoms) differentiation represents an important part of SiRMS. We consider that specification of atoms only by their nature (actually reflects atom identity, for example, C, N, O) realized in many QSAR methods limits the possibilities of pharmacophore fragment selection. For example, if the –NH– group has been selected as the fragment (pharmacophore) determining activity and the ability of H-bond formation is a factor determining its activity, H-bonds donors, for example, the OH-group will be missed. The use of atom differentiation using H-bond marks mentioned above avoids this situation. One can make analogous examples for other atomic properties (lipophilicity, partial charge, refraction, etc.).

Thus, the SD at the 2D level is a number of simplexes of fixed composition and topology. It is necessary to note that, in addition to the simplex descriptors, other structural parameters, corresponding to molecular fragments of different size, can be used for 1D and 2D-QSAR analysis. The use of 1–4 atomic fragments is preferable because further extension of the fragment length could increase the probability of the model overfitting and decrease its predictivity and DA.

2.5D models. It's well known that the stereochemical moieties of the investigated compounds could affect biological activity to at least at the same level as their topology. Although the most adequate description of stereochemistry of compounds is possible only on 3D and 4D levels of molecular structure modeling, 2D models of molecules can also provide stereochemical information. In the case when a compound contains a chiral center on the atom X ($X = C$, Si, P, etc.), the special marks X^A, X^R, X^S (A – achiral X atom, R – "right" surrounding of X atom, S – "left" surrounding of X atom) can be used to reflect the stereochemistry information of such a center. In each case, the configuration (R or S) of a chiral center can be determined by the Kahn–Ingold–Prelog rule [43]. For example, in the situation where atom X has been differentiated to three different types depending on its stereochemical surroundings, i.e., X^A, X^R, X^S, the different types are analyzed in the molecular model as separate atoms. Conventionally, such molecular models can be considered as 2.5D because not only topological (molecular graph) but also stereochemical information has been taken into account. If simplex vertexes (X atoms) have been differentiated by some physical–chemical properties (e.g., partial charges, lipophilicity) then the differences between atoms X^A, X^R, X^S will be leveled as in normal 2D models. For subsequent QSAR analysis, the simplexes differentiated by atom individuality have been used separately and in combination with those differentiated by physical–chemical properties.

3D models. At the 3D level, not only the topology but also the stereochemistry of molecule is taken into account. It is possible to differentiate all the simplexes as right (R), left (L), symmetrical (S), and plane (P) achiral. For example:

(R)

(L)   (S)   (P)

The stereochemical configuration of simplexes is defined by modified Kahn–Ingold–Prelog rules [39]. A SD at this level is a number of simplexes of fixed composition, topology, chirality, and symmetry.

4D models. For the 4D-QSAR models, each *SD* is calculated by the summation of the products of descriptor values for each conformer (SD_k) and the probability of the realization of the corresponding conformer Eq. (5-2) (P_k).

$$SD = \sum_{k=1}^{N} (SD_k \cdot P_k), \qquad (5\text{-}2)$$

where N is a number of conformers being considered.

As is well known [44], the probability of conformation P_k is defined by its energy equation (5-3):

$$P_k = \left\{ 1 + \sum_{i \neq k} \text{EXP} \left(\frac{-(E_i - E_k)}{RT} \right) \right\}^{-1}, \qquad \sum_k P_k = 1, \qquad (5\text{-}3)$$

where E_i and E_k are the energies of conformations i and k, respectively.

The conformers are analyzed within an energy band of 5–7 kcal/mol. Thus, the molecular SD at the 4D level takes into account the probability of the realization of the 3D-level SD in the set of conformers. At the 4D level the other 3D whole-molecule parameters, which are efficient for the description of spatial forms of the conformer (e.g., characteristics of inertia ellipsoid, dipole moment), can be used along with SD. An example of the representation of a molecule as sets of simplexes with different levels of structure detailed (1D–4D) is depicted in Figure 5-2.

Double nD models. The interaction of a mixture with a biological target cannot normally be described simply as the average between interactions of its parts, since the last interactions have different reactivity. It is also applicable for mixtures of compounds with synergetic or anti-synergetic action [45]. Because of these issues, the SiRMS approach has been developed and improved in order to make this method suitable for the execution of QSAR analysis for molecular mixtures and ensembles. With this purpose it's necessary to indicate whether the parts of unbound simplexes belong to the same molecule or to a different one. In the latter case, such unbound simplex will reflect the structure not of a single molecule, but will characterize a pair of different molecules. Simplexes of this kind are structural descriptors of the mixtures of compounds (Figure 5-3). Their usage allows for the analysis of synergism, anti-synergism, or competition in the mixture's interaction with the biological target. Obviously, such an approach is suitable for different nD-QSAR models, where $n = 1$–4[1]. If in the same task both mixtures and single compounds have been considered, it's necessary to represent individual compounds as the mixture of two similar molecules for the correct description of such systems [46]. Thus, this approach has been named by authors as "double nD-QSAR." Although such methodic is suitable only for binary mixtures, it can be easily extended to more complicated tasks. For molecular ensembles (associates), it is necessary to use one more simplex type – simplexes with intermolecular bonds.

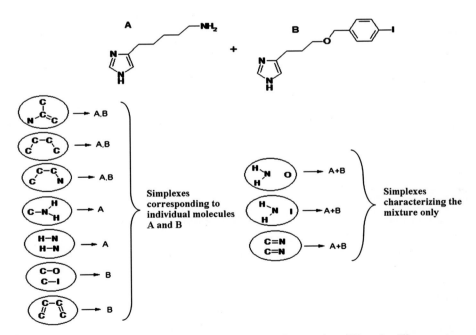

Figure 5-3. Example of structure description of the mixture of antagonists of histamine H3 – receptors (A-imphetamine, B-iodoproxiphane)

---

[1] For 1D-QSAR models unbounded simplexes characterize only the mixtures.

In this chapter, the application of the "double nD-QSAR" approach is demonstrated with the example of chiral AChE inhibitors [46] (see Section 5.4.4).

5.2.2.2. Lattice Model

The lattice model (LM) approach has been developed by the authors [19] using similar principles as CoMFA and CoMSIA (see Chapter 4), which utilize a more elaborated description of the molecules and consider parameters reflecting peculiarities of the intermolecular interaction of the compounds analyzed and their spatial structure. However, in addition, molecular properties are described with a variety of complementary parameters. The whole set of parameters generated ranges from the most simple, such as the presence or absence of particular atoms in the molecule, to more sophisticated parameters that could be used for the consideration of the stereochemistry of the analyzed molecule and its interaction with the environment.

The description of compounds includes several steps. In the first, the spatial structure of the analyzed molecules is obtained from experimental data (i.e., X-ray analysis) or from quantum mechanical calculations. In the case of flexible molecules, it is necessary to select one of the stable conformations. This may be achieved using a conformational search [47] or some complementary information regarding the biologically active conformation of the molecule. The conformation of each molecule is placed into a lattice of cubic cells. The size of a cell can be varied, by default it equals 2 Å, that corresponds approximately to the average van der Waals radius of an organogenic atom. The invariant disposition of the molecule in the lattice is achieved by the superposition of the center of mass of the molecules with the origin of the coordinates. In addition, the principal axes of inertia of the molecule are also superimposed with the coordinate axes of the lattice. If the analyzed structures contain a large common structural fragment, their alignment is carried out mainly according to this fragment.

All structural parameters in the LM can be classified as follows:

- Integral parameters describing properties of the whole molecular structure;
- Local parameters describing the separate fragments of the molecule;
- Field parameters describing the influence of the molecule on the enclosing space.

Integral parameters are characteristics of inertia ellipsoid, dipole moment, molecular refraction, lipophilicity, parachor, and average polarizability. If available, some information about the environment and mutual disposition of the pharmacophores can be also included into the analysis [48].

Local parameters were used to describe the properties of cells occupied by atoms. They include parameters corresponding to the presence or absence of some atoms in the cell (i.e., the presence of C or O), average lipophilicity, refraction, polarizability, electrostatic charge, and electronegativity of fragments and atoms. All charge characteristics were calculated using the Jolly-Perry [40, 49] method of smoothing of electronegativity.

Field parameters describe the characteristics of vacant cells. They include

(1) An electrostatic potential in the vacant cell [Eq. (5-4)]:

$$EP_i = \sum_j^{n_1} \frac{q_j}{r_{ij}} \qquad (5\text{-}4)$$

where i is the number of the cell, j is the number of the atom, q_j is the charge of the atom j [40, 49], and r_{ij} is the distance between the atom j and the cell i;

(2) A lipophilicity potential [50] in the vacant cell [Eq. (5-5)]:

$$LP_i = \sum_j \frac{f_j}{(1 + r_{ij})} \qquad (5\text{-}5)$$

where i is the number of the cell, j is the number of the atom, f_j is the lipophilicity of the atom (group), and r_{ij} is the distance between the atom j and the cell i;

(3) A probability of an occupancy of a vacant cell by different atoms i, k ("probe-atoms") or probability of it to be empty [Eq. (5-6)]:

$$P_k = \left\{ 1 + \sum_{i \neq k} EXP \left(\frac{-(E_i - E_k)}{RT} \right) \right\}^{-1}, \quad \sum_k P_k = 1, \qquad (5\text{-}6)$$

where E_i or E_k is the energy of interaction between the molecule and the corresponding probe-atom i or k in the analyzed cell.

A set of atoms C_{sp}^3, N_{sp}^3, O_{sp}^3, C_{sp}^2, N_{sp}^2, O_{sp}^2 Cl, H and the absence of any atom ("vacuum") were used as probes. If CoMFA [16] uses energy attributes to characterize the analyzed cells, in LM the probabilities of the occupancy of a cell represents a different approach for the description of interactions between the molecule and the biological target. It might be argued that a probability-based scheme offers improvements over an energy-based method.

(4) A possibility of the presence of hydrogen bond donor or acceptors in the cell. It is assumed that such a hydrogen bond can be formed between this donor or this acceptor and the analyzed molecule.

All structural parameters, i.e., integral, local, and field parameters contain an exhaustive description of the molecular structure. Thousands of descriptors (their exact number depends on the characteristics of the lattice) are generated within the proposed approach for each analyzed molecule. This reduces the probability of missing the most significant parameters required to correlate activity of the analyzed molecules with their structure.

The efficiency of the LM approach has been demonstrated on different tasks, e.g., [19, 48, 51, 52].

5.2.2.3. *Whole-Molecule Descriptors and Fourier Transform of Local Parameters*

SDs at all levels of differentiation (1D–4D) are the fragmentary parameters which describe not a molecule as a whole, but its different parts. In order to reflect the structural features of a whole molecule, it is necessary to carry out the Fourier transformation [53] for the spectrum of structural parameters. The spectrum of structural parameters is the discrete row of values arranged in a determined order. The mode of ordering is not crucial (frequently descriptors are lexicographically ordered), but it must be the same for all compounds of an investigated task. As a result of the Fourier transformation, the high-frequency harmonics characterize small fragments while the low-frequency harmonics correspond to the global molecule properties. The Fourier transformation of a discrete function of parameters $P(i)$ can be presented as Eq. (5-7):

$$P(i) = \frac{a_0}{2} + \sum_{k=1}^{M-1} \left(a_k \cos \frac{2\pi k(i-1)}{N} + b_k \sin \frac{2\pi k(i-1)}{N} \right) + a_{N/2} \cos(\pi(i-1))$$

(5-7)

where

$$a_k \frac{2}{N} \cdot \sum_{i=1}^{N} P_i \cdot \cos \left(\frac{2\pi \cdot k \cdot (i-1)}{N} \right), \quad b_k = \frac{2}{N} \cdot \sum_{i=1}^{N} P_i \cdot \sin \left(\frac{2\pi \cdot k \cdot (i-1)}{N} \right)$$

(5-8)

or in an alternative form [Eq. (5-9)]

$$p(i) = \frac{q_0}{2} + \sum_{k=1}^{M-1} \left(q_k \sin \left[\frac{2\pi k(i-1)}{N} + \psi_k \right] \right) + q_{n/2} \cos[\pi(i-1)], \quad (5\text{-}9)$$

The amplitudes and phase angle in Eq. 5-9 are defined as follows:

Amplitudes: $q_k = \sqrt{a_k^2 + b_k^2}$, Phase angle: $\psi_k = \arctan((a_k)/b_k)$. (5-10)

where k is the number of harmonics, N is the total number of simplex descriptors, $M = \text{int}(N{-}1)/2$ is the total number of harmonics, a_k and b_k are the coefficients of expansion procedure, $q_{n/2} = 0$ for even N.

Values of amplitudes (a_k, b_k, q_k) can be used as the parameters for the solution of QSAR tasks [19, 54]. PLS equations containing amplitudes a_k and b_k can be mechanistically interpreted, because they can be represented as a linear combination of source structural parameters (5-7). Amplitudes q_k have poor mechanistic interpretation because of the more complex dependence from the source structural parameters (5-9). However, all the amplitudes (a_k, b_k, q_k) separately or together allow for well-fitted, robust, and predictive models to be obtained; hence, they can be used as an additional (completely different) tool for the virtual screening.

Such whole-molecule parameters, such as characteristics of inertia ellipsoid (moments of inertia I_X, I_Y, I_Z and its ratio I_X/I_Y, I_Y/I_Z, I_X/I_Z), dipole moment, molecular refraction, lipophilicity, also can be used for different levels of representation of the molecular structure.

All mentioned integral parameters can be united with SD which usually leads to the most adequate model that unites the advantages of molecular descriptors of every mentioned type.

5.2.3.   Hierarchy of Statistical Methods

As was mentioned above, different statistical methods have been used in HiT QSAR to establish the structure–activity relationship depending on the scale of the investigated property (binomial $\rightarrow$ nominal $\rightarrow$ ordinal $\rightarrow$ continual).

5.2.3.1.   Classification Trees

The classification tree (CT) approach is a non-parametric statistical method of analysis [55]. It allows for the analysis of data sets regardless of the number of investigated compounds and the number of their characteristics (descriptors). In the CT approach, the models obtained represent the hierarchical sets of rules based on descriptors selected for the description of the investigated property. The rule represents "IF-THEN" logical construction. For example, a simple rule can be "IF lipophilicity > 3 THEN compound is active." In fact, such model is presented by a set of consecutive nodes, and each of them contains certain sets of compounds which correspond to this node rule. The CT method has several advantages: obtaining of intuitively understandable models using natural language, quick learning and predicting processes, non-linearity of obtained models, and the ability to develop models using ranked values of the activity (it allows for the analysis of sets of compounds with heterogeneous experimental activity values).

The usage of CT methods for QSAR analysis is limited due to the poor mechanistic interpretation of the models. It is difficult to make quantitative estimation of the influence of descriptors used in the model and to determine structural fragments interfering or promoting activity.

A new approach for the interpretation of CT models, based on a trend-vector procedure (see Section 5.2.3.3), has been proposed to solve this problem. It allows for the determination of the quantitative influence of descriptors used in the model built on the investigated property [Eq. (5-11)]:

$$T_j = \frac{1}{m} \sum_{i=1}^{m} [(A_i - A_{\mathrm{mean}})] \qquad (5\text{-}11)$$

where T_j is the relative influence of jth descriptor on investigated property, m is the number of compounds in the certain node, A_i is the activity rank of ith compound, and A_{mean} is the mean value of activity rank for the whole set of compounds.

The relative influence (T_j) of each descriptor used in the CT model are calculated by applying Eq. (5-11) to each node of the model (excepting the root node). Furthermore, each calculated influence has a corresponding range of descriptor values (D) according to node rule, within which this influence has been implemented. As a result of such analyses, ranges of descriptor values and corresponding relative influences can be determined. When descriptor has several overlapping ranges of values then the relative influence values should be summarized in the overlapping interval.

The approach described is valid only for models with classification scale of activity. It can be considered as a restriction of the method. However, estimation of activity level is an appropriate result in many cases relating to the investigation of biological activity. In the case of the usage of simplex (fragmentary) descriptors for the representation of molecular structure, T_j values obtained in this manner are the cumulative influences of all simplexes of a certain type in the molecule. It allows for the calculation of the relative atomic influences for each investigated compound according to Eq. (5-12).

$$T_a = \frac{T_j}{4N_j} \qquad (5\text{-}12)$$

where T_a is the relative influence of each atom included in the jth simplex of certain molecule, T_j is the relative influence of the jth simplex, $4N_j$ is the number of jth simplexes (value of jth descriptor) in certain molecules multiplied by four (number of atoms in a simplex).

Calculated relative atom influences can be visualized on the investigated compounds. They allow for the determination of the relative influences of separate molecular fragments by summarizing the influences of individual atoms included in certain fragments.

5.2.3.2.   Trend-Vector

The trend-vector (TV) procedure [19, 56, 57] does not depend on the form of corresponding dependence and can use many structural parameters. This method can predict the properties of analyzed molecules only in a rank scale and can be used if biological data are represented in an ordinal scale (see Figure 5-1). Similar to a dipole moment vector, TV characterizes a division of "conventional charges" (corresponding to active and inactive classes) in the multi-dimensional space of structural parameters S_{ij} ($i = \overline{1,n}$ – number of molecules, $j = \overline{1,m}$ – number of structural parameters). Each component of a TV is determined by Eq. (5-13)

$$T_j = \frac{1}{n} \cdot \sum_{i=1}^{n} (A_i - \bar{A}) \cdot S_{ij}, \qquad (5\text{-}13)$$

and reflects a degree and direction of influence of the jth structural parameter on the magnitude of a property A. The prediction of activity is obtained using the following relation:

$$\text{rank}(A_i) = \text{rank}\left(\sum_{j=1}^{m} T_j S_{ij}\right) \qquad (5\text{-}14)$$

It is important to note that each component of the TV is calculated independently from the others and its contribution to a model is not adjusted. Thus, the influence on the reliability of the model of the number of structural parameters used is not so critical, as in the case of the regression methods. The quality of the structure–property relationship can be estimated by the Spearman rank correlation coefficient calculated between ranks of the experimental and calculated activities A_i.

The search for models using the TV method in HiT QSAR is achieved by the methods of exhaustive or partial search after the removal of mutual correlations. It was discovered by the authors [10, 32] that descriptors involved in the best TV models (several decades of models with approximately identical quality) form a good subset for the subsequent usage in PLS. Noise elimination can be one of the probable explanations of the success of the TV procedure.

5.2.3.3.   Multiple Linear Regression

The greatest number of QSAR/QSPR investigations has been made using linear statistic methods [58]. In such approaches, the investigated property is represented as a linear function of calculated descriptors [Eq. (5-15)]:

$$y' = a_0 + \sum_{i=1}^{n} a_i x_i \qquad (5\text{-}15)$$

where y' is the calculated values of investigated property (y), x_i is the structural descriptors (independent variables), a_i is the regression coefficients determined during the analysis by the least squares method, n is the number of variables in the regression equation.

The use of linear approaches is very convenient for investigations because the theory of selection of the most important attributes and obtaining of the final equations is well developed for such methods. The quality of the obtained model is estimated by the correlation coefficient R between the observed values of the investigated property (y) and those predicted by Eq. (5-15) (y'). The R^2 value is explained by regression measure of the part of common scatter relative to average y. The term of adequacy of the obtained regression model with the chosen level of risk α will be F [Eq. (5-16)] [58]:

$$F = \frac{R^2(m - n - 1)}{(1 - R^2) \cdot n} \geq F_{xp} \ (n - 1, m - n, \alpha), \tag{5-16}$$

where m is the number of molecules in the training set and F_{xp} (n–1, m–n, 1–α) is the percent points of the F-distribution for given level of significance 1–α.

The relative simplicity of regression approaches is also their shortcoming; they show poor results during the extrapolation of complicated structure–activity relationships. Their usage is further hampered in the case of large numbers of descriptors, since the total number of descriptors in a MLR equation must be at least ten times fewer than the number of training set compounds [59].

5.2.3.4. *Partial Least Squares or Projection to Latent Structures (PLS)*

A great number of simplex descriptors have been generated in HiT QSAR. The PLS-method has proved efficient for working with a great number of variables [60–62]. The PLS regression model may be written as Eq. (5-17) [62]:

$$Y = b_0 + \sum_{i=1}^{N} b_i x_i, \tag{5-17}$$

where Y is an appropriate activity, b_i are the PLS regression coefficients, x_i is the ith descriptor value, and N is the total number of descriptors.

This is not apparently different from MLR (see Section 5.2.3.3), except that the values of the coefficients b are calculated using PLS. However, the assumptions underlying PLS are radically different from those of MLR. In PLS one assumes the x-variables to be collinear and PLS estimates the covariance structure in terms of a limited number of weights and loadings. In this way, PLS can analyze any number of x-variables (K) relating to the number of objects (N) [62].

5.2.4. Data Cleaning and Mining

The removal of highly correlated and constant descriptors, the use of genetic algorithms (GA) [63], trend-vector methods [56, 57], and automatic variable selection (AVS) strategies that are similar to interactive variable selection (IVS) [61] and evolutionary variable selection (EVS) [60] have been used for selection of descriptors in PLS. The removal of highly correlated descriptors is not necessary for PLS analysis, since descriptors are reduced to series of uncorrelated latent variables. However, this procedure frequently helps to obtain more adequate models and reduce a number of used variables up to five times. During this procedure one descriptor from each pair having a pair correlation coefficient r satisfying $|r| > 0.90$ has been eliminated.

5.2.4.1.   *Automatic Variable Selection (AVS) Strategy in PLS*

The AVS strategy in PLS is used to obtain highly adequate models by removing the "noise" data, i.e., systematic variations in X (descriptors space) that are orthogonal to Y (investigated property). This strategy is similar to IVS [61], EVS [60], OSC [64], and O-PLS [65] and has the same objective but uses different means.

The essence of AVS consists of the following: at the first step of the AVS the model containing all descriptors is obtained. Then variables with the smallest normalized regression coefficients (b_i, Eq. (5-17)) are excluded from the X-matrix and in the next step the PLS model is obtained. This procedure has been repeated stepwise until the amount of variables equals 1. The AVS strategy can be used either for all structural parameters or after different variable selection procedures (e.g., removal of highly correlated descriptors, TV procedure, GA). An application of the AVS procedure resulted in the decreasing of the model complexity (number of descriptors and latent variables) and an increase in model predictivity and robustness.

5.2.4.2.   *Genetic Algorithms*

GA imitates such properties of living nature as natural selection, adaptability, heredity. The use of the heuristic organized operations of "reproduction," "crossing," and "mutation" from casual or user-selected starting "populations" generates the new "chromosomes" – or models. The utility of the GA is its flexibility. With adjustment of the small set of algorithm parameters (number of generations, crossover and mutation type, crossover and mutation probability, and type of selection), it is possible to find a balance between the time for search and the quality of decision. In the HiT QSAR, GA is used as a tool for the selection of adequate PLS, MLR, and TV models. Descriptors from the best model obtained by the preliminary AVS procedure have usually been used as the starting "population." GA is not a tool for the elucidation of the global maximum or minimum, and very often a subsequent AVS procedure and different enumerative techniques allow one to increase the quality of the obtained PLS models.

5.2.4.3.   *Enumerative Techniques*

As mentioned above, the usage of the methods of exhaustive or partial searching (depending on the number of selected descriptors) after AVS or GA very often allow one to increase the quality of the obtained models (PLS, MLR, and TV). After the statistical processing model or models with the best combinations of statistic characteristics (R^2, Q^2) have been selected from the obtained resulting list, and they may be submitted for subsequent validation using an external test set. The general scheme

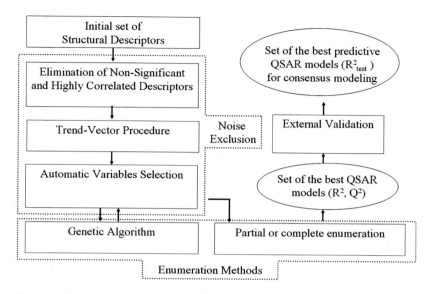

Figure 5-4. General scheme of the PLS models generation and selection applied in the HiT QSAR

of the PLS model generation and selection applied in the HiT QSAR is presented in Figure 5-4. This procedure can be repeated several times using as input an initial set of SD of different levels of molecular structure representation (usually 2D–4D) and/or with various kinds of atom differentiation (see above) with the purpose to develop several resulting "predictive" QSAR models for consensus modeling. This approach is believed to yield more accurate predictions.

5.2.5.   Validation of QSAR Models

To have any practical utility, up-to-date QSAR investigations must be used to make predictions [66]. The statistical fit of a QSAR can be assessed in many easily available statistical terms (e.g., correlation coefficient R^2, cross-validation correlation coefficient Q^2, standard error of prediction S).

Cross-validation is the statistical practice of partitioning a sample of data into subsets such that the analysis is initially performed on a single subset, while the other subset(s) are retained for subsequent use to confirm and validate the initial analysis. The initial subset of data is called the training set; the other subset(s) are called validation sets. In QSAR analysis, only two types of cross-validation are used:

(1) *K-fold cross-validation.* In K-fold cross-validation, the original sample is partitioned into K subsamples. Of the K subsamples, a single subsample is retained as the validation data for testing the model and the remaining K – 1 subsamples are used as training data. The cross-validation process is then repeated K times (the folds), with each of the K subsamples used exactly once as the validation data.

(2) *Leave-one-out cross-validation.* As the name suggests, leave-one-out cross-validation (LOOCV) involves using a single observation from the original sample as the validation datum and the remaining observations as the training data. This is repeated such that each observation in the sample is used once as the validation data. This is the same as a K-fold cross-validation with K being equal to the number of observations in the original sample.

The determination coefficient (Q^2) calculated in cross-validation terms is the main characteristic of model robustness. Q^2 is calculated by the following formula:

$$Q^2 = 1 - \frac{\sum_Y (Y_{\text{pred}} - Y_{\text{actual}})^2}{\sum_Y (Y_{\text{actual}} - Y_{\text{mean}})^2} \tag{5.18}$$

where Y_{pred} is a predicted value of activity, Y_{actual} is an actual or experimental value of activity, and Y_{mean} is the mean activity value.

The shortfalls of cross-validation are the following:

(1) The training task must be solved N times leading to substantial calculative expenses in time and resources.
(2) The estimation of cross-validation assumes that the training algorithm is already given. It has no idea how to obtain "good" algorithms and which properties must be inherent to them.
(3) An attempt to use cross-validation for training as an optimizable criterion leads to loss of its unbiasedness property and there is a risk of overfitting.

At the same time statistical fit should not be confused with the ability of a model to make predictions. The only method to obtain a meaningful assessment of statistical fit is to utilize the so-called "test set". During this procedure a certain proportion of the data set molecules (10–85%) are removed to form the test set before the modeling process begins (remaining molecules form the training set). Once a model has been developed, predictions can be made for the test set. This is the only method by which the validity of a QSAR can be more or less truly assessed. However, one must understand that sometimes it means only the model ability to predict the certain test set. It is important that both training and test sets cover the structural space of the complete data set as much as possible.

In the HiT QSAR, the following procedure has been used for the formation of the test set: a dissimilarity matrix for all initial training set molecules has been developed on the basis of relevant structural descriptors. Such a descriptor set can be obtained using different procedures for descriptor selection (for example, see Chapter 4) or directly from the model generated for all investigated compounds. In our opinion the use of the whole set of descriptors generated at the very beginning is not completely correct, because during QSAR research we are interested not in structural similarity by itself, but from the point of view of the investigated activity and the descriptor selection will help the avoidance of some distortions caused by the insignificance of structural parameters from the initial set for this task.

A dissimilarity matrix is based on the estimation of structural dissimilarity between all investigated molecules. A measure of the structural dissimilarity for molecules M, M' can be calculated using the Euclidean distance in the multidimensional space of structural parameters S [Eq. (5-19)]:

$$SD(M, M') = \sqrt{\sum_{i=1}^{n} (S_i - S'_i)^2,}$$

(5.19)

where n is the a number of molecules in data set.

Thus, total structural dissimilarity toward the rest of initial training set compounds could be calculated for every molecule from a sum of the corresponding Euclidean distances. In the meanwhile, all the compounds were divided into groups depending on their activity, where the number of groups equals the number of molecules that one wants to include into test set. Then one compound from each group has been chosen to go to the test set according to its maximal (or minimal) total Euclidean distance from the other molecules in this group, or by random choice. Most likely, the use of several (three is the enough minimum) test sets constructed by different principles and subsequent comparison and averaging of the obtained results is more preferable than the use of only one set for the model validation. In that way, the first test set has been constructed to maximize its diversity from the training set, i.e., the compounds with maximal dissimilarity were chosen. This is the most rigorous estimation, sometimes it can lead to the elimination of all of the dissimilar compounds from the training set, i.e., such splitting of the training set when the test set structures would not be predicted correctly by the developed model and would be situated outside of DA. The second test set is created in order to minimize its diversity from the training set, i.e., less dissimilar compounds from each group were removed. The last test set has been chosen in random manner taking into account activity variation only.

5.2.6. Hierarchy of Aims of QSAR Investigation

HiT QSAR provides not only hierarchy of molecular models, systems of descriptors, and statistical models, but also the hierarchy of the aims of QSAR investigation (Figure 5-1). Targets of the first level are activity prediction or virtual screening. Any descriptors could be used here, even those that are only poorly interpretable or noninterpretable, e.g., different topological indices, informational-topological indices, eigenvalues of various structural matrices. In other words, at this level descriptors which are not expected to be used for subsequent analysis of structural factors promoting or interfering with activity can be used.

The aims of the second level must include the interpretability of obtained QSAR models. Only descriptors which have clear physico-chemical meaning, e.g., reflecting such parameters of the molecule such as dipole moment, lipophilicity, polarizability, van der Waals volume, can be used at this level. Analysis of

QSAR models corresponding to this level allows one to reveal structural factors promoting or interfering with the investigated property. Such information can be useful for the generation of hypotheses about mechanisms of biological action and assumptions about the structure of biological target. Finally, the presence of information useful for molecular design is expected from QSAR models corresponding to the third level of purposes. As a rule, fragmentary descriptors have been used in such models. In this case, the analysis of the degree and direction of influence of such descriptors on activity can give immediate information for the optimization of known structures and design of novel substances with desired properties.

5.2.6.1.   *Virtual Screening (Including Consensus Modeling and DA)*

As mentioned above, QSAR investigations must be used to make predictions for compounds with unknown activity values (so-called "virtual screening"). In order to increase the quality of predictions, these authors recently started to apply consensus QSAR modeling which has become more and more popular [67]. It also represents one of the crucial concepts of HiT QSAR [31, 36] and can be briefly described by the statement "More models that are good and different." The efficiency of this technique can be easily explained by the fact that nearly the same predictions obtained by different and independent methods (either statistical or descriptors generation) are more reliable than single prediction made by even the best fitted and predictable model.

From another aspect, in order to analyze the predictivity of PLS models and according to the OECD QSAR principles [38], different DA procedures have been included in the HiT QSAR. The first procedure is an integral DA called "ellipsoid" developed by the authors [11]. It represents a line at the 1D level; an ellipse at the 2D level; an ellipsoid at the 3D level; and multidimensional ellipsoids in more complicated n-dimensional spaces. Its essence consists of the following: the distribution of training set molecules in a space of latent variables T_1–T_A (axes of coordinates) can be obtained from PLS. For each coordinate axis (T_1 and T_2 in our case) the root-mean-square deviations S_{T1} and S_{T2} have been determined. DA represents an ellipsoid that is built from the molecules of the training set distribution center ($T_1 = 0$; $T_2 = 0$) with the semi-axes length $3S_{T1}$ and $3S_{T2}$, respectively [11] (Figure 5-5). Further, the correct positions in relation to this center have been calculated for every molecule (including molecules from prediction set). If a work set molecule does not correspond to the DA criteria, it is termed "influential," i.e., it has unique (for given training set) structural features that distinguish it from the other compounds. If a new molecule from the prediction set is situated out of the DA (region outside ellipsoid), its prognosis from the corresponding QSAR model is less reliable (model extrapolation). And, naturally, the prognoses for molecules nearest to the center of the DA are most reliable.

The second approach – the integral DA rectangle has been also developed by the authors [11]. Two extreme points (so-called virtual activity and inactivity etalons)

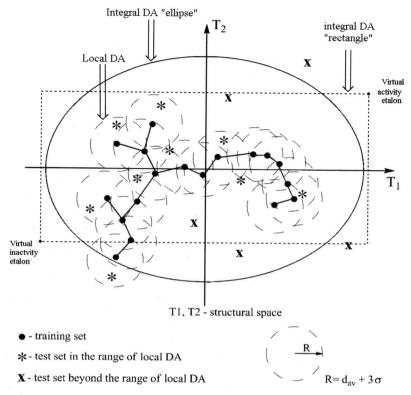

Figure 5-5. Different domain applicability procedures in the HiT QSAR: integral (*ellipsoid, rectangle*) and local

are determined in a space of structural features. The first one has maximal values of descriptors (training set data) promoting activity and minimal interfering. This point corresponds to a hypothetic molecule – the peculiar activity etalon. The second point, analogically, is an inactivity etalon, i.e. contains maximal values of descriptors interfering activity and minimal promoting. Vectors that unite these points (directed from inactive to active) depict the tendency of activity change in the variable space. This vector is a diagonal for the rectangle that determines DA [11] (Figure 5-5). All the mentioned trends concern the "influential" points from the training set and model extrapolation for new molecules from the prediction set remain and for the DA rectangle approach.

The third method is based on the estimation of leverage value h_i [68]. It has been visualized as a Williams plot [69] and is described in detail in [70]. For leverage, a value of 3 is commonly used as a cut-off value for accepting predictions, because points that lie ±3 standard deviations from the mean cover 99% of the normally distributed data. For training set molecules high leverage values do not always indicate outliers from the model, i.e., points that are outside the model domain. If high leverage points ($h_i > h_{cr}$, separated by vertical bold line) fit the model well (i.e., have small residuals), they are called "good high leverage points" or good influence points. Such points stabilize the model and make it more precise. High leverage points, which do not fit the model (i.e., have large residuals) are called "bad high leverage points" or bad influence points. They destabilize the model [70]. A new molecule is situated out of the DA (model extrapolation) if it has $h_i > h_{cr} = 3(A+1)/M$, where A – number of the PLS latent variables and M – number of molecules in a work set.

Recently, a local (Tree) approach for DA estimation has been developed by authors in order to avoid the inclusion of hollow space into the DA that is the lack of integral DA methods. The following are required for its realization:

(1) Obtaining of a distance matrix between the training set molecules in the structural space of descriptors of the QSAR model. The molecules in the given approach have been analyzed in the coordinates of the latent variables of the PLS model considered.
(2) Detection of the shortest distances between molecules using the above-mentioned matrix. Building of an extreme short distance tree for all training set molecules.
(3) Finding of average distance (d_{av}) and its root-mean-square deviation (σ) for inclusion in the tree average values. Such a distance is the characteristic of average density of molecules distribution in the structural space.

Following this procedure, all the points corresponding to test set molecules have been taken into account in the structural space. If any of test set molecules have been situated on the distance bigger than $d_{av}+3\sigma$ from the nearest training set point, it means that this test set molecule is situated outside DA. Respectively, molecules belonging to the DA are situated on the distance less than $d_{av}+3\sigma$ from the training set points. The scheme of DA estimation has been depicted in Figure 5-5.

Such an approach for DA estimation is similar, to some extent, to methods described in [70]. As opposed to integral approaches, e.g. [11], where the convex region (polyhedron, ellipsoid) which could contain vast cavities has been determined in the structural space, the approach presented here is local. The space of the structural parameters has been analyzed locally, i.e., regions around every training set point are analyzed. The presence of cavities in the structural space which correspond to DA is undesirable and it has been eliminated in the given approach.

Summarizing, it's necessary to note that if a new structure is lying inside the DA, it is not a final argument for a correct prediction; rather, it is an indication of the reduced uncertainty of a prediction. In exactly the same way, the situation of the compound outside the DA does not lead to the rejection of the prediction; it is just an indication of the increased uncertainty of the subsequent virtual screening prediction. Naturally, such compounds could be predicted (by model extrapolation) with great accuracy, but it will be more by co-incidence than design. Unfortunately, there is currently no unbiased estimation of prognosis reliability, and the relative character of any DA procedure was reflected in [11, 70]. Thus, it should be remembered that the DA is not a guide to action but only a probable recommendation.

All of the mentioned DA procedures together, or separately, are applied to selected single models before being averaged in consensus model. The accuracy of the DA consensus model has been compared with the adequacy of consensus models without DA consideration. The authors recommend the use of consensus DA models for subsequent virtual screening excepting the case of substantial loss of coverage of training and prediction sets with only a limited benefit in predictivity.

5.2.6.2. *Inverse Task Solution and Interpretation of QSAR Models*

Using Eq. (5-15) it is not difficult to make the inverse analysis (interpretation of QSAR models) in the frameworks of the SiRMS approach. The contribution of each j-atom (C_j) in the molecule can be defined as the ratio of the sum of the PLS regression coefficients (b_i) of all simplexes this atom contains (M) to a number of atoms (n) in the simplex (or fragment) [Eq. (5-20)]:

$$C_j = \frac{1}{n} \sum_{i=1}^{M} b_i, \quad \text{(for simplex } n = 4) \tag{5-20}$$

According to this formula, the atom contribution depends on the number of simplexes which include this atom. This value (number of simplexes) is not constant; it varies in different molecules and depends on other constituents (surroundings), and hence, this contribution is non-additive. Atoms that have a positive or negative influence on the studied biological activity of compounds can be colored. It helps to present the results and to determine visually (additionally to the automate search) the groups of atoms affecting the activity in different directions and with varying strength. The example of the representation of the obtained results on the molecule using color-coding according to the contribution of atoms into antirhinoviral activity [11] is represented in Figure 5-6. Atoms and structural fragments reducing antiviral activity are colored in *red* (*dark gray* in printed version) and that enhance antiviral

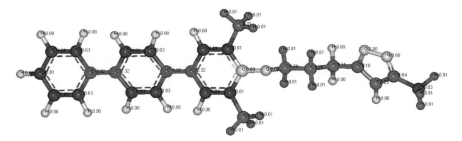

Figure 5-6. Color-coded structure according to atoms contributions to activity against HRV-2 [11]. Atoms and structural fragments reducing antiviral activity are colored in *dark gray* and that enhancing antiviral activity in *light gray* and *white*

activity in *green* (*light gray and white* in printed version). Atoms and fragments with no effect are colored in *gray*.

The automatic search procedure for pre-defined fragments from the data set and their relative effect on activity has been realized in HiT QSAR. The procedure of the fragment searching in molecule is based on a fast algorithm for solving the maximum clique problem [71]. Some molecular fragments promoting and interfering anti-influenza activity [12, 29, 34] are represented in Table 5-1 as well as their average relative influence on it.

Table 5-1. Molecular fragments governing the anti-influenza activity change (Δ lgTID$_{50}$) and their average relative influence on it [12, 29, 34]

Enhance the activity

3.0	2.4	1.9
	$-(CH_2)_2-O-$	
1.7	1.4	0.8

Decrease the activity

$-(CH_2)_n-NH-$		$-CO-NH_2$
$n = 2-3$		
−0.3 to −0.4	−0.2	−0.2

5.2.6.3.   *Molecular Design*

It is possible to design compounds with a desired activity level from the SiRMS via the generation of allowed combinations of simplexes determining the investigated property. The simplest way is soft drug design [72] that consists of replacing of undesired substituents by more active ones, or by the insertion of fragments, promoting the activity instead of non-active parts of molecule or hydrogen atoms. The use of this technique allows one to retain newly designed compounds in the same region of structural space as the training set compounds. The accuracy of prognosis can be estimated using the DA techniques (see below). However, the use of soft drug design keeps within the limits of the initial chemical class of training set compounds. More drastic drug design is, certainly, more risky, but it allows for much more dramatic results. Almost certainly, new structures would lie outside the DA region. That, however, does not mean uncertainty of prediction, but extrapolation of the model predictivity and a certain lack of any DA procedure. However, at the same time, we can receive compounds of completely different (from initial training set) chemical classes as the output of such design. It was demonstrated in [12, 28, 29], where, in searching for a new antiviral and anticancer agents, we started our investigations from macrocyclic pyridinophanes and through several convolutions of QSAR analyses came out with nitrogen analogues of crown ethers in the first and acyclic aromatic structures with the azomethine fragment in the second case.

5.2.7.   HiT QSAR Software

The HiT QSAR software for Windows has been designed and developed as an instrument for high-value QSAR investigations including the solution of the following tasks:

- Creation of QSAR projects;
- Calculation of lipophilicities and partial atom charges;
- Molecules superposition in the lattice approaches;
- Generation of different integral, simplex, lattice (local and field), and harmonic descriptors;
- Data mining (see Section 5.2.4);
- Obtaining of statistical models by PLS, MLR, and TV approaches with the usage of total and partial enumeration methods, GA, AVS strategy, etc.
- Inverse task solution – interpretation of the equations developed as color-coded diagrams for the molecules or their fields;
- Determination of the contributions (increments) of the fragments in the property investigated;
- Consensus modeling of the property investigated taking into account the DA of the model.

Graphic visualization of molecules, the atoms' influence on the investigated properties, lattice models, different fields, etc. was implemented using the open graphic language (OpenGL) library from Silicon Graphics©. HiT QSAR software is accessible on your request. Please contact the authors if you have any questions about its usage. Summarizing the information above, the HiT QSAR workflow (Figure 5-7) has recently been developed and used by authors for the solution of different QSAR/QSPR tasks.

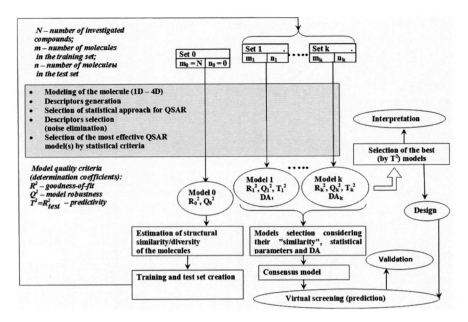

Figure 5-7. HiT QSAR workflow

As was mentioned above, the proposed technology operates on a set of different models. At the preliminary stage "Model 0" (Figure 5-7) is generated for the initial division of investigated molecules into training and test sets. Subsequent generation of sets 1–K is required for the development of consensus QSAR models. In all cases, such statistical characteristics as R^2, Q^2, R^2_{test} have been taken into account as well as the model DA.

5.3.   COMPARATIVE ANALYSIS OF HiT QSAR EFFICIENCY

The HiT QSAR based on SiRMS has proved efficient in numerous studies to solve different "structure–activity/property" problems [3, 10–12, 25–30, 33, 35] and it has been interesting to compare it with the other successful QSAR approaches and software. The results of a comparative analysis are shown in Table 5-2. Obviously, HiT QSAR does not have the problem of the optimal alignment of the set of molecules considered that is inherent to CoMFA and its analogues [16–19]. The SiRMS approach is similar to HQSAR [20] in certain ways, but has none of its restrictions (only topological representation of molecular structure an ambiguity of descriptor formation during the molecular hologram hashing). In addition, contrary to HQSAR, different physical and chemical properties of atoms (charge, lipophilicity, etc.) can be taken into account in SiRMS (Table 5-2).

Table 5-2. Comparison of different QSAR methods

Criterion		HiT QSAR	CoMFA CoMSIA HASL GRID	CODESSA DRAGON	HQSAR
Adequacy of representation of molecular structure	1D-4D	1D - 4D	3D	2D 3D	2D
Absence of "molecular alignment" problem		Yes	No	Yes	Yes
Explicit con sideration of stereochem istry and chirality		Yes	Partly	No	No
Consideration of physical-chemical properties of atoms	charge, lipophilicity, polarizability etc.	Yes	Partly	Partly	No
Possibility of molecular design		Yes	Partly	No	Partly

Thus, main advantages of the HiT QSAR are the following:

- The use of different (1D–4D) levels of molecular modeling;
- The absence of the "molecular alignment" problem;
- Explicit consideration of stereochemical features of molecules;
- Consideration of different physical and chemical properties of atoms;
- Clear methods (rules) for molecular design.

5.3.1. Angiotensin Converting Enzyme (ACE) Inhibitors

After such a theoretical comparative analysis, it was logical to test the efficiency of the proposed HiT QSAR on real representative sets of compounds. All such sets only contain structurally similar compounds to avoid the "molecular align-ment" problem and, therefore, to facilitate the usage of the "lattice" approaches (CoMFA and CoMSIA). One hundred and fourteen angiotensin converting enzyme (ACE) inhibitors [73] represent the first set. Different statistic models obtained by

HiT QSAR have been compared with those published in [73]. The structure of enalaprat – a representative compound from the ACE data set is displayed below:

The ability of ACE inhibition (pIC$_{50}$) has been investigated. The training set consists of 76 compounds and 38 structures were used in a test set [73]. In the given work, we have compared the resulting PLS-models built with the use of descriptors generated from the following QSAR approaches:

(a) CoMFA – comparative molecular fields analysis [16];
(b) CoMSIA – comparative molecular similarity indexes analysis [18];
(c) EVA – eigenvalue analysis [74];
(d) HQSAR – hologram QSAR [20];
(e) the Cerius 2 program (Accelrys, Inc., San Diego, CA) – method of traditional integral (whole-molecule) 2D and 2.5D^2 descriptors generation;
(f) HiT QSAR based on SiRMS [3, 11, 32].

Because all the mentioned approaches compare parameters generated at 2D or 3D levels of molecular structure representation, the corresponding SD, the Fourier parameters, and united models with mixed (simplex + Fourier) parameters were taken for comparison. The advantage of HiT QSAR over other methods is revealed by the comparison of such statistical descriptions of the QSAR models, as the determination coefficient for training (R^2) and test (R^2_{test}) sets; the determination coefficient calculated in the cross-validation terms (Q^2) as well as the standard errors of prediction for both sets (see Table 5-3). For example, for SiRMS $Q^2 = 0.81$–0.87, for the Fourier models $Q^2 = 0.73$–0.80, and for the other methods $Q^2 = 0.65$–0.72. It is necessary to note that the transition to 3D level allows for the improvement of the quality of the QSAR models obtained. At the same time, the usage of the Fourier parameters does not lead to good predictive models ($R^2_{\text{test}} = 0.37$–0.51) for this task. United models (simplex + Fourier) have the same predictive power as the simplex ones, but, because of the presence of integral parameters, they are sufficiently different to provide another aspect of the property.

5.3.2.   Acetylcholinesterase (AChE) Inhibitors

The second set used for comparative analysis consisted of 111 acetylcholinesterase (AChE) inhibitors. The structure of E2020 – a representative compound from the AChE data set is displayed below:

---

^{2} This classification is offered by the authors of Cerius2.

The ability to model AChE inhibition (pIC_{50}) has been investigated. The training set consists of 74 compounds and 37 structures were used as a test set [73]. The methods compared and the principles of comparison are similar to the ones described above. The main trends revealed for the ACE set were also the same for the AChE inhibitors. The advantage of HiT QSAR over other methods have been observed with all statistical parameters (Table 5-3), but especially on predictivity of the models: for SiRMS $R^2_{test} = 0.74$–0.82, for the Fourier models $R^2_{test} = 0.59$–0.61, and for the other methods $R^2_{test} = 0.16$–0.47. As in the previous case, consideration of the spatial structure of investigated compounds improved the quality of the models obtained.

Table 5-3. Statistical characteristics of the QSAR models obtained for ACE and AChE data sets by different methods

QSAR method	R^2 ACE	R^2 AChE	Q^2 ACE	Q^2 AChE	R^2_{test} ACE	R^2_{test} AChE	S_{ws} ACE	S_{ws} AChE	S_{test} ACE	S_{test} AChE	A ACE	A AChE
CoMFA*	0.80	0.88	0.68	0.52	0.49	0.47	1.04	0.41	1.54	0.95	3	5
CoMSIA(basic)*	0.76	0.86	0.65	0.45	0.52	0.44	1.15	0.45	1.48	0.98	3	6
CoMSIA(extra)*	0.73	0.86	0.66	0.46	0.49	0.44	1.22	0.45	1.53	0.98	2	4
EVA*	0.84	0.96	0.70	0.41	0.36	0.28	0.93	0.23	1.72	1.11	4	4
HQSAR*	0.84	0.72	0.72	0.33	0.30	0.37	0.95	0.64	1.80	1.01	4	5
Cerius 2*	0.82	0.38	0.72	0.3	0.51	0.16	1.00	0.95	1.50	1.2	4	1
Simplex 2D	0.87	0.81	0.81	0.65	0.73	0.74	0.86	0.53	1.13	0.67	2	2
Simplex 3D	0.92	0.89	0.87	0.84	0.85	0.82	0.68	0.41	0.85	0.56	2	2
Fourier 2D	0.83	0.71	0.80	0.61	0.37	0.61	0.96	0.66	1.7	0.82	5	4
Fourier 3D	0.78	0.81	0.73	0.71	0.51	0.59	1.1	0.53	1.5	0.84	4	4
Mix** 2D	0.86	0.81	0.80	0.69	0.75	0.74	0.9	0.53	1.07	0.67	2	2
Mix** 3D	0.90	0.89	0.88	0.84	0.85	0.82	0.74	0.4	0.83	0.56	2	2

where
R^2 – correlation coefficient
Q^2 – cross-validation correlation coefficient (10-fold, see Chapter 5)
R^2_{test} – correlation coefficient for test set
S_{ws} – standard error of a prediction for training set
S_{test} – standard error of a prediction for test set
A – number of PLS latent variables
*Statistic characteristics from [73] were shown
**Mix = Simplex + Fourier descriptors

Summarizing, it is necessary to note that we understand that the advantage of simplex descriptors generated in HiT QSAR may be partially a result of some of the differences in the statistical approaches applied (e.g., in addition to GA, TV and AVS procedures have been used). However, these mathematical differences are not responsible for all the improvements in the investigated approaches. Thus, it is obvious from the results obtained that HiT QSAR simplex models are well-fitted, robust and, in the main, they are much more predictive than QSAR models developed by other approaches.

5.4.   HiT QSAR APPLICATIONS

The application of HiT QSAR for the solution of different QSAR/QSPR tasks on different levels of representation of molecular structure is highlighted briefly below. The PLS method has been used for the development of QSAR models in all the cases described below.

5.4.1.   Antiviral Activity

Because a lot of different viral serotypes and strains exist, vaccine development for prevention of a wide variety of viral infections is considered to be impracticable. The present treatment options for such infections are unsatisfactory [75–77]. However, there are ongoing attempts to develop antiviral drugs [78–84]. That is why computational approaches, which can distinguish highly active inhibitors from less useful compounds and predict more potent substances, have been used for the analysis of antiviral activity for many years [4, 6, 7, 12, 13, 29].

5.4.1.1.   Antiherpetic Activity of N,N'-(bis-5-nitropyrimidyl) Dispirotripiperazine Derivatives[3] (2D)

HiT QSAR was applied to evaluate the influence of the structure of 48 N,N'-(bis-5-nitropyrimidyl)dispirotripiperazines (see structures below) on their antiherpetic activity, selectivity, and cytotoxicity with the purpose to understand the chemicobiological interactions governing their activities, and to design new compounds with strong antiviral activity [3].

[3]   The authors express sincere gratitude to Dr. M. Schmidtke, Prof. P. Wutzler, Dr. V. Makarov, Dr. O. Riabova, Mr. N. Kovdienko and Mr. A. Hromov for fruitful cooperation that made the development of this task possible.

The common logarithms of 50% cytotoxic concentration (CC_{50}) in GMK cells, 50% inhibitory concentration (IC_{50}) against HSV-1, and the selectivity index (SI = CC_{50}/IC_{50}) were used to develop 2D-QSAR models. Spirobromine – a medicine with a nitrogen-containing dispiro structure possessing anti-HSV-1 activity was included in the training set. The statistic characteristics of QSAR models obtained are quite high ($R^2 = 0.84$–0.91; $Q^2 = 0.61$–0.68; $R^2_{test} = 0.68$–0.71) and allow for the prediction of antiherpetic activity, cytotoxicity, and selectivity of new compounds. Electrostatic factors (38%) and hydrophobicity (25%) were the most important determinants of antiherpetic activity (Figure 5-8). The results of the QSAR analysis demonstrate a high impact of individual structural fragments for antiviral activity. Molecular fragments that promote and interfere with antiviral activity were defined on the basis of the models obtained. Thus, for example, the insertion of non-cationic linkers such as *N*-(2-aminoethyl)ethane-1,2-diamine, ethylenediamine, or piperazine instead of dispirotripiperazine leads to a complete loss of activity while the presence of methyloxirane leads to a strong increase. Using the established results and observations, several new dispirotripiperazine derivatives – potential antiviral agents – were computationally designed. Two of these new compounds (**1** and **2**, Table 5-4) were synthesized. The results of biological tests confirm the predicted high values of antiviral activity and selectivity (they are about two logarithmic units more active and one order more selective than spirobromine) as well as low toxicity of these compounds.

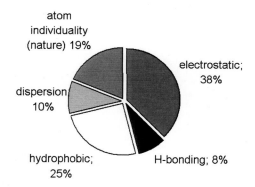

Figure 5-8. Relative influence of some physico-chemical factors on variation of anti-HSV-1 activity estimated on the basis of QSAR models

5.4.1.2. Antiherpetic Activity of Macrocyclic Pyridinophanes[4]

The antiherpetic data set was similar to that for the anti-influenza study and was also characterized by essential structural variety: different macrocyclic pyridinophanes and their acyclic analogues plus well-known antiviral agents including acyclovir as a reference compound:

---

[4] Anti-influenza and antiherpetic investigations described below were carried out as a result of fruitful cooperation with Dr. V.P. Lozitsky, Dr. R.N. Lozytska, Dr. A.S. Fedtchouk, Dr. T.L. Gridina, Dr. S. Basok, Dr. D. Chikhichin, Mr. V. Chelombitko and Dr. J.-J. Vanden Eynde. The authors express sincere gratitude for all mentioned above colleagues.

Table 5-4. Perspective potent compounds – results of computer-assisted molecular design

The antiherpetic activity against HSV-1 strain US was expressed as a percentage of the inhibition of HSV reproduction in treated cell cultures (Hep-2) in comparison with untreated ones. As in previous cases, the antiherpetic study has a multistep cyclic character: synthesis – biological tests – QSAR analysis – virtual screening and computer-assisted drug design – synthesis –, etc. [25, 28, 29, 34]. Initially, 14 compounds (mostly macrocyclic pyridinophanes and their acyclic analogues) have been investigated for antiherpetic activity [29]. At the present stage [25], after the several QSAR convolutions, 37 compounds were divided between training and test sets (26 and 11 compounds respectively) and the set of QSAR models with different adequacy levels (2D, 4D, and 3D) has been obtained as a result of the investigations. All the obtained QSAR models were well fitted, robust, predictive ($R^2 = 0.82$–0.90, $Q^2 = 0.60$–0.65, $R^2_{test} = 0.70$–0.78), and have a defined DA and clear mechanistic interpretation. For the 3D-QSAR investigations the set of "productive" conformers has been used. They were determined as the most active from the results of 4D-QSAR modeling.

All the models developed (2D–4D) indicate the impact of hydrophobic (~50%) and electrostatic (~20%) factors on the variation of antiherpetic activity. The strong promotion of antiherpetic activity by aminoethylene fragments was revealed. It was also discovered that an important factor for the HSV inhibition is the presence of an amino group connected to aliphatic fragment. A tendency of antiviral activity increasing with the strengthening of acceptor properties of compound's aromatic rings was revealed. This information was used for the design of potent antiherpetic agent **1** (Table 5-4). The use of SiRMS allows to progress in searching for new antiherpetic agents starting from macrocyclic pyridinophanes [29] and finishing in symmetric piperazine containing macroheterocycle 1,4,7,10,13,16,19,22,25,28-Decaaza-tricyclo[26.2.2.2*13,16*]tetratriacontane (**1**).

5.4.1.3. [(Biphenyloxy)propyl]isoxazole Derivatives – Human Rhinovirus 2 Replication Inhibitors[5] (2D)

QSAR analysis of antiviral activity of [(biphenyloxy)propyl]isoxazole derivatives

was developed using HiT QSAR based on SiRMS to reveal chemico-biological interactions governing their activities as well as their probable mode of action, and to design new compounds with a strong antiviral activity [11]. The common logarithms

---

[5] The authors express sincere gratitude to Dr. M. Schmidtke, Prof. P. Wutzler, Dr. V. Makarov, Dr. O. Riabova and Ms. Volineckaya for fruitful cooperation that made possible the development of this task.

of 50% cytotoxic concentration (CC_{50}) in HeLa cells, the 50% inhibitory concentration (IC_{50}) against human rhinovirus 2 (HRV-2), and the selectivity index (SI = CC_{50}/IC_{50}) of [(biphenyloxy)propyl]isoxazole derivatives were used as cytotoxicity, antiviral activity, and selectivity assessments, respectively. The set of molecules consists of 18 compounds including pleconaril as a reference compound. They have not been divided into training and test sets because of the low number of compounds (i.e., the structural information contained in each molecule in this case is unique and useful). The statistic characteristics of the resulting 2D-QSAR models are quite satisfactory ($R^2 = 0.84$–0.92; $Q^2 = 0.70$–0.87) for the prediction of CC_{50}, IC_{50}, and SI values and permit the virtual screening and molecular design of new compounds with high anti-HRV-2 activity. The results indicate the high influence of atom's individuality on all the investigated properties (~40%), electrostatic factors on selectivity (~50%), where these factors along with atom individuality play the determining role, and hydrophobic interactions on the antiviral activity (~40%). The presence of terminal 5-trifluoromethyl-1,2,4-oxadiazole and p-fluorophenyl fragments in a molecule leads to strong enhancement of its useful properties, i.e., increase of activity toward HRV-2 as well as selectivity and decrease of cytotoxicity. An additional terminal aromatic ring – naphthalene or phenyl – strongly reduces activity toward HRV-2 and, to a lesser degree, SI. The virtual screening and molecular design of new well-tolerated compounds with strong anti-HRV-2 activity has been performed on the basis of QSAR results. Three different DA approaches (DA rectangle and ellipsoid as well as leverage) give nearly the same results for each QSAR model and additionally allow for the estimation of the quality of the prediction for all designed compounds. A hypothesis to the effect that external benzene substituent must have negative electrostatic potential and definite length L (approximately 5.5–5.6 A) to possess strong antiviral activity has been suggested. Most probably, the fluorine atom in the *para*-position of terminal aromatic ring (compounds **2-4**, Table 5-4) is quite complementary ($L = 5.59$ A) to the receptor cavity for such an interaction. It is necessary to note that pleconaril ($L = 5.54$ A) completely satisfies the indicated criteria. In the case of nitroaromatics, the accumulation of nitro groups in the region of receptor cavity will lead to strengthening of electrostatic interactions with the biological target and, therefore, to an increase in activity.

Several new compounds have been designed computationally and predicted as having high activity and selectivity. Three of them (**2–4**, Table 5-4) were synthesized. Subsequent experimental testing revealed a strong coincidence between experimental and predicted anti-HRV-2 activity and SI. Compounds **2–4** are similar in their cytotoxicity level to plecanoril, but they are more active and selective.

5.4.1.4.   Anti-influenza Activity of Macrocyclic Pyridinophanes[4] (2D–4D)

All the advantages of HiT QSAR were demonstrated during the investigation of anti-influenza activity on the data set possessing structural variety: different macrocyclic pyridinophanes, their acyclic analogues, and well-known antiviral agents (deiteforin, remantadine, ribavirin, ambenum, and others) [12, 29]:

remantadine

ribavirin

Anti-influenza activity (virus A/Hong Kong/1/68 (H3N2)) was expressed in lgTID$_{50}$ and reflected the suppression of viral replication in "experimental" samples in comparison with "controls." The structures investigated were divided between training and test sets (25 and 6 compounds, respectively).

In accordance with the hierarchical principles of the approach offered, the QSAR analysis was solved sequentially on the 2D, 4D, and 3D levels.[6] The set of QSAR models with different adequacy levels (2D, 4D, and 3D) was obtained as a result of the investigations. All the obtained QSAR models were well fitted, robust, predictive ($R^2 = 0.94$–0.98, $Q^2 = 0.85$–0.95, and $R^2_{test} = 0.98$–0.99)[7], and have defined DA and clear mechanistic interpretation. For 3D-QSAR investigations the set of "productive" conformers has been used. They were determined as the most active from the results of 4D-QSAR modeling. The results indicate the great impact of atom individuality on the variation of anti-influenza activity (37–50%). Hydrophobic/hydrophilic and electrostatic interactions also played an important role (15–22%). The shape of molecules (4D and 3D models) also effects anti-influenza activity but has the smallest influence (11 and 16%, respectively). The cylindrical form of molecules ($I_X/I_Y \to 1$) with small diameters ($I_Y \to$ min) promotes anti-influenza activity. The molecular fragments governing the change of anti-influenza activity and their average relative influence (Table 5-1) were determined. For example, the presence of oxyethylene or 2-iminomethylphenol fragments promotes antiviral activity and aminoethylene fragments decreases it.

The purposeful design of new molecules **5–7** (Table 5-4) with adjusted activity level was developed by obtained results. The high level of all predicted (all the resulting 2D–4D models show the strong coincidence of predictions) values of anti-influenza activity was confirmed experimentally. Thus, during the QSAR investigations [12, 29] the search for active compounds began from macrocyclic pyridinophanes and finally results in benzene derivatives containing the 2-iminomethyl-phenol fragment (**5–7**, Table 5-4).

---

[6] In this and antiherpetic research 1D modeling were not performed.

[7] We are aware that these models can approximate not only variation of activity but also variation of experimental errors. The high values of R^2_{test} can be explained by the fact that test compounds are very similar to those in the training set, that there are only few compounds in test set, by high quality of obtained models, by simple good luck or by combination of all mentioned factors.

5.4.2.   Anticancer Activity of MacroCyclic Schiff Bases[8] (2D and 4D)

The investigation of influence of the molecular structure of macrocyclic Schiff bases (see structures below) on their anticancer activity has been carried out by

means of the 4D-QSAR SiRMS approach [10]. The panel of investigated human malignant tumors includes 60 lines of the following nine cell cultures: leukemia, CNS cancer, prostate cancer, breast cancer, melanoma, non-small cell lung cancer, colon cancer, ovarian cancer, and renal cancer. Anticancer activity was expressed as the percent of the corresponding cell growth. The training set is very structurally dissimilar and consists of 30 macrocyclic pyridinophanes, their analogues, and some other compounds.

The use of simple topological models generated by EMMA [85] allows the description of the anticancer activity of macrocyclic pyridinophanes (MCP) for only five cell cultures [86]. These studies show that even within the simple topological model it is possible to detect some patterns of the relationship between the structure of MCP and their activity. The consideration of spatial structure improves the situation, but only at the 4D level reliable QSAR models ($R^2 = 0.74$–0.98; $Q^2 = 0.54$–0.84) were obtained for all of the investigated cells (except leukemia, where $Q^2 < 0.5$; however, even in this case the designed compound was predicted correctly) and averaged activity (most of lines and cells are highly correlated) that indicate the importance of not the most active or favorable single conformer but the set of interacting conformers within the limits of energy gap of 3 kcal/mol. It was discovered that the presence of the N^1,N^3-dimethylenepropane-1,3-diamine fragment strongly promotes anticancer activity. This fragment was used as a linker between two naphthalen-2-oles that leads to the creation of universal anticancer agent active against all mentioned tumors except prostate cancer. It is necessary to note that the use of SiRMS allow one starting from 12 macrocyclic pyridinophanes [86] in the search for anticancer agents to finally result in symmetric open-chained aromatic compounds connected by above-mentioned linker [10].

[7] The authors express sincere gratitude to Dr. V.P. Lozitsky, Dr. R.N. Lozytska and Dr. A.S. Fedtchouk for fruitful cooperation during the development of this task.

5.4.3.   Acute Toxicity of Nitroaromatics

5.4.3.1.   Toxicity to Rats[8] (1D–2D)

HiT QSAR based on 1D and 2D simplex models and some other approaches for the description of molecular structure have been applied for (i) evaluation of the influence of the characteristics (constitutional and structural) on the toxicity of 28 nitroaromatic compounds (some of them belonging to a widely known class of explosives, see structures below); (ii) prediction of the toxicity of new nitroaromatic derivatives; (iii) analysis of the effects of substitution in nitroaromatic compounds on in vivo toxicity

R = H, F, Cl, OH, NO$_2$, COOH, CH$_3$, CH$_2$Cl

The 50% lethal dose to rats (LD$_{50}$) has been used to develop the QSAR models based on simplex representation of molecular structure. The preliminary 1D-QSAR results show that even the information on the composition of molecules reveals the main characteristics for the variations in toxicity [87].

A novel 1D-QSAR approach that allows for the analysis of the non-additive effects of molecular fragments on toxicity has been proposed [87]. The necessity of the consideration of substituents' impact for the development of adequate QSAR models of nitroaromatics' toxicity was demonstrated.

The statistic characteristics for all the 1D-QSAR models developed, with the exception of the additive models, were quite satisfactory ($R^2 = 0.81$–0.92; $Q^2 = 0.64$–0.83; $R^2_{test} = 0.84$–0.87). Successful performance of such models is due to their non-additivity, i.e., the possibility of taking into account the mutual influence of substituents in a benzene ring which governs variations in toxicity and could be mediated through the different C–H fragments of the ring.

The passage to 2D level, i.e., consideration of topology, allows for the improvement of the quality of the obtained QSAR models ($R^2 = 0.96$–0.98; $Q^2 = 0.91$–0.93; $R^2_{test} = 0.89$–0.92) to predict the activity for 41 novel compounds designed by the application of new combinations of substituents represented in the training set [37]. The comprehensive analysis of variations in toxicity as a function of the position and nature of the substituent was performed. Among the contributions analyzed in this work are the electrostatic, hydrophobic, and van der Waals interactions of toxicants to biological targets. Molecular fragments that promote and interfere with toxicity were defined on the basis of models obtained. In particular, it was found that in most cases, insertion of fluorine and hydroxyl groups into nitroaromatics increases toxicity, whereas insertion of a methyl group has the opposite effect. The influence

[8] The authors express sincere gratitude to Prof. J. Leszczynski, Dr. L. Gorb and Dr. M. Quasim for fruitful cooperation during the development of this task.

of chlorine on toxicity is ambiguous. Insertion of chlorine at the *ortho*-position to the nitro group leads to substantial increase in toxicity, whereas the second chlorine atom (at the *para*-position to the first) results in a considerable decrease in toxicity. The mutual influence of substituents in the benzene ring is substantially non-additive and plays a crucial role regarding toxicity. The influence of different substituents on toxicity can be mediated via different C–H fragments of the aromatic ring.

The correspondence between observed and predicted toxicity obtained by the 1D and 2D models was good. The single models obtained were summarized in the most adequate consensus model that allows for an improved accuracy of toxicity prediction and demonstrate its ability to be used as a virtual screening tool.

5.4.3.2.   Toxicity to Tetrahymena Pyriformis[9] (2D)

The present study applies HiT QSAR to evaluate the influence of the structure of 95 various nitroaromatic compounds (including some widely known explosives, see structures below) to the toxicity to the ciliate *T. pyriformis* (QSTR – quantitative structure–toxicity relationship); for the virtual screening of toxicity of new nitroaromatic derivatives; analysis of the characteristics of the substituents in nitroaromatic compounds as to their influence on toxicity.

R = H, F, Cl, OH, NO$_2$,
COOH, CH$_3$, OAlc,
CHO, CN, NH2, etc.

The negative logarithm of the 50% inhibition growth concentration (IGC$_{50}$) was used to develop 2D simplex QSTR models.

During the first part of the work the whole initial set of compounds was divided into three overlapping sets depending on the possible mechanism of action [88]. The 2D-QSTR PLS models obtained were quite satisfactory ($R^2 = 0.84$–0.95; $Q^2 = 0.68$–0.86). The predictive ability of the QSTR models was confirmed through the use of three different test sets (maximal similarity with training set, also minimal one and random choice, taking into account toxicity range only) for any obtained model ($R^2_{test} = 0.57$–0.85).

The initial division into different sets was confirmed by the QSTR analysis, i.e., the models developed for structures with one mechanism (e.g., redox cyclers) cannot satisfactorily predict the others (e.g., those participating in nucleophilic attack). However, the reliable predictive model can be obtained for all the compounds, regardless of mechanism, when structures of different modes of action are sufficiently represented in the training set.

In addition, the classification and regression trees (CRT) algorithm has been used to obtain models that can predict possible mechanism of action. The quality of the

---

[9] The authors express sincere gratitude to Prof. J. Leszczynski, Dr. L. Gorb, Dr. M. Quasim and Prof. A. Tropsha for fruitful cooperation during the development of this task.

CRT models obtained is also quite good. The final models had only 15–20% misclassification errors. The obtained models have correctly predicted mechanism of action for compounds of the test set (76–81%).

The comparative analysis of similarity/difference of all nine selected QSAR models has been carried out using the correlation coefficient and Euclidean distance between the sets of toxicity predicted values. It has been shown that all of them are quite close between themselves and the vector of observed activity values. Hence, *T. pyriformis* toxicity by nitroaromatic compounds is complicated and multifactorial process where, most probably, factors determining penetration and delivery of toxicant to biological target play the most important role. Reactivity of nitroaromatics, seemingly, only has an auxiliary role. This was confirmed by the absence of any correlation between toxicity and Hammett constants of substituents. In this regard, the difference in the mechanisms of toxicant interaction with biomolecules (reactions of nucleophylic substitution or radical reduction of nitro group) is important but do not determine for the value of its toxicity.

Molecular fragments that promote and interfere with toxicity were defined using the interpretation of the PLS models obtained. For example, oxibutane and aminophenyl substituents promote the toxicity of nitroaromatics to *T. pyriformis* but carboxyl groups interfere with toxicity. It was also shown that substituent interference in the benzene ring plays the determining role for toxicity. Contributions of the substituents to toxicity are substantially non-additive. Substituents interference effects the activation of aromatic C–H fragments with regard to toxicity.

The structural factors of nitroaromatics which characterize their hydrophobicity and ability to form electrostatic interactions are the most important for the toxic action of the compounds investigated; local structural characteristics (presence of one or other fragments) are more important than integral (whole-molecule) ones.

All the nine selected models were used for consensus predictions of toxicity of an external test set which consists of 63 nitroaromatics. PLS models based on compounds from one mechanism of action were used for consensus predictions only in the case when the CRT model was able to predict such a mechanism. Thus, the predictivity of the consensus model on the external test set was quite satisfactory ($R^2_{test} = 0.64$).

5.4.4. AChE Inhibition[10] (2.5D, Double 2.5D, and 3D)

HiT QSAR has been used for the consensus QSAR analysis of AChE inhibition by various organophosphate compounds. SiRMS and LM QSAR approaches have been used for descriptor generation. Different chiral organophosphates represented by their (R)- and (S)-isomers, racemic mixtures, and achiral structures (totally 42 points) have been investigated. A successful consensus model ($R^2 = 0.978$) based on 14 best QSAR models ($R^2 = 0.91$–0.99; $Q^2 = 0.86$–0.98; $R^2_{test} = 0.82$–0.97),

---

[10] The authors express sincere gratitude to Prof. J. Leszczynski, Dr. L. Gorb and Dr. J. Wang for fruitful cooperation during the development of this task.

obtained using different QSAR approaches and training sets for several levels and methods of molecular structure representation (2.5D, double 2.5D, and 3D), was used for the prediction of AChE inhibition of new compounds. The trend established on the training set compounds [(S)-isomers are more active than (R)-ones] applies to all new predicted structures.

Atom individuality (including stereochemistry of the chiral surroundings of the asymmetric phosphorus atom) plays the determining role in the variation of activity and is followed by the dispersion and electrostatic characteristics of the OPs. The molecular fragments promoting or interfering with the activities investigated were determined. Identical fragments in the achiral compounds have smaller contributions to activity in comparison with their role in chiral molecules. The influence of phosphorus on the AChE inhibition has a wide range of variation and is very dependent on its surroundings. The substitution of oxygen in $\geq P = O$ by sulfur leads to decreasing AChE inhibition. The presence of the 2-sulphanylpropane fragment facilitates a decrease in activity. Oxyme-containing fragments are actively promoting with activity. The most active predicted compound (2-[(E)-({[cyano(cyclopentyloxy)phosphoryl]oxy}imino)methyl]-1-methylpyridinium) contains oxyme and cyclopentyl parts and is more toxic than oxyme-containing OPs from the training set.

It was also shown in the given work that the topological models of molecular structure (2.5D and double 2.5D) with the identification of stereochemical center of investigated compounds allow for the description of the OPs' ability to inhibit AChE.

5.4.5.   5-HT$_{1A}$ Affinity (1D–4D)[11]

This work was devoted to the analysis of the influence of the structure of N-alkyl-N'-arylpiperazine derivatives (see structures below) on their affinity for the 5-HT$_{1A}$ receptors (5-HT$_{1A}$R).

Several PLS and MLR models have been obtained for the training set containing 42 ligands of 5-HT$_{1A}$R represented on the 1D–4D levels by SiRMS [32]. All the models obtained have acceptable statistical characteristics ($R^2 = 0.71$–0.96, $Q^2 = 0.66$–0.88). There is improvement in the models from 1D $\rightarrow$ 2D $\rightarrow$ 4D $\rightarrow$ 3D. Molecular fragments which have an influence on the affinity for 5-HT$_{1A}$R have been identified. Analysis of the spatial structure of "productive" conformers determined according to 4D-QSAR model shows considerable similarity to the existing pharmacophore models [89–91] and has allowed for improvement.

---

[11] The authors express sincere gratitude to Academician S.A. Andronati and Dr. S.Yu. Makan for fruitful cooperation during the development of this task.

The 2D-QSAR classification task has been solved using the PLS and CRT methods for the set of 364 ligands of 5-HT$_{1A}$R (284 in the training set and 62 in the test set) [92]. The PLS model showed a 65% accuracy for the prediction of test set compounds and the CRT model – 74%. The results of these models have a considerable correspondence between each other that additionally confirmed their validity. It has been shown that, in general, a polymethylene chain comprising three or fewer CH$_2$ groups has a negative influence on affinity for 5-HT$_{1A}$R and a chain comprising four or more CH$_2$ groups has a positive influence. Electron-donating substituents (*o*–OCH$_3$, *o*–OH, *o*–Cl) at the *ortho*-position of phenyl ring strongly promoted affinity. A 2,3-dihydrobenzodioxin-5-yl residue has a similar influence on affinity. Electron-accepting substituents (*m*-CF$_3$) in phenyl have high affinity. Electron-accepting substituents at the *para*-position of the phenyl ring (*p*-NO$_2$, *p*-F) have a stronger negative influence on affinity to 5-HT$_{1A}$R than electron-donating ones (*p*–OCH$_3$). The following conclusions have been made about the influence of the terminal fragments (substituents of *N*-alkyl group) on affinity. Saturated polycyclic fragments and small aromatic residues demonstrated positive influence on affinity and larger aromatic fragments show a negative effect. According to the following analysis, the optimal van der Waals volume for the terminal moiety must be approximately 500 Å^3 or less.

Molecular design and virtual screening of new potential ligands of 5-HT$_{1A}$R has been developed on the basis of the obtained results. Several most promising compounds have been chosen for subsequent investigations, two of them are represented in Table 5-4 (**8** and **9**).

5.4.6. Pharmacokinetic Properties of Substituted Benzodiazepines (2D)

The influence of the structure of substituted benzodiazepines (27 compounds, see below)[12] on the variation of their pharmacokinetic properties including bioavailability, semi-elimination period, clearance, and volume of distribution in the organism of man has been studied [94].

X=O,S,NH, 2H
R1=H, Alc, etc.
R2=H, OH, COOH, OCOAlc
R3,R4=H, halogen

Simplex descriptors in addition to some integral parameters generated by the Dragon software [93] were used for the development of statistic models.

---

[12] The authors express sincere gratitude to Dr I.Yu. Borisyuk and Acad. N.Ya. Golovenko for a fruitful collaboration.

Reasonably adequate quantitative "structure-pharmacokinetic properties" relationships were obtained using the PLS and MLR statistical approaches ($R^2 = 0.91$–0.95, $Q^2 = 0.81$–0.94) [94]. Structural factors affecting the change of pharmacokinetic properties of substituted benzodiazepines were revealed on the basis of the obtained models.

Bioavailability. Although there is no correlation between absolute bioavailability (F) and lipophilicity (R$\approx$0), the trend of increasing of molecular fragments' contribution to common bioavailability alongside with increasing of its lipophilicity is observed quite clearly. This trend is the most evident in case of aromatic fragments. Pentamerous aromatic heterocycles have the greatest influence on bioavailability.

Thus, the presence of benzene rings in a molecule increases its bioavailability in a series of substituted benzodiazepines and substitution on the aromatic rings leads to a decrease in bioavailability. Also one can note that the more oxygen atoms in a molecule, the lower the bioavailability. It has been determined that the oxygen atoms are hydrogen bond acceptors. This is in agreement with Lipinski's "rule of five" [24], whereby good bioavailability is observed when the drug corresponds to the following physico-chemical characteristics: molecular weight < 500; $\log P \leq 5$; number of groups – proton donors ≤ 5; number of groups – proton acceptors ≤ 10.

Clearance. For clearance (Cl) of the investigated series, the trend is opposite to that for bioavailability. Thus, the presence of H-donors in a molecule, substitution in aromatic rings as well as an increase of molecule saturation leads to an increase in clearance.

Time of semi-elimination. The influence of structural fragments on the variation of the time of semi-elimination is similar to that described for bioavailability. Thus, all lipophilic aromatic fragments have high values for increasing semi-elimination time.

Volume of distribution. During the analysis of the influence of structure of benzodiazepines on their volume of distribution, the same trends as for clearance were revealed. Thus, refraction (electronic polarizability) increases the volume of distribution and high aromaticity and hydrophilicity decrease it.

The resulting PLS models have been used for the development of virtual screening of pharmacokinetic properties of novel compounds belonging to bezdiazepines family [94].

5.4.7. Catalytic Activity of Crown Ethers[13] (3D)

HiT QSAR was applied to develop the QSPR analysis of the phase-transfer catalytic properties of crown ethers in the reaction of benzyl alcohol oxidation by potassium chlorochromate:

$$3PhCH_2OH + 2KCrO_3Cl \xrightarrow[CH_2Cl_2]{Crown\ Ether} 3PhCHO + 2KCl + Cr_2O_3 + 3H_2O \quad (5\text{-}21)$$

[13] The authors express sincere gratitude to Prof. G.L. Kamalov, Dr. S.A. Kotlyar and Dr. G.N. Chuprin for fruitful cooperation during the development of this task.

The objects of the investigation were 66 structurally dissimilar crown ethers, their acyclic analogues and related compounds. The compounds were not divided into training and test sets. Catalytic activity was expressed as the percentage of conversion acceleration.

The distinctive feature of this study is the absence of any reliable relationship between topololgical (2D) structure of crown ethers and their catalytic properties. At the 4D level a not very robust ($Q^2 = 0.46$) relationship was obtained and, only at the 3D level, after the selection of the conformations with the most acceptable formation of complexes with potassium, was a reliable model formed ($R^2 = 0.87$; $Q^2 = 0.66$). Alongside the positive effect of biphenyl and diphenyloxide fragments on catalytic activity of the investigated compounds, the slight preference of "transoid" on *cis*-conformations of crown ethers containing mentioned fragments was shown. The undesirability of the cyclohexyl fragment was determined as well as the certain limits of crown ether dentacy (4–8). These findings, as well as the predominant role of electrostatic factors in investigated process (~50%), correspond to the known mechanisms of catalytic action of the crown ethers. Two potent catalysts **10** and **11** (Table 5-4) were designed and introduced as a result of the QSPR analysis.

5.4.8.   Aqueous Solubility[14] (2D)

This work was devoted to the development of new QSPR equations which will accurately predict S_w for compounds of interest to the US Army (explosives and their metabolites) using the SiRMS approach with subsequent validation of the obtained results using a broad spectrum of available experimentally determined data.

The series of the different QSPR models that supplement each other excludes the application of additive schemes and provides a solution to the problems of virtual screening, the evaluation of influence of the structural factors on solubility, etc., have been developed and used with the consensus part of hierarchical QSAR technology.

The training set consists of 135 compounds and the test set includes 156 compounds. Two-dimensional simplex and derived from them Fourier integral descriptors have been used to obtain the set of well-fitted, robust, and predictive (internally and externally) QSPR models ($R^2 = 0.90$–0.95; $Q^2 = 0.85$–0.91; $R^2_{test} = 0.78$–0.87). External validation using four different test sets also reflects a high level of predictivity ($R^2_{test1} = 0.7$–0.87; $R^2_{test2} = 0.82$–0.88; $R^2_{test3} = 0.66$–0.76; $R^2_{test4} = 0.86$–0.91). Here test$_1$ – mixed set of 27 compounds from different chemical classes; test$_2$ – set of 100 pesticides; test$_3$ – McFarland set of 18 drugs and pesticides; and test$_4$ – Arthursson set of 11 drugs. When all 156 compounds have been united in one external set, $R^2_{testU} = 0.87$ has been reached. The application of DA estimated by the two different approaches (Ellipsoid DA and Williams Plot) leads to a loss of coverage but does not improve the quality of the prediction ($R^2_{testU} = 0.87$).

---

[14] The authors express sincere gratitude to Prof. J. Leszczynski, Dr. L. Gorb and Dr. M. Quasim for fruitful cooperation during the development of this task.

Special attention was paid to the accurate prediction of the solubility of polynitro military compounds, e.g., HMX, RDX, CL-20. Comparison of the solubility values for such compounds predicted by our QSPR results and EPI SuiteTM and SPARC techniques indicates that both DoD and Environmental Protection Agency will have considerable advantage using the SiRMS models developed here.

5.5.   CONCLUSIONS

In summary, it can be concluded that the QSAR technology considered is a universal instrument for the development of effective QSAR models which provide reliable enough virtual screening and targeted molecular design of various compounds with desired properties. This is a result of its hierarchical structure and wide descriptor system.

The comparative analysis of HiT QSAR with the most popular modern QSAR approaches reflects its advantage, especially in predictivity. The efficiency of HiT QSAR was demonstrated on various QSAR/QSPR tasks at different (1D–4D) levels of molecular modeling. HiT QSAR is under permanent development and improvement. Currently the system of descriptors devoted to adequate description of structure of nanomaterials on the basis of carbon polyhedrons (fullerenes, nanotubes, etc.), algorithms of consensus modeling, and procedures for QSAR analysis of complex mixtures are under development. The technology developed has been realized as a complex of computer programs "HiT QSAR." The trial version is available on request for everyone who is interested in it.

REFERENCES

1. Ooms F (2000) Molecular modeling and computer aided drug design. Examples of their applications in medicinal chemistry. Curr Med Chem 7:141–158
2. Thomas G (2008) Medicinal chemistry: An introduction, 2nd edn John Wiley & Sons Inc, New York
3. Artemenko AG, Muratov EN, Kuz'min VE et al. (2007) Identification of individual structural fragments of N,N'-(bis-5-nitropyrimidyl)dispirotripiperazine derivatives for cytotoxicity and anti-herpetic activity allows the prediction of new highly active compounds. J Antimicrob Chemother 60:68–77
4. Bailey TR, Diana GD, Kowalczyk PJ et al. (1992) Antirhinoviral activity of heterocyclic analogs of win 54954. J Med Chem 35:4628–4633
5. Butina D, Gola JMR (2004) Modeling aqueous solubility. J Chem Inf Comp Sci 43:837–841
6. de Jonge MR, Koymans LM, Vinkers HM et al. (2005) Structure based activity prediction of HIV-1 reverse transcriptase inhibitors. J Med Chem 48:2176–2183
7. Jenssen H, Gutteberg TJ, Lejon T (2005) Modelling of anti-HSV activity of lactoferricin analogues using amino acid descriptors. J Pept Sci 11:97–103
8. Kovatcheva A, Golbraikh A, Oloff S et al. (2004) Combinatorial QSAR of ambergris fragrance compounds. J Chem Inf Comp Sci 44:582–595
9. Kubinyi H (1990) Quantitative structure–activity relationships (QSAR) and molecular modeling in cancer research. J Cancer Res Clin Oncol 116:529–537
10. Kuz'min VE, Artemenko AG, Lozitska RN et al. (2005) Investigation of anticancer activity of macrocyclic Schiff bases by means of 4D-QSAR based on simplex representation of molecular structure. SAR QSAR Environ Res 16:219–230

11. Kuz'min VE, Artemenko AG, Muratov EN et al. (2007) Quantitative structure–activity relationship studies of [(biphenyloxy)propyl]isoxazole derivatives – human rhinovirus 2 replication inhibitors. J Med Chem 50:4205–4213

12. Muratov EN, Artemenko AG, Kuz'min VE et al. (2005) Investigation of anti-influenza activity using hierarchic QSAR technology on the base of simplex representation of molecular structure. Antivir Res 65:A62–A63

13. Verma RP, Hansch C (2006) Chemical toxicity on HeLa cells. Curr Med Chem 13:423–448

14. Zhang S, Golbraikh A, Tropsha A (2006) The development of quantitative structure–binding affinity relationship (QSBR) models based on novel geometrical chemical descriptors of the protein–ligand interfaces. J Med Chem 49:2713–2724

15. Selassie CD (2003) History of QSAR. In: Abraham DJ (ed) Burger's medicinal chemistry and drug discovery. Wiley, New York, p 960

16. Cramer RD, Patterson DI, Bunce JD (1988) Comparative molecular field analysis (CoMFA). 1. Effect of shape binding to carrier proteins. J Am Chem Soc 110:5959–5967

17. Doweyko AM (1988) The hypothetical active site lattice. An approach to modeling active sites from data on inhibitor molecules. J Math Chem 31:1396–1406

18. Klebe G, Abraham U, Mietzner T (1994) Molecular similarity indeces in comparative analysis (CoMSIA) of molecules to correlate and predict their biological activity. J Med Chem 37:4130–4146

19. Kuz'min VE, Artemenko AG, Kovdienko NA et al. (2000) Lattice model for QSAR studies. J Mol Model 6:517–526

20. Seel M, Turner DB, Wilett P (1999) HQSAR – a highly predictive QSAR technique based on molecular holograms. QSAR 18:245–252

21. Pavan M, Consonni V, Gramatica P et al. (2006) New QSAR modelling approach based on ranking models by genetic algorithms – variable subset selection (GA-VSS). In: Brüggeman R, Carlsen L (eds) Partial order in environmental sciences and chemistry. Springer Berlin Heidelberg, Berlin, pp 181–217

22. Kuz'min VE, Muratov EN, Artemenko AG et al. (2008) The effect of nitroaromatics composition on theirs toxicity in vivo. 1D QSAR research. Chemosphere 72:1373–1380

23. Baurin N, Mozziconacci JC, Arnoult E et al. (2004) 2D QSAR consensus prediction for high-throughput virtual screening. An application to COX-2 inhibition modeling and screening of the NCI database. J Chem Inf Model 44:276–285

24. Vedani A, Dobler M (2000) Multi-dimensional QSAR in drug design. Progress in Drug Res 55: 107–135

25. Artemenko A, Kuz'min V, Muratov E et al. (2007) Molecular design of active antiherpetic compounds using hierarchic QSAR technology. Antivir Res 74:A76

26. Artemenko A, Muratov E, Kuz'min V et al. (2006) Molecular design of novel antimicrobial agents on the base of 4-thiazolidone derivatives. Clin Microbiol Infec 12:1557

27. Artemenko A, Muratov E, Kuz'min V et al. (2006) Influence of artifical ribonucleases structure on their anti-HIV activity. Antivir Res 70:A43

28. Artemenko AG, Kuz'min VE, Muratov EN et al. (2005) Investigation of antiherpetic activity using hierarchic QSAR technology on the base of simplex representation of molecular structure. Antivir Res 65:A77

29. Kuz'min VE, Artemenko AG, Lozitsky VP et al. (2002) The analysis of structure-anticancer and antiviral activity relationships for macrocyclic pyridinophanes and their analogues on the basis of 4D QSAR models (simplex representation of molecular structure). Acta Biochim Polon 49: 157–168

30. Kuz'min VE, Artemenko AG, Muratov EN et al. (2007) QSAR analysis of anti-coxsackievirus B3 nancy activity of 2-amino-3-nitropyrazole[1,5-α]pyrimidines by means of simplex approach. Antivir Res 74:A49–A50

31. Kuz'min VE, Artemenko AG, Muratov EN et al. (2005) The hierarchical QSAR technology for effective virtual screening and molecular design of the promising antiviral compounds. Antivir Res 65:A70–A71

32. Kuz'min VE, Artemenko AG, Polischuk PG et al. (2005) Hierarchic system of QSAR models (1D-4D) on the base of simplex representation of molecular structure. J Mol Model 11:457–467

33. Muratov E, Artemenko A, Kuz'min V et al. (2006) Computational design of the new antimicrobials based on the substituted crown ethers. Clin Microbiol Infec 12:1558

34. Muratov EN (2004) Quantitative evaluation of the structural factors influence on the properties of nitrogen-, oxygen- and sulfur-containing macroheterocycles. National Academy of Sciences of Ukraine, A.V. Bogatsky Physical-Chemical Institute, Odessa, p 202

35. Muratov EN, Kuz'min VE, Artemenko AG et al. (2006) QSAR studies demonstrate the influence of structure of [(biphenyloxy)propyl]isoxazole derivatives on inhibition of coxsackievirus B3 (CVB3) replication. Antivir Res 70:A77

36. Kuz'min VE, Artemenko AG, Muratov EN (2008) Hierarchical QSAR technology on the base of simplex representation of molecular structure. J Comp Aid Mol Des 22:403–421

37. Kuz'min VE, Muratov EN, Artemenko AG et al. (2008) The effects of characteristics of substituents on toxicity of the nitroaromatics: HiT QSAR study. J Comp Aid Mol Des 22:747–759. doi:10.1007/s10822-10008-19211-x

38. QSAR, Expert, Group (2004) The report from the expert group on (quantitative) structure–activity relationships [(Q)SARs] on the principles for the validation of (Q)SARs. In: OECD series on testing and assessment. Organisation for Economic Co-operation and Development, Paris, p 206

39. Kuz'min VE (1995) About homo- and heterochirality of dissymetrical tetrahedrons (chiral simplexes). Stereochemical tunneling. Zh Strucur Khim (in Russ) 36:873–878

40. Jolly WL, Perry WB (1973) Estimation of atomic charges by an electronegativity equalization procedure calibration with core binding energies. J Am Chem Soc 95:5442–5450

41. Wang R, Fu Y, Lai L (1997) A new atom-additive method for calculating partition coefficients. J Chem Inf Comp Sci 37:615–621

42. Ioffe BV (1983) Chemistry refractometric methods, 3 ed. Himiya, Leningrad

43. Cahn RS, Ingold CK, Prelog V (1966) Specification of molecular chirality. Angew Chem Int Ed 5:385–415

44. Burkert U, Allinger N (1982) Molecular mechanics. ACS Publication, Washington, DC

45. Hodges G, Roberts DW, Marshall SJ et al. (2006) Defining the toxic mode of action of ester sulphonates using the joint toxicity of mixtures. Chemosphere 64:17–25

46. Kuz'min VE, Muratov EN, Artemenko AG et al. (2009) Consensus QSAR modeling of phosphor-containing chiral AChE inhibitors. J Comp Aid Mol Des 28:664–677

47. Hyperchem 7.5 software. Hypercube, Inc. 1115 NW 4th Street, Gainesville, FL 32601, USA

48. Kuz'min VE, Artemenko AG, Kovdienko NA et al. (1999) Lattice models of molecules for solution of QSAR tasks. Khim-Pharm Zhurn (in Russ) 9:14–20

49. Kuz'min VE, Beresteckaja EL (1983) The program for calculation of atom charges using the method of orbital electronegativities equalization. Zh Struct Khimii (in Russ) 24:187–188

50. Croizet F, Langlois MH, Dubost JP et al. (1990) Lipophilicity force field profile: An expressive visualization of the lipophilicity molecular potential gradient. J Mol Graphics 8:53

51. Artemenko AG, Kovdienko NA, Kuzmin VE et al. (2002) The analysis of "structure-anticancer activity" relationship in a set of macrocyclic pyridinophanes and their acyclic analogues on the basis of lattice model of molecule using fractal parameters. Exp Oncol 24:123–127

52. Lozitsky VP, Kuz'min VE, Artemenko AG et al. (2000) The analysis of structure–anti-influenza relationship on the basis molecular lattice model for macrocyclic piridino-phanes and their analogs. Antivir Res 50:A85

53. Marple SL Jr (1987) Digital spectral analysis with applications. Prentice-Hall Inc., Englewood Cliffs, NJ

54. Kuz'min VE, Trigub LP, Shapiro YE et al. (1995) The parameters of shape of peptide molecules as a descriptors in the QSAR tasks. Zh Struct Khimii (in Russ) 36:509–517

55. Breiman L, Friedman JH, Olshen RA et al. (1984) Classification and regression trees. Wadsworth, Belmont

56. Carhart RE, Smith DH, Venkataraghavan R (1985) Atom pairs as molecular features in structure–activity studies. Definition and application. J Chem Inf Comput Sci 25:64–73

57. Vitiuk NV, Kuz'min VE (1994) Mechanistic models in chemometrics for the analysis of multidimensional data of researches. Analogue of dipole-moments method in the structure(composition)–property relationships analysis. ZhAnalKhimii 49:165–167

58. Ferster E, Renz B (1979) Methoden der Korrelations und Regressionanalyse. Verlag Die Wirtschaft, Berlin

59. Topliss JG, Costello RJ (1972) Chance correlations in structure–activity studies using multiple regression analysis. J Med Chem 15:1066–1068

60. Kubinyi H (1996) Evolutionary variable selection in regression and PLS analyses. J Chemometr 10:119–133

61. Lindgren F, Geladi P, Rannar S et al. (1994) Interactive variable selection (IVS) for PLS. Part 1: Theory and algorithms. J Chemometr 8:349–363

62. Rannar S, Lindgren F, Geladi P et al. (1994) A PLS kernel algorithm for data sets with many variables and fewer objects. Part 1: Theory and algorithm. J Chemometr 8:111–125

63. Rogers D, Hopfinger AJ (1994) Application of genetic function approximation to quantitative structure–activity relationships and quantitative structure–property relationships. J Chem Inf Comp Sci 34:854–866

64. Wold S, Antti H, Lindgren F et al. (1998) Orthogonal signal correction of nearinfrared spectra. Chemometrics Intell Lab Syst 44:175–185

65. Trygg J, Wold S (2002) Orthogonal projections to latent structures (O-PLS). J Chemometr 16:119–128

66. Cronin MTD, Schultz TW (2003) Pitfalls in QSAR. J Mol Struct (Theochem) 622:39–51

67. Zhang S, Golbraikh A, Oloff S et al. (2006) A novel automated lazy learning QSAR (ALL-QSAR) approach: Method development, applications, and virtual screening of chemical databases using validated ALL-QSAR models. J Chem Inf Model 46:1984–1995

68. Neter J, Kutner MH, Wasseman W et al. (1996) Applied linear statistical models. McGraw-Hill, New York

69. Meloun M, Militku J, Hill M et al. (2002) Crucial problems in regression modelling and their solutions. Analyst 127:433–450

70. Jaworska J, Nikolova-Jeliazkova N, Aldenberg T (2005) QSAR applicability domain estimation by projection of the training set in descriptor space: A review. Altern Lab Anim 33:445–459

71. Östergard PRJ (2002) A fast algorithm for the maximum clique problem. Discrete Appl Math 120:195–205

72. Bodor N, Buchwald P (2000) Soft drug design: General principles and recent applications. Med Res Rev 20:58–101

73. Sutherland JJ, O'Brien LA, Weaver DF (2004) A comparison of methods for modeling quantitative structure–activity relationships. J Med Chem 47:5541–5554

74. Heritage TV, Ferguson AM, Turner DB et al. (1998) EVA: A novel theoretical descriptor for QSAR studies. Persp Drug Disc Des 11:381–398

75. Barnard DL (2006) Current status of anti-picornavirus therapies. Curr Pharm Des 12:1379–1390

76. Patick AK (2006) Rhinovirus chemotherapy. Antivir Res 71:391–396

77. Rotbart HA (2002) Treatment of picornavirus infections. Antivir Res 53:83–98

78. Binford SL, Maldonado F, Brothers MA et al. (2005) Conservation of amino acids in human rhinovirus 3C protease correlates with broad-spectrum antiviral activity of rupintrivir, a novel human rhinovirus 3C protease inhibitor. Antimicrob Agents Chemother 49:619–626

79. Conti C, Mastromarino P, Goldoni P et al. (2005) Synthesis and anti-rhinovirus properties of fluoro-substituted flavonoids. Antivir Chem Chemother 16:267–276

80. Cutri CC, Garozzo A, Siracusa MA et al. (2002) Synthesis of new 3-methylthio-5-aryl-4-isothiazolecarbonitriles with broad antiviral spectrum. Antiviral Res 55:357–368

81. Diana GD, Cutcliffe D, Oglesby RC et al. (1989) Synthesis and structure–activity studies of some disubstituted phenylisoxazoles against human picornavirus. J Med Chem 32:450–455

82. Dragovich PS, Prins TJ, Zhou R et al. (2002) Structure-based design, synthesis, and biological evaluation of irreversible human rhinovirus 3C protease inhibitors. 6. Structure-activity studies of orally bioavailable, 2-pyridone-containing peptidomimetics. J Med Chem 45:1607–1623

83. Gaudernak E, Seipelt J, Triendl A et al. (2002) Antiviral effects of pyrrolidine dithiocarbamate on human rhinoviruses. J Virol 76:6004–6015

84. Kaiser L, Crump CE, Hayden FG (2000) In vitro activity of pleconaril and AG7088 against selected serotypes and clinical isolates of human rhinoviruses. Antiviral Res 47:215–220

85. Suchachev DV, Pivina TS, Shliapochnikov VA et al. (1993) Investigation of quantitative "structure-shock-sensitivity" relationships for organic polynitrous compounds. Dokl RAN (in Russ) 328:50–57

86. Kuz'min VE, Lozitsky VP, Kamalov GL et al. (2000) The analysis of "structure–anticancer activity" relationship in a set of macrocyclic 2,6-bis (2- and 4-formylaryloxymethyl) pyridines Schiff bases. Acta Biochim Polon 47:867–875

87. Kuz'min VE, Muratov EN, Artemenko AG et al. (2008) The effect of nitroaromatics' composition on their toxicity in vivo: Novel, efficient non-additive 1D QSAR analysis. Chemosphere 72(9):1373–1380. doi:10.1016/j.chemosphere.2008.1004.1045

88. Katritzky AR, Oliferenko P, Oliferenko A et al. (2003) Nitrobenzene toxicity: QSAR correlations and mechanistic interpretations. J Phys Org Chem 16:811–817

89. Chilmonczyk Z, Szelejewska-Wozniakowska A, Cybulski J et al. (1997) Conformational flexibility of serotonin$_{1A}$ receptor ligands from crystallographic data. Updated model of the receptor pharmacophore. Archiv der Pharmazie 330:146–160

90. Hibert MF, Gittos MW, Middlemiss DN et al. (1988) Graphics computer-aided receptor mapping as a predictive tool for drug design: Development of potent, selective, and stereospecific ligands for the 5-HT1A receptor. J Med Chem 31:1087–1093

91. Hibert MF, Mcdermott I, Middlemiss DN et al. (1989) Radioligand binding study of a series of 5-HT1A receptor agonists and definition of a steric model of this site. Eur J Med Chem 24:31–37

92. Kuz'min VE, Polischuk PG, Artemenko AG et al. (2008) Quantitative structure–affinity relationship of 5 HT1A receptor ligands by the classification tree method. SAR & QSAR in Env Res 19:213–244

93. Todeschini R, Consonni V (2000) Handbook of molecular descriptors, 1st ed. Wiley-VCH, Weinheim

94. Artemenko AG, Kuz'min VE, Muratov EN et al. (2009) The analysis of influence of benzodiazepine derivatives structure on its pharmacocinetic properties. Khim-Pharm Zhurn 43:36–45 (in Russ)

CHAPTER 6

ROBUST METHODS IN QSAR

BEATA WALCZAK, MICHAŁ DASZYKOWSKI, AND IVANA STANIMIROVA

Department of Chemometrics, Institute of Chemistry, The University of Silesia, 40-006, Katowice, Poland, e-mail: beata@us.edu.pl

Abstract: A large progress in the development of robust methods as an efficient tool for processing of data contaminated with outlying objects has been made over the last years. Outliers in the QSAR studies are usually the result of an improper calculation of some molecular descriptors and/or experimental error in determining the property to be modelled. They influence greatly any least square model, and therefore the conclusions about the biological activity of a potential component based on such a model are misleading. With the use of robust approaches, one can solve this problem building a robust model describing the data majority well. On the other hand, the proper identification of outliers may pinpoint a new direction of a drug development. The outliers' assessment can exclusively be done with robust methods and these methods are to be described in this chapter

Keywords: Outliers, Robust PCA, Robust PLS

6.1.   INTRODUCTION

The chemical behavior of a given molecule and its ability to interact with the surrounding environment are determined by its molecular properties. Most of the molecular properties related to the electronic configuration of the molecule, to its different conformational and steric effects, and to its physico-chemical and topological properties cannot be measured directly. Therefore, molecular descriptors are developed as a numerical expression of the molecular properties (Chapter 3). The properties that can be obtained experimentally, e.g., biological activity or toxicity are another expression of the molecular properties. These observed properties are related back to the intrinsic properties in order to predict the behavior of a molecule from its structure and physico-chemical properties. Construction of a quantitative/qualitative model that describes this relationship is the main goal of any quantitative/qualitative structure–activity relationship (QSAR) study. In this context, the chemometric calibration techniques are highly valued. Specifically principal component regression (PCR) and partial least squares regression (PLS) [1, 2]

177

T. Puzyn et al. (eds.), Recent Advances in QSAR Studies, 177–208.
DOI 10.1007/978-1-4020-9783-6_6, © Springer Science+Business Media B.V. 2010

have become the usual methods of choice where a large number of descriptors are used. In addition to the quantitative issues, one is interested in identifying groups of molecules with similar properties as quantified by a set of molecular descriptors or by a certain observed property (e.g., biological activity, toxicity). Cluster analysis and principal component analysis have proven to be excellent methods for the exploration and visualization of the huge numbers of descriptor data generated, whereas with classification and/or discriminant methods one can create logic rules for the classification of molecules.

In this chapter, we will focus on the principles of "robust" data exploration and modeling. In the standard QSAR applications, the term "robust" is used to describe a model with good predictive properties for a relatively broad collection of new molecules. However, in the context of robust statistics the term "robust" is reserved for a group of methods that provide good estimates for the majority of data. The statistical term "robust" will be used exclusively in this chapter and the robust variants of some classic methods applied for the exploration and modeling of QSAR data will be introduced.

The motivation for using the robust over the classic methods stems from the nature of QSAR data, where atypical observations are often present. Such atypical observations are called outliers. They are usually the result of the improper calculation of some molecular descriptors and/or experimental error in determining the property to be modelled. The use of three-dimensional descriptors requires an alignment of molecules, which is another potential source of error in the calculated descriptor data. The presence of outliers in the QSAR data matrix influences the performance of classic statistical methods with the so-called least squares loss function [3]:

$$V = \min \sum_{i=1}^{m} r_i^2 \qquad (6\text{-}1)$$

where r_i is the residual of the ith object obtained from the least squares model.

Classic principal component analysis (PCA) and partial least squares regression [4] are typical methods with such a loss function.

Two strategies are usually followed to handle outliers. The first strategy consists of the identification and subsequent elimination of outliers from the data matrix. Then a classic approach can be used to model the "clean" data. Alternatively, one may perform diagnostics on the residuals obtained from the classic model to identify outliers. However, such an investigation is rather misleading, since the outliers strongly influence the fit of any classic model [5]. The problem is even further complicated when the data contain many outliers and the so-called masking and swamping effects (see the explanation given in Section 6.6.3) may take place. Therefore, a less popular, but a more appropriate and efficient strategy in the QSAR field is to apply robust methods directly to the data.

For instance, down-weighting the influence of outliers can diminish the risk of distorting the least squares model constructed for the complete data set to a great extent.

Even though outliers are considered as to be unique molecules that cause considerable difficulties during data processing, they can identify important information about an eventual source of error or about possible reasons for their unique behavior, which is especially valued in the QSAR studies. Let us first describe the types of outlying objects that may be present in the QSAR data.

6.2.   OUTLIERS AND THEIR GENESIS IN THE QSAR STUDIES

Several types of outliers can be distinguished depending on the QSAR problem that is being investigated. Molecules with structures that are different compared to the structures of the remaining molecules are outliers in the space of molecular descriptors (**X**-space). The **X**-data characterize in a specific way either the structure of the molecule or its binding strength to the specific target, e.g., to the active site of an enzyme. A number of descriptors can be used for this purpose [6], including those which characterize the three-dimensional nature of molecules. Calculation of three-dimensional descriptors requires minimizing the molecules' energy using docking or alignment procedures, as is performed in the comparative molecular field analysis (CoMFA, see Chapter 4) [7]. An example of such a philosophy can be found in [8], where a set of selected HIV reverse transcriptase inhibitors were docked into the binding pocket of the reverse transcriptase using the pharmacophore-based docking algorithm [9]. To define the inhibition strength, the non-bond interaction energies between the candidate inhibitor and 93 amino acid residues forming 16 different reverse transcriptase binding pockets were then computed. The procedures used for minimizing the molecules' energy, as well as docking and alignment methods, are considered as time-consuming and error-prone approaches. Their optimal performance is crucial in obtaining high-quality data. In practice, the presence of outliers in **X**-data caused by the sub-optimal performance of these methods is not an exception. The molecules can be aligned differently with respect to their shape and charge distribution [10]. As pointed out in [11], the large flexibility of the binding site can also be a possible source of outliers.

Docking of molecules is not an easy task. The difficulties are mainly associated with the choice of the crystallographic structure of a target protein. At present, the protein data bank contains more than ten X-ray crystallographic structures of the HIV reverse transcriptase and choosing which of them should be used for docking of putative inhibitors is not straightforward. Moreover, docking of inhibitors in all known target structures may be unfeasible [8]. In practice, the docking procedure is performed either by the use of a "compromise" target structure (average of all known target structures) or by subsequent docking of molecules to the target structures followed by averaging the computed interaction energies. Both approaches possess some docking uncertainties due to which outliers can be introduced into **X**-data. An inappropriate choice of molecular descriptors might be another reason that **X**-data contain outliers.

Furthermore, some molecules may have a specific mechanism of chemical behavior [12] associated with their unique binding properties [13]. For instance, some molecules may have multiple binding modes resulting in higher overall interaction energies in comparison with molecules with a single binding mode. Such a molecule can be attractive as a potential drug and its identification as an outlier can pinpoint new directions for future research.

Outliers identified according to the dependent variable (e.g., biological activity, toxicity), **y**, belong to another type of outliers. These are molecules that have generally high or low values of the determined property due to typographical or experimental error.

In summary, two types of outliers may be present in the collected data:

(i) outliers in the **X** data and
(ii) outliers in calibration or experimental determination.

The choice of methodology for analysis depends strongly on the aim of the QSAR study and the collected data. Supervised models such as principal component regression (PCR) and PLS are constructed when the goal of the study is to model the relationship between **X** and **y**. In this case, one is interested in finding all types of outliers bearing in mind that the molecules found as outliers in the **X**-space are not necessarily outliers in **y**. Outliers in calibration are objects that do not follow the model appropriate for data majority [3]. Three types of calibration outliers can be distinguished, i.e., good and bad leverage objects as well as high residual objects (see Figure 6-1). Compared to the good leverage objects that are far from the data center in the **X**-space and in **y**, the bad leverage observations do not fit the model. The high residual outliers are objects with large absolute differences between the observed and predicted values of **y**. One can also look for the same types of outliers in the descriptor **X**-data, which are explored by the unsupervised chemometric approach. Here the residuals from the constructed robust PCA model and the distances of objects from the robust data center in the space of robust scores are considered.

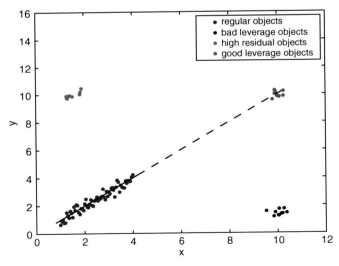

Figure 6-1. Illustration of the regular and outlying objects in calibration: good leverage, bad leverage, and high residual objects

Before introducing the robust versions of PCA and PLS and discussing their use in the QSAR studies, some major concepts of robustness will be presented. In the following section, the term "estimator" is used in a somewhat general sense to refer to any function that aims to estimate a value characterizing the data (e.g., data location, scatter).

6.3. MAJOR CONCEPTS OF ROBUSTNESS

Robust properties of estimators are evaluated in different ways, but using a single measure of robustness is usually insufficient to obtain a complete picture of their performance. Here only the concepts of the breakdown point [14], influence function [15], efficiency, and equivariance properties [3], which are probably the most popular and frequently used measures to assess the robustness of an estimator, will be discussed briefly. A more detailed description of various robustness measures can be found in [3, 16, 17].

6.3.1. The Breakdown Point of an Estimator

The breakdown point, which is defined as the smallest fraction of outliers that can make the estimator useless, seems the most intuitive and appealing robustness measure. It is said that the estimator has a 0% breakdown point when a single outlier completely distorts the result obtained from the estimator. For example, two well-known classic estimates of data location and scatter, namely data mean and standard deviation perform well when data are normally distributed. However, if even one data sample (data object) has a very different value compared to the remaining ones, these two classic estimators "breakdown." Similar to the data mean and standard deviation, any least squares estimator has a breakdown point of 0%. Estimators having the highest possible breakdown point of 50% are called high breakdown point estimators. Many estimators have a breakdown point somewhere between the two extremes (0 and 50%) and its exact value depends on the type of estimator used and its properties.

6.3.2. Influence Function of an Estimator

Another quantitative measure of the robust properties of an estimator is the influence function. As the name of this function suggests, it measures the influence of a single observation on the outcome of the estimator. When the influence function of an estimator is bounded, i.e., takes values only in a certain interval, then the estimator is robust, otherwise it is not. A few commonly used influence functions, Ψ, bounded and unbounded ones, are presented in Figure 6-2.

6.3.3. Efficiency of an Estimator

The efficiency of an estimator is a measure of its performance with contaminated and uncontaminated data. It is defined as the ratio of the mean squared error of the robust estimator to the mean squared error of the classic estimator when they are

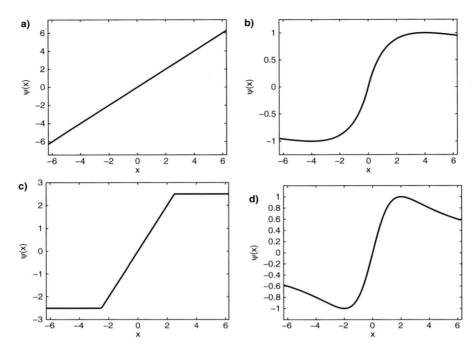

Figure 6-2.   Examples of: (**a**) unbounded influence function of the least squares estimator and bounded influence functions such as (**b**) "fair," (**c**) Huber, and (**d**) Cauchy

both applied to uncontaminated data (normally distributed data). In other words, an efficient estimator neither ignores outliers nor treats regular observations as if they were outliers.

6.3.4.   Equivariance Properties of an Estimator

The estimator is said to be equivariant when a systematic change in the data causes an analogical impact on the estimator [3]. In general, one speaks about the equivariance properties of location, scale, and the regression estimators.

Affine equivariant estimators are the most desired ones. They are independent of affine data transformations such as rotation, scaling, and translation, which are linear data transformations. The affine transformation preserves the collinearity between objects as well as the ratio of distances. Three types of equivariance properties are discussed for regression estimators: regression, scale, and affine equivariance. A regression estimator is the regression equivariant when an additional linear dependence results in an appropriate modification of the regression coefficients. Regression estimators belonging to the family of scale equivariant estimators are independent of the measurement scale. Last, but not the least, the affine equivariance of the regression estimator is the most difficult to fulfill. In this context, any affine data transformation causes a corresponding change in the regression coefficients. Practically, the affine equivariance of the regression estimators is not always required. For instance, the scale and orthogonal equivariance are sufficient to obtain a robust PLS model [18].

6.4. ROBUST ESTIMATORS

Over the years, different families of robust estimators have been proposed for an estimation of data location and scatter. They can differ greatly with respect to their robust properties. Here, for the sake of brevity, several estimators will be briefly reviewed. In general, the usual problems that are the direct focus in various QSAR studies rely on a robust estimation of data location and scatter (covariance) under the presence of outliers. An appropriate estimation of data location and covariance is of great importance and is required in many methods which actively use the covariance matrix. To emphasize the scale of this problem, it is probably sufficient to mention at this point that principal components are the eigenvectors either of the covariance or correlation matrix and that the covariance matrix is required for the construction of the PLS model.

6.4.1. Robust Estimators of Data Location and Scatter

Among different robust data location estimators, the median is the simplest to define the robust center of a univariate distribution [3]. Even though the median is a highly robust estimator with the maximum breakdown point of 50%, it is not very efficient. In multivariate settings, several median generalizations can be used, including the L_1-median, also called the "spatial" median. The problem of estimating the L_1-median center of the data relies on finding a point, μ_{L1}, in the multivariate data space that minimizes the sum of Euclidean distances between this point and all of the data points [3]. The differences among the mean (*magenta dot*), the robust data center estimated using the coordinate-wise median (*blue dot*), and the L_1-median center (*red dot*) are illustrated in Figure 6-3 for a contaminated set of 30 objects described by two variables. The data set contains five outlying objects.

The L_1-median center can be obtained using an iterative procedure in which the following criterion is minimized:

$$\min_{\mu_{L1}} \sum_{i=1}^{n} \|\mathbf{x}_i - \mu_{L1}\| \tag{6-2}$$

where $\|\bullet\|$ is the L_1-norm and n is the number of variables.

The consecutive steps of the iterative approach can be summarized as follows:

(i) set the L_1-median as the median of the parameters, e.g., use the coordinate-wise median as the initial estimate;
(ii) center the data using the current estimates of the L_1-median, μ_{L1};
(iii) compute the weight for each data object as [Eq. (6-3)]:

$$w_i = \frac{\mathbf{x}_i}{\|\mathbf{x}_i\|} \tag{6-3}$$

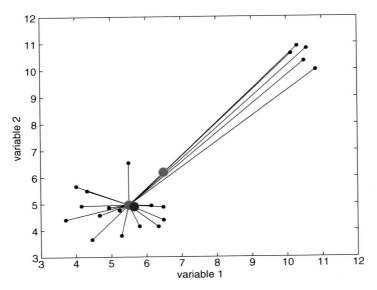

Figure 6-3. Illustration of the difference among three data location estimators applied to two-dimensional data set with five outlying observations: classic mean (*magenta dot*), median (*blue dot*), and L_1-median (*red dot*)

(iv) compute the new data center for *m* objects as [Eq. (6-4)]:

$$\hat{\mathbf{c}} = \frac{\text{diag}(\mathbf{w}) \cdot \mathbf{X}}{\sum\limits_{i=1}^{m} w_i} \qquad (6\text{-}4)$$

(v) check the convergence limit, *d* [Eq. (6-5)]:

$$d = \sum_{j=1}^{n} \left| \mu_{\text{L1}j} - \hat{c}_j \right| \qquad (6\text{-}5)$$

(vi) calculate new estimates of the L_1-median using Eq. (6-6):

$$\boldsymbol{\mu}_{\text{L1}} = \boldsymbol{\mu}_{\text{L1}} + \hat{\mathbf{c}} \qquad (6\text{-}6)$$

(vii) return to step (ii) if *d* is larger than the predefined limit.

The L_1-median estimator and the standard median have the highest possible breakdown point of 50%. It is worth noting that the L_1-median is reduced to the standard median in a one-dimensional space.

A parameter that describes data distribution or scale is the standard deviation. This estimator is also non-robust because it is highly influenced by arithmetic mean.

There are several robust variants of this estimator. The median of absolute deviation about the median, also called the median absolute deviation (MAD), and the Qn estimators are usually applied [19]. The robust scale estimator MAD is defined as:

$$\sigma_{MAD} = c_{MAD} \cdot \text{median}_i \cdot \left| x_i - \text{median}_j \left(x_j \right) \right| \qquad (6\text{-}7)$$

where c_{MAD} is the correction factor which is equal to 1.483. The correction factor is required to increase the efficiency of the MAD estimator for the uncontaminated data following the normal distribution. The MAD estimator is highly robust and relatively simple to compute, but in some applications its efficiency can be unsatisfactory. Therefore, another robust scale estimator called the Qn estimator [19] was proposed. It has a relatively high efficiency for normally distributed data and can deal with up to 50% of outliers in the data. The Qn estimator for a single variable is defined as the value of the element corresponding to the first quartile of the sorted absolute pair-wise differences between objects. More formally it can be expressed as:

$$\sigma_{Qn} = 2.2219 \cdot c_{Qn} \cdot \left\{ \left| x_i - x_j; \ i < j \right| \right\}_{(k)} \qquad (6\text{-}8)$$

where $k = \binom{h}{2} \approx \binom{n}{2}/4$ and $h = [n/2]+1$.

Similar to the MAD estimator, the value obtained is modified to achieve better efficiency with a normal distribution. In general, the constant factor, c_{Qn}, depends on the number of objects and tends to 1 with the increasing number of elements.

In contrast to the estimators already discussed, there are robust estimators that estimate data location and scatter simultaneously. These are the M estimator, MVT (multivariate trimming), MVE (minimum volume ellipsoid), the Stahel-Donoho estimator, and MCD (minimum covariance determinant) [3]. They provide robust estimates of data covariance.

6.4.2.   Robust Estimators for Multivariate Data Location and Covariance

Historically, the Stahel-Donoho estimator of multivariate data location and scatter was proposed first. This estimator, known as the "outlyingness-weighted median" [20, 21], is a highly robust and affine equivariant estimator. As its name suggests, the influence of outliers is discarded by down-weighting each object, $\mathbf{x}_i$, by a weight defined from the outlyingness measure, o_i, using a positive and decreasing weight function. The outlyingness of an object is determined as the maximum value obtained from the projection of an object onto a set of normalized directions $\mathbf{p}$:

$$o_i = \max_{\|p\|=1} \frac{\left| \mathbf{x}_i \mathbf{p}^T - \text{median} \left(\mathbf{x}_i \mathbf{p}^T \right) \right|}{\sigma_{MAD} \left(\mathbf{x}_i \mathbf{p}^T \right)} \qquad (6\text{-}9)$$

In the original version of the Stahel-Donoho estimator, MAD was used to estimate the scale of projections. However, MAD can be replaced by a more efficient estimator of scale, e.g., the Qn estimator. The concept of the Stahel-Donoho estimator is closely related to the principle of the projection pursuit method [22]. With the Stahel-Donoho estimator, one assumes that the outlying observations should be uncovered on some univariate projections. This assumption is very attractive, but in practice the computation of this estimator requires solving an optimization problem, which is a very time-consuming task. Consequently, only an approximated solution of the Stahel-Donoho estimator given by Eq. (6-9) is obtained [23]. The breakdown point of the estimator attains 50% when $m > 2n+1$. Using the outlyingness measure, a weight can be attributed to each object and in this way, a robust estimation of the multivariate data location and scatter [24, 25] is derived.

Another concept of a robust data covariance estimator was introduced in [26]. The MVT estimator, known as the multivariate trimming approach, is an iterative procedure of computing the Mahalanobis distance for each object to obtain a so-called "clean" subset of objects. The clean subset contains a specified fraction of objects with the smallest Mahalanobis distances that is used to obtain robust estimates of the data mean and covariance. The iterative procedure is continued while the mean of retained objects changes. The MVT estimator reaches convergence relatively quickly and is affine equivariant. In [21], it was reported that the MVT estimator has a breakdown point of at most $1/n$ and its robust properties greatly depend on the data dimensionality which is a serious drawback. When the number of data variables outnumbers the number of objects, which is often the case in the QSAR studies, the MVT estimator cannot be applied directly.

The minimum volume estimator (MVE) proposed in [27] is also a robust estimator of data location and covariance. With this affine equivariant estimator, the "clean" subset of p objects is found as a population of objects that define an ellipsoid of the smallest volume (where $p = m/2 + 1$). There are several algorithms for MVE [28–30]. The breakdown point of 50% of the MVT estimator is expected when the number of samples tends toward infinity. To achieve a better efficiency at the normal distribution, the covariance estimate obtained for the "clean" subset is multiplied by a suitable correction factor. The MVE estimator is computationally demanding for large data sets.

The minimum covariance determinant (MCD) estimator is yet another highly robust estimator of data covariance [3], which has gained much attention in recent years. MCD was frequently used to make robust variants of methods used in chemometrics [31]. Using MCD, the "clean" subset of objects with the covariance matrix of the smallest possible determinant is determined. The estimator has a breakdown point of 50%. It has a relatively good efficiency compared to its predecessors and its solution can be found relatively quickly using the FAST-MCD algorithm [25]. The MATLAB code of the FAST-MCD algorithm is available from [32]. Moreover, other estimators, which are more efficient than MCD, have been proposed in [33, 34].

The main steps of the MCD algorithm can be summarized as follows:
The following process is repeated through 500 iterations:

(i) select p objects randomly. The p value can be set by default as $0.5 \cdot (m+n+1)$, where m is the number of objects in the data and n is the number of variables in the data;

(ii) compute mean, covariance, and the Mahalanobis distances using a subset of p objects and perform the next step two times;

(iii) select p objects with the smallest Mahalanobis distances and on the basis of these objects compute the data mean, covariance, and the Mahalanobis distances for all of the objects;

(iv) retain ten subsets of objects, for which the determinant of the covariance matrix is the smallest;

(v) after 500 iterations, perform step (iii) on the best "clean" subset of objects as long as convergence is not reached;

(vi) use the "clean" subset of objects in order to detect outliers using diagnostics based on the Mahalanobis distance.

Similar to MVT, where the Mahalanobis distances are used to define the "clean" subset, the MCD estimator can only be computed if the number of objects in the "clean" subset exceeds the number of data variables. Otherwise, a data dimensionality reduction is required. A further increase of MCD efficiency can be gained using a weighting scheme [4]. In the re-weighted variant of MCD, only objects with the Mahalanobis distances below a definite cut-off value $\left(\sqrt{\chi^2_{n,\,0.975}} \right)$ receive weights equal to one, and thus, only they are considered in defining the final estimate of robust covariance matrix:

$$
w_i = \begin{cases} 1 & \text{if } MD_i \leq \sqrt{\chi^2_{n,0.975}} \\ 0 & \text{otherwise} \end{cases}
\tag{6-10}
$$

6.5. EXPLORING THE SPACE OF MOLECULAR DESCRIPTORS

The aim of the exploration of the space of molecular descriptors, **X**-space, is to reveal the similarities/differences among the molecules studied, to obtain information about the correlation among various descriptors, and eventually to investigate whether the data contain molecules that are very different in terms of the selected descriptors, in comparison with the remaining ones. Principal component analysis (PCA) and its robust variants can be applied successfully for the exploration of **X**-data. Therefore, the following section is devoted to these methods.

6.5.1. Classic Principal Component Analysis

PCA [35, 36] allows for a representation of multivariate data into a low-dimensional space spanned by new orthogonal variables, which are obtained as linear combinations of the original variables by maximizing the description of data variance. For any centered data matrix $\mathbf{X}_c$ (m, n), the PCA decomposition can be presented as

$$\mathbf{X_c} = \mathbf{TP}^\mathrm{T} \tag{6-11}$$

where $\mathbf{T}$ (m, r) is the matrix containing the data scores, $\mathbf{P}$ is the loading matrix of dimension (n, r), and r is the mathematical rank of the data that is equal to min (m, n).

Each orthogonal component (eigenvector) constructed is associated with its corresponding eigenvalue. The total sum of eigenvalues is equal to the total data variance. The eigenvalue of the ith eigenvector can be calculated as the sum of its squared score elements when the data are centered. Principal components are ordered according to the magnitude of the corresponding eigenvalues. The first component describes the largest part of the data variance and the consecutive ones account for a smaller part of the data variance than the preceding principal component. For data exploration, a plot of eigenvalues or a plot of the cumulative data variance described by the models of increasing complexity is a standard tool which helps to distinguish the significant components from the non-significant ones associated only with the data noise. The number of significant components, denoted as f, is usually much lower than the number of original variables, n, and expresses the degree of data compression. The higher the degree of correlation among the original variables the better the compression of the studied data set, which results in a smaller number of significant principal components. The number of significant PCs can be determined on the basis of a cross-validation procedure [36].

Projection of objects onto the plane defined by the main principal components (e.g., PC1 and PC2 or PC1 and PC3), the so-called score plot, allows the distribution of objects and their similarity or dissimilarity in the space of parameters to be studied. Analogous projection of data variables enables sub-groups of correlated and independent variables to be found. Simultaneous interpretation of score and loading projections gives the possibility to draw conclusions for the objects' structure in terms of the variables studied. In QSAR studies, the score matrix, $\mathbf{T}$, contains information about the studied molecules, while the loading matrix, $\mathbf{P}$, holds information about the studied molecular descriptors.

6.5.2. Robust Variants of Principal Component Analysis

Similar to any least square method, this basic tool of data compression and visualization is very sensitive to outliers (see Figure 6-4). One or a few outlying objects can greatly influence the results of PCA and the final data interpretation.

There are several robust variants of PCA that fall into one of the following three categories:

- Methods that make use of a robust covariance matrix. The classic covariance matrix is replaced by its robust estimate obtained using the robust estimates of data location and spread. A PCA decomposition of the robust covariance matrix provides a set of robust eigenvectors and eigenvalues.
- Methods that provide robust estimates of eigenvectors and eigenvalues directly without the need to obtain the robust estimate of the covariance matrix. Generally,

these methods are based on a projection pursuit search for a data structure in high-dimensional data. In fact, the experimental data are projected from a high-dimensional onto a lower-dimensional space by maximizing a robust measure of data spread called the projection index.

- Hybrid methods that combine the two above listed procedures for dealing with outliers.

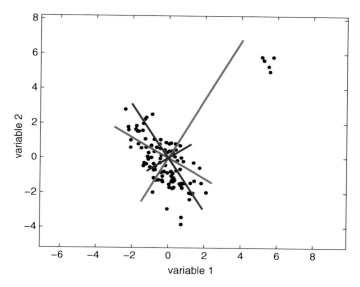

Figure 6-4. A projection of simulated data containing five outliers onto the two-dimensional space spanned by original variables, x_1 and x_2. The first two principal components constructed for the data are shown with *red lines* and the PCs obtained after removing the outliers are illustrated with *blue lines*

Some of the robust PCA methods can efficiently handle only a restricted number of variables. A comparison of different robust PCA methods can be found in [37, 38]. We will limit our presentation to three robust PCA procedures that are suitable for processing high-dimensional data. These PCA methods are selected to demonstrate the properties of one approach from each category.

6.5.2.1.　Spherical and Elliptical PCA

Spherical PCA (sPCA) [39] belongs to the first category of robust PCA methods. It is the simplest and most intuitively appealing approach. With sPCA, all objects receive weights proportional to the inverse of their distances to the robust center of the data. In this way, the potential influence of outliers is diminished. This is equivalent to a projection of all objects onto a sphere of unit radius measured from the robust data center (see Figure 6-5). The robust scores, $\mathbf{T}^w$, are then found by projecting centered $\mathbf{X}$-data onto the loadings obtained from the standard PCA applied to the weighted data, $\mathbf{X}^w$.

More formally, the sPCA algorithm can be summarized as follows:

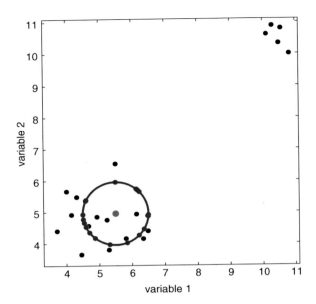

Figure 6-5. Projection of a two-dimensional data set containing 20 objects onto a sphere of unit radius centered at the L_1-median

(i) center the data, $\mathbf{X}$ (m, n), using the L_1-median estimator of data location;

(ii) calculate weights for all objects, defined as in Eq. (6-12):

$$w_i = \frac{1}{\sum_{j=1}^{n} (x_{ij})^2} \tag{6-12}$$

(iii) perform PCA on the weighted data matrix:

$$\mathbf{X}^{\text{w}} = \text{diag}(\mathbf{w})\,\mathbf{X} \tag{6-13}$$

$$\mathbf{X}^{\text{w}} = \mathbf{T}^{\text{w}}\mathbf{P}^{\text{wT}} \tag{6-14}$$

where operation "diag" transforms the vector of weights $\mathbf{w}$ into a diagonal matrix, the diagonal elements of which are the weights;

(iv) use loadings $\mathbf{P}^{\text{w}}$ to calculate robust scores:

$$\mathbf{T}^{\text{r}} = \mathbf{X}\mathbf{P}^{\text{wr}} \tag{6-15}$$

A variant of this method called elliptical PCA (ePCA) in which different scales of the data variables were taken into the account, has also been proposed. In this approach, the objects are projected onto a hyperellipse, the radii of which are proportional to the Qn scale estimator of each variable. The MATLAB code for estimation of the Qn scale was implemented in the LIBRA toolbox [40], which is available from [41].

6.5.2.2. *Projection Pursuit with the Qn Scale*

Many approaches based on projection pursuit (PP) have been presented, e.g., [42–47]. The main difference among them is in the type of projection index used. For instance, Li and Chen [43] applied the M estimator of scale, whereas Xie et al. [46] as well as Galpin and Hawkins [45] proposed optimizing the L_1-norm of the data variance as a robust measure of data spread. On the other hand, Xie et al. [46] applied the generalized simulated annealing method as an optimization algorithm to identify the global minimum, while Croux and Ruiz-Gazen [47] developed the C-R algorithm with the L_1-median for estimation of data center and the robust Qn scale estimator as a projection index. The main idea of the PP-Qn method is based on finding directions in the experimental space that maximize the Qn scale (a robust equivalent of the standard deviation). In the algorithm, first the data are centered about their L_1-median center and all objects are then projected onto a set of directions defined by the normalized vectors passing through the objects and data origin. Next the Qn scale is estimated for each projection and the direction with the maximum Qn value is selected as the first loading vector. The **X**-residuals are further analyzed as long as the desired number of factors is constructed. Since the considered directions are restricted to pass through the center and data points, a suboptimal solution might be obtained when the number of objects in the studied data set is small. A remedy for this problem is to add random directions to the data. The method described is easy to implement, the estimates are defined explicitly, good efficiency with a smooth and bounded influence function and the maximum breakdown point are achieved as well as a quick estimation of the first q eigenvectors without the need to compute them all.

6.5.2.3. *ROBPCA – A Robust Variant of PCA*

The ROBPCA method proposed in [48] belongs to the third category of robust PCA approaches. It is a hybrid procedure combining the idea of projection pursuit with the robust estimation of data location and covariance in a low-dimensional space. The PCA method is used in the preliminary step for data dimensionality reduction. Then the robust data center and covariance is found using the re-weighted MCD estimator in the reduced space of the projected samples. Finally, the estimates of data location and covariance are transformed back to the original data space and the robust estimates of multivariate data location and scatter are calculated. The ROBPCA method can be summarized in the following four steps:

(i) perform PCA for preliminary data dimensionality reduction;

(ii) compute the outlyingness measure (i.e., the projection index) for every object and construct the initial H-subset (H_0) containing h objects with the smallest outlyingness measure (the choice of h determines the robustness of the method and its efficiency; the default value of h is set to 75% of the total number of objects in the data);

(iii) perform a further data dimensionality reduction by projecting the data onto k-dimensional subspace spanned by the first k eigenvectors of the empirical covariance matrix obtained for objects in H_0;

(iv) compute the robust data center and covariance in the k-dimensional subspace and apply the re-weighted MCD estimator to the projected data.

The MATLAB implementation of the ROBPCA algorithm is available from [49]. The original algorithm of ROBPCA is designed to construct an optimal PCA subspace of a definite dimensionality, f. The solutions obtained are not nested, which means that the model with $f+1$ components should be recalculated. A faster version called ROBPCA-fmax, which handles data sets of dimension up to 100 and $f_{max} = 10$, was proposed by Engelen et al. [38].

6.6.   CONSTRUCTION OF MULTIVARIATE QSAR MODELS

In the majority of the QSAR studies, the number of descriptors used greatly outnumbers the number of available samples thereby increasing the possibility of obtaining a high correlation among descriptors. Construction of a model that describes the relationship of the highly correlated **X**-data with a property **y** (toxicity, biological activity) is then problematic when applying the classic multiple linear regression (MLR) approach, since the regression coefficients cannot be calculated. A possible remedy for this problem is to select several orthogonal variables either using some preliminary knowledge or using a variable selection scheme, e.g., the stepwise MLR approach. Another more general and efficient strategy to deal with the multicollinearity in **X**-data is to obtain a few orthogonal variables that describe the covariance between **X**-data and **y**. The partial least squares (PLS) regression has proved to be a successful tool for this purpose. There are several algorithms that can be used to construct a PLS model among which the non-iterative partial least squares (NIPALS) method [1] is the oldest. An improved variant of the classic PLS algorithm called SIMPLS [50] allows for the quick and efficient processing of a large number of descriptors by performing the calculations on the economic size of the **X**-matrix.

To make the presentation easier to understand, a description of the classic PLS algorithm is presented in the next section.

6.6.1.   Classic Partial Least Squares Regression

With the partial least squares (PLS) model, one aims to describe the linear relationship between a set of explanatory variables, **X** [1, 2] and a response variable, **y**. As mentioned before, PLS is capable of providing a solution when variables in the data are highly correlated. This is possible because the original variables are represented by a few new orthogonal latent factors, **T**, obtained by maximizing the covariance of **X** with **y**. The model can be presented mathematically as follows:

$$\mathbf{X} = \mathbf{TP}^{\mathrm{T}} + \mathbf{E} \qquad (6\text{-}16)$$

$$\mathbf{y} = \mathbf{Tq} + \mathbf{r} = \mathbf{Xb} + \mathbf{r} \tag{6-17}$$

In these equations, $\mathbf{X}$ is the original centered data matrix with m rows (molecules) and n columns (descriptors), $\mathbf{q}$ holds f regression coefficients associated with f PLS factors, $\mathbf{T}$. The residual matrix $\mathbf{E}$ represents the differences between observed and predicted $\mathbf{X}$, while $\mathbf{r}$ holds the differences between observed and predicted $\mathbf{y}$. The regression coefficients, $\mathbf{b}$, are obtained according to Eq. (6-18):

$$\mathbf{b} = \mathbf{W}(\mathbf{P}^{\mathrm{T}}\mathbf{W})^{-1}\mathbf{q} \tag{6-18}$$

in which $\mathbf{W}$ is the matrix of loadings maximizing the covariance criterion and $\mathbf{P}$ is the product of $\mathbf{X}$ and $\mathbf{T}$. As was already mentioned, the classic PLS regression estimator has a breakdown point of 0% and it provides an inadequate solution when outliers are present in the data. Therefore, robust versions of PLS are of great value.

6.6.2. Robust Variants of the Partial Least Squares Regression

There are several versions of robust PLS which differ in the way outliers are handled. Some of the first proposals for the robust PLS method [51] are considered as partially robust [52] since only some of the steps in the algorithm are made robust. This does not entirely guarantee the dealing with multivariate outliers properly. The first robust PLS method based on a robust sample covariance matrix and cross-covariance of $\mathbf{X}$ and $\mathbf{y}$ was presented in [52]. In order to derive the robust covariance matrices, the authors adopted the robust Stahel-Donoho estimator of data scatter with the Huber's weight function [16]. However, the method is computationally demanding which strongly limited its use. Therefore, the method did not gain popularity and was rather neglected. The next proposal was the iteratively re-weighted PLS (IRPLS) method [53]. As the name of the method suggests, the objects are iteratively weighted according to their residuals from the model. The authors of the method used various weight functions and evaluated their robust properties. They showed that with the use of the "fair" function, the highest breakdown point of 44% was obtained. The main drawback of this method is that only the outliers with respect to $\mathbf{y}$ are down-weighted, diminishing their influence on the model. In some applications this could be sufficient, but in general abnormal samples can also be found in the generated or experimentally obtained $\mathbf{X}$-data. A natural and necessary continuation in development of the robust PLS method is a method that is (i) capable of handling outliers in both $\mathbf{X}$ and $\mathbf{y}$ data, (ii) computationally fast, (iii) statistically efficient, and (iv) highly robust in terms of breakdown point. Several approaches possessing such properties are described in the next section.

6.6.2.1. Partial Robust M-Regression

The robust properties of partial robust M-regression (PRM) introduced in [54] are obtained by weighting the objects in a way which guarantees that the calibration model built is representative for the majority of the data. Two types of continuous weights are considered, namely the leverage and residual weights. For the ith object, the leverage weight w_i^x is defined for the f robust factors in the following way:

$$w_i^x = \Phi \left(\frac{\|\mathbf{t}_i - \boldsymbol{\mu}_{L1}(\mathbf{T})\|}{\text{median}_i \|\mathbf{t}_i - \boldsymbol{\mu}_{L1}(\mathbf{T})\|}, c \right) \qquad (6\text{-}19)$$

and

$$\Phi(z,c) = \frac{1}{\left(1 + \left|\frac{z}{c}\right|\right)^2} \qquad (6\text{-}20)$$

where $\|\bullet\|$ is the Euclidean norm, $\mathbf{t}_i$ represents f PLS scores for the ith object, "median" denotes the median estimate, μ_{L1} is the L_1-median robust estimator of the location [19], and c is the tuning constant which is equal to four for majority of applications [54].

The residual weights, w_i^r, are found according to Eq. (6-21):

$$w_i^r = \Phi \left(\frac{r_i}{\sigma}, c \right) \qquad (6\text{-}21)$$

In this Eq. (6-21), r_i is the residual element, i.e., the squared differences between observed and predicted response values for the ith object that are obtained from the model with f factors and σ is the MAD estimator of data scale [3] defined as:

$$\sigma = \underset{i}{med} \left| r_i - \underset{j}{med}(r_j) \right| \qquad (6\text{-}22)$$

Finally, the objects are weighted by the inverse of global weights, w_i, which are obtained as a combination of both weights:

$$w_i = \sqrt{w_i^x w_i^r} \qquad (6\text{-}23)$$

Similar to IRPLS, in PRM the weights are estimated iteratively as long as the algorithm convergence is not reached. The convergence criterion is fulfilled when the difference between the norm of the regression coefficients associated with the robust PLS factors, $\mathbf{q}$, of two consecutive steps is negligible, e.g., 10^{-2}. The main steps of the PRM algorithm can be summarized in the following way:

(i) initialize the global weights, w_i, according to Eqs. (6-19, 6-20, 6-21, 6-22, and 6-23);

(ii) build a PLS model for the $\mathbf{X}$ rows weighted using w_i;

(iii) calculate the residuals, r_i, for all objects according to the interim PLS model and update the values of the global weights; and

(iv) repeat the estimation from step (ii) to (iii) as long as the convergence criterion is not met.

The PRM routine is implemented in the recently presented toolbox for multivariate calibration (TOMCAT). The toolbox developed in the MATLAB environment is available from [55].

6.6.2.2. *Robust Version of PLS via the Spatial Sign Preprocessing*

Another concept to make the PLS method robust has been proposed by Serneels et al. [56]. The idea is to project the multivariate data objects onto a sphere of unit radius in the direction passing through the robust data center as illustrated in Figure 6-6 and then to construct a classic PLS model for the transformed data. This method is known as a spatial sign preprocessing PLS. Figure 6-6 shows an object (*black dot*) and its projection (*red dot*) onto the sphere with a center in the robust center of the data.

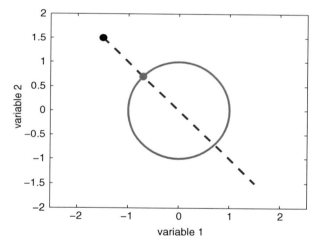

Figure 6-6. Projection of an object (*black dot*) with coordinates [−1.5 1.5] onto a unit sphere with center in the L_1-median. The *red dot* represents the object transformed with new coordinates given as [−0.71 0.71]

A spatial sign transformation is a method to obtain a robust covariance estimate, $\hat{\Sigma}$, and this can be expressed as

$$\hat{\Sigma} = \frac{1}{m-1} \sum_{i=1}^{m} \text{sgn}\,(\mathbf{x}_i - \mu_{L1})\,\text{sgn}\,(\mathbf{x}_i - \mu_{L1})^{T} \tag{6-24}$$

with the sign function defined as

$$\text{sgn}\,(\mathbf{x}_i) = \begin{cases} \mathbf{x}_i / \|\mathbf{x}_i\| & \text{if } \mathbf{x}_i \neq 0 \\ 0 & \text{if } \mathbf{x}_i = 0 \end{cases} \tag{6-25}$$

In Eq. (6-24) the L_1-median is used to center the data, but generally different estimators of data location can be applied. Compared to PRM and RSIMPLS, the method proposed is only moderately robust in terms of the breakdown point as this is confirmed by the simulation study presented in [57]. Nevertheless, this simple modeling approach has a number of very useful features. It can easily be implemented within the PLS framework with no extra computational cost. The model obtained is relatively efficient providing satisfactory estimates for the normally distributed data. Therefore, it can be considered as an attractive alternative to more computationally demanding robust PLS variants for data with a moderate data contamination.

6.6.2.3.   RSIMPLS and RSIMCD – Robust Variants of SIMPLS

The popularity of the SIMPLS algorithm [50] is due to its fast performance for wide-type data (where the number of variables exceed the number of samples) as found in many QSAR analyses. Two empirical covariance matrices are used in the algorithm. One is computed for the input $\mathbf{X}$ data, and the other one is the cross-covariance of $\mathbf{X}$ and $\mathbf{y}$ which makes the classic SIMPLS algorithm sensitive to outliers. To obtain the robust variant of SIMPLS, both covariance matrices have to be replaced with their robust counterparts. Two robust approaches, RSIMPLS and RSIMCD that consist of two steps were proposed. The step common to both methods is the use of ROBPCA to obtain robust scores. Specifically, this step is important to derive the robust cross-covariance matrix. Then the robust regression is carried out. This step is performed differently in both methods. Re-weighted multiple linear regression is used in RSIMPLS, while the MCD-based regression [57] is applied in RSIMCD. Both approaches are characterized by a breakdown point of 50% which makes them highly robust, but RSIMPLS is computationally faster in comparison with the RSIMCD approach.

6.6.3.   Outlier Diagnostics Using Robust Approaches

Another important issue to be discussed is how to identify the outlying samples. The simplest univariate approach is to compare the z-transformed values of objects x_i with a definite cut-off value, e.g., 2.5 [4]:

$$z_i = \frac{|x_i - median(\mathbf{x})|}{\sigma_{Qn}} \tag{6-26}$$

In this Eq. (6-26), the robust Qn estimator of data scale is used.

In the multivariate case, detection of outliers is based on the robust version of the Mahalanobis distance which is defined as the distance of each object to the robust center of the multivariate data. The robust variant of the Mahalanobis distance can be obtained when the location and scatter of multivariate data, $\mathbf{C}_R$, are derived using the robust estimator of data location and scatter, e.g., MCD.

$$MD_i = \sqrt{(\mathbf{x}_i - \boldsymbol{\mu})^{\mathrm{T}} \mathbf{C}_R (\mathbf{x}_i - \boldsymbol{\mu})} \tag{6-27}$$

The chi-square distribution with two degrees of freedom and a 97.5% confidence level is used as the cut-off value, e.g., $c = \sqrt{\chi^2_{2,0.975}}$. The confidence level defines the proportion of objects with the Mahalanobis distances below the cut-off level, c.

As was pointed out before, outlier identification is not an easy task because the masking and swamping effects can take place. When the data contain many outliers they can hide each other's influence and it can happen that some of them may be unnoticed if a non-robust approach is applied. On the other hand, some regular objects may be recognized as outliers as a result of the swamping effect.

Let us consider data presented in [58] containing 188 measured values of log K_{OW} and toxicity (log LC_{50}). A total of 13 outliers (*red dots* in Figure 6-7) were deliberately introduced into the data by setting log LC_{50} values to zero for those compounds for which LC_{50} was undetermined. Their negative influence upon the data covariance can be noticed immediately since the covariance ellipse (*red line*) drawn for a confidence level of 97.5% has changed its orientation toward them. However, after down-weighting the influence of outliers, the orientation of the covariance ellipse (*green line*) indicates a correct correlation.

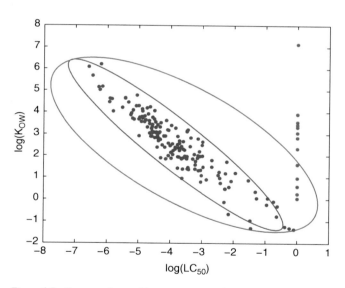

Figure 6-7. Two covariance ellipses *green* and *red lines* with a confidence level of 97.5% constructed for clean data (*green objects*) and contaminated data (*green and red objects*), respectively. The data contain 188 measured values of log LC_{50} and log K_{OW}

The identification of outliers is usually performed on the basis of the so-called distance–distance plot or outliers' map. Regardless of the technique used, robust PCA or robust PLS, both plots display the residuals from the robust model as a function of the corresponding Mahalanobis distance (a robust distance) computed in the space of robust scores. The cut-off values are defined differently for the residuals and for the robust distances. The cut-off line of the residuals is defined for their robust z-transformed values (see Eq. 6-26) and is set to 2.5 or 3, which corresponds to a confidence level of 97.5 or 99.9%. The cut-off line for the robust distances is determined using the chi-squared test with f degrees of freedom (models' complexity) and a confidence level, p, $c = \sqrt{\chi^2_{f,p}}$. Four types of objects can be distinguished with respect to their position in the space of f robust factors and residuals from the robust model in the plot (see Figure 6-8):

- regular objects that are below the corresponding cut-off lines, i.e., objects with short robust distances and small absolute residuals;
- high residual objects that are above the cut-off line of absolute residuals, i.e., objects with large absolute residuals from the model and short robust distances;
- good leverage objects are characterized by long robust distances, but small absolute residuals from the model; and
- bad leverage objects are those with long robust distances and large absolute residuals.

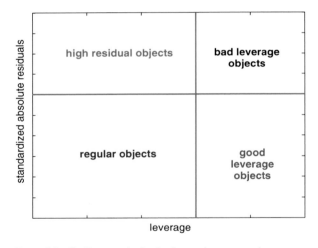

Figure 6-8. Outlier map obtained using a robust approach

6.7. EXAMPLES OF APPLICATIONS

6.7.1. Description of the Data Sets Used to Illustrate Performance of Robust Methods

Data set 1: inhibitors of HIV reverse transcriptase. Data set 1 contains the van der Waals and Coulomb interactions energies calculated between 208 inhibitors and the side chain and backbone parts of the selected 93 amino acid residues forming the reverse transcriptase binding pocket. Prior to calculation of the interaction energies, the inhibitors were docked into 16 binding pockets using a pharmacophore-based docking algorithm [9]. The crystal structures of HIV-RT complexed with several inhibitors were mostly taken from the Protein Data Bank [59]. A total of 208 inhibitors successfully bind in 10 of the 16 binding pocket structures with interaction energies below -30 kcal·mol^{-1}. The van der Waals and Coulomb interaction energies were averaged resulting in a data table with the dimensions 208×372. The weakest interactions in terms of energies were removed and the final data set contains 54 columns. The data set was described in [8] in detail and can be obtained from [60].

Data set 2: group of baseline toxicity aquatic pollutants. Data set 2 contains 50 compounds that are a subset of a large collection of xenobiotics known as aquatic pollutants [58]. The aquatic pollutants belong to the same groups of chemical compounds including phenols, anilines, and mononitrobenzenes. Each compound is described by 11 molecular descriptors, including an energy level of the highest and the lowest occupied orbital, electronegativity, hardness, a dipole moment, polarizability, solvent accessible molecular surface area, molecular volume, the most negative charge and the most positive charge on any non-hydrogen atom, and log K_{OW}. The acute toxicity for all of the compounds, a dependent variable, was determined and reported as log LC$_{50}$.

6.7.2. Identification of Outlying Molecules Using the Robust PCA Model

Projection of inhibitors on the plane defined by the first two robust scores obtained using the C-R algorithm is presented in Figure 6-9a. Two groups of inhibitors and a few inhibitors that are relatively far away from the groups can be distinguished. The natural grouping of inhibitors is due to a different mechanism of binding to the active site of the HIV reverse transcriptase. Inhibitor no. 83 is close to the blue-type inhibitors (e.g., inhibitors nos. 149, 160, and 187), while inhibitors nos. 149, 160, and 187 are more distant from the majority of the blue-type inhibitors. Using the robust loading plot, see Figure 6-9b, the contribution of original data variables to the construction of robust principal components can also be studied. Variable no. 20, which describes the van der Waals interaction energies computed between the side chain of amino acid no. 318 and inhibitors, has the largest absolute loading value and consequently plays an important role in explaining the docking behavior of the blue- and the red-type of inhibitors along the first robust PC. Compared with the red-type inhibitors, the blue-type inhibitors interact with the side chain of amino acid no. 318 via the van der Waals interaction more strongly.

In general, some of the inhibitors may be incorrectly docked into the reverse transcriptase binding pocket, since the docking of molecules is not an easy task. However, such information can hardly be deduced using the classic score plot. Therefore, construction of an outlier map that takes into account the location of an object in the space of the robust PCA model and its residuals from this model is required. The outlier map is obtained for a definite number of robust principal components which can be selected in various ways [61]. Here, the complexity of the robust PCA model was decided on the basis of a scree plot of robust eigenvalues (see Figure 6-10a). Eight robust principal components can be considered for the

B. Walczak et al.

example studied. The outlier map shown in Figure 6-10b indicates the presence of six high residual objects, i.e., inhibitors (nos. 2, 80, 83, 179, 185 and 198) that do not fit the robust PCA model well. All of them have high residuals from the robust PCA model. Inhibitor no. 140 can be regarded as a good leverage object because it

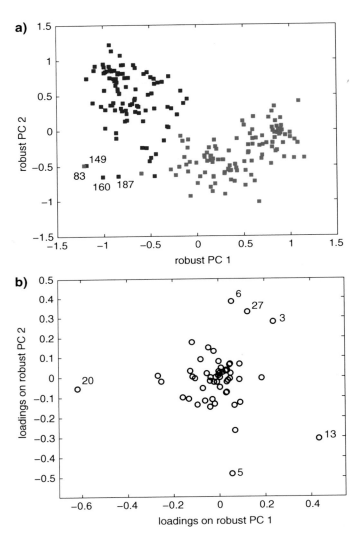

Figure 6-9. Results of the robust principal component analysis: (**a**) projection of two types of inhibitors (*blue and red*) on the plane defined by the first two robust principal components and (**b**) projection of variables onto the first two robust PCs

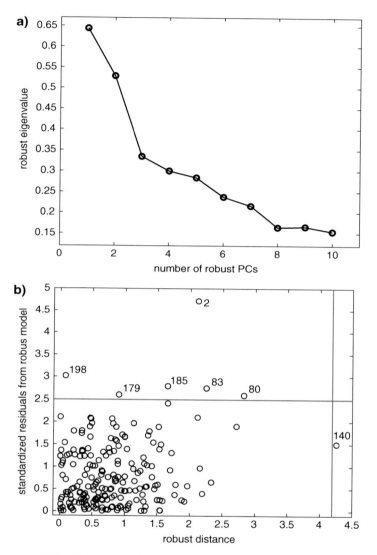

Figure 6-10. (a) Scree plot of the first ten robust eigenvalues and (b) outlier map constructed for eight robust PCs

is located relatively far from the data majority in the robust score space, but it still fits the model having a standardized absolute residual value below the cut-off line.

It should be emphasized that many outliers may be neglected due to possible masking and swamping effects, but their presence will be revealed when the outlier map is built for the robust model.

6.7.3.   Construction of the Robust QSAR Model with the PRM Approach

Let us consider that a calibration model is required for data set 2. First, the data were split into a calibration and a test set. The Kennard and Stone algorithm [62] was adopted in order to include all sources of the data variance into the calibration set. Using this algorithm, the samples of the calibration set are selected so that they cover the experimental domain uniformly. A total of 40 objects were included into the calibration set and the remaining ten samples formed the test set used to test the predictive power of the model. Here the performance of the PRM method will be presented due to its conceptual simplicity. To demonstrate the efficiency of PRM, it is compared with the classic PLS model. From a practical point of view, the performance of the robust model should be virtually the same as the classic approach for uncontaminated data. In this context, classic PLS and its robust counterpart PRM perform equally well in terms of prediction error for the studied data set with four factors as shown in Figure 6-11.

The optimal complexity of the PLS and PRM models is found by the Monte Carlo cross-validation approach. With the Monte Carlo cross-validation procedure,

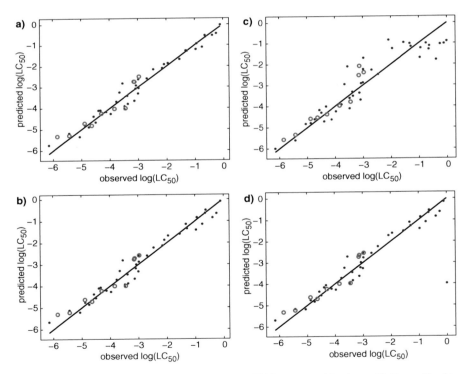

Figure 6-11. Two calibration models: (**a**) PLS and (**b**) PRM constructed for data set 2. Two calibration models: (**c**) PLS and (**d**) PRM constructed for contaminated data set 2 with calibration samples colored in blue and the test set samples shown with *red circles*

a number of objects is selected randomly at each step and omitted. Then the values for the omitted objects are predicted using the PLS model built for the remaining objects. In this way, calibration models of different complexities are built. This procedure is repeated k times, and the k prediction errors obtained for models of definite complexity are averaged. To estimate the prediction error of contaminated data with PRM, the so-called trimming procedure is adopted, which means that the error estimate is obtained after removing the largest fraction of residuals, e.g., 5%.

For the studied data, the Monte Carlo procedure was repeated 160 times. The leave-10-objects-out scheme was used in the case of the PLS model, while the leave-20-objects-out procedure with the trimming fraction of 5% was used in PRM. The root mean square error (RMSE) of calibration and the root mean square error of prediction (RMSEP) obtained from both models are given in Table 6-1.

Table 6-1. Results obtained from PLS and PRM models of f factors which are constructed for clear and contaminated data set 2. The trimming fraction of PRM is assumed to be 5%

Data	Method	f	RMSE	RMSECV	RMSEP
Clean data	PLS	4	0.27	0.27	0.35
	PRM	4	0.28	0.36	0.35
Contaminated data	PLS	2	0.61	0.63	0.47
	PRM	4	0.29	0.40	0.36

Of course when the calibration data contain outliers, the robust modeling technique outperforms the classic approach. To illustrate this property, a single outlier was deliberately introduced into the calibration set and the PLS and PRM models were again constructed. The same input settings were used as before, i.e., the same number of Monte Carlo iterations and the same number of objects to be left out. However, the complexity of PLS was now found to be two. Figure 6-11c illustrates a clear deterioration of the classic model properties in terms of fit and prediction power, while better results are obtained with the robust approach. The PRM model has again complexity of four. Figure 6-11d shows that compound no. 2 has a very large residual from the model, and it can be easily distinguished from the remaining compounds. This distinction is impossible when the classic model is constructed (see Figure 6-11c) since the model is highly influenced by this compound.

With the PRM model, it is possible to distinguish among the different types of molecules with respect to their potential influence upon the model. This can be done either using the weights attained during the model's construction or by analyzing the outlier map. Even though the plots are constructed differently, the conclusions about the outlying character of objects are similar.

The leverage and residual weights of objects that are used to construct the model are shown in Figure 6-12a. Object no. 2 has the smallest weight since its highest influence on the model is the most diminished. The outlying character of this object is mainly apparent in the space of the model's residuals. Using the outlier

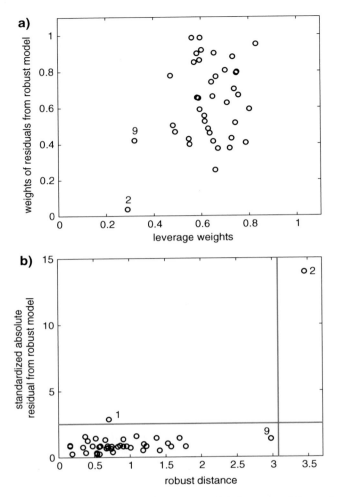

Figure 6-12. (**a**) Plot representing the inverse of the leverage weights vs. the inverse of residual weights used to down-weight the negative influence of outliers in the four-factor PRM model constructed for contaminated data set 2 and (**b**) an outliers map with the horizontal and vertical cut-off lines set for the normal and χ^2 distribution with four degrees of freedom and a confidence level of 97.5%

map shown in Figure 6-12b for the PRM model, it can also be deduced that object no. 2 has the largest outlying character since its robust distance and its standardized value of the residual from the robust model are very large and their values exceed the defined cut-off lines. According to the outlier categorization, object no. 2 is a bad leverage object, while object no. 1 is a high residual object. The standardized absolute residual value for the latter object exceeds only the horizontal cut-off line. The presence of good leverage objects in the data has a positive impact on the model since they can reinforce the model extending its calibration range (see Figure 6-1). For the studied data, object no. 9 is a good leverage observation because it is relatively far away from majority of the data, but still fits the robust model (see Figure 6-12b).

6.8. CONCLUDING REMARKS AND FURTHER READINGS

In general, modeling of multivariate contaminated data and the identification of unique molecules are not easy tasks. As has been presented in this chapter, a successful identification of outliers can only be expected with robust techniques that are specially designed to describe the majority of data well. The robust approach chosen for this purpose should be highly efficient, which guarantees that such methods have virtually the same performance as the classic approaches for uncontaminated data. Owing to the attractive properties of robust techniques, they are highly valued and well-suited for handling various QSAR data in which outliers are expected to be present. In particular, robust data exploration of molecular descriptors space, identification of outlying molecules, and construction of robust calibration and classification/discrimination models are of great interest. Some of these applications have been demonstrated in this chapter.

To date many handbooks [3, 16, 17] have presented the principles of robust statistics and robust methods. During the last decade, the robust techniques have been intensively popularized in the chemical sciences and their usefulness seems to be acknowledged already. For further reading, there are a number of tutorial papers and book chapters [4, 5, 63–65] that present a detailed overview of the robust methods applied in chemical sciences.

REFERENCES

1. Martens H, Næs T (1989) Multivariate calibration. John Wiley & Sons, Chichester
2. Næs T, Isaksson T, Fearn T, Davies T (2002) Multivariate calibration and classification. NIR Publications, Chichester
3. Rousseeuw PJ, Leroy AM (1987) Robust regression and outlier detection. John Wiley & Sons, New York
4. Rousseeuw PJ, Debruyne M, Engelen S et al. (2006) Robustness and outlier detection in chemometrics. Crit Rev Anal Chem 36:221–242
5. Walczak B, Massart DL (1998) Multiple outlier detection revisited. Chemom Intell Lab Syst 41:1–15
6. Todeschini R, Consonni V (2000) Handbook of molecular descriptors. Wiley, New York
7. Cramer RD, Patterson DE, Bunce JD (1988) Comparative molecular field analysis (CoMFA). Effect of shape on binding of steroids to carrier proteins. J Am Chem Soc 110:5959–5967
8. Daszykowski M, Walczak B, Xu QS et al. (2004) Classification and regression trees – Studies of HIV reverse transcriptase inhibitors. J Chem Inf Comput Sci 44:716–726
9. Daeyaert F, de Jonge M, Heeres J et al. (2004) A pharmacophore docking algorithm and its application to the cross-docking of 18 HIV-NNTI's in their binding pockets. Protein Struct Funct Genet 54:526–533
10. Wehrens R, de Gelder R, Kemperman GJ et al. (1999) Molecular challenges in modern chemometrics. Anal Chim Acta 400:413–424

11. Kim KH (2007) Outliers in SAR and QSAR: 2. Is a flexible binding site a possible source of outliers. J Comput Aided Mol Design 21:421–435

12. Lipnick RL (1991) Outliers: their origin and use in the classification of molecular mechanisms of toxicity. Sci Tot Environ 109/110:131–153

13. Kim KW (2007) Outliers in SAR and QSAR: Is unusual binding mode a possible source of outliers. J Comput Aided Mol Design 21:63–86

14. Hampel FR (1971) A general definition of qualitative robustness. Ann Mat Stat 42:1887–1896

15. Hampel FR (1974) The influence curve and its role in robust estimation. Annal Stat 69:383–393

16. Huber PJ (1981) Robust statistics. John Wiley & Sons, New York

17. Maronna RA, Martin RD, Yohai VJ (2006) Robust statistics. John Wiley & Sons, Chichester

18. Croux C, Ruiz-Gazen A (2005) High breakdown estimators for principal components: The projection-pursuit approach revisited. J Mul Anal 95:206–226

19. Rousseeuw PJ, Croux C (1993) Alternatives to Median Absolute Deviation. J Am Stat Assoc 88:1273–1283

20. Stahel WA (1981) Robust estimation: infinitesimal optimality and covariance matrix estimators. PhD Thesis, ETH, Zürich

21. Donoho DL (1982) Breakdown properties of multivariate location estimators. PhD Qualifying paper, Harvard University

22. Friedman JH, Tukey JW (1974) A projection pursuit for exploratory data analysis. IEEE Trans Comput 23:881–889

23. Maronna RA, Yohai VJ (1995) The behaviors of the Stahel-Donoho robust multivariate estimator. J Am Stat Assoc 90:330–341

24. Croux C, Haesbroeck G (1999) Influence function and efficiency of the minimum covariance determinant of scatter matrix estimator. J Mul Anal 71:161–190

25. Rousseeuw PJ, Van Driessen K (1999) A fast algorithm for minimum covariance determinant estimator. Technometrics 41:212–223

26. Gnanadesikan R, Kettenring JR (1972) Robust estimates, residuals, and outlier detection with multiresponse data. Biometrics 28:81–124

27. Rousseeuw PJ (1985) Multivariate estimation with high breakdown point. In: Grossmann W, Pflug G, Vinche I (eds) Mathematical statistics and applications, Vol. B. Reidel, Dordrecht

28. Woodruff DL, Rocke DM (1993) Heuristic search algorithms for the minimum volume ellipsoid. J Comput Graph Stat 2:69–95

29. Cook RD, Hawkins DM, Weisberg S (1992) Exact iterative computations of the robust multivariate minimum volume ellipsoid estimator. Stat Prob Lett 16:213–218

30. Agulló J (1996) Exact iterative computation of the multivariate minimum volume ellipsoid estimator with a branch and bound algorithm. In: Prat A (ed) Computational statistics. Physica-Verlag, Heidelberg

31. Hubert M (2006) Robust calibration. In: Gemperline P (ed) Practical guide to chemometrics. Taylor & Francis, London

32. ftp://ftp.win.ua.ac.be/pub/software/agoras/newfiles/fastmcdm.gz. Accessed on the 16th of August 2009

33. Rousseeuw PJ, Yohai VJ (1984) Robust regression by means of S-estimators. In: Franke J, Härdle W, Martin D (eds) Robust and nonlinear time series. Lecture notes in statistics, vol 26. Springer, New York, pp 256–272

34. Yohai VJ (1987) High breakdown-point and high efficiency robust estimates for regression. Annal Stat 15:642–656

35. Wold S, Esbensen K, Geladi P (1987) Principal component analysis. Chemom Intell Lab Syst 2:37–52

36. Malinowski ER (1991) Factor analysis in chemistry. John Wiley & Sons, New York

37. Stanimirova I, Walczak B, Massart DL et al. (2004) A comparison between two robust PCA algorithms. Chemom Intell Lab Syst 71:83–95
38. Engelen S, Hubert M, Vanden Branden K (2005) A comparison of three procedures for robust PCA in high dimensions. Austrian J Stat 34:117–126
39. Locantore N, Marron JS, Simpson DG et al. (1999) Robust principal component analysis for functional data. Test 8:1–73
40. Verboven S, Hubert M (2005) LIBRA: a MATLAB library for robust analysis. Chemom Intell Lab Syst 75:127–136
41. http://wis.kuleuven.be/stat/robust/Libra.html. Accessed on the 16th of August 2009
42. Ruymgaart FH (1981) A robust principal analysis. J Mul Anal 11:485–497
43. Li G, Chen ZL (1985) Projection-pursuit approach to robust dispersion matrices and principal components: Primary theory and Monte Carlo. J Am Stat Assoc 80:759–766
44. Amman LP (1993) Robust singular value decompositions: A new approach to projection pursuit. J Am Stat Assoc 88:505–514
45. Galpin JS, Hawkins DM (1987) Methods of L1 estimation of a covariance matrix. Comput Stat Data Anal 5:305–319
46. Xie YL, Wang JH, Liang YZ et al. (1993) Robust principal component analysis by projection pursuit. J Chemometr 7:527–541
47. Croux C, Ruiz-Gazen A (1996) A fast algorithm for robust principal components based on projection pursuit. In: Prat A (ed) Compstat: Proceedings in computational statistics. Physica-Verlag, Heidelberg
48. Hubert M, Rousseeuw PJ, Vanden Branden K (2005) ROBPCA: A new approach to robust principal component analysis. Technometrics 47:64–79
49. http://wis.kuleuven.be/stat/robust/Libra.html. Accessed on the 16th of August 2009
50. de Jong S (1993) SIMPLS: An alternative approach to partial least squares. Chemom Intell Lab Syst 42:251–263
51. Wakeling IN, Macfie HJH (1992) A robust PLS procedure. J Chemometr 6:189–198
52. Gil JA, Romera R (1998) On robust partial least squares (PLS) methods. J Chemometr 12:365–378
53. Cummins DJ, Andrews CW (1995) Iteratively reweighted partial least squares: a performance analysis by Monte Carlo simulation. J Chemometr 9:489–507
54. Serneels S, Croux C, Filzmoser P et al. (2005) Partial Robust M-regression. Chemom Intell Lab Syst 79:55–64
55. Daszykowski M, Serneels S, Kaczmarek K et al. (2007) TOMCAT: A MATLAB toolbox for multivariate calibration techniques. Chemom Intell Lab Syst 85:269–277
56. Serneels S, De Nolf E, Van Espen PJ (2006) Spatial sign preprocessing: a simple way to impart moderate robustness to multivariate estimators. J Chem Inf Model 3:1402–1409
57. Hubert M, Vanden Branden K (2003) Robust methods for partial least squares regression. J Chemometr 17:537–549
58. Verhaar HJM, Ramos EU, Hermens JLM (1996) Classifying environmental pollutants. 2: separation of class 1 (baseline toxicity) and class 2 ('polar narcosis') type compounds based on chemical descriptors. J Chemometr 10:149–162
59. http://www.rcsb.org/pdb/home/home.do. Accessed on the 16th of August 2009
60. http://pubs.acs.org/doi/suppl/10.1021/ci034170h/suppl_file/ci034170hsi20031126_085339.txt Accessed on the 16th of August 2009
61. Hubert M, Engelen S (2004) Fast cross-validation of high-breakdown resampling methods for PCA. In: Antoch J (ed) Proceedings in computational statistics. Springer-Verlag, Heidelberg
62. Kennard RW, Stone LA (1969) Computer aided design of experiments. Technometrics 11:137–148
63. Liang YZ, Kvalheim OM (1996) Robust methods for multivariate analysis—a tutorial review. Chemom Intell Lab Syst 32:1–10

64. Møller SF, von Frese J, Bro R (2005) Robust methods for multivariate data analysis. J Chemometr 19:549–563
65. Daszykowski M, Kaczmarek K, Vander Heyden Y et al. (2007) Robust statistics in data analysis – a review. Basic concepts. Chemom Intell Lab Syst 85:203–219

CHAPTER 7

CHEMICAL CATEGORY FORMATION AND READ-ACROSS FOR THE PREDICTION OF TOXICITY

STEVEN J. ENOCH

School of Pharmacy and Chemistry, Liverpool John Moores University, Liverpool L3 3AF, England, e-mail: s.j.enoch@ljmu.ac.uk

Abstract: The aim of this chapter is to outline the principles of chemical category formation and the use of read-across methods to fill data gaps to aid regulatory toxicological decision making. The chapter outlines the Organisation for Economic Co-operation and Development (OECD) principles for the design of a chemical category. This section aims to give a flavour of the steps that need to be considered when forming a chemical category. This is followed by a description of the advantages that considering chemicals within categories bring in risk assessment. The importance of how to define chemical similarity and several commonly used methods is discussed. Finally a brief review of the limited literature available showing actual examples of read-across methods is presented

Keywords: Chemical categories, Read-across

7.1.   INTRODUCTION

Chemical category formation and subsequent read-across analysis have been suggested as being essential if the objectives of REACH are going to be achieved without the excessive use of animals [1, 2]. The use of the chemical category approach is already common in a number of regulatory environments outside of the European Union namely in the United States and Canada. In terms of the Organisation for Economic Co-operation and Development (OECD) a chemical category has been defined as "a group of chemicals whose physiochemical and toxicological properties are likely to be similar or follow a regular pattern as a result of structural similarity, these structural similarities may create a predictable pattern in any or all of the following parameters: physicochemical properties, environmental fate and environmental effects, and human health effects" [2]. On a practical level, this process involves treating a closely related (or similar) group of chemicals as a category. Within the category toxicological data will exist for some, but not all of the chemicals for the endpoints of interest. Thus data gaps are likely to exist for some of the properties or endpoints for each chemical, with it being likely that differing

209

T. Puzyn et al. (eds.), Recent Advances in QSAR Studies, 209–219.
DOI 10.1007/978-1-4020-9783-6_7, © Springer Science+Business Media B.V. 2010

data gaps will exist for different chemicals within the category. It is for these data gaps that structure–activity relationship methods (such as read-across) will have to be utilised to make predictions for the missing toxicological data.

7.2.   BENEFITS OF THE CATEGORY FORMATION

The recent OECD documentation detailing category formation highlighted a number of key benefits of the approach when applied to regulatory decision making about the safety of chemicals [2]. These can be summarised as follows:

1. Animal testing is reduced by the interpolation and/or extrapolation to other chemicals in the category. The use of existing data further reduces the need for additional testing.
2. Evaluation of chemicals using a category approach involves the use of a greater volume of data than assessing chemicals individually (as has been carried out in the past).
3. Development of a category aids the evaluation of chemicals which otherwise might not be assessed.
4. Chemicals which might not be able to be assessed in standard animal protocols can be investigated using the category approach [3, 4].
5. The category approach has the potential to aid in the risk assessment of chemicals for which animal tests do not reliably predict effects in humans [4].

As a practical benefit of the utilisation of such category approaches, the US EPA needed to conduct new animal tests for only 6% of 1257 chemicals assessed as part of the High Production Volume Challenge (HPVC) Program [5]. In this programme, existing data were available for 50% of the chemicals; a further 44% of the data required was estimated using methods such as read-across.

7.3.   CHEMICAL SIMILARITY

The fundamental requirement for category formation is the ability to assess how similar a group of chemicals are that might form a category. Unfortunately no single measure of chemical similarity exists which can be universally applied across any endpoint. Instead one can consider a number of general approaches that have been suggested to be beneficial in the formation of a chemical category, with each one of them trying to ensure that for differing scenarios the resulting category contains chemicals acting via the same mechanism of action [2].

The first of these methods, and perhaps the simplest, is based upon forming a category around a common functional group such as an aliphatic aldehyde or aromatic ketone, the so-called "common functional group approach". The second approach, generally suitable for categories dealing with physicochemical properties such as boiling point, aims to make use of simple counts of carbon chains lengths.

The third and fourth methods are more complex and aim to deal with category formation for complex mixtures and metabolically related chemicals. In terms of complex substances or biological material in which a single chemical substance does not exist, it has been suggested that common constituents, similar carbon ranges or chemical class are likely to be useful in the formation of suitable categories. Such substances are referred to as "substances of unknown or variable composition, complex reaction products or biological material" (UVCB). Finally, chemicals can be grouped into a common category if they have a common precursor and/or common breakdown products; this can be thought of as the metabolic pathway category approach.

A related approach to chemical similarity that has been suggested to form useful chemical categories is the "mechanism-based approach", with it being suggested that a number of toxicological endpoints can be understood in terms of a common initialising event, usually a chemical reaction between an electrophilic chemical and a nucleophilic side chain in either amino or nucleic acids. A number of authors have documented such approaches [6–8].

Finally, it has been suggested that chemoinformatic approaches are able to form useful categories, especially in the identification of less obvious analogues from larger data sets [9]. Such methods rely upon the use of computational indices to encode structural information about chemicals; these indices can then be compared and chemicals within a certain distance located [10].

Given the numerous methods for developing chemical categories, it is unlikely a single method will always be the most appropriate, in contrast it being likely that more than a single method will be utilised in the formation of a single category. For example, a suitable category might be formed by the combination of assigning chemicals to a single electrophilic mechanism and then further restricting the chemicals within the category by the length of the carbon chains. Such decisions need to be made based on category by category basis with constant reference to the available experimental data. The aims of the remainder of this chapter are to highlight a general method by which chemical categories can be formed. In addition, the chapter will draw several examples from the literature to illustrate the differing ways of forming a chemical category, highlighting examples in which read-across has been used to fill data gaps.

7.4. GENERAL APPROACH TO CHEMICAL CATEGORY FORMATION

The recently published OECD guidelines for chemical category formation outlined nine steps required for the robust definition of a chemical category [2]. The first of these is to consider whether the chemical/chemicals of interest have already been assigned a category by other workers. A number of organisations provide resources for existing chemical categories for high volume chemicals, including the US EPA, OECD and the UN [11–13]. Assuming that the chemical of interest has not already been assigned to a category, eight further steps are suggested by the OECD; briefly these are

1. Development of a category hypothesis as the basis for the grouping of the chemicals. This definition should fully document the chemicals (names, structures)

and the endpoints that the category is applicable to. Care should be taken to fully document the structural domain that the category is applicable to. This definition covers molecular features such as chain lengths, molecular weight ranges, and the types of chemicals which should be included or excluded.

2. Gather data for each category member. This step involves the acquisition of all available toxicological and physicochemical data for each of the category members.
3. Evaluation of the quality and adequacy of the data available for each category member.
4. Construction of a data matrix showing the available data and crucially identifying gaps in the available data.
5. Evaluate the category hypothesis and if possible perform read-across to fill data gaps. This step aims to ensure that the hypothesis put forward in step 1 is fully valid and if so, and provided sufficient data exist, then the missing data in the data matrix be filled using appropriate read-across methods. Crucially if the data gathered in step 3 cannot or do not support the hypothesis proposed in step 1, then an alternate category might be required.
6. Should the data in step 5 support the category hypothesis but be insufficient for one or more of the endpoints covered by the category, further testing might need to be undertaken. Such testing should be designed in order to minimise animal usage whilst maximising information content.
7. If additional testing has been undertaken then a further assessment of the category should be undertaken. This is essentially a repeat of step 5.
8. If the category assessment is found to acceptable then the new category should be fully documented according to the OECD guidelines [2].

A common way to view the data matrix and how read-across methods might be used to fill any gaps in the data matrix is shown in Table 7-1.

Table 7-1. Schematic representation of data matrix required for a chemical category (X represents data points which are known and O represents missing data)

	Chemical 1	Chemical 2	Chemical 3	Chemical 4
Property 1	X	O	X	X
Property 2	O	X	X	O
Endpoint 1	X	O	X	X
Endpoint 2	X	X	O	X

7.5. EXAMPLES OF CATEGORY FORMATION AND READ-ACROSS

The above guidelines show the idealised methodology that should be employed in a regulatory environment for the formation of a chemical category. The remainder of the chapter will highlight studies in the literature into the development of categories of chemicals and then, in some cases, to perform read-across within these categories.

The focus of these sections is the illustration of three types of similarity method that can be used to aid category formation: these being chemical class, common mechanism of action and chemoinformatic approaches.

7.5.1. Chemical Class-Based Categories

A recent study utilised the category approach to assess the developmental toxicity of a group of phthalates esters with varying side carbon chain lengths [14]. The study used a further five phthalate esters of differing benzene substitution patterns and chain lengths to test the category hypothesis. The authors showed that differences in physicochemical properties, absorption rates or metabolism between the phthalate esters could not explain the differing reproductive toxicity profiles. The analysis of the chemicals in the study enabled a strict definition of the applicability domain of the category to be made, this being *ortho*-phthalate esters with carbon chain lengths between four and six carbons. The authors suggested that such chemicals acted via a common mechanism of action, most likely through binding to the anti-androgenic receptor. The study highlighted the use of both a chemical class and chain length restrictions in the formation of a suitable category. In addition, it showed that a clear mechanistic rationale could be offered for a complex endpoint within a well-defined chemical category.

7.5.2. Mechanism-Based Categories

A number of authors have demonstrated the use of mechanistic categories (rather than chemical class-based categories) for skin sensitisation and acute fish toxicity [6–8, 15–18]. Research has suggested that five principle organic chemistry mechanisms can be used as the basis for categorisation [15]. Briefly these mechanisms involve the attack by nucleophilic amino acid side chains (typical sulphur or nitrogen) on electrophilic fragments of potentially toxic chemicals; the mechanisms are summarised in Figure 2-1. Methods to enable chemicals to be assigned to these so-called reactive mechanisms have been published in the literature [19, 20] and included in the OECD (Q)SAR Application Toolbox which is freely available from the OECD website.

Additional studies have highlighted the ability of both QSAR and read-across methods to fill data gaps within these reactive mechanisms for both skin sensitisation and acute fish toxicity [18, 21–24]. One recent study demonstrated the utility of a computational measure of electrophilicity in making quantitative read-across predictions for a series of skin sensitising chemicals within the Michael mechanistic domain [21]. The study suggested the following methodology should be used to make a prediction for a "query chemical":

1. Calculate the electrophilicity for a database of chemicals in the Michael mechanistic domain with known EC3 values. The database was ranked based on electrophilicity (Table 7-1).

2. Select the two closest chemicals to the "query chemical" in terms of electrophilicity, one with a lower electrophilicity value, the other a higher electrophilicity value. Given that the database was ranked by electrophilicity the closest chemical with lower electrophilicity would be the chemical immediately preceding the "query chemical", whilst the closest chemical with greater electrophilicity would be the one immediately following the "query chemical". For example, to make the prediction for chemical 3 in Table 7-2, chemicals 2 and 4 would be chosen.

3. Linear extrapolation between electrophilicity and pEC3 using the two closest chemicals selected in step 1 allows a prediction to be made for the "query chemical". This step is equivalent to plotting electrophilicity against pEC3 for the two closest chemicals and using the electrophilicity value of the "query chemical" to predict its pEC3 value.

4. The predicted pEC3 value is then converted into an EC3 value.

Examples of the predictions possible from this methodology are shown in Table 7-2.

Table 7-2. Examples of read-across predictions made using the method described in the text. NP means a prediction has not been made as there is not a chemical more electrophilic (larger ω) or less electrophilic (smaller ω) in this small, four-chemical, example database

Name	Structure	Experimental EC3	Predicted EC3	ω
trans-2-hexenal		5.5	NP	1.608
1-(4-methoxyphenyl)-1-penten-3-one		9.3	9.87	1.734
Safranal		7.5	5.29	1.796
diethyl maleate		5.8	NP	1.804

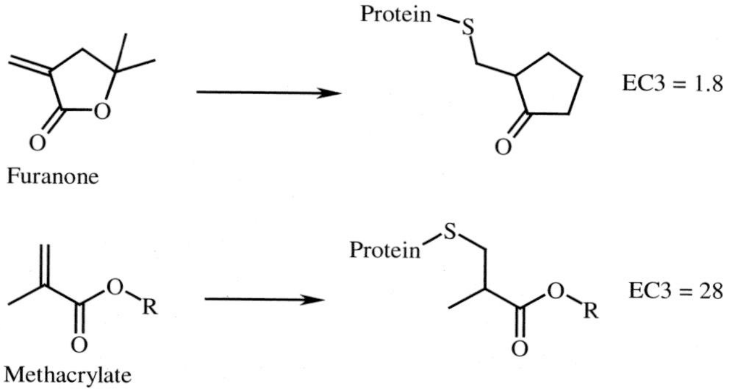

Figure 7-1. Ring strain release leading to increased skin sensitisation in 5,5-dimethyl-3-methylene-dihydro-2(3H)-furanone

Also highlighted was the need for sub-categories within the Michael domain, as 5,5-dimethyl-3-methylene-dihydro-2(3H)-furanone was found to be a significantly more potent skin sensitiser than would be suggested from its calculated electrophilicity. The authors suggested that upon reaction with a skin protein the furanone ring undergoes release of ring strain energy and thus is more reactive than the equivalent aliphatic molecules (Figure 7-1). It is therefore likely that for chemicals such as these, in which additional factors such as the release of ring strain energy are important, separate categories within the Michael domain will be required.

The use of calculated electrophilicity to make read-across predictions demonstrated that for good quality, interpretable predictions to be made requires subtle mechanistic understanding and appropriate categories and sub-categories to be formed. This suggested use of sub-categories within a mechanistic category is in keeping with the phthalates study in which sub-categories were used within a larger chemical class-based category [14].

Another study [25] grouped compounds containing α,β-unsaturated carbonyl compounds together. Such compounds are believed to be able to interact covalent with proteins, enzymes and DNA through various mechanisms. As such, they are able to stimulate a range of environmental toxicities and adverse health effects. Koleva et al. [26] assume that compounds in this category (aldehydes and ketones) act by a common mechanism of action (Michael-type addition). The acute aquatic toxicities to *Tetrahymena pyriformis* of compounds within the category were obtained in an effort to develop approaches for (qualitative) read-across. In addition, *Salmonella typhimurium* (strain TA100) mutagenicity data were analysed to establish the structural differences between mutagenic and non-mutagenic compounds. These structural differences were compared with the structural characteristics of molecules associated with acute aquatic toxicity in excess of narcosis as well as other end points, for example, skin sensitisation. The results indicate that a category can be formed that allows structural information and boundaries to be elucidated.

7.5.3.   Chemoinformatics-Based Categories

Chemoinformatics-based similarity measures have also been suggested for its use in the development of chemical categories [9, 26]. The primary example of this approach in the scientific literature makes use of a range so-called fingerprint methods. Such methods involve encoding the structural information within a molecule as a bit string in which each "bit" indicates the presence (if the bit is set as 1) or absence (if the bit is set as 0) of a particular molecular feature. These methods have been widely used in the drug discovery paradigm for locating similar chemicals from large chemical inventories [10, 27].

A recent study highlighted the usefulness of such approaches by using the freely available Toxmatch software (freely available from http://ecb.jrc.it/qsar/qsar-tools) to develop a small category of chemicals starting from a query chemical of interest [9]. The starting point for the study was a Schiff base chemical whose pEC3 was not known. By using the in-built fingerprint and similarity functions the software was able to locate three analogues from the 210 chemical local lymph node assay database [28] (Table 7-3).

Table 7-3.   Schiff base category formation and subsequent read-across predictions using similarity indices

Chemical	EC3 (% wt)	Similarity
	1.07 (predicted)	"query chemical"
	3.00	0.60
	6.30	0.60
	1.30	0.87

The authors then used the similarity measures to perform linear extrapolation between the similarity measures and pEC3 values of the three most similar chemicals. It was then possible to use this relationship to obtain a predicted pEC3 value for the query chemical (Table 7-3). Additional category formation and subsequent read-across examples were also presented using the bioaccumulation and fathead minnow data sets. A further study [10] has illustrated the use of the Toxmatch to form groupings of compounds from which it is possible to make assessment of teratogenicity.

7.6.   CONCLUSIONS

This chapter has demonstrated the general concepts that are required for the regulatory usage of chemical categories. It is clear from the material presented that the formation of a chemical category is a complex process requiring expert knowledge about both the physicochemical properties of the suggested group of chemicals and crucially their mechanisms of action across the endpoints of interest. In addition, the chapter has highlighted a number of read-across examples from the literature. Whilst examples of read-across predictions in the wider literature are currently limited, those presented in this chapter show that given a well-defined category (or indeed sub-category) good quality read-across predications can be made. These publications support the category hypothesis and help show that within these categories simple read-across methods enable mechanistically interpretable predictions to be made for complex toxicological endpoints.

ACKNOWLEDGEMENT

The funding of the European Union Sixth Framework CAESAR Specific Targeted Project (SSPI-022674-CAESAR) is gratefully acknowledged.

REFERENCES

1. van Leeuwen K, Schultz TW, Henry T et al. (2009) Using chemical categories to fill data gaps in hazard assessment. SAR QSAR Environ Res 20:207–220
2. OECD (2007) Guidance on Grouping of Chemicals. Series on Testing and Assessment No. 80. ENV/JM/MONO (2007)28, pp. 72–77. http://appli1.oecd.org/olis/2007doc.nsf/linkto/env-jm-mono(2007)28
3. Comber M, Simpson B (2007) Grouping of petroleum substances. In: Worth AP, Patlewicz G (eds) A compendium of case studies that helped to shape the REACH guidance on chemical categories and read across. EUR report no 22481 EN. European Chemicals Bureau, Joint Research Centre, European Commission, Ispra, Italy. Available from ECB website: http://ecb.jrc.it/qsar/background/index.php?c=CAT
4. Hart J (2007) Nickel compounds – a category approach for metals in EU Legislation. A report to the Danish Environmental Protection Agency. http://cms.mim.dk/NR/rdonlyres/07DB028E-134E-4796-BF6D-97B9AD5F9E82/0/Nikkel.pdf

5. Worth AP, van Leeuwen CJ, Hartung T (2004) The prospects for using (Q)SARs in a changing polit-
    ical environment—high expectations and a key role for the European Commission's Joint Research
    Centre. SAR QSAR Environ Res 15:331–343
6. Aptula AO, Roberts DW (2006) Mechanistic applicability domains for non-animal based prediction
    of toxicological end points: general principles and application to reactive toxicity. Chem Res Toxicol
    19:1097–1105
7. Roberts DW, Aptula AO, Patlewicz G (2006) Mechanistic applicability domains for non-animal
    based prediction of toxicological endpoints. QSAR analysis of the Schiff base applicability domain
    for skin sensitization. Chem Res Toxicol 19:1228–1233
8. Schultz TW, Sinks GD, Cronin MTD (1997) Identification of mechanisms of toxic action of phe-
    nols to *Tetrahymena pyriformis* from molecular descriptors. In: Chen F, Schuurmann G (eds)
    Quantitative Structure-Activity Relationships in Environmental Sciences VII, SETAC, Florida
9. Patlewicz G, Jeliazkova N, Gallegos Saliner A et al. (2008) Toxmatch – a new software tool to aid in
    the development and evaluation of chemically similar groups. SAR QSAR Environ Res 19:397–412
10. Enoch SJ, Cronin MTD, Madden JC et al. (2009) Formation of structural categories to allow for
    read-across for reproductive toxicity. QSAR Comb Sci 28:696–708
11. US EPA (2002) Chemical categories report (http://www.epa.gov/opptintr/newchems/pubs/chemcat.htm)
12. OECD (2004) OECD Integrated HPV Database (http://cs3-hq.oecd.org/scripts/hpv/)
13. UNEP Chemicals Screening Information Dataset (SIDS) for High Volume Chemicals
    (http://www.chem.unep.ch/irptc/sids/OECDSIDS/sidspub.html)
14. Fabjan E, Hulzebos E, Mennes W et al. (2006) A category approach for reproductive effects of
    phthalates. Crit Rev Toxicol 36:695–726
15. Aptula AO, Patlewicz G, Roberts DW (2005) Skin sensitization: reaction mechanistic applicability
    domains for structure-activity relationships. Chem Res Toxicol 18:1420–1426
16. Aptula AO, Roberts DW, Schultz TW et al. (2007) Reactivity assays for non-animal based prediction
    of skin sensitisation potential. Toxicol 231:117–118
17. Russom CL, Bradbury SP, Broderius SJ (1997) Predicting modes of toxic action from chemical
    structure: acute toxicity in the fathead minnow (*Pimephales promelas*). Environ Toxicol Chem
    16:948–967
18. Seward JR, Hamblen EL, Schultz TW (2002) Regression comparisons of *Tetrahymena pyriformis*
    and *Poecilia reticulata* toxicity. Chemosphere 47:93–101
19. Schultz TW, Yarbrough JW, Hunter RS et al. (2007) Verification of the structural alerts for Michael
    acceptors. Chem Res Toxicol 20:1359–1363
20. Enoch SJ, Madden JC, Cronin MTD (2008) Identification of mechanisms of toxic action for skin
    sensitisation using a SMARTS pattern based approach. SAR QSAR Environ Res 19:555–578
21. Enoch SJ, Cronin MTD, Schultz TW et al. (2008) Quantitative and mechanistic read across for
    predicting the skin sensitization potential of alkenes acting via Michael addition. Chem Res Toxicol
    21:513–520
22. Roberts DW, Aptula AO, Patlewicz G (2007) Electrophilic chemistry related to skin sensitization.
    Reaction mechanistic applicability domain classification for a published data set of 106 chemicals
    tested in the mouse local lymph node assay. Chem Res Toxicol 20:44–60
23. Patlewicz GY, Basketter DA, Pease CKS et al. (2004) Further evaluation of quantitative structure-
    activity relationship models for the prediction of the skin sensitization potency of selected fragrance
    allergens. Contact Derm 50:91–97
24. Patlewicz GY, Wright ZM, Basketter DA et al. (2002) Structure-activity relationships for selected
    fragrance allergens. Contact Derm 47:219–226
25. Koleva YK, Madden JC, Cronin MTD (2008) Formation of categories from structure-activity rela-
    tionships to allow read-across for risk assessment: toxicity of α, β-unsaturated carbonyl compounds.
    Chem Res Toxicol 21:2300–2312

26. Jaworska J, Nikolova-Jeliazkova N (2007) How can structural similarity analysis help in category formation? SAR QSAR Environ Res 18:195–207

27. Leach AR (2001) Molecular modelling: Principles and applications. Pearson Education Limited, Harlow

28. Gerberick GF, Ryan CA, Kern PS et al. (2005) Compilation of historical local lymph node data for evaluation of skin sensitization alternative methods. Dermatitis 16:157–202

Part II
Practical Application

CHAPTER 8

QSAR IN CHROMATOGRAPHY: QUANTITATIVE STRUCTURE–RETENTION RELATIONSHIPS (QSRRs)

ROMAN KALISZAN AND TOMASZ BĄCZEK

Department of Biopharmaceutics and Pharmacodynamics, Medical University of Gdańsk, Gen. J. Hallera 107, 80416 Gdańsk, Poland

Abstract: To predict a given physicochemical or biological property, the relationships can be identified between the chemical structure and the desired property. Ideally these relationships should be described in reliable quantitative terms. To obtain statistically significant relationships, one needs relatively large series of property parameters. Chromatography is a unique method which can provide a great amount of quantitatively precise, reproducible, and comparable retention data for large sets of structurally diversified compounds (analytes). On the other hand, chemometrics is recognized as a valuable tool for accomplishing a variety of tasks in a chromatography laboratory. Chemometrics facilitates the interpretation of large sets of complex chromatographic and structural data. Among various chemometric methods, multiple regression analysis is most often performed to process retention data and to extract chemical information on analytes. And the methodology of quantitative structure–(chromatographic) retention relationships (QSRRs) is mainly based on multiple regression analysis. QSRR can be a valuable source of knowledge on both the nature of analytes and of the macromolecules forming the stationary phases. Therefore, quantitative structure–retention relationships have been considered as a model approach to establish strategy and methods of property predictions.

Keywords: QSRR, Retention predictions, Characterization of stationary phases

8.1.   INTRODUCTION

8.1.1.   Methodology of QSRR Studies

At the current state of development of chemistry, it appears easier to synthesize a compound with a definite chemical structure than with a required property. Usually, reaction pathways can correctly be estimated for established chemical structures, whereas predicting properties of specific product(s) of the reaction is still a matter of scientific guesswork [1].

Chemical reactivity, in the sense of forming the new chemical bonds or breaking of the existing ones, seems to depend mostly on the compound's structure itself. On the other hand, valid predictions from chemical formula of even the simplest

223

T. Puzyn et al. (eds.), Recent Advances in QSAR Studies, 223–259.
DOI 10.1007/978-1-4020-9783-6_8, © Springer Science+Business Media B.V. 2010

properties, such as anesthetic potency, boiling point, or chromatographic retention, can only be obtained within series of homologues or otherwise closely congeneric compounds by extrapolation or interpolation of the measured property of several representatives of the series.

A compound's properties depend strongly on the environment in which it is placed. It is not only the molecular structure of compounds, but also their interactions with molecules forming their environment, which justifies them being called drugs, toxins, cosmetics, hormones, pheromones, odorants, detergents, pesticides, herbicides, environmental pollutants, conductors, building materials, and so on. Unlike chemical reactions, the interactions of molecules which form the environment in which the molecules are placed cause neither the breaking of existing bonds nor the formation of new bonds.

To predict a given biological or physicochemical property, the relationships must be identified between the chemical structure and the desired property. Optimally, these relationships should be described in reliable quantitative terms. To obtain statistically significant relationships, one needs relatively large series of property parameters. Chromatography (especially high-performance liquid chromatography, HPLC) is a unique method which can yield a great amount of quantitatively comparable, precise, and reproducible retention data for large sets of structurally diversified compounds (analytes). These data can be mutually related because all of them are determined at the same experimental conditions (or can be standardized by simple interpolation or extrapolation). Therefore, quantitative structure–(chromatographic) retention relationships (QSRRs) have been considered as a model approach to establish strategy and methods of property predictions.

In 1977 the first three publications appeared on what is now termed QSRR [2–4]. A monograph on QSRR published in 1987 considered several hundred publications [5]. Since then, reviews [6–8] and several books [9–13] have dealt with the topic.

Reliable QSRR methods have been established to predict retention and to elucidate molecular mechanism of retention on diverse stationary phases. These could be useful during HPLC method development and to rationally design new HPLC stationary phases of required properties. QSRR analysis has also been applied to facilitate protein identification in proteomics. The QSRR-processed chromatographic data have been proposed to preselect the most promising drug candidates from a multitude of synthesized or computer-designed structures. All these issues deserve a comprehensive review to better understand and employ in practice the rules of chemistry.

First, QSRR reports resulted from the application of the methodology used for studies of quantitative structure–(biological) activity relationships (QSARs) –so-called Hansch approach [14] –to the analysis of chromatographic data. The presently applied methodology and goals of QSRR studies is schematically presented in Figure 8-1.

To perform a QSRR study, one needs a set of quantitatively comparable retention parameters for a sufficiently large series of analytes and a set of their structural descriptors. Through the use of computerized statistical and chemometric techniques, retention parameters are characterized in terms of various analyte

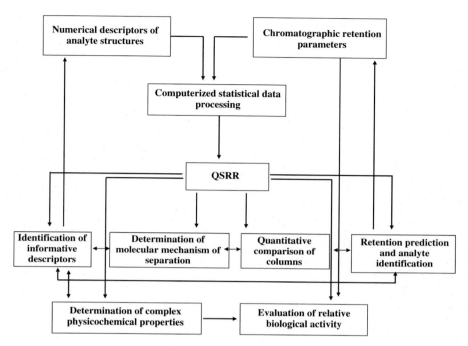

Figure 8-1. Methodology and goals of QSRR studies

descriptors. If statistically significant and physically meaningful QSRR are obtained then they can be applied to

(i)   identify the most useful (regarding properties) structural descriptors;
(ii) predict retention for a new analyte and to identify unknown analytes;
(iii) gain insight into molecular mechanism of separation operating in a given chromatographic system;
(iv) quantitatively compare separation properties of individual types of chromatographic columns;
(v)   evaluate other than chromatographic physicochemical properties of analytes, such as lipophilicity and dissociation constants; and
(vi) estimate relative bioactivities within sets of drugs and other xenobiotics as well as material properties of members of a family of chemicals.

To obtain reliable QSRRs appropriate input data are necessary and statistical analysis must be carried out. Chromatography can provide large amounts of suitable input data. That is in a chromatographic analysis conditions may be kept constant for many separated analytes. Thus, the analyte structure is the single variable in the system.

QSRR analysis seems to be especially attractive from the general chemometric point of view. That is because QSRRs provide the best testing of the applicability of individual structural parameters for property description. The skill and knowledge gained from QSRR studies may be applicable to other chemometric studies.

A number of QSRR reports deserve the interest of physical, analytical, medicinal, and environmental chemists. The fact is, however, that not every published QSRR provides worthy information. Some QSRRs are statistically invalid, and occasionally statistically valid correlations are developed for chemically invalid principles.

8.1.2. Intermolecular Interactions and Structural Descriptors of Analytes

First, it must be emphasized that intermolecular interactions governing chromatographic separation are not the interactions causing definite chemical alterations of the analyte molecules. These are not the protonation, oxidation, reduction, complex formation nor other stoichiometric processes. Only in ion-exchange chromatography, where the separation determining forces are ionic in nature, can discrete chemical alterations be said to occur. In other chromatographic techniques and modes only the forces that can occur between closed-shell molecules are involved. The known intermolecular interaction types are given in Table 8-1.

Formulae defining the potential energy of individual types of interactions between two molecules are described in physical chemistry textbooks. However, their application to actual chromatographic separation processes is not that straightforward.

Table 8-1. Binding types and binding energies with example systems. Compiled after Albert [15] and Seydel and Schaper [16]

Type of interaction	Energy of interaction (kJ/mol)	Example system
Covalent bonding	$-(170\text{--}460)$	CH_3-OH
Strong ionic bonding	-40	$HRN-H--O$ $H^{\oplus}---^{\ominus}O$ $\diagdown C-R'$
Ionic bonding	-20	$R_4N^{\oplus}---Cl^{\ominus}$
Ion – dipole interactions	$-(4\text{--}17)$	$R_4N^{\oplus}---:NR_3$
Dipole-dipole and dipole induced dipole interactions (Keesom effect and Debye effect, respectively)	$-(4\text{--}17)$	$O=C^{\delta+}---:NR_3$
Hydrogen bonding interactions	$-(4\text{--}17)$	$-OH---O=$
Electron pair donor- Electron pair acceptor interactions (charge transfer effect)	$-(4\text{--}17)$	$-OH--- \overset{\diagdown C \diagup}{\underset{\diagup C \diagdown}{\|}}$
Dispersive interactions (London – Hall effect)	$-(2\text{--}4)$	$\overset{\|}{\underset{\|}{C}}----\overset{\|}{\underset{\|}{C}}$
Hydrophobic interactions (a hybride of nonpolar and polar interactions)	-4	$R \diagup CH_2 \diagdown CH_2 \diagup R$ $R \diagup CH_2 \diagdown CH_2 \diagup R$

Let's consider, for example, the so-called reversed-phase high-performance liquid chromatography (RP HPLC) on a chemically bonded hydrocarbon-silica stationary phase. One has to consider the following mutually interacting entities: the analyte, the hydrocarbon bonded to silica matrix, the mobile phase components preferentially adsorbed on the stationary phase, the silanol groups of the silica support, and all the components of the eluent. In view of system complexity, no satisfactory model is known which would permit quantitative prediction of retention. A rational explanation of the observed retention differences in terms of intermolecular interactions appears possible, however.

An increase in logarithm of retention factor, *log k*, of an analyte with an increasing number of carbon atoms for homologous series is typically observed in RP HPLC. Considering first the interactions of analytes with the hydrocarbon chain bond to stationary phase, one will identify the dispersive forces (London-Hall effect) as differentiating the homologues. At the same time, one can assume the input to separation due to the orientation interactions (Keesom effect) of homologues as negligible, because the polarity (dipole moment) of the hydrocarbon part of the molecules is practically negligible. Also the dipole-induced dipole interactions (Debye effect) should be similar for all homologues because the dipole moments within homologues series are similar. On the other hand, the magnitude of dispersion interactions (London-Hall effect) increases with increasing polarizability of the analytes, which actually reflects their molecular size ("bulkiness"). In the case of analyte interactions with the eluent (which is polar in RP HPLC), for homologous analytes the orientation effects and the inductive effects are undoubtedly stronger than the interactions of the analyte with the non-polar hydrocarbon of stationary phase. Of course, the dipole–dipole and dipole-induced dipole interactions with the eluent are similar for all homologues due to the similarity of their dipole moments.

The attraction of homologous analytes by a mobile phase resulting from the dispersive and dipole inductive interactions is mostly affected by the analyte polarizability. Because the dispersive interactions usually prevail among intermolecular interactions and the polarizability of hydrocarbon, e.g., octadecyl chains of the stationary phase, is greater than the polarizability of the small molecules of typical eluents used in RP HPLC, the net effect of all the van der Waals interactions (Keesom, Debye, and London-Hall effects) will be the increased retention of larger homologues. Of course, the interactions of analytes with the components of a chromatographic system are normally further complicated by the non-van der Waals interactions listed in Table 8-1, in particular hydrogen bonding and charge transfer interactions.

The observed decreasing of analytes' retention in RP HPLC with their increasing degree of ionization can also be explained in terms of known intermolecular interactions. One can assume that the dispersive interactions of analyte ions with both phases do not differ significantly from these interactions for non-ionized analyte molecules. However, in the case of analyte ions the ion–dipole interactions became dominating. Such attractive interactions are of practical importance between analyte ions and the polar molecules of the eluent used in RP HPLC as opposed to interactions of ions with non-polar hydrocarbon moieties of stationary phase. That is true especially since the ion–dipole long-distance interactions are stronger than regular short-range van der Waals interactions and decrease with the second power of the distance between the interacting species. Thus, eluent pH strongly affects retention of weak acids and bases in RP HPLC.

In 1937 Hammett formulated the well-known relationships for the calculation of substituent effects on reaction rates and chemical equilibria. By analogy to the Hammett electronic substituent constant in 1964, Hansch and Fuijta [14] introduced the substituent hydrophobic constant π. Already by 1965 Iwasa et al. [17] reported the correlation between π and a substituent linked to thin-layer chromatographic retention, ΔR_M. Another linear free-energy related substituent constant defined from chemical reactivities was the Taft steric constant, E_s. These classical empirical structural descriptors found little application in QSRR analysis [12, 13, 18].

Furthermore, Taft, Carr, Abraham and co-workers [19–21] studied the nature of RP HPLC separations and developed an approach based on solvatochromic comparison method, the so-called linear solvation energy relationships (LSERs). That approach is based on a general Eq. (8-1) describing logarithm of analyte retention factor, $\log k$:

$$\log k = \text{constant} + M(\delta_m^2 - \delta_s^2)V_x/100 + S(\pi_s^* - \pi_m^*)\pi_x^* + A(\beta_s - \beta_m)\alpha_x + B(\alpha_s - \alpha_m)\beta_x$$

$$(8\text{-}1)$$

where the subscript x designates an analyte property such as molar volume, V_x, polarizability/dipolarity, π_x^*, hydrogen bonding acidity, α_x, and hydrogen bonding basicity, β_x. Each analyte property is multiplied by a term, which represents the difference in complementary solvent properties of the mobile (subscript m) and the stationary (subscript s) phases. Thus, α_m and α_s are the abilities of the phases (bulk or bonded) to donate a hydrogen bond. These properties complement the analyte's ability to accept a hydrogen bond, β_x. Similarly, δ_m^2 and δ_s^2, the squares of the Hildebrand solubility parameter or cohesive energies of the two phases, complement the analyte intrinsic molar volume, V_x.

However, solvatochromic parameters are empirically obtained and therefore available only for a limited number of compounds. There is no such limitation for structural parameters, which can be derived by computational chemistry based solely on the structural formula of a compound. The constitutive–additive parameters such as molar refractivity, n-octanol–water partition coefficients calculated by fragmental methods (CLOGP), quantum-chemical and molecular mechanics indexes, and parameters derived from molecular graphs can also be treated in this manner. Examples of structural descriptors employed in QSRR analysis are given in Table 8-2.

The number of structural descriptors which can be found for an individual analyte is practically unlimited. The first commercially available software introduced by Hasan and Jurs [22] processed some 200 different structural descriptors. In their comprehensive review published in 1996, Katritzky and co-workers [23] described numerous quantum-chemical descriptors. Further reports from Katritzky's laboratory enlarged the number of methods of quantifying the structural information about the molecule [24–26].

Table 8-2. Exemplary structural descriptors used in QSRR studies

Molecular bulkiness-related descriptors	*Molecular polarity-related (electronic) descriptors*
Carbon number	Dipole moments
Molecular mass	Atomic and fragmental electron excess charges
Refractivity	Orbital energies of HOMO and LUMO
Polarizability	Partially charged areas
van der Waals volume and area	Local dipoles
Solvent-accessible volume and area	Sub-molecular polarity parameters
Total energy	
Calculated partition coefficient (clog P)	
Molecular geometry-related (shape) descriptors	*Combined molecular shape/polarity parameters*
Lenght-to-breadth ratio	Comparative molecular field analysis (CoMFA) parameters
STERIMOL parameters	Comparative molecular surface (CoMSA) parameters
Moments of inertia	
Shadow area parameter	*Molecular graph-derived (topological) descriptors*
	Molecular connectivity indices
Physicochemical empirical and semiempirical parameters	Kappa indices
Hammett constants	Information content indices
Hansch constants	Topological electronic index
Taft steric constants	
Hydrophobic fragmental parameters	*Indicator variables*
Solubility parameters	Zero-one indices
Linear solvation energy relationship (LSER) parameters	
Partition coefficienct (log P)	Ad hoc designed descriptors
Boiling temperatures	
pK_a values	

A tremendous work was completed in 2000 by Todeschini and Consonni [27] who thoughtfully analyzed 3300 references and collected about 1800 known descriptors in a form of encyclopedia. That valuable monograph gives detailed characteristics of the descriptors known in the chemical literature, whether they are physicochemical or topological in their nature. Appropriate software [28] for the calculation of individual descriptors is currently widely used all over the world. After the Todeschini and Consonni monograph [28] some new descriptors have been proposed [29, 30] but still there is a space for the invention and imagination of the individual researcher.

When proposing specific ad hoc descriptors the requirement is that these descriptors are well defined and identified, even if their physical meaning is unclear. An example may be the so-called topological electronic index T^E proposed in 1986 [31]. That index gave rise to a number of modifications compiled by Todeschini and Consonni [27]. Its originality consisted in accounting for electron excess charge distribution in three-dimensional space. To get T^E for an energy-optimized structure, the distances $r_{i,j}$ between each pair of atoms are calculated based on the atom coordinates. Then, for every pair of atoms, the absolute value of the excess charge difference is divided by the square of the respective interatomic distance. The resulting numbers are summed for all possible atomic pairs in the molecule.

Descriptors calculated from molecular formula or molecular graph appear attractive. The problem is if they are actually related to a molecular property of a compound or are only casually related. From a chemistry point of view, excepting perhaps chemical documentation, only the mathematical properties of chemical structures are of interest, which are related to the physicochemical or biological properties of compounds. Although several non-empirical structural descriptors were reported to contribute to numerous multivariable QSRR equations, it cannot be said that any outstandingly reliable, property-specific structural descriptor was found that was universally applicable to various chemical families.

Generally, chemical formulae represent molecules as sets of balls (atoms) connected by stronger or weaker, longer or shorter springs (bonds and interatomic interactions). In a collision of one ball and spring system with another, the crackings and fusions leading to a new entity can be easily understood. Hence, reactivity seems to emerge as an innate feature of a molecule. However, a molecule is a definite physical entity different from the component atoms. Molecular properties are constitutive rather than additive regarding the composing atoms. That constitutiveness remains a kind of mystery not accounted for neither by conventional structural formula nor by quantum chemistry models. One can subsequently invent thousands of descriptors based on established chemical coding without improving the design of materials with desired properties.

8.2.   CHROMATOGRAPHIC RETENTION PREDICTIONS

Although the physical meaning of the applied descriptors is often disputable, there are QSRRs in literature, which are able to predict retention well. These QSRRs are derived statistically. There are advanced statistical techniques available for proper descriptor selection, precluding at the same time deriving of formally invalid models.

A good prediction of retention of structurally defined analytes by QSRR has been documented in the literature. Even if "predictions" are demonstrated retrospectively and applied to members of closely congeneric families of analytes only, there is no doubt that at least those reported more recently are not artifacts. Since Randić's original paper [32] several proposed structural descriptors have permitted the prediction of the gas–liquid chromatographic (GLC) retention of saturated hydrocarbons, including isomers. More recent multivariable QSRR cannot be questioned formally as regards the retention prediction potency within series of related compounds.

The problem is that complex predictive QSRRs are of limited value for the actual design of compounds with a desired property. Of course, any structure can be drawn and its property calculated. However, there is no way to identify the specific structure with a set of dozen or more descriptors to provide a definite retention factor or other more valuable property, such as bioactivity. What is worse, such a structure may not exist at all because adjusting one or more descriptors leads to unfavorable changes of other QSRR descriptors.

8.2.1. Retention Predictions in View of Optimization of HPLC Separations

QSRRs derived for high-performance liquid chromatographic (HPLC) retention data are generally of lower statistical quality than those reported in GLC. That is due to a greater complexity of phase systems in LC than in GLC. The structure–retention relationships in LC are often semi-quantitative rather than quantitative. Regardless, they may as such be useful for the confirmation of the identification of the proper analyte, separation conditions optimization, and elucidation of mechanism of retention at molecular level. However, oversimplification such as plotting retention vs. molecular mechanics calculated energy of analyte-stationary phase model interaction [33] shows nothing but a poor correlation and certainly does not mean that solvent effects are unimportant in LC.

The QSRRs in HPLC published up until 1997 have been reviewed previously [12]. None had retention prediction potency of actual practical value. That also applies to later published QSRRs, such as those relating RP HPLC retention and micellar electrokinetic capillary chromatographic (MECC) retention of 32 steroid hormones to sets of complex topological indexes [34]. Contrary to GLC, the poor performance of both topological and quantum-chemical descriptors in prediction of HPLC retention was found by Makino [35]. The QSRR reported [36] for as few as eight hydroxybenzoic acid derivatives subjected to RP HPLC cannot be treated as reliable, because as many as three sophisticated topological indexes had to be used. Also, the QSRR equations [37] describing RP HPLC retention of O-aryl, O-(methylthioethylideneamino)-phosphates (13 compounds chosen of total 20 studied) in terms of calculated molecular refractivity *(CMR)*, and energy of core–core repulsion *(ECCR)*, are of highly disputable value for any prediction. At first sight, a strong intercorrelation is evident between *CMR* and *ECCR* (Table 3 in reference 37) precluding their joint use in the same regression equation.

The QSRR derived by Ledesma and Wornat [38] makes good physical sense and has correct statistics. The RP HPLC retention of 12 polycyclic aromatic hydrocarbons (PAH) was related to the maximum excess charge difference between two atoms in a molecule, Δ [39], and analyte polarizability, α. That way two types of intermolecular interactions which govern retention are accounted for polar forces resulting from permanent or induced dipoles from the analyte, stationary phase, and mobile phase molecules and non-polar forces resulting from dispersive interactions.

However, the QSRR obtained by Ledesma and Wornat [38] cannot be extended beyond closely the series of congeneric PAHs. Long ago the effect of molecular shapes (degree of elongation) of individual PAH on chromatographic retention of isomers was noted and quantified, first in GLC [40], then also in HPLC [41]. Recently, Wise and co-workers [42] derived a predictive QSRR to identify PAH and methyl-substituted PAH chromatographed on monomeric and polymeric C18 HPLC stationary phases. A large set of calculated electronic, topological spatial, and thermodynamic molecular descriptors was examined through the use of the partial least squares (PLS) statistical technique. A reduced set of abstract descriptors (the PLS components) was derived from the initial series of molecular descriptors to maximize their correlation with the retention properties while keeping them mutually orthogonal. The obtained QSRR model accounts well for retention differences among analytes and discerns differences in analyte shape selectivity between the two stationary phases studied.

Considering RP HPLC data, PLS modeling was also applied to 17 chalcones. The authors [43] derived a QSRR with five PLS components obtained after the processing of 20 molecular graph-derived descriptors. Such QSRRs are claimed to provide chalcone retention prediction on RP HPLC stationary phases of different polarity. A poor description of retention was obtained when using calculated octanol–water partition coefficient, *CLOGP*, in combination with a count of O–H and N–H bonds. Also, QSRRs of moderate quality were obtained by Luco and co-workers [44] for amino acids chromatographed on three reversed-phase columns. The molar volume of the side chain, a polarity factor, and the energy of the lowest unoccupied molecular orbital, E_{LUMO}, were the descriptors employed. Using PLS analysis of the RP HPLC data, the same authors attempted to model amino acid transport across the blood–brain barrier and in pigeon erythrocytes.

Another multivariate data processing method, i.e., principal component analysis (PCA) was used in QSRR studies on liquid chromatographic (TLC and HPLC) retention data of ditetrazolium salts [45]. The first two principal components obtained after PCA of 14 topological and quantum-chemical analyte descriptors accounted for the retention differences observed in the liquid chromatographic systems studied.

An original approach in terms of QSRRs was reported by Åberg and Jacobsson [46]. Three-dimensional images of molecules with a pulse-coupled neural network (PCNN), thus obtaining a short time series representation of the molecules, were processed. Such a representation appeared suitable for QSRR modeling with PLS of a series of 24 steroids. No report was noted to develop the approach further, however.

On the other hand, artificial neural networks (ANN) were employed to obtain a QSRR to predict RP HPLC retention in the gradient mode of phenylthiocarbamoyl amino acids derivatives [47]. The molecular structure of each amino acid was encoded with 36 descriptors calculated from the molecular formula. A lipophilicity parameter, along with three molecular size and shape parameters, was found to be important for analyte retention. However, dominating for differences in retention data analyzed was the effect of mobile phase composition.

ANN have also been reported to predict the retention of a series of herbicides in RP HPLC at various pHs and compositions of water–methanol mobile phases [48]. Within congeneric herbicides the lipophilicity parameter, log P, and the dipole moment, μ, accounted for the effect of molecular structure of the analyte. That series was too short (four analytes) to draw any conclusion, however. The effect of methanol content on logarithm of retention coefficient is linear and pH was kept within narrow range providing a full analyte ionization. Therefore, the findings are as expected. The claimed advantage of ANN over multilinear regression (MLR) in deriving QSRR model cannot be considered as proven.

In comparative studies performed in the authors' laboratory [49, 50], the predictive power of ANN was observed to be similar to that of MRA. For a structurally diverse series of predesigned test analytes gradient retention times were determined on three modern columns. Then, based on linear solvent strength (LSS) theory of Snyder [51, 52], retention parameters were related to eluent composition. Both isocratic and gradient retention times for any structurally defined analyte could be calculated based on LSS and on the QSRR models derived for the test series of analytes. Three such QSRR models were considered: one based on linear solvation energy relationships (LSERs) [19–21], another relating retention to logarithm of n-octanol–water partition coefficient, log P, and third describing retention in terms of water accessible molecular surface area, dipole moment, and minimum atomic excess charge [53]. Using both ANN and MRA, the combined LSS/QSRR approach was demonstrated to provide approximate, yet otherwise unattainable, a priori predictions of retention of analytes based solely on their chemical formulae. That way a rational chemometric basis for a systematic optimization of conditions of chromatographic separations has been elaborated as an alternative to the trial-and-error method normally applied before.

A paper reflecting the current stage of development of predictive QSRR is provided by Schefzick et al. [54]. These authors chromatographed 62 structurally diverse analytes on 15 modern reversed-phase columns at five acetonitrile concentration gradient conditions. The LC/MS data obtained were related to 2419 molecular descriptors of the analytes calculated by currently available commercially software (identified in the report). A genetic algorithm selected the 20 most predictive descriptors. According to the original authors, all the descriptors selected can be explained physically. That is somewhat surprising, considering the large number of descriptors available for the analysis and the considerable variable reduction due to removing of the correlated and invariant descriptors. The claim might be generally true but the physical sense of a descriptor such as the "absolute value of the difference between $CASA^+$ (positive charge weighted surface area, ASA^+ times max $\{q_i > 0\}$), and $CASA^-$ (MOE)" lacks interpretability. The paper by Schefzick et al. [54] might be recommended as methodologically instructive and representing the state of the art in predictive QSRR. The only problem is that the verification of its reliability is almost impossible as 48 of the 62 analytes studied have unknown structures.

Whereas the physical meaning of the structural descriptors used in the previously discussed QSRR were considered as disputable, the parameters from the so-called comparative molecular field analysis (CoMFA) [55] can be considered to be even more abstract (see Chapter 4). Luo and Cheng [56] described the RP HPLC retention data of 33 purine compounds. First, a moderate QSRR ($R^2 = 0.790$) was obtained by relating retention to dipole moment, molecular moment of inertia, polar surface area, and molecular volume. Partial atomic charges for the optimized structures were imported into the Tripos Sybyl 6.9.1 modeling software suite [57] for CoMFA-QSRR analysis. The steric (Lennard-Jones) and electrostatic (Coulombic) interaction energies between the purine molecule and the interaction probe were computed at each lattice point. A model based on various partial atomic charge formalisms resulted in a correlation superior to the classical QSRR.

QSRR models are usually built using multiple linear regression (MLR) methods. However, MLR can be used only when the number of analytes is larger than the number of descriptor variables and the descriptors included in a given MLR equation are not highly correlated (actually, they should be orthogonal). Since thousands of molecular descriptors are readily available [27] either variable selection methods have to be applied prior to MLR or other chemometric modeling methods have to be used, such as artificial neural networks, principal component regression, or partial least squares [58–60]. These latter methods use combinations of the original variables (latent variables), that makes the understanding of the resulting QSRR equations rather impossible. Therefore, MLR with feature selection is preferred in QSRR.

Recently published advanced methods of variable selection used in QSRR studies of HPLC data are genetic algorithms [61], classification and regression tree (CART) analysis [62], and multivariate adaptive regression splines (MARS) [63]. These methods provide selection of predictive descriptors. However, they do not guarantee that the statistically correct model, automatically produced, will make good physical sense. Statistics themselves cannot improve the model beyond the limits of the actual chemical information contained in structural descriptors. Evidently, a suitable translation which would reveal the properties of compounds encoded into their structure in a reliable manner is still lacking.

There is an attitude to QSRR when emphasis is put more strongly on the physical meaning of the equations rather than on their statistical quality. The advocates of this approach suggest that approximate predictions can also be of help to confirm analyte identification, to optimize HPLC separation conditions, and to differentiate mechanism of separation dominating in specific chromatographic systems. Earlier reports on that approach have been discussed in details in references [5] and [12].

In a series of papers, Abraham and co-workers [64–68] developed a QSRR model beginning with from the solvatochromic theory of Taft and Kamlet [19–21]. Solvatochromic descriptors of analytes are derived from spectroscopic experiments and are not linearly related to free-energy changes, however. In his approach Abraham had to redefine and determine a series of descriptors of analytes to replace solvatochromic descriptors in LFER models. Abraham et al. [69] used an experimental equation (8-2):

$$\log k = \log k_0 + rR_2 + vV_x + s\pi_2^H + aS\alpha_2^H + bS\beta_2^H \tag{8-2}$$

where R_2 – excess molar refraction (E), V_x – characteristic McGowan volume (V), p_2^H – dipolarity/polarizability (S), $S\alpha_2^H$ – effective hydrogen bonding acidity (A), $S\beta_2^H$ – effective hydrogen bonding basicity (B), and r, v, s, a, b – regression coefficients characterizing corresponding properties of the chromatographic system.

The model shows a good retention prediction potency but requires empirically determined structural parameters that are available for several thousand compounds [70] but certainly not for all possible analytes. In 2002 Wilson et al. [71] elaborated a LSER-based procedure for retention prediction based on limited number of experimental measurements. The approach, however comprehensive, appears rather complex for routine retention prediction purposes.

Recently, Abraham et al. [70] provided a review on the principles and methodology of the determination of solvation parameters (analyte descriptors) on the basis of chromatographic analyses. The applicability of Abraham model in QSRR studies of RP HPLC data has been confirmed by many other authors [72–78]. It was found useful for the prediction of the retention of specific analytes and even more for the elucidation of mechanisms of retention operating on individual RP HPLC stationary phases. A very recent review by Vitha and Carr [79] clearly demonstrates the advantages of Abraham model in the chemical interpretation of the mechanism of chromatographic retention at molecular (sub-molecular) level.

A full Abraham model comprises terms which often appear insignificant when applied to the defined sets of analytes. Abraham et al. [70] insist on the use of new symbols (E, V, S, A, and B) for parameters of analytes in LSER model for simplification of the notation. However, such simplification obscures the physical meaning of individual parameters.

Individual researchers report QSRR equations accounting for RP HPLC retention differences among more or less structurally diversified series of analytes comprising only the V_x (or R_2), $\Sigma\beta_2^H$, and (less frequently) $\Sigma\alpha_2^H$ terms as statistically significant. For specific analytes and chromatographic systems, the π_2^H term is also reported as being significant [73, 79–81]. A reduced form (8-3) of the LSER-based QSRR equation has been derived for a designed series of structurally diverse test analytes and proposed for quantitative comparison of RP HPLC columns [53]:

$$\log k = k_1 + k_2 S\alpha_2^H + k_3 S\beta_2^H + k_4 V_x \tag{8-3}$$

where k_1–k_4 are regression coefficients.

Another, novel LSER-based QSRR model is the hydrophobic-subtraction model of retention [71, 82, 83]. In this approach, a subtraction of the hydrophobicity contribution to the RP HPLC retention is done to better see the remaining contributions to retention: those which are due to interactions other than the hydrophobic analyte-stationary/mobile phase. The resulting general equation (8-4) describing column selectivity, α, is

$$\log \alpha = \log \left(\frac{k}{k_{EB}} \right) = H\eta - S^*\delta + A\beta + B\alpha + C\kappa \tag{8-4}$$

where k is the retention factor of a given analyte and k_{EB} is the value of k for a non-polar reference analyte (e.g., ethylbenzene) determined on the same column under the same conditions. The remaining symbols represent either eluent- and temperature-dependent properties of the analyte (η, δ, β, α, κ) or eluent- and temperature-independent properties of the column (H, S^*, A, B, C). The column parameters denote the following column properties: H – hydrophobicity, S^* – steric resistance to insertion of bulky analyte molecules into the stationary phase, A – column hydrogen-bond acidity, mainly attributable to non-ionized silanols, B – column hydrogen-bond basicity, hypothesized to result from sorbed water in the stationary phase, C – column cation-exchange activity due to ionized silanols. The parameters η, δ, β, α, and κ denote complementary properties of the analyte: η – hydrophobicity, δ – molecular "bulkiness" or resistance of the analyte to its insertion into the stationary phase, β – hydrogen-bond basicity, α – hydrogen-bond acidity and κ – approximate charge (either positive or negative) on the analyte molecule. The values of each analyte parameter, η, δ, β, α, κ, are relative to the values for ethylbenzene, the reference analyte for which all the analyte parameters are zero.

A simple and robust QSRR model (8-5) applicable to RP HPLC was built [49, 84–86] employing the following analyte descriptors: μ – total dipole moment, δ_{Min} – electron excess charge of the most negatively charged atom, A_{WAS} – water-accessible molecular surface area:

$$\text{Retention Parameter} = k_1 + k_2\mu + k_3\delta_{Min} + k_4A_{WAS} \qquad (8\text{-}5)$$

where *retention parameter* may be either isocratic $\log k$ or gradient retention time and t_R and k_1–k_4 are regression coefficients.

From equations in the form of Eq. (8-5), retention parameters for a structurally representative and sufficiently large (for meaningful statistics) model series of analytes chromatographed in a given RP HPLC system can be evaluated. Such a model series of 18 test analytes has been designed [53, 85]. Later on, it was found [49] that the model series could have been shortened to 15 compounds without meaningful loss of statistical significance of the resulting general QSRR equations.

Changes in RP HPLC retention parameters accompanying the changes in composition of the mobile phase are usually in good agreement with the linear solvent strength (LSS) model [51, 52]. In that model (8-6) the logarithm of retention factor for a given analyte, $\log k$, is related linearly to the volume fraction of organic modifier in a binary aqueous eluent, φ:

$$\log k = \log k_w - S\varphi \qquad (8\text{-}6)$$

where $\log k_w$ is the value of $\log k$ extrapolated to 100% water as the mobile phase, i.e., to $\varphi = 0$; S is a constant characteristic for a given analyte and a given RP HPLC system.

Based on Eq. (8-6) one can determine $\log k_w$ by extrapolation to $\varphi = 0$. However, the procedure requires a number of chromatographic runs and therefore is time-consuming. The LSS model allows for the calculation of $\log k_w$ (and S) from retention data obtained in two gradient RP HPLC experiments [51, 52]. The calculations can be performed by commercially distributed software. Based on results of two gradient runs carried at different gradient times, t_G, one can also calculate gradient retention time, t_R, at pre-designed gradient conditions. Of course, having $\log k_w$ and S one can also calculate retention coefficients corresponding to any chosen isocratic conditions. All that is possible according to the fundamental LSS equation (8-7):

$$t_R = (t_0/b) \log(2.3k_0b + 1) + t_0 + t_D \qquad (8\text{-}7)$$

where t_0 is dead time, k_0 is isocratic value of k at the start of gradient elution, t_D is gradient delay time (dwell time), and b is gradient steepness parameter which is determined as follows (8-8):

$$b = t_0 \Delta\varphi S/t_G = V_m \Delta\varphi S/t_G F \qquad (8\text{-}8)$$

where V_m is column dead volume, $\Delta\varphi$ is the change in the mobile phase composition, t_G is gradient time, and F is flow-rate as S is as defined earlier. A more comprehensive deriving of Eq. (8-8) than done by original authors [51, 52] can be found in reference [87].

Pairs of general QSRR equations in the form of Eq. (8-5) were derived to describe two gradient retention times of model series of analytes for six columns. Either methanol or acetonitrile was used as organic modifiers of the eluent. Thus, the QSRR models obtained served to calculate predicted gradient retention times of other analytes, as well as the parameters log k_w and S required for calculating isocratic retention parameters. Calculated and experimental retention times served to determine the relative errors in the predicted retention coefficients.

The predictive potency of the approach is depicted in Figure 8-2. Clearly, it allows for only a first approximation of retention for a new analyte [49]. The relative error in gradient retention factor is 14–32%. On the other hand, the predictions

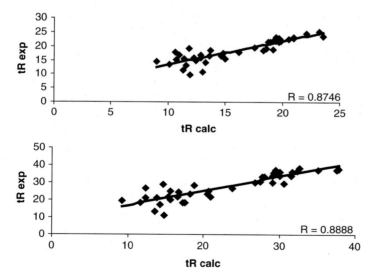

Figure 8-2. Correlations between the calculated and the experimental gradient retention times obtained on XTerra MS column (Waters Corporation, Milford, MA, USA) for a set of test analytes: 5–100% methanol gradient developed within $t_G = 20$ min (**a**) and $t_G = 40$ min (**b**) [49]

obtained for compounds for which only structural formula is known are better than guessing, even by experienced chromatographer.

A subsequent report [86] confirmed the retention prediction capability of the approach based on Eq. (8-5) and LSS theory for new, previously untested, analytes of a known molecular structure. Actually, better predictions than provided by Eq. (8-5) were observed employing Eq. (8-9):

$$\text{retention parameter} = k_1 + k_2 \, \log P \qquad (8\text{-}9)$$

where $\log P$ are values of the logarithm of octanol–water partition coefficient calculated from structural formulae by commercially available software; k_1 and k_2 are regression coefficients. Deriving Eq. (8-9) for a series of model analytes [53, 85] and employing it for any compound, a good prediction of gradient RP HPLC retention was obtained. The quality of the prediction seems to depend on the software used for calculation of $\log P$: ACD (Advanced Chemistry Development, Toronto, Canada) gave more predictive $\log P$ than HYPERCHEM (Hypercube, Waterloo, Canada) and CLOGP (Biobyte, Claremont, CA, USA). However, for such compounds, which had not been included in the partitioning database employed for deriving the $\log P$ data by the calculation methods, the universal QSRR based on the molecular modeling-derived parameters of Eq. (8-5) might be more reliable (Figure 8-3).

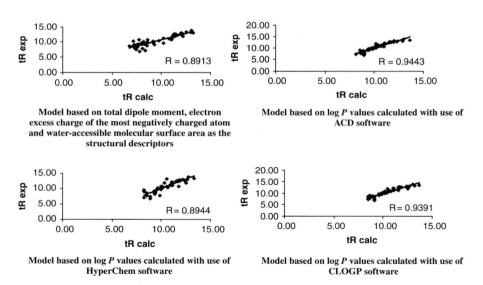

Model based on total dipole moment, electron excess charge of the most negatively charged atom and water-accessible molecular surface area as the structural descriptors

Model based on log *P* values calculated with use of ACD software

Model based on log *P* values calculated with use of HyperChem software

Model based on log *P* values calculated with use of CLOGP software

Figure 8-3. Prediction of retention from molecular modeling parameters and from calculated log *P* [87]

8.2.2.   Retention Predictions in Proteomics Research

One of the preliminary steps of proteomic analysis is peptide separation. However, little information from HPLC, usually employed for the separation, is utilized in proteomics. Meanwhile, prediction of the retention time for a given peptide, combined with routine MS/MS data analysis, could help to improve the confidence of peptide identifications.

The chromatographic retention behavior of peptides is usually related to their amino acid composition [88]. However, different values of retention coefficients of the same amino acid in different peptides in different neighborhoods are commonly observed. Specific retention coefficients were derived by regression analysis [89] or by artificial neural networks [90] with the use of a set of peptides retention.

Recently, a QSRR model has been proposed [91] to predict the peptide gradient RP HPLC retention at given separation conditions. The following structural descriptors of peptides were employed: the logarithm of the sum of gradient retention times of the amino acids composing the individual peptide, log Sum_{AA}, the logarithm of the peptide van der Waals volume, log VDW_{Vol}, and the logarithm of its calculated *n*-octanol–water partition coefficient, $CLOGP$. The general QSRR equation has the following form (8-10):

$$t_R = k_1 + k_2 \log Sum_{AA} + k_3 \log VDW_{Vol} + k_4\,C\log P \qquad (8\text{-}10)$$

where t_R is the gradient HPLC retention time and k_1–k_4 are regression coefficients.

To employ Eq. (8-10) one first needs retention coefficients of the 20 natural amino acids determined at given HPLC separation conditions. Then, a set of structurally diverse peptides must be available to determine t_R values in the same HPLC conditions and which will then allow the derivation of an appropriate QSRR equation. In the original paper [91], the training set comprised 35 peptides. The QSRR model describing gradient HPLC retention time, t_R, of these 35 peptides on an XTerra MS column (Waters, Milfors, MA, USA) with 0–60% acetonitrile gradient developed within 20 min was as follows:

$$t_R = 7.52(\pm 3.12) + 15.24(\pm 1.54) \log Sum_{AA} - 5.83(\pm 1.84) \log VDW_{Vol}$$
$$+ 0.26(\pm 0.08)\,C\log P$$
$$(8\text{-}11)$$

The statistics of Eq. (8-11) were very good: correlation coefficient $R = 0.966$, standard error of estimate $s = 1.06$, F-test value $F = 144$ and significance level $p < 10^{-17}$.

After deriving Eq. (8-11) an additional 66 peptides were chromatographed under the same conditions and a QSRR was then derived for all 101 peptides:

$$t_R = 8.02(\pm 2.04) + 14.86(\pm 0.93) \log Sum_{AA} - 5.77(\pm 1.16) \log VDW_{Vol}$$
$$+ 0.28(\pm 0.06)\,C\log P$$
$$(8\text{-}12)$$

For Eq. (8-12) $R = 0.963$, $s = 0.97$, $F = 411$ and $p < 10^{-54}$.

Equation (8-12) is not only of very high statistical quality, but it comprises the same peptide variables as Eq. (8-11) with very similar regression coefficients. Therefore, both the physical sense and predictive potency of Eq. (8-11) has been confirmed.

The good quality of QSRR model equation (8-10) to predict the retention of variable sets of peptides chromatographed on different reversed-phase columns at diverse gradient HPLC conditions has been confirmed in [92–94].

Retention data for 90 peptides of 98 reported in [92] as determined in one RP HPLC system of 19 originally tested were applied in a QSRR study by Put et al. [95]. These authors considered 1726 calculated molecular descriptors of the peptides. Retention of the peptides was modelled with so-called uninformative variable elimination partial least squares (UVE-PLS) method as well as the classical partial least squares regression. A subset of 63 peptides was used to build the QSRR models. The remaining 27 compounds served to evaluate their predictive power. Correlations between the measured relation data and those calculated from the author's [95] model are better than originally reported [92] for the QSRR based on Eq. (8-10). At the same time, empirical data for amino acids to calculate log Sum_{AA} are unnecessary. However, the latent variables extracted from 1726 molecular descriptors are not interpretable physically. In addition, a question arises whether the same set of descriptor variables would be selected for QSRR describing retention data if any other of 19 HPLC systems studied was considered or another set of peptides used. Certainly, it would be annoying if small changes in HPLC conditions required a radical change in the QSRR model, even if the statistics were perfectly correct. The QSRR model defined in Eq. (8-10) holds for all the 19 HPLC systems studied. Differences are observed only between regression coefficients k_1–k_4 depending rationally on the column and the separation conditions applied.

Another approach to protein identification supported by QSRR was developed by Cramer and co-workers [96–100]. These authors studied protein retention in cation-exchange chromatography or hydrophobic interaction chromatography on various stationary phases. QSRR models were based on a support vector machine (SVM) regression technique of Song et al. [101]. Molecular descriptors were computed from protein crystal structure and primary sequence information. The predictive power of the QSRR models was demonstrated with proteins not included in the derivation of the models. The authors claim that they obtained physically interpretable models of protein retention, thus providing insight into the factors influencing protein affinity in different HPLC systems. The physical meaning of such descriptors (listed in the original paper [101]) like *SIEPIA*, i.e., surface integral of the integral average of the electrostatic potential is, however, not clear.

8.3.   CHARACTERIZATION OF STATIONARY PHASES

For the prediction of retention and hence for confirmation of proper analyte iden-tification, QSRR models which possess a high statistical quality can be used (however uncomfortable a chemist may feel using them without understanding the physical meaning behind them). Ideally, a QSRR to be used in comparative studies of stationary phases and mechanism of retention must be physically interpretable.

A detailed review of QSRRs for the quantitative characteristics of retention properties of stationary phases published up until 1997 can be found in reference [12]. Readers who are interested in the classification of stationary phases based on principal component analysis (PCA) of retention data of test series of analytes are referred to that monograph. The demonstration of similarities and dissimilarities of the retention properties of individual GLC, TLC, and HPLC stationary phases, by showing their relationships in the plot of the first two principal components, is now a routine procedure in the studies of new phases. An instructive methodologically more recent example of grouping of 30 stationary phases for GLC by PCA and relating rationally that grouping to phases polarity indicators (McReynolds constants, Snyder's selectivity parameters) was given by Heberger [102]. In HPLC, the explanation of PCA grouping was performed [103, 104] in terms of regression coefficients of the molecular descriptors in the QSRR equations of the form of Eqs. (8-2), (8-5) and (8-9).

Following [12] several reports appeared to be related to the evaluation by QSRRs of the retention properties of reversed-phase materials used in HPLC. The conclusions drawn by individual authors as regards the molecular mechanism operating in individual types of HPLC systems were mutually consistent [8, 73, 80, 105–109]. However, quantitative comparisons were difficult, because the QSRR reported have been derived for different sets of test analytes. Therefore, it appeared advisable to design a model series of test analytes that could be recommended for individual types of QSRR analysis for comparative studies of the mechanism of chromatographic separation.

The requirements of high quality and meaningful statistics were taken into consideration during the designing of the test series of analytes. The analytes were selected such that, within the series, the intercorrelations were minimized among the individual analyte structural descriptors. At the same time, the selection of test analytes was designed to provide as wide a range as possible and even distribution of individual structural descriptor values. Moreover, the structural descriptors induced in the final QSRR equations must all be significant at the 99.9% significance level. The multiple correlation coefficient, R, should be around 0.99. Additionally, the series of test analytes must be large enough to exclude chance correlations but not too big to save work required for chromatographic and structural analysis. There is a rule of thumb [110] that at least five data points of regressand (retention parameter) should fall per regressor (structural descriptor). Therefore, if one describes the retention coefficients of a set of analytes by a three-descriptor QSRR equation, then the number of the analytes should be 15 or greater. And finally, the test analytes should be readily accessible and cause no operational problems.

Having all the above in mind, three model series of analytes for three types of QSRR analysis have been proposed [53]. One for the reduced LSER-based model (Eq. 8-3), another for the model expressed by Eq. (8-9), and a third for the model employing calculated chemistry descriptors, Eq. (8-5). All the models were demonstrated to provide reliable QSRRs for five different sets of RP HPLC retention data, accounting for differences in both stationary phase material (C8, C18, and IAM columns) and the organic modifier used (methanol or acetonitrile). These equations discriminated quantitatively individual chromatographic systems and were interpretable in straightforward chemical categories. In view of such QSRRs obtained, the retention processes clearly emerged as the net effects of fundamental intermolecular interactions involving the analyte and the components of the chromatographic system.

In the following papers, the performance of the approach described above was considered in the objective evaluation of chromatographic columns [77, 85, 111]. Independently, in 1999 Rohrschneider [112] proposed a set of 26 prototypical test substances for characterization of RP HPLC columns by QSRR, the performance of which has not been further confirmed.

In Table 8-3, the regression coefficients of Eq. (8-5) describing the retention parameter, log k_w, of the test series of analytes proposed in [81] on seven modern reversed-phase columns with both methanol and acetonitrile as organic modifiers of aqueous mobile phases are presented [85]. Log k_w is a standardized retention parameter preferred in QSRR studies instead of individual isocratic log k. It is intercept of the linear relationship between the isocratic log k values and the corresponding contents of the organic modifier in the eluent.

The physical meaning of equations characterized in Table 8-3 is clearly recognized. The net positive input to retention is because of the A_{WAS} parameter. This parameter is evidently related to the ability of analytes to take part in dispersive interactions. Obviously, these attractive London-type interactions are stronger between an analyte and the bulky ligand of the stationary phase than between the same analyte and the small molecules of the eluent. Therefore, there is a positive sign at the k_4 regression coefficient in Eq. (8-5).

Table 8-3. Coefficients of a general QSRR equation, $\log k_w = k_1 + k_2\mu + k_3\delta_{Min} + k_4 AWA_S$, for test series of analytes relating retention parameters, log k, to structural descriptors of analytes from molecular modeling

log k_w (column, organic solvent)	k_1	k_2	k_3	k_4	R	s	F
log k_w (Aluspher, MeOH)	−3.3477	−0.3490	−2.3799	0.0180	0.9938	0.1775	375
log k_w (Hisep, MeOH)	−2.6759	−0.1100	−1.4997	0.0140	0.9847	0.1974	148
log k_w (Nova-Pak, MeOH)	−1.9956	−0.3054	−3.0880	0.0171	0.9955	0.1471	513
log k_w (Luna, MeOH)	−1.2007	−0.2477	−3.0623	0.0149	0.9960	0.1211	579
log k_w (Discov.Amide, MeOH)	−1.9653	−0.2284	−2.3735	0.0151	0.9946	0.1364	431
log k_w (Discov.Cyano, MeOH)	−2.3211	−0.1128	−1.2506	0.0114	0.9952	0.0886	505
log k_w (Mix-Chol-AP, MeOH)	−2.7928	−0.2563	−2.3630	0.0172	0.9894	0.2153	217
log k_w (Aluspher, ACN)	−1.8284	−0.3111	−2.8433	0.0126	0.9856	0.2131	158
log k_w (Hisep, ACN)	−1.5144	−0.1398	−1.5771	0.0103	0.9819	0.1694	125
log k_w (Nova-Pak, ACN)	−0.1658	−0.2328	−4.1331	0.0096	0.9881	0.1683	192
log k_w (Luna, ACN)	0.1658	−0.2223	−3.2134	0.0081	0.9881	0.1424	192
log k_w (Discov.Amide, ACN)	−0.3196	−0.1828	−3.1903	0.0086	0.9888	0.1373	205
log k_w (Discov.Cyano, ACN)	−1.6697	−0.1346	−1.5310	0.0095	0.9922	0.1032	292
log k_w (Mix-Chol-AP, ACN)	−1.4724	−0.2112	−2.3100	0.0115	0.9855	0.1819	157s

R – correlation coefficient, s – standard error of estimate, F – value of F-test of significance [85]

The inputs to retention through the specific polar intermolecular interactions are reflected by the coefficients k_2 and k_3 in Table 8-3. Coefficient k_2 describes that the net effect to retention provided by dipole moment, μ, is negative. It appears reasonable because the dipole–dipole and dipole-induced dipole attractions are obviously stronger between an analyte and the polar molecules of the eluent than between the same analyte and the non-polar ligands (mainly hydrocarbons) of the stationary phase.

An analogous explanation is valid with regards to the coefficient k_3 in Table 8-3. The greater the electron distribution on the most charged atom in the molecule the less retained is the analyte.

Finally, the coefficient k_2 in Table 8-3 differentiates the phases from each other. It also clearly differentiates the methanol–water from the acetonitrile–water system. According to k_2, the most polar of the phases studied would be Hisep and Discovery Cyano. The Discovery Amide, Luna, and Mix-Cholesterol-AP stationary phases would be of lesser dipolarity. The least polar would be the Aluspher column followed by Nova-Pak. This arrangement agrees well with observations drawn from QSRRs based on the reduced LSER model equation (8-3), also studied in the work.

8.4. ASSESSMENT OF LIPOPHILICITY BY QSRR

The processes of drug absorption, distribution, and excretion in pharmacokinetic phase of drug action, as well as drug-receptor interactions in the pharmacodynamic phase (Figure 8-4), are dynamic in nature. One can state that the analyte's distribution processes in chromatography are also equally dynamic.

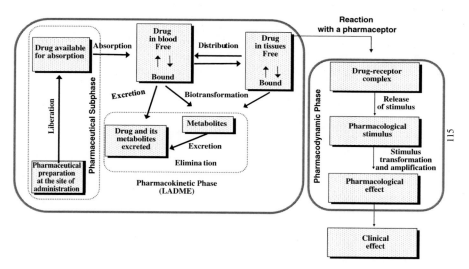

Figure 8-4. Fundamental processes at the basis of drug action

The partition of a compound between organic and aqueous phases determines both its permeation through biological membranes and retention in reversed-phase liquid chromatographic systems. Therefore, QSRRs might be of use in the fast and convenient assessment of lipophilicity (hydrophobicity) of xenobiotics [113].

The expressions "hydrophobicity" and "lipophilicity" are very often considered to be synonymous. In the scientific use, however, their meaning may be different [114]. The following operational definitions have been suggested by the IUPAC:

Hydrophobicity is the association of non-polar groups or molecules in an aqueous environment which arises from the tendency of water to exclude non-polar molecules.

Lipophilicity represents the affinity of a molecule or a moiety for a lipophilic environment. It is commonly measured by its distribution behavior in a biphasic system, either liquid–liquid (e.g., partition coefficient in 1-octanol–water) or solid–liquid (e.g., retention in TLC or reversed-phase HPLC) systems.

The partition coefficient was first introduced in 1872 by Berthelot and Jungfleisch [115]. In 1941 Martin [116] demonstrated that retention coefficient obtained from TLC could be related to partition coefficient. There are hundreds of reports in scientific literature on the chromatographic determination of lipophilicity and relevant QSRRs. Most have been systematically reviewed in earlier monographs [5, 8, 10, 114, 117, 118]. Among them special recommendation deserves the exhaustive critical review by Poole and Poole [118]. Also worth of noting are reviews emphasizing specific aspects of lipophilicity determination by TLC [119], HPLC [120], and MEKC (micellar electrokinetic chromatography) [118, 121, 122].

Although there are many reports on very good correlations between the logarithm of retention coefficient, log k, and the slow-equilibrium n-octanol–water partition coefficient, log P, dissimilarities between the two partition systems are evident.

For QSRR studies, the retention parameter is usually applied which corresponds to a given composition of mobile phase. In reversed-phase HPLC it normally is log k_w, which corresponds to pure water as the mobile phase.

Extrapolation to pure water is a useful and convenient means of standardization of chromatographic lipophilicity parameters. The question is, however, which is the most appropriate description of the dependence of retention parameters on the composition of the eluent.

There are two models used in practice. A linear one, expressed by Eq. (8-6), and quadratic one described by Eq. (8-13):

$$\ln k = A\varphi^2 + B\varphi + C \qquad (8\text{-}13)$$

where φ is the volume fraction of the organic modifier in binary aqueous eluent and A, B, and C are regression coefficients.

Statistical evaluation of linear and quadratic models was performed [123]. The relationships describing retention were derived for a set of 23 selected test analytes chromatographed on 18 HPLC columns using methanol–water and acetonitrile–water solutions as mobile phase. It was ascertained whether the square term in

Eq. (8-13) improves the description of retention in a statistically significant manner. It was also checked whether the retention data extrapolated to a hypothetical neat water eluent, log k_w, obtained with the two models and the two organic modifiers are equivalent or should be considered different. The study proved that both the models give similar results and the extrapolated log k_w values do not differ significantly statistically in the case of methanol-containing mobile phases. In the case of acetonitrile–water systems, the log k_w values obtained with linear and quadratic models are statistically different for most columns studied. Such differences are underlined in the relevant review by Poole et al. [124].

The best correlations with log P were obtained employing log k_w data derived with the linear model for methanol–water systems in agreement with earlier suggestions [125, 126]. For a series of test analytes, the log k_w values achieved by isocratic elution and by organic solvent gradient elution at pH suppressing ionization were observed to correlate very well [127, 128]. The values of log k_w produced by the two methods are closely similar for most analytes for two columns (Inertsil ODS-3 packed with octadecyl-bonded silica and Aluspher 100 RP-select B packed with polybutadiene-coated alumina) and the two organic modifiers (methanol, acetonitrile) studied (Figure 8-5). However, a higher correlation was noted in the case of methanol-containing eluents on Inertsil ODS-3 and Aluspher 100 RP-select B columns than on the same columns when acetonitrile-containing mobile phases were used. For the Aluspher 100 RP-select B column not only is the correlation between log k_w determined isocratically and in gradient mode is higher, but also the slope of the relationship is close to unity and the intercept is close to zero. In other words, the log k_w values determined on the alumina-based stationary phase by gradient method are strictly equivalent to those obtained by standard isocratic procedure. That might be due to the absence of specific interactions of analytes with stationary phase support material of alumina, as opposed to the not fully controllable interactions of analytes with free silanol groups present on the surface of typical silica-based reversed-phase material.

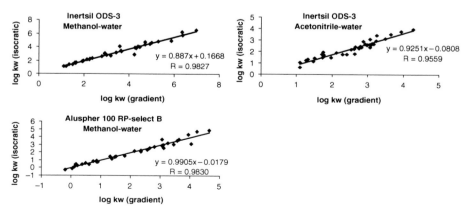

Figure 8-5. Correlations between log k_w values determined by gradient vs. isocratic RP HPLC mode on two columns with two organic modifiers of binary aqueous eluent [128]

It has been demonstrated on a structurally diverse series of test analytes that the log k_w values determined by both isocratic and organic gradient mode correlated very well with literature log P data [128]. Again, the correlation is better when methanol is used as the organic modifier in the mobile phase. A better log k_w vs. log P correlation was obtained on alumina-based column than on the conventional silica-based column.

A high-throughput procedure for the determination of lipophilicity and acidity, applicable to multicomponent mixtures of analytes and based on gradient RP HPLC, has recently been theoretically and experimentally established [87, 129–132]. The approach is based on the simultaneous development of the programmed gradient of pH and of organic modifier concentration in the mobile phase. The derived comprehensive theoretical model of the pH/organic modifier double-gradient RP HPLC allows a rapid assessment of both acidity, pK_a, of weak acids and bases and their lipophilicity. Verification of the reliability of the parameters determined by the new method was demonstrated [132] on a series of 93 drug analytes with acidic and basic properties for which reliable literature pK_a and log P data were accessible. Figure 8-6 presents a correlation between chromatographic and literature pK_a data. It must be noted here that very recently Subirats et al. [133] reported a better correlation between the chromatographically determined and standard pK_a values of ionizable analytes. However, the method applied is time-consuming and inconvenient for practical use, because it requires a series of controlled isocratic measurements.

Obviously, chromatographic processes cannot directly model the bulk octanol–water partitioning process, because the non-polar stationary phase in liquid chromatography is an interphase (immobilized at one end) and not a bulk medium. In RP HPLC, the partition of analytes occurs between the mobile phase and the stationary zone formed by preferential adsorption of the organic component of the eluent on the stationary phase. Differences between RP HPLC systems and the regular liquid–liquid partition system become obvious if one realizes that in

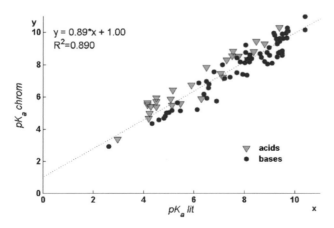

Figure 8-6. Chromatographically determined acidity parameter vs. literature pK_a data for 93 drug analytes studied in double-gradient RP HPLC [133]

chromatography possible equilibrium involves the following types of intermolecular interactions: analyte–stationary phase, analyte–solvent, solvent–stationary phase, solvent–solvent and mutual interactions between the flexible fragments of the stationary phase. Therefore, the relationships between log P and the RP HPLC retention parameters are often less than moderate [12, 134, 135].

To facilitate lipophilicity determination by RP HPLC, Donovan and Pescatore [136] proposed applying a fast methanol–water gradient and a short polymeric octadecyl–polyvinyl alcohol (ODP) column. Gulyaeva et al. [137] used that method to determine lipophilicity of 63 drugs of diverse structure. The partitioning of the drugs was also studied at the same pH 7.4 in aqueous dextran-polyethylene glycol two-phase system and in the octanol-buffer system. A rather poor correlation ($R = 0.918$) was found between the lipophilicity parameters determined in the standard octanol-buffer system and when employing the RP HPLC method. Of the three analytical techniques employed, no single lipophilicity parameter correlated with blood–brain barrier permeability of the drugs studied.

A fast gradient HPLC procedure was used in comparative studies on the lipophilicity of 11 basic drugs [138]. Two polar-embedded phases were studied: alkylamide SG-AP and cholesterolic SG-CHOL. Two other phases considered were octadecyl-bonded silica: monolithic Chromolith RP-18e and silica gel Supelcosil C_{18}-DB. Methanol and acetonitrile were used as mobile phase modifiers. The logarithms of retention factors from gradient elution, log k_g, were related to calculated log P values (*CLOGP*) of analytes. As was expected, better linearity was found in the case of methanol-buffer than acetonitrile-buffer eluents.

Determinations of lipophilicity of non-dissociated forms of basic analytes caused specific problems when using silica-based materials because of instability of silica support at higher pH. It was partially solved after introducing polymer encapsulated silica stationary phases. The optimization of the stationary phase with respect to the preparation of silica support and the manufacturing of modern columns for the analysis of basic pharmaceuticals is discussed in detail by Vervoort et al. [139].

Buszewski et al. [140] prepared a chemically bonded silica stationary phase (SG-MIX) with different non-polar and polar functional groups, such as hydroxyl, amino, cyano, phenyl, octadecyl, and octyl on the surface of the silica support. Correlation between log k_w from that column and log P was significantly higher than for the previously discussed alkylamide column. Evidently SG-MIX stationary phase has retention characteristics closer to that of standard reversed-phase materials.

For a long time, the reduction of silanophilic interaction by the addition of variable masking agents to the mobile phase has been suggested to improve the correlations between RP HPLC retention parameters and log P [5, 12]. Lombardo et al. [141] confirmed older observations that the presence of decylamine in mobile phase used for lipophilicity determination of neutral and basic compounds improves the reliability of the method.

Recently, ionic liquids have been demonstrated [142, 143] as the effective residual free silanol blocking agents. For example, addition of 1.5% (v/v) 1-ethyl-3-methylimidazolium tetrafluoroborate to methanol–water eluent markedly improved correlation of log k_w with log P of basic drugs.

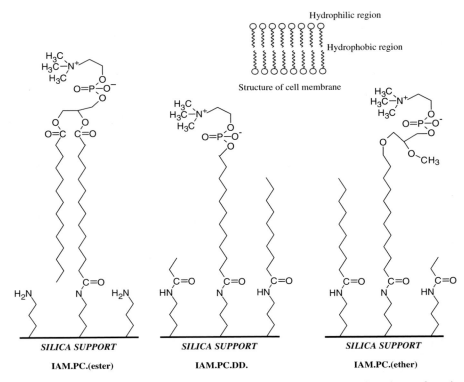

Figure 8-7. Structure of immobilized artificial membrane stationary phases and a scheme of a unit biomembrane

The chromatographic approach seems to appear as an effective tool in duplicating log *P* data. However, the lipophilicity measuring system is not designed to model octanol–water partition system but the processes occurring in the biophase. Therefore, the similarity of the components of the chromatographic and the biological system appeared rational. An original chromatographic stationary phase was introduced by Pidgeon and co-workers [144, 145] to immobilize enzymes under the name of immobilized artificial membrane (IAM). The phases of that type contain a phosphatidylcholine ligand, having a polar head groups and non-polar hydrocarbon chains, which is covalently bound to aminopropylsilica, and thus form a cofluent monolayer of immobilized lipid-like entities (Figure 8-7). As with biomembranes, the polar head groups protrude away from the organized hydrophobic layer and are the first contact sites between the analytes and the surface of IAM. Since a correlation was reported [146] between retention coefficient determined on IAM columns and log *P*, the columns called wide interest of medicinal chemists. Generally, the numerous correlations reported were not very significant and depended on chemical structure of the chromatographed compounds [12, 120, 122, 147–150].

8.5.   QSRR IN AFFINITY CHROMATOGRAPHY

Modern techniques and procedures of HPLC allow for the inclusion of various biomacromolecules as active components of separation systems. Protein stationary phases for HPLC were introduced in the early 1980s for enantioselective determination of chiral drugs. Among them were bovine serum albumin, human serum albumin, α_1-acid glycoprotein, ovomucoid, flavoprotein, avidin, pepsin, trypsin, α-chymotrypsin, cellulase, lysozyme, keratin, collagen, melanin, amylose tris(3,5-dimethylphenylcarbamate), cellulose triacetate, cellulose tribenzoate, zein, riboflavin-binding proteins and basic fatty acid-binding proteins from chicken liver. Protein phases were subjects of systematic reviews [151, 152].

Since Wainer and co-workers [153–155] proposed the use of HPLC, employing human serum albumin chemically bound to silica as the stationary phase, to quantitative the protein binding of several classes of drugs, chromatography became a convenient tool for studying drug–protein interactions. That way a substantial amount of precise and reproducible drug binding-related data can be readily produced, unlike in the standard methodology in which the drug and the protein are incubated in a buffer solution and the free fraction is separated from the bound fraction by dialysis or ultrafiltration.

For example, an α_1-acid glycoprotein (AGP) bound to a silica HPLC column was demonstrated to derive QSRRs and then characterize structurally the binding site for basic drugs on the protein [156, 157]. For a series of 49 basic drugs of diverse chemical structures and pharmacological activities, the log k_{AGP} was determined. The QSRR equation is given by

$$\log k_{\text{AGP}} = 1.688(\pm 0.245) + 0.658(\pm 0.040) \log k_{\text{IAM}} + 3.342(\pm 0.841)N_{ch} \\ -0.0081(\pm 0.0030)S_T$$

$$(8\text{-}14)$$

where log k_{IAM} was a lipophilicity parameter determined on an IAM column, N_{ch} was electron excess charge on the most highly charged aliphatic nitrogen atom, and S_T was surface area of a triangle drawn with one vertex located on the aliphatic nitrogen and the two remaining vertexes placed on the extreme atoms of the aromatic substituents in the geometry optimized molecular structure (Figure 8-8). The statistical quality of Eq. (8-14) is good: $R = 0.929$, $s = 0.163$, $F = 92$.

The above QSRR equation (8-14) provides an indication of the qualitative characteristics of the binding mode of xenobiotics to AGP [158] and allows for an indirect identification of structural features of the binding site of basic drugs on that protein. The site is assumed to possess a form of an open conical pocket. Its internal surface contains lipophilic regions at the base of the cone. There is an anionic region close to the spike of the cone. The hypothetical mechanism of binding is as follows: protonated aliphatic nitrogen guides drug molecules toward the anionic region. Lipophilic hydrocarbon fragments of the drug fix the molecule in the lipophilic regions of the binding site. There is a steric restriction for the molecule to plunge into the binding site.

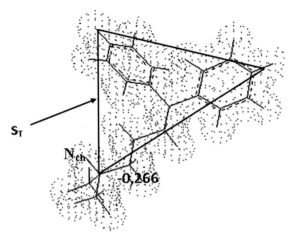

Figure 8-8. Graphical representation of structural descriptors N_{ch} and S_T of drugs used in QSRR equation (8-14) with the example of pheniramine

8.6.   CONCLUSIONS

In relating structure and chromatographic retention, the statistical approach is applicable and lacks the rigor of thermodynamics, but provides otherwise inaccessible information. Currently, QSRRs are the most extensively studied manifestations of linear free-energy relationships. QSRRs are also the most often reported results of chemometric data processing. At the same time, the performance in QSRR verifies the reliability of numerous proposed chemometric methods as well as structural descriptors of chemical compounds. Although all QSRRs have been derived retrospectively, their potential for the identification of chromatographed analytes must be recognized, even if the physical meaning of structural parameters employed is occasionally unclear. Due to QSRRs, an optimization of chromatographic conditions can be rationally guided to provide a good separation of given structurally defined analytes. Likewise, intermolecular interactions determining separations on specific stationary phases may be identified and quantitatively compared. Including biomacromolecules as components of separation systems allows for the ready in vitro determination of measures of their interactions with drugs and other xenobiotics. Then, structural requirements for compounds to participate in these interactions can be identified and the most effective agents rationally designed. Relating conveniently chromatographically derived lipophilicity parameters to pharmacokinetic properties facilitates the selection of those drug candidates which are likely to be absorbed to and distributed within a living organism. Systematic information extracted chemometrically from the readily produced large sets of diverse chromatographic data might be used for pharmacologically consistent classification of analytes. When applied to sets of metabolites determined in biological fluids that approach may result in diagnostic differentiation of donors of these fluids.

Prediction of retention changes, accompanying the changes in composition of the eluent, might be achieved by applying QSRR-based models. However, the main limitation of a priori prediction of retention (as well as other physicochemical and biological properties) of chemical compounds from their structure is the overall difficulty of translation of structural formulae into sets of numerical descriptors. There is a lack of such translation that would reveal the properties encoded in the structure in a sufficiently reliable manner considering applications for all analytes available in laboratory practice. Hopefully, further progress in molecular modeling and theoretical chemistry will bring better characterization of chemical compounds and make the predictions of their properties more feasible. Quantitative structure–retention relationships (QSRRs) are an efficient and reliable tool to test the performance of such new solutions.

ACKNOWLEDGEMENT

The work was supported by the Polish State Committee for Scientific Research Project N N405 1040 33.

REFERENCES

1. Kaliszan R (1992) Quantitative structure–retention relationships. Anal Chem 64:619A–631A
2. Kaliszan R, Foks H (1977) The relationship between the Rm values and the connectivity indices for pyrazine carbothioamide derivatives. Chromatographia 10:346–349
3. Kaliszan R (1977) Correlation between the retention indices and the connectivity indices of alcohols and methyl esters with complex cyclic structure. Chromatographia 10:529–531
4. Michotte Z, Massart DL (1977) Molecular connectivity and retention indexes. J Pharm Sci 66:1630–1632
5. Kaliszan R (1987) Quantitative structure–chromatographic retention relationships. Wiley, New York
6. Karelson M, Lobanov VS, Katritzky AR (1996) Quantum-chemical descriptors in QSAR/QSPR studies. Chem Rev 96:1027–1043
7. Kaliszan R (2000) Recent advances in quantitative structure–retention relationships. In: Valko K (ed) Separation methods in drug synthesis and purification. Elsevier, Amsterdam
8. Kaliszan R (2007) QSRR: Quantitative structure–(chromatographic) retention relationships. Chem Rev 107:3212–3246.
9. Smith RM (1995) Retention and selectivity in liquid chromatography. Elsevier, Amsterdam
10. Jurs PC (1996) Computer software applications in chemistry. Wiley, New York
11. Forgacs E, Cserhati T (1997) Molecular bases of chromatographic separations. CRC Press, Boca Raton
12. Kaliszan R (1997) Structure and retention in chromatography. A chemometric approach. Harwood, Amsterdam
13. Jinno K (ed) (1997) Chromatographic separations based on molecular recognition. Wiley, New York
14. Hansch C, Fujita T (1964) ρ-σ-π analysis. A method for the correlation of biological activity and chemical structure. J Am Chem Soc 86:1616–1626
15. Albert A (1965) Selective toxicity. Wiley, New York
16. Seydel JK, Schaper K-J (1979) Chemische struktur und biologische aktivität von wirkstoffen. Verlag Chemie, Weinheim

17. Iwasa J, Fuijta T, Hansch CJ (1965) Substituent constants for aliphatic functions obtained from partition coefficients. J Med Chem 8:150–153

18. Wang QS, Zhang L (1999) Review of research on quantitative structure–retention relationships in thin-layer chromatography. J Liq Chrom Rel Techn 22:1–14

19. Sadek PC, Carr PW, Doherty RM et al. (1985) Study of retention processes in reversed-phase high-performance liquid chromatography by the use of the solvatochromic comparison method. Anal Chem 57:2971–2978

20. Carr PW, Doherty RM, Kamlet MJ et al. (1986) Study of temperature and mobile-phase effects in reversed-phase high-performance liquid chromatography by the use of the solvatochromic comparison method. Anal Chem 58:2674–2680

21. Snyder LR, Carr PW, Rutan SC (1993) Solvatochromically based solvent-selectivity triangle. J Chromatogr A 656:537–547

22. Hasan MN, Jurs PC (1990) Prediction of gas and liquid chromatographic retention indices of polyhalogenated biphenyls. Anal Chem 62:2318–2323

23. Katritzky AR, Karelson M, Lobanov VS (1997) QSPR as a means of predicting and understanding chemical and physical properties in terms of structure. Pure Appl Chem 69:245–248

24. Lucic B, Trinajstic N, Sild S et al. (1999) A new efficient approach for variable selection based on multiregression: Prediction of gas chromatographic retention times and response factors. J Chem Inf Comput Sci 39:610–621

25. Katritzky AR, Chen K, Maran U et al. (2000) QSPR correlation and predictions of GC retention indexes for methyl-branched hydrocarbons produced by insects. Anal Chem 72:101–109

26. Ignatz-Hoover F, Petrukhin R, Karelson M et al. (2001) QSRR correlation of free-radical polymerization chain-transfer constants for styrene. J Chem Inf Comput Sci 41:295–299

27. Todeschini R, Consonni V (2000) Handbook of molecular descriptors. Wiley-VCH, Weinheim

28. Todeschini R, Consonni V, Pavan M (http://www.disat.unimib.it/chm/dragon.htm) Dragon software version 5.0, 2004

29. Ivanciuc O, Ivanciuc T, Cabrol-Bass D et al. (2000) Comparison of weighting schemes for molecular graph descriptors: Application in quantitative structure–retention relationship models for alkylphenols in gas-liquid chromatography. J Chem Inf Comput Sci 40:723–743

30. Junkes BS, Amboni RDMC, Yunes RA et al. (2003) Prediction of the chromatographic retention of saturated alcohols on stationary phases of different polarity applying the novel semi-empirical topological index. Anal Chim Acta 477:29–39

31. Ośmiałowski K, Halkiewicz J, Kaliszan R (1986) Quantum chemical parameters in correlation analysis of gas–liquid chromatographic retention indices of amines. J Chromatogr 361:63–69

32. Randić M (1975) On characterization of molecular branching. J Am Chem Soc 97:6609–6615

33. Hanai T (2005) Chromatography in silico, basic concept in reversed-phase liquid chromatography. Anal Bioanal Chem 382:708–717

34. Salo M, Sirén H, Volin P et al. (1996) Structure–retention relationships of steroid hormones in reversed-phase liquid chromatography and micellar electrokinetic capillary chromatography. J Chromatogr A 728:83–88

35. Makino M (1999) An analysis of the chromatographic behavior of the environmental contaminants polychlorinated naphthalenes using electrical and topological descriptors. Toxicol Environ Chem 73:117–128

36. Dimov N, Stoev S (1999) A new approach to structure–retention relationships. Acta Chromatogr 9:17–18

37. Zhang L, Zhang M, Tang GZ et al. (2000) Retention prediction system of O-aryl,O-(1-methylthioethylidene-amino)phosphates on RP-HPLC. HRC J High Resol Chromatogr 23:445–448

38. Ledesma EB, Wornat MJ (2000) QSRR prediction of chromatographic retention of ethynyl-substituted PAH from semiempirically computed solute descriptors. Anal Chem 72:5437–5443

39. Kaliszan R, Ośmiałowski K, Tomellini SA et al. (1986) Quantitative retention relationships as a function of mobile and C18 stationary phase composition for non-cogeneric solutes. J Chromatogr 352:141–155

40. Kaliszan R, Lamparczyk H, Radecki A (1979) A relationship between regression of dimethylnitrosamine-demethylase by polycyclic aromatic hydrocarbons and their shape. Biochem Pharmacol 28:123–125

41. Wise SA, Bonnett WJ, Guenther FR et al. (1981) A relationship between reversed-phase C18 liquid chromatographic retention and the shape of polycyclic aromatic hydrocarbons. J Chromatogr Sci 19:457–465

42. Lippa KA, Sander LC, Mountain RD (2005) Molecular dynamics simulations of alkylsilane stationary-phase order and disorder. 1. Effects of surface coverage and bonding chemistry. Anal Chem 77:7852–7861

43. Montaña MP, Pappao B, Debattista NB et al. (2000) High-performance liquid chromatography of chalcones: Quantitative structure–retention relationships using partial least-squares (PLS) modeling. Chromatographia 51:727–735

44. Silva MF, Chipre LF, Raba J et al. (2001) Amino acids characterization by reversed-phase liquid chromatography. Partial least-squares modeling of their transport properties. Chromatographia 53:392–400

45. Csiktusnádi-Kiss GA, Forgács E, Markuszewski M (1998) Application of multivariate mathematical–statistical methods to compare reversed-phase thin-layer and liquid chromatographic behaviour of tetrazolium salts in Quantitative Structure–Retention Relationships (QSRR) studies. Analusis 26:400–406

46. Åberg MK, Jacobsson SP (2001) Pre-processing of three-way data by pulse-coupled neural networks – An imaging approach. Chemom Intell Lab Sys 57:25–36

47. Tham SY, Agatonovic-Kustrin S (2000) Application of the artificial neural network in quantitative structure-gradient elution retention relationship of phenylthiocarbamyl amino acids derivatives. J Pharm Biomed Anal 28:581–590

48. Ruggieri F, D'Archivio AA, Carlucci G et al. (2005) Application of artificial neural networks for prediction of retention factors of triazine herbicides in reversed-phase liquid chromatography. J Chromatogr A 1076:163–169

49. Bączek T, Kaliszan R (2002) Combination of linear solvent strength model and quantitative structure–retention relationships as a comprehensive procedure of approximate prediction of retention in gradient liquid chromatography. J Chromatogr A 962:41–55

50. Kaliszan R, Bączek T, Buciński A et al. (2003) Prediction of gradient retention from the linear solvent strength (LSS) model, quantitative structure–retention relationships (QSRR), and artificial neural networks (ANN). J Sep Sci 26:271–282

51. Snyder R, Dolan JW (1998) The linear-solvent-strength model of gradient elution. Adv Chromatogr 38:115–185

52. Haber P, Bączek, T, Kaliszan R et al. (2000) Computer simulation for the simultaneous optimization of any two variables and any chromatographic procedure. J Chromatogr Sci 38:386–392

53. Al-Haj MA, Kaliszan R, Nasal A (1999) Test analytes for studies of the molecular mechanism of chromatographic separations by quantitative structure–retention relationships. Anal Chem 71:2976–2985

54. Schefzick S, Kibbey C, Bradley MP (2004) Prediction of HPLC conditions using QSPR techniques: An effective tool to improve combinatorial library design. J Comb Chem 6:916–927

55. Cramer RD III, Patterson DE, Bunce JD (1988) Comparative molecular field analysis (CoMFA). 1. Effect of shape on binding of steroids to carrier proteins. J Am Chem Soc 110:5959–5967

56. Luo H, Cheng Y-K (2005) Quantitative structure–retention relationship of nucleic-acid bases revisited. CoMFA on purine RPLC retention. QSAR Comb Sci 24:969–975

57. Sybyl (Linux) 6.9.1 (2003) Tripos, Inc. St. Louis, MO, USA

58. Sutter JM, Peterson TA, Jurs PC (1997) Prediction of gas chromatographic retention indices of alkylbenzenes. Anal Chim Acta 342:113–122

59. Massart DL, Vandeginste BGM, Buydens LMC et al. (1997) Handbook of chemometrics and qualimetrics: Part A. Elsevier, Amsterdam

60. Vandeginste BGM, Massart DL, Buydens LMC et al. (1998) Handbook of chemometrics and qualimetrics: Part B. Elsevier, Amsterdam

61. Ros F, Pintore M, Chrétien JR (2002) Molecular descriptor selection combining genetic algorithms and fuzzy logic: Application to database mining procedures. Chemometr Intell Lab Sys 63:15–26

62. Put R, Perrin C, Questier F et al. (2003) Classification and regression tree analysis for molecular descriptor selection and retention prediction in chromatographic quantitative structure–retention relationship studies. J Chromatogr A 988:261–276

63. Put R, Xu QS, Massart DL, Vander Heyden Y (2004) Multivariate adaptive regression splines (MARS) in chromatographic quantitative structure–retention relationship studies. J Chromatogr A 1055:11–19

64. Abraham MH, Chadha HS, Whiting GS et al. (1994) Hydrogen bonding. 32. An analysis of water–octanol and water–alkane partitioning and the Δlog P parameter of seiler. J Pharm Sci 83:1085–1100

65. Abraham MH, Treiner C, Rosés M et al. (1996) Linear free energy relationship analysis of microemulsion electrokinetic chromatographic determination of liophilicity. J Chromatogr A 752:243–249

66. Abraham MH, Chadha HS, Leitao RAE et al. (1997) Determination of solute lipophilicity, as log P(octanol) and log P(alkane) using poly(styrene-divinylbenzene) and immobilised artificial membrane stationary phases in reversed-phase high-performance liquid chromatography. J Chromatogr A 766:35–47

67. Abraham MH (1997) On characterization of some GLC chiral stationary phases: LFER analysis. Anal Chem 69:613–617

68. Abraham MH (2004) The factors that influence permeation across the blood–brain barrier. Eur J Med Chem 39:235–240

69. Abraham MH, Poole CF, Poole SK (1999) Classification of stationary phases and other materials by gas chromatography. J Chromatogr A 842:79–114

70. Abraham MH, Ibrahim A, Zissimos AM (2004) Determination of sets of solute descriptors from chromatographic measurements. J Chromatogr A 1037:29–47

71. Wilson NS, Nelson MD, Dolan JW et al. (2002) Column selectivity in reversed-phase liquid chromatography: I. A general quantitative relationship. J Chromatogr A 961:171–193

72. Zhao J, Carr PW (2000) A comparative study of the chromatographic selectivity of polystyrene-coated zirconia and related reversed-phase materials. Anal Chem 72:302–309

73. Park JH, Yoon MH, Ryu YK et al. (1998) Characterization of some normal-phase liquid chromatographic stationary phases based on linear solvation energy relationships. J Chromatogr A 796:249–258

74. Rosés M, Bolliet D, Poole CF (1998) Comparison of solute descriptors for predicting retention of ionic compounds (phenols) in reversed-phase liquid chromatography using the solvation parameter model. J Chromatogr A 829:29–40

75. Li J, Cai B (2001) Evaluation of the retention dependence on the physicochemical properties of solutes in reversed-phase liquid chromatographic linear gradient elution based on linear solvation energy relationships. J Chromatogr A 905:35–46

76. Li J, Sun J, Cui Z et al. (2006) Quantitative structure–retention relationship studies using immobilized artificial membrane chromatography I: Amended linear solvation energy relationships with the introduction of a molecular electronic factor. J Chromatogr A 1132:174–182

77. Vonk EC, Lewandowska K, Claessens HA et al. (2003) Quantitative structure–retention relationships in reversed-phase liquid chromatography using several stationary and mobile phases. J Sep Sci 26:777–792

78. Poole CF, Poole SK (2002) Column selectivity from the perspective of the solvation parameter model. J Chromatogr A 965:263–299

79. Vitha M, Carr PC (2006) The chemical interpretation and practice of linear solvation energy relationships in chromatography. J Chromatogr A 1126:143–194

80. Jackson PT, Schure MR, Weber TP et al. (1997) Intermolecular interactions involved in solute retention on carbon media in reversed-phase high-performance liquid chromatography. Anal Chem 69:416–425

81. Sandi A, Bede A, Szepesy I et al. (1997) Characterization of different RP-HPLC columns by a gradient elution technique. Chromatographia 45:206–214

82. Tan LC, Carr PW (1998) Study of retention in reversed-phase liquid chromatography using linear solvation energy relationships II. The mobile phase. J Chromatogr A 799:1–19

83. Snyder LR, Dolan JW, Carr PW (2004) The hydrophobic-subtraction model of reversed-phase column selectivity. J Chromatogr A 1060:77–116

84. BĄczek T, Kaliszan R, Novotná K et al. (2005) Comparative characteristics of HPLC columns based on quantitative structure–retention relationships (QSRR) and hydrophobic-subtraction model. J Chromatogr A 1075:109–115

85. Kaliszan R, van Straten MA, Markuszewski M et al. (1999) Molecular mechanism of retention in reversed-phase high-performance liquid chromatography and classification of modern stationary phases by using quantitative structure–retention relationships. J Chromatogr A 855:455–486

86. Al-Haj MA, Kaliszan R, Buszewski B (2001) Quantitative structure–retention relationships with model analytes as a means of an objective evaluation of chromatographic columns. J Chromatogr Sci 39:29–38

87. BĄczek T, Kaliszan R (2003) Predictive approaches to gradient retention based on analyte structural descriptors from calculation chemistry. J Chromatogr A 987:29–37

88. Kaliszan R, Wiczling P, Markuszewski MJ (2004) pH gradient reversed-phase HPLC. Anal Chem 76:749–760

89. Meek JL (1980) Prediction of peptide retention times in high-pressure liquid chromatography on the basis of amino acid composition. Proc Natl Acad Sci USA 77:1632–1636

90. Palmblad M, Ramström M, Markides KE et al. (2002) Prediction of chromatographic retention and protein identification in liquid chromatography/mass spectrometry. Anal Chem 74:5826–5830

91. Petritis K, Kangas LJ, Ferguson PL et al. (2003) Use of artificial neural networks for the accurate prediction of peptide liquid chromatography elution times in proteome analyses. Anal Chem 75:1039–1048

92. Kaliszan R, BĄczek T, Cimochowska A et al. (2005) Prediction of high-performance liquid chromatography retention of peptides with the use of quantitative structure–retention relationships. Proteomics 5:409–415

93. BĄczek T, Wiczling P, Marszałł M et al. (2005) Prediction of peptide retention at different HPLC conditions from multiple linear regression models. J Proteome Res 4:555–563

94. BĄczek T (2006) Chemometric evaluation of relationships between retention and physicochemical parameters in terms of multidimensional liquid chromatography of peptides. J Sep Sci 29:547–554

95. BĄczek T (2005) Improvement of peptides identification in proteomics with the use of new analytical and bioinformatic strategies. Curr Pharm Anal 1:31–40

96. Put R, Daszykowski M, BĄczek T et al. (2006) Retention prediction of peptides based on uninformative variable elimination by partial least squares. J Proteome Res 5:1618–1625

97. Mazza CB, Sukumar N, Breneman CM et al. (2001) Prediction of protein retention in ion-exchange systems using molecular descriptors obtained from crystal structure. Anal Chem 73:5457–5461

98. Ladiwala A, Rege K, Breneman CM et al. (2003) Investigation of mobile phase salt type effects on protein retention and selectivity in cation-exchange systems using quantitative structure–retention relationship models. Langmuir 19:8443–8454

99. Tugcu N, Song M, Breneman CM et al. (2003) Prediction of the effect of mobile-phase salt type on protein retention and selectivity in anion-exchange systems. Anal Chem 75:3563–3572

100. Ladiwala A, Xia F, Luo Q et al. (2006) Investigation of protein retention and selectivity in HIC systems using quantitative structure–retention relationship models. Biotechnol Bioeng 93:836–850

101. Chen J, Luo Q, Breneman CM et al. (2007) Classification of protein adsorption and recovery at low salt conditions in hydrophobic interaction chromatographic systems. J Chromatogr A 1139:236–246

102. Song MH, Breneman CM, Bi JB et al. (2002) Prediction of protein retention times in anion-exchange chromatography systems using support vector regression. J Chem Inf Comput Sci 42:1347–1357

103. Heberger K (1999) Evaluation of polarity indicators and stationary phases by principal component analysis in gas–liquid chromatography. Chemom Intell Lab Sys 47:41–49

104. Markuszewski M, Krass JD, Hippe T et al. (1998) Separation of nitroaromatics and their transformation products in soil around ammunition plants: New high performance liquid chromatographic charge transfer stationary phases. Chemosphere 37:559–575

105. Jiskra J, Claessens HA, Cramers CA et al. (2002) Quantitative structure–retention relationships in comparative studies of behavior of stationary phases under high-performance liquid chromatography and capillary electrochromatography conditions. J Chromatogr A 977:193–206

106. Cserháti T, Forgacs E, Payer K et al. (1998) Quantitative structure–retention relationships in separation mechanism studies on polyethylene-coated silica and alumina stationary phases. LCGC Int 4:240–252

107. Abraham MH, Rosés M, Poole CF et al. (1997) Hydrogen bonding. 42. Characterization of reversed-phase high-performance liquid chromatographic C18 stationary phases. J Phys Org Chem 10:358–368

108. Sandi A, Szepesy L (1998) Characterization of various reversed-phase columns using the linear free energy relationship. I. Evaluation based on retention factors. J Chromatogr A 818:1–17

109. Valko K, Plass M, Bevan C et al. (1998) Relationships between the chromatographic hydrophobicity indices and solute descriptors obtained by using several reversed-phase, diol, nitrile, cyclodextrin and immobilised artificial membrane-bonded high-performance liquid chromatography columns. J Chromatogr A 797:41–55

110. Baranowska I, Zydroń, M (2002) Quantitative structure–retention relationships of xanthines in RP HPLC systems with the new Chromolith RP-18e stationary phases. Anal Bioanal Chem 373:889–892

111. Topliss JG, Edwards RP (1979) Chance factors in studies of quantitative structure–activity relationships. J Med Chem 22:1238–1244

112. Jiskra J, Claessens HA, Cramers CA et al. (2002) Quantitative structure–retention relationships in comparative studies of behavior of stationary phases under high-performance liquid chromatography and capillary electrochromatography conditions. J Chromatogr A 977:193–206

113. Rohrschneider L (1999) Prototypical test substances for a reversed phase in a retention parameter model. HRC J High Resol Chromatogr 22:454–458

114. Kaliszan R (2000) Quantitative structure–retention relationships (QSRR) in chromatography. In: Wilson ID, Adlard ER, Cooke M (eds) et al. Encyclopedia of separation science, vol 9. Academic Press, San Diego

115. Pliška V, Testa B, van de Waaterbemd H (1996) Lipophilicity in drug action and toxicology. VCH, Weinheim

116. Berthelot M, Jungfleisch E (1872) Sur les lois qui pre´sident au partage d'un corps entre deux dissolvants (expe´riences). Ann Chim Phys 26:396–407

117. Martin AJP (1941) A new form of chromatogram employing two liquid phases. Biochem J 35:1358–1368

118. Cserháti T, Valko K (1994) Chromatographic determination of molecular interactions. Applications in biochemistry, chemistry and biophysics. CRC Press, Boca Raton

119. Poole SK, Poole CF (2003) Separation methods for estimating octanol–water partition coefficients. J Chromatogr B 797:3–19

120. Wang QS, Zhang L, Yang HZ (1999) Lipophilicity determination of some potential photosystem ii inhibitors on reversed-phase high-performance thin-layer chromatography. J Chromatogr Sci 37:41–44

121. Nasal A, Siluk D, Kaliszan R (2003) Chromatographic retention parameters in medicinal chemistry and molecular pharmacology. Curr Med Chem 10:381–426

122. Kaliszan R, Nasal A, Markuszewski MJ (2003) New approaches to chromatographic determination of lipophilicity of xenobiotics. Anal Bioanal Chem 377:803–811

123. Nasal, A, Kaliszan R (2006) Progress in the use of HPLC for evaluation of lipophilicity. Curr Comp-Aided Drug Design 2:327–340

124. BÄczek T, Markuszewski M, Kaliszan R, van Straten, MA, Claessens HA (2000) Linear and quadratic relationships between retention and organic modifier content in eluent in reversed phase high-performance liquid chromatography: A systematic comparative statistical study. HRC J High Resol Chromatogr 23:667–676

125. Poole CF, Gunatilleka AD, Poole SK (2000) In search of a chromatographic model for biopartitioning. Adv Chromatogr 40:159–230

126. Kaliszan R (1986) Quantitative relationships between molecular structure and chromatographic retention. Implications in physical, analytical, and medicinal chemistry. CRC Crit Rev Anal Chem 16:323–383

127. Braumann T (1986) Determination of hydrophobic parameters by reversed-phase liquid chromatography: Theory, experimental techniques, and application in studies on quantitative structure–activity relationships. J Chromatogr 373:191–225

128. Kaliszan R, Haber P, BÄczek T et al. (2001) Gradient HPLC in the determination of drug lipophilicity and acidity. Pure Appl Chem 73:1465–1475

129. Kaliszan R, Haber P, BÄczek T et al. (2002) Lipophilicity and pK_a estimates from gradient high-performance liquid chromatography. J Chromatogr A 965:117–127

130. Wiczling P, Markuszewski MJ, Kaliszan R (2004) Determination of pKa by pH gradient reversed-phase HPLC. Anal Chem 76:3069–3077

131. Wiczling P, Markuszewski MJ, Kaliszan M et al. (2005) pH/organic solvent double-gradient reversed-phase HPLC. Anal Chem 77:449–458

132. Kaliszan R, Wiczling P (2005) Theoretical opportunities and actual limitations of pH gradient HPLC. Anal Bioanal Chem 382:718–727

133. Wiczling P, Kawczak P, Nasal A et al. (2006) Simultaneous determination of pKa and lipophilicity by gradient RP HPLC. Anal Chem 78:239–249

134. Subirats X, Bosch E, Rosés M (2007) Retention of ionisable compounds on high-performance liquid chromatography XVII. Estimation of the pH variation of aqueous buffers with the change of the methanol fraction of the mobile phase. J Chromatogr A 1138:203–215

135. Novotny L, Abdel-Hamid M, Hamza H (2000) Inosine and 2′-deoxyinosine and their synthetic analogues: Lipophilicity in the relation to their retention in reversed-phase liquid chromatography and the stability characteristics. J Pharm Biomed Anal 24:125–132

136. Dai J, Jin S, Yao S, Wang LS (2001) Prediction of partition coefficient and toxicity for benzaldehyde compounds by their capacity factors and various molecular descriptors. Chemosphere 42:899–907

137. Donovan SF, Pescatore MC (2002) Method for measuring the logarithm of the octanol–water partition coefficient by using short octadecyl-poly(vinyl alcohol) high-performance liquid chromatography columns. J Chromatogr A 952:47–61

138. Gulyaeva N, Zaslavsky A, Lechner P et al. (2003) Relative hydrophobicity and lipophilicity of drugs measured by aqueous two-phase partitioning, octanol-buffer partitioning and HPLC. A simple model for predicting blood–brain distribution. Eur J Med Chem 38:391–396

139. Welerowicz T, Buszewski B (2005) The effect of stationary phase on lipophilicity determination of β-blockers using reverse-phase chromatographic systems. Biomed Chromatogr 19:725–736

140. Vervoort RJM, Debets AJJ, Claessens HA et al. (2000) Optimisation and characterisation of silica-based reversed-phase liquid chromatographic systems for the analysis of basic pharmaceuticals. J Chromatogr A 897:1–22

141. Buszewski B Gadzała-Kopciuch RM Kaliszan R et al. (1998) Polyfunctional chemically bonded stationary phase for reversed phase high-performance liquid chromatography. Chromatographia 48:615–622

142. Lombardo F, Shalaeva MY, Tupper KA et al. (2001) ElogDoct: A tool for lipophilicity determination in drug discovery. 2. Basic and neutral compounds. J Med Chem 44:2490–2497

143. Kaliszan R, Marszałł MP, Markuszewski MJ et al. (2004) Suppression of deleterious effects of free silanols in liquid chromatography by imidazolium tetrafluoroborate ionic liquids. J Chromatogr A 1030:263–271

144. Marszałł MP, BĄczek T, Kaliszan R (2006) Evaluation of the silanol-suppressing potency of ionic liquids. J Sep Sci 29:1138–1145

145. Pidgeon C, Venkataram UV (1989) Immobilized artificial membrane chromatography: Supports composed of membrane lipids. Anal Biochem 176:36–47

146. Thurnhofer H, Schnabel J, Betz M et al. (1991) Cholesterol-transfer protein located in the intestinal brush-border membrane. Partial purification and characterization. Biochim Biophys Acta 1064:275–286

147. Kaliszan R, Kaliszan A, Wainer IW (1993) Deactivated hydrocarbonaceous silica and immobilized artificial membrane stationary phases in high-performance liquid chromatographic determination of hydrophobicities of organic bases: Relationship to log P and CLOGP. J Pharm Biomed Anal 11:505–511

148. Ducarme A, Neuwels M, Goldstein S et al. (1998) IAM retention and blood–brain barrier penetration. Eur J Med Chem 33:215–223

149. Kępczyńska E, Bojarski J, Haber P et al. (2000) Retention of barbituric acid derivatives on immobilized artificial membrane stationary phase and its correlation with biological activity. Biomed Chromatogr 14:256–260

150. Vrakas D, Hadjipavlou-Litina D, Tsantili-Kakoulidou A (2005) Retention of substituted coumarins using immobilized artificial membrane (IAM) chromatography: A comparative study with n-octanol partitioning and reversed-phase HPLC and TLC. J Pharm Biomed Anal 39:908–913

151. Kaliszan R (1998) Retention data from affinity high-performance liquid chromatography in view of chemometrics. J Chromatogr B 715:229–244

152. Bertucci C, Bartolini M, Gotti R et al. (2003) Drug affinity to immobilized target bio-polymers by high-performance liquid chromatography and capillary electrophoresis. J Chromatogr B 797:111–129

153. Domenici E, Bertucci C, Salvadori P et al. (1990) Synthesis and chromatographic properties of an HPLC chiral stationary phase based upon human serum albumin. Chromatographia 29:170–176

154. Domenici E, Bertucci C, Salvadori P et al. (1991) Use of a human serum albumin-based high-performance liquid chromatography chiral stationary phase for the investigation of protein binding: Detection of the allosteric interaction between warfarin and benzodiazepine binding sites. J Pharm Sci 80:164–166

155. Noctor TAG, Pham CD, Kaliszan R et al. (1992) Stereochemical aspects of benzodiazepine binding to human serum albumin. I. Enantioselective high performance liquid affinity chromatographic examination of chiral and achiral binding interactions between 1,4-benzodiazepines and human serum albumin. Mol Pharmacol 42:506–511

156. Kaliszan R, Nasal A, Turowski M (1995) Binding site for basic drugs on α1-acid glycoprotein as revealed by chemometric analysis of biochromatographic data. Biomed Chromatogr 9:211–215

157. Kaliszan R, Nasal A, Turowski M (1996) Quantitative structure–retention relationships in the examination of the topography of the binding site of antihistamine drugs on α1-acid glycoprotein. J Chromatogr A 722:25–32

158. Goolkasian DL, Slaughter RL, Edwards DJ et al. (1983) Displacement of lidocaine from serum α1-acid glycoprotein binding sites by basic drugs. Eur J Clin Pharmacol 25:413–417

CHAPTER 9

THE USE OF QSAR AND COMPUTATIONAL METHODS IN DRUG DESIGN

FANIA BAJOT

School of Pharmacy and Chemistry, Liverpool John Moores University, Liverpool, L3 3AF, England
e-mail: f.bajot@ljmu.ac.uk

Abstract: The application of quantitative structure–activity relationships (QSARs) has significantly impacted the paradigm of drug discovery. Following the successful utilization of linear solvation free-energy relationships (LSERs), numerous 2D- and 3D-QSAR methods have been developed, most of them based on descriptors for hydrophobicity, polarizability, ionic interactions, and hydrogen bonding. QSAR models allow for the calculation of physicochemical properties (e.g., lipophilicity), the prediction of biological activity (or toxicity), as well as the evaluation of absorption, distribution, metabolism, and excretion (ADME). In pharmaceutical research, QSAR has a particular interest in the preclinical stages of drug discovery to replace tedious and costly experimentation, to filter large chemical databases, and to select drug candidates. However, to be part of drug discovery and development strategies, QSARs need to meet different criteria (e.g., sufficient predictivity). This chapter describes the foundation of modern QSAR in drug discovery and presents some current challenges and applications for the discovery and optimization of drug candidates

Keywords: Blood–brain barrier, Cyclooxygenase inhibitor, Drug design, hERG, QSAR

9.1.   INTRODUCTION

The discovery and development of a new drug is an expensive and time-consuming process. Therapeutic effects and hazards to health are assessed using a series of experimental and in vivo tests. However, usage of animal models is often subject to ethical (and financial) considerations. Therefore, alternative methods are being developed to reduce the requirement of animals in testing. In silico methods are often implemented due to their lower cost; an added bonus is their significant contribution to the identification and development of effective drugs from new chemical entities (NCEs) (see Chapter 1). The computational tools are principally used for

  (i)   the conformational analysis of molecular structure (e.g., molecular dynamics);
 (ii)   the characterization of drug–target interactions (e.g., molecular docking); and
(iii)   the assessment and optimization of drug activity using quantitative structure-activity relationships (QSARs).

261

T. Puzyn et al. (eds.), Recent Advances in QSAR Studies, 261–282.
DOI 10.1007/978-1-4020-9783-6_9, © Springer Science+Business Media B.V. 2010

QSAR methods in drug design are used particularly for the estimation of physico-chemical properties, biological effects as well as understanding the physicochemical features governing a biological response. As a result, QSAR provides low-cost tools for the selection of novel "hits" and for "lead" optimization during drug discovery and development.

The foundation of modern QSAR was pioneered by Prof. Corwin Hansch and co-workers during the early 1960s [1]. Since these studies, numerous QSAR models have been proposed for the evaluation and the understanding of biological and physicochemical properties of NCEs. A further milestone in QSAR development was the introduction of 3D-QSAR which has contributed to advances in the usage of QSAR for selecting drug candidates (Chapter 4). This strong interest in QSAR for design and development of new drugs is demonstrated by the plethora of publications and the numerous companies proposing QSAR-based software (Chapters 1, 4, and 10).

The aim of this chapter is to consider the application of (Q)SAR in drug design. After a brief introduction on the drug discovery and development process, this chapter will cover the foundation of modern QSAR in drug discovery and present some current challenges and applications for the discovery and optimization of drug candidates.

9.2.   FROM NEW CHEMICAL ENTITIES (NCEs) TO DRUG CANDIDATES: PRECLINICAL PHASES

The development of new chemical entities (NCEs) as new effective drugs is conducted under stringent conditions in order to ensure the therapeutic effect and the safety of the new compounds. To achieve this challenge, the benefits (therapeutic effects) and the risk (toxic effects) of the NCEs are evaluated, respectively, during the preclinical and the clinical phases of development [2, 3]. After the selection of a validated target relating to a particular disease state, preclinical drug development aims to gather relevant data in order to propose a drug candidate for clinical test. During the preclinical stage, the pharmacological profile and the acute toxicity of the drug candidate are assessed using in silico, in vitro methods and animal models. The three-stage procedure allowing the selection of the most effective NCEs (the drug candidates) is presented in Figure 9-1 and the three stages are described in detail below.

9.2.1.   Stage 1: Hit Finding

For a given target (receptor, enzyme, etc.), this stage aims to identify "hit" compounds from diverse libraries (corporate, commercial, etc.) and/or by medical observations. High-throughput screening (HTS) and in silico evaluations are used to screen NCEs with suitable pharmacodynamic (PD) activity. The PD properties of a given molecular entity are defined as the biochemical and physiological effects of the entity on the body.

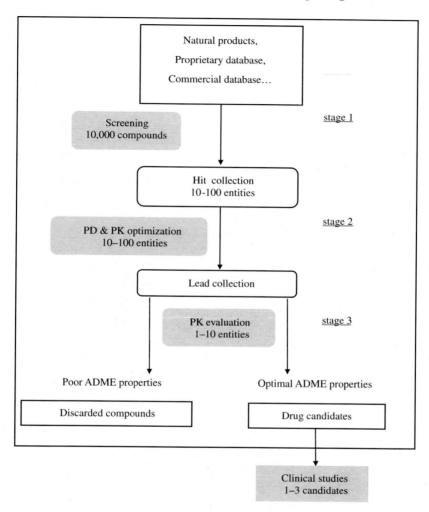

Figure 9-1. Preclinical testing strategy for a validated pharmaceutical target. Drug candidates are selected following three stages of development: (1) the selection of "hit" compounds, (2) the optimization of "lead" compounds, and (3) the choice of the drug candidates

9.2.2. Stage 2: Lead Finding

This stage is a key milestone of the drug candidate discovery process. The pharmacokinetic (PK) properties govern the bioavailability of the NCEs and, therefore, the correct delivery of the drug to its target site. The PK properties are represented by the processes of absorption, distribution, metabolism, and elimination (ADME) undergone by the NCEs in the organism (Chapter 10). During this stage, "hit" molecules presenting good ADME and physicochemical properties are identified and taken further as lead compounds.

9.2.3.   Stage 3: Lead Optimization

This stage includes the evaluation of various properties of lead analogs in order to propose the drug candidates. Accordingly, lead analogs are generated by operating different structural modifications around the lead's molecular scaffold. The chemical structures with the optimal potency, solubility, and ADME profile are selected as drug candidates. Computational strategies for lead finding and optimization include molecular docking and (Q)SAR.

At the successful completion of the preclinical phase, the selected drug candidates can progress into phases I, II, and III of clinical development.

9.3.   FAILURE IN DRUG CANDIDATES' DEVELOPMENT

In spite of the stringent procedure and substantial financial investment of drug development (the development of a typical drug may cost up to one billion US dollars) of 5000 molecules tested in the preclinical phase, only one reaches the market [4]. Kennedy et al. [5] identified the factors associated with failures during clinical assessments (Figure 9-2). Poor PK properties (39%), lack of efficacy (30%), toxicity in animals (11%), and adverse effects to man (10%) are the most common reasons to explain the attrition of molecules in pharmaceutical research.

Currently, poor ADME properties are considered as the main reason for failure during drug development [6, 7] (Chapter 10). Besides these ADME issues, the selection of potent NCEs needs to be improved to reduce failure associated with a lack of efficacy. Numerous computational tools, with variable success in their application, have been proposed to address ADME and potency during the early phase of drug discovery and development [8–10]. These tools range from very trivial "rules of thumb," e.g., Lipinski's rule of 5 [11] to more complex and multivariate

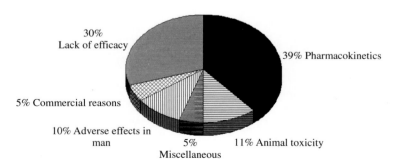

Figure 9-2. Reasons for the failure of the development of NCEs in clinical development. Adapted from [5]

QSAR models, including the use of models based on neural networks [12, 13], etc.) (Chapter 10).

9.4. CLASSIC QSAR IN DRUG DESIGN

A brief overview of some of the "classical" techniques to develop QSARs, particularly those that have been applied for drug design, is provided.

9.4.1. Hansch Analysis

In the early 1960s, Prof. Corwin Hansch proposed the use of linear multiple regression in order to predict the biological response of compounds yet to be synthesized [14, 15]. In this approach, each chemical structure is represented by several parameters which describe hydrophobicity, steric properties, and electronic effects and is usually formalized into the following equation:

$$\log 1/C = a(\text{hydrophobic parameter}) + b(\text{electronic parameter})$$
$$+ c(\text{steric parameter}) + \text{constant} \tag{9-1}$$

where C is molar concentration producing the biological effect; a, b, and c are the regression coefficients.

As an example, the antiadrenergic activities of *meta*- and *para*-substituted N,N-dimethyl-α-bromo-phenethylamines (the general structure is shown in Figure 9-3) have been analyzed by Hansch et al. [16] (9-2):

$$\log 1/C = 1.22\pi - 1.59\sigma + 7.82$$
$$n = 22; r = 0.918; s = 0.238 \tag{9-2}$$

where π is the hydrophobic substituent bonding constant [17]; σ is Hammett's substituent constant for electronic effects [18]; n is the number of compounds; r is the correlation coefficient, and s is the standard error of the estimate.

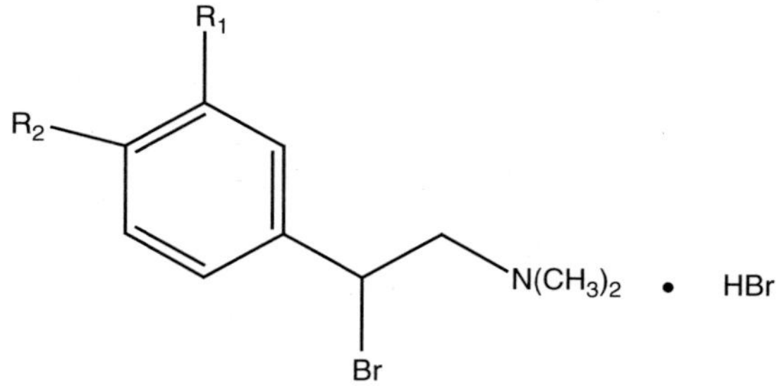

Figure 9-3. *Meta-* and *para*-substituted N, N-dimethyl-α-bromo-phenethylamines. The *meta-* and *para*-substituents R_1 and R_2 include H, F, Cl, Br, I, and Me

It should be noted that steric properties (i.e., molecular size and shape) were not found to be relevant to the modeling of this effect; hence a steric descriptor is not included in this equation.

Hansch's approach has been largely applied to analyze different biological data (e.g., affinity data, inhibition constant, pharmacokinetic parameters) in the framework of diverse therapeutics areas (e.g., antibacterial, anticancer, antimalarial drugs) [14, 19].

9.4.2.   Non-parametric Methods: Free-Wilson and Fujita-Ban

Independent to the equation developed by Hansch, non-parametric methods have been proposed by Free-Wilson [20, 21] and Fujita-Ban [22]. These are summarized by the following equation:

$$\text{Biological Activity} = \sum_{ij} \alpha ij \bullet Rij + \mu \qquad (9\text{-}3)$$

For a series of chemical analogs i-j, the biological activity is assumed to be the sum of intrinsic activity of the skeleton (μ) and the additive contribution of the Rij substituents (αij).

The method is simple and has the advantage of being independent of the possible problems associated with the calculation of molecular descriptors. However, each substituent is assumed not to interact (e.g., intra-molecular hydrogen bonding) which is considered to be a major limitation. Numerous examples of the Free-Wilson approach are available [19].

9.4.3.   Linear Solvation Free-Energy Relationships (LSERs)

Derived directly from the work of Hansch, linear solvation free-energy relationship (LSER) analysis [23, 24] is a powerful QSAR approach which analyzes a property's dependence on solute–solvent interactions [25]. LSER analysis decomposes the given molecular property to its solvatochromic descriptors. Three different terms derived from solvatochromic properties are involved in a LSER equation (9-4):

Property = cavity/bulk term + polarizability term + H-bond term(s) + constant
(9-4)

These terms represent the three main interaction forces operating between the solute and the solvent, namely

- The hydrophobic forces represented by the bulk term. This parameter is related, in part, to the energy needed to break the water–water interactions to form a cavity for the solute in the polar phase and in part to the stabilizing interactions with non-polar phases. The van der Waals volume (V_w) is generally used to represent the bulk parameter.
- The polarizability term π^* measures the dipolar interaction (induced or not) between the solute and the environment.

- The H-bonding terms measure the H-bonding interactions of the solute with the solvent. H-bonding terms are represented by the acceptor capacity (β) and the donor capacity (α).

The relation between the property and the solvatochromic descriptors is established using multiple regression analysis (9-5):

$$\text{Property} = a\alpha + b\beta + c\pi^* + dV_w + \text{constant} \tag{9-5}$$

where a, b, c, and d are the coefficients of regression for each parameter.

Currently, molecular modeling offers the computation of the theoretical solvatochromic parameters with the aim to derive in silico solvatochromic models. The theoretical linear solvation energy relationships (TLSER) method comprises the calculated volume of van der Waals (V_w), the theoretical solvatochromic parameters, namely the dipolarity/polarizability π^*, the H-bond donor acidity α, and the H-bond acceptor basicity β [26, 27]. Numerous (T)LSER models for endpoints such as bioavailability, physicochemical properties, and biological responses have been published [28–30]. Two successful applications of the method are described critically below.

Example 1: LSER analysis of lipophilicity. The lipophilicity of a drug is its capacity to be partitioned between a polar phase and a non-polar environment. Lipophilicity is often expressed as the logarithm of the partition coefficient of a compound distributed between immiscible phases of n-octanol and water (log P) (Eq. 9-6). Experimental methods for log P evaluations include the shake-flask method, reversed-phase high-performance liquid chromatography (RP-HPLC), and the potentiometric method [31]. Besides these experimental techniques, there is a plethora of free and commercial computational tools for log P prediction [32]. Lipophilic parameters are often highly significant in (Q)SAR models. ADME properties of molecules and drug–target interactions are also partly governed by their lipophilic properties [33–36].

$$\log P = \log \left(\frac{C_{\text{octanol}}}{C_{\text{water}}} \right) \tag{9-6}$$

Where C_{octanol} and C_{water} are the substance concentrations at equilibrium in n-octanol and aqueous phases.

Some important efforts have been made toward a better characterization of the intermolecular forces underlying lipophilicity [37]. LSER analysis is suitable for decomposing lipophilicity in terms of polar and hydrophobic intermolecular forces [38]. For instance, using solvation descriptors for 613 drug-like compounds, Lombardo et al. [39] derived the following LSER for log P (9-7):

$$\log P = -0.034\,\alpha - 3.46\,\beta - 1.05\,\pi + 3.81\,V_w + 0.56\,R^2 + 0.088$$
$$n = 613;\ r^2 = 0.995;\ s = 0.116;\ F = 23{,}100 \tag{9-7}$$

where R^2 is the excess molar refraction and F is the Fisher Statistic.

This LSER analysis indicates the positive contribution of steric descriptors in octanol/water partitioning. Polar interactions represented by hydrogen-bonding descriptors and the dipolarity/polarizabillity terms contribute negatively to octanol/water partitioning.

Example 2: TLSER Analysis of Blood–Brain Barrier Permeation. Targeting the brain represents an important challenge in pharmaceutical research [40]. The central nervous system (CNS) is the site of disruption in biochemical pathways, which leads to numerous brain disorders [41, 42]. Furthermore, some adverse affects can arise from the cerebral activity of drugs developed for peripheral targets, as observed with the sedative effect by the first generation of histamine H1 antagonists [43, 44]. Thus, the ability of a compound to reach the CNS may be of crucial importance. This is commonly assumed to be controlled by the blood–brain barrier (Chapter 10). Abraham et al. [45] successfully applied LSER to analyze the blood–brain permeation capacity (BB) of a set of 30 diverse chemicals (9-8):

$$\log BB = -1.80\alpha - 1.60\beta - 0.97\pi^* + 1.89V_w - 0.72$$
$$n = 30; \; r^2 = 0.87; \; s = 0.52; \; F = 41 \tag{9-8}$$

Polar interactions, namely acceptor (β) and donor (α) hydrogen-bonding interactions and polarizability (π^*) are unfavorable to blood–brain barrier permeation, whereas hydrophobicity, encoded by the volume V, is favorable. These results have confirmed the interest of an in-depth description and decomposition of the intermolecular interaction forces.

9.5.   QSAR METHODS IN MODERN DRUG DESIGN

The synthesis of prospective new compounds has been revolutionized by combinatorial chemistry, which allows for the easy synthesis of many tens or hundreds of diverse compounds. Among these chemicals, only a small proportion will have a suitable PK and PD profile to be a drug candidate. Accordingly, in combination with HTS technology, (Q)SAR is considered as a worthwhile in silico tool for rapidly filtering out non-optimal compounds. Various physicochemical and PK properties (e.g., bioavailability and solubility) are usually addressed in the first instance using (Q)SAR tools.

Drug-likeness of NCEs is particularly evaluated during the early stages of drug discovery. Simple rules of thumb, such as the "rule of 5" (also known as Lipinski's rules [11]) can be used to filter molecules which are likely to be only weakly bioavailable. The "rule of 5" is routinely used as a guideline during the drug development. This rule derives from the studies of 2245 compounds from the World Drug Index and comprises four criteria noted by Madden (Chapter 10). Poor oral bioavailability is expected if a NCE satisfies two elements of the four criteria.

In order to avoid future failure during drug development phases, certain biological processes and activities are systemically assessed during the preclinical phase. For instance, ability of drug to induce cardiotoxicity (Chapter 11) is now carefully examined during the preclinical phase. However, experimental evaluation can be time-consuming, expensive, or be subject to ethical barriers. In this context, QSAR methods are useful for assessing large sets of molecules. In addition, by including interpretable descriptors, QSAR analyses allow the identification of the structural and physicochemical features modulating the activities of compounds. This contributes to assisting and guiding the design and the selection of potent and safe drug candidates.

9.5.1. Tools for QSAR

Since the definition of the Hansch approach, QSAR has evolved considerably both in terms of its methodology as well as its application in drug design. A plethora of molecular descriptors are now available and new methods, such as 3D-QSAR methods, have contributed to leverage the role of QSAR in drug discovery. The following sections deals with current tools, methods, and concepts which play a role in QSAR development for drug design.

9.5.1.1. Data and Databases

The first stage in deriving a (Q)SAR model is to gather and select the molecules with activity data to include in the training set (Chapter 12). The number of published data obtained from the same experimental conditions is often too much small to perform an in silico model; therefore the training set commonly contains data coming from different sources. The main sources of information for constituting a training set are internal (corporate) data, publicly available data, as well as free and commercial chemical library providers.

Among the numerous informatics tools for supporting the computational chemist, knowledge management tools and, in particular, databases are of use to construct a training set. Indeed, chemical structures, physicochemical properties, and more occasionally biological (pharmacology) activity are accessible via free or commercial database providers, some of which are summarized in Table 9-1 (see also Chapter 11, Table 11-2). The user can generally explore the database using different input formats such as chemical identifiers (chemical name, CAS registry number), structure, and sub-structures as well as physicochemical properties values. The resulting information can be downloaded in different formats (e.g., SMILES, sdf file, txt file).

A few specialized websites provide free "ready to use" training set(s) (chemical structures associated with activity data). An example of such a website is http://www.cheminformatics.org/. Several projects aim to structure and share information associated with compounds, disease states, targets, and activity data. The ChemBank website is a good example of this concept (http://chembank.broad. harvard.edu/). In addition to these free tools, some companies specialized in pharmaceutical knowledge management have centralized data drawn from scientific journals, patents, or drug monographs. These commercial tools allow for the retrieval of different information (chemical structures, in vitro or in vivo activity data) using different inputs (e.g., target, molecular structure, disease).

Table 9-1. Sources of data for QSAR studies for drug design

Name	Website	Examples of the type of records
Aureus Pharma	http://www.aureus-pharma.com	Chemical structure, biological activities (in vitro, in vivo)
ChemBank	http://chembank.broad.harvard.edu	
ChemIDPlus	http://chem.sis.nlm.nih.gov/chemidplus	Chemical structure, toxicity, physicochemical properties, etc.
Chemoinformatics.org	http://www.cheminformatics.org	Data sets
Chemspider	http://www.chemspider.com	Physicochemical properties
Developmental therapeutics program (DTP)	http://dtp.nci.nih.gov/webdata.html	Chemical structure (2D and 3D), biological data (oncology)
QSAR World	http://www.qsarworld.com	Literature, data sets
Scifinder (CAS Database)	http://www.cas.org/	Literature, chemical structure

9.5.1.2.   *Novel Molecular Descriptors*

Molecular descriptors can be calculated from the chemical formula (1D descriptors), the 2D structure (2D descriptors), and the 3D conformation (3D descriptors) using a large number of methods (Chapters 3, 5, 12, and 14). These methods are based on atom types, molecular fragments, and the three-dimensional structure, respectively. The degree of information encoded in 1D descriptors is very low. For example, the formula of morphine, $C_{17}H_{19}NO_3$, corresponds to more than 3649 compounds referenced by the Chemical Abstract Service (September 12, 2006). Thus, 1D descriptors are rarely included in QSAR approaches. A QSAR model can, in part, be characterized by the relative contributions obtained from the molecular descriptors. This contribution can assist in the identification and optimization of drug candidates. Consequently, a judicious choice of molecular descriptors contributes to obtain a QSAR with satisfactory model interpretability. The following section describes some of the descriptors applied in drug design and elsewhere since the formalization of the Hansch approach.

Topological descriptors are calculated from the molecular graph and encode the molecular connectivity into numerical values called topological indexes. The first topological index developed by Wiener [46] was very simple and did not take into account the atom types and the bond orders; thereafter, other topological indexes have been proposed (e.g., Randic index X [47], Balaban index J [48], Kier and Hall indexes [49]). A full discussion of topological indexes is provided by Todeschini and Consonni (Chapter 3).

Steric descriptors relate to the size and shape of molecules. These are some of the most crucial structural properties which modulate the biological activity of compounds [50, 51]. In effect, an efficient and a specific drug–target interaction can be ensured if there is a sufficient shape and surface complementarity between the drug and the target [52, 53]. Steric descriptors, namely the volume, the surface, and the shape are useful to describe the global structural features of a molecule. The volume is often represented by the van der Waals volume, the Connolly volume, or molar refractivity. The van der Waals volume can be calculated simply using the van der Waals radii [54]. The Connolly volume [55] is a better simulation of reality since it represents the solvent accessible volume. Different to volume and other descriptors of molecular bulk, translating molecular shape into a single numerical value is a real challenge [56]. The steric influence of a substituent can be described by using the four geometrical parameters of Verloop – the so-called Sterimol descriptors [57]. The general form of a molecule can be obtained with the evaluation of its spheroidal properties. The VolSurf software [58] provides a method to estimate molecular globularity G and rugosity R with the aim of measuring the drug's ellipsoidal characteristics; the calculation of these parameters derives from the computed molecular volume – see the next section for a more complete description on VolSurf.

The lipophilicity, polarity, and hydrophobicity are often described as the main properties which govern the bioavailability of drugs [59, 60]. Currently, semi-empirical procedures are able to evaluate the lipophilicity, the hydrogen-bonding abilities, and the polarizability with reasonable accuracy. The advantage of semi-empirical procedures is their experimental background which confers a better prediction of the property. In this way, it is possible, from the 2D or 3D structure of a molecule, to estimate its physicochemical properties profile using some computational tools. The fragmental methods of Rekker [61] and the ClogP algorithm from Hansch and Leo [62] are the most famous methods to provide a theoretical log P value. However, most of the pharmaceutical compounds are ionizable and, thus, the pH-dependent distribution coefficient (log D) is preferentially used. Generally, in the case of ionizable compounds, it is assumed that the partition of the ionized species can be neglected, meaning that the partition coefficient of the neutral species log P can easily be calculated from the log D taking into account the amount of ionized fraction at the given pH [63, 64]. These fragmental methods for computing lipophilicity are 2D approaches and cannot take into account the effects of the 3D conformation. This must be accounted for by different methods, some of which are described in the next section.

9.5.1.3.   3D-QSAR

The size and the 3D conformation of a molecule are some of the crucial structural properties which modulate the biological activity of a compound [50, 51]. Moreover, numerous molecules of pharmaceutical interest present a degree of flexibility allowing in some case changes in expected hydrophobic or hydrophilic properties. As described previously, relevant models can be obtained using 2D-QSAR methods;

however, 2D descriptors are not able to take into account the important 3D specificities described above [65]. The 3D-QSAR method, Comparative Molecular Field Analysis (CoMFA) was introduced by Cramer in 1988 and represents an important step in the progression of QSAR [66]. CoMFA and other 3D-QSAR methods are described by Sippl (Chapter 4).

The GRID force field is possibly the most widely used to map the physicochemical properties in 3D molecular space, perhaps because of its unspecificity toward the types of molecular structures (e.g., organic molecules and macromolecules) [67–70]. The GRID fields describe the variation of the interaction energy between a target molecule and a chemical probe (e.g., water probe, hydrophobic probe, or amide probe), placed in a 3D grid constructed around the target [67]. The GRID force field is based on three types of interaction forces, namely the induction and dispersion interactions, the charged interactions (measured by an electrostatic field), and the hydrogen-bonding acceptor and donor interactions.

An interesting alternative to CoMFA and GRID is the VolSurf software proposed by Cruciani et al. [58]. In VolSurf, no molecular alignment is required to compare the molecular interaction fields (MIF). VolSurf extracts numerical descriptors of MIFs, and multivariate statistics (e.g., partial least squares) are used to relate the variation of MIF descriptors to the variation of compounds' activity. VolSurf converts the information contained in the 3D fields into numerical values depending on eight 3D isopotential contours [8]. Among the MIFs' descriptors, the volume (V), the INTEGY moment (I), and the capacity factor (CF) are three main parameters which are used particularly to characterize each isopotential contour. The INTEraction enerGY (INTEGY) moments express the degree of MIFs delocalization around the molecule. The capacity factor is the MIFs surface per surface unit. In addition to MIFs descriptors, four-shape parameters such as the molecular volume V, the molecular surface (S), the rugosity (R), and the globularity (G) are also available. A number of VolSurf descriptors are summarized in Table 9-2. The successful applications of the VolSurf procedures are principally in predicting pharmacokinetic properties [58, 71]; nevertheless, several models have also been proposed for activities such as bactericidal activity of quinolones [72].

Table 9-2. Summary of some VolSurf descriptors describing a MIF, the number of energetic level n varies from 1 to 8. Adapted from [71]

VolSurf descriptors	Definition
Volume regions V_n	The molecular size of the MIFs computed at the energetic level n
INTEGY moments I_n	The imbalance between the center of mass of a molecule and the barycentre of its MIF computed at the energetic level n
Capacity factors CF_n	The ratio of the MIF surface over the total molecular surface. In other words, it is the MIFs surface per surface unit
Local interaction energy minima	The energy interaction (in kcal mol^{-1}) of the three local energy minima
Energy minima distances	The distances between the best three local energy minima

Recently, a 3D linear solvation energy model (3D solvatochromic) method based on molecular interaction fields and VolSurf has been proposed [73]. The physico-chemical properties are represented by four MIFs associated with, respectively, the hydrophobic part of the molecular lipophilicity potential (MLP) [74]; the acceptor and donor parts of the molecular hydrogen-bonding potential (MHBP) [75]; and the GRID DRY field. This method has allowed an in-depth analysis of the intermolecular interaction forces involved in pharmacokinetic mechanisms, such as skin permeation, blood–brain barrier permeation [73], and affinity of flavonoid derivatives toward P-glycoprotein (Pgp).

9.5.1.4. *Applicability Domain in QSARs*

The composition of the data matrix used to develop a (Q)SAR contributes directly to the model accuracy. The set of molecules used to build the model (training set), the molecular descriptors, and the biological parameters must be rigorously selected. Indeed, the molecular diversity, the quality of the biological values (e.g., standard errors), as well as the range of biological activity covered are some factors influencing the quality of the model.

Many models developed are "local" models, i.e., based on homologous series of compounds [76]; "global" models are based on larger and more structurally heterogeneous data sets (Chapter 11). The current tendency is to introduce diversity into the training set in order to assist in the prediction of activity for diverse compounds. This gradual change from local model to global model has necessitated the better characterization of the applicability domain of an individual model or (Q)SAR. The concept of the applicability domain is defined by Netzeva et al. [77, 78] (see also Chapter 12). Figure 9-4 provides a simple schematic representation of the applicability domain of a local model in the whole of a defined chemical

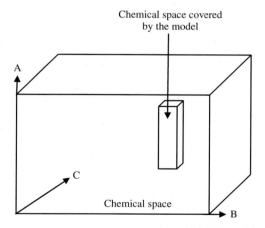

Figure 9-4. Schematic representation of a chemical space delimited by three descriptors A, B, and C; and the chemical space (applicability domain) covered by the molecular structures included in a given model

space. For a user to have confidence in a predicted value from a model, the compound (whose activity is to be predicted) should be within the applicability domain of the model.

9.6.   QSAR IN MODERN DRUG DESIGN: EXAMPLES

In the last decade, non-optimal ADMET drug profiles have been closely related to lack of clinical efficacy and adverse effects, which in some cases have resulted in the withdrawal of drugs [6, 7]. At the same time, assessing pharmacodynamic (PD) profiles of NCEs remains an important task during drug discovery. Experimental assays (e.g., in vitro or in vivo) are time-consuming and expensive for screening large chemical libraries. As a result, in silico approaches have increasingly gained relevance and interest to obtain the ADMET and pharmacodynamic profiles of chemicals. In this section, examples are provided for the development of QSARs for current issues in drug discovery ADMET properties and pharmacological activity.

9.6.1.   Example 1: Application of QSAR to Predict hERG Inhibition

QT prolongation observed on an electrocardiogram is closely associated with torsade de pointes and cardiac arrhythmias. In the past few years, a number of drugs have been removed from the market due to their ability to produce QT prolongation [79, 80]. Blockade of the human ether-a-go-go (hERG) potassium channel is assumed to be the mechanism leading to torsade de pointe. Therefore, early identification of potent hERG blockers represents a central task in the safety assessment of drug candidates [79].

The hERG channel is a plasma membrane with a large pore cavity able to bind a range of inhibitor sizes [81]. The 3D structure of hERG is not available due to the plasma membrane characteristics of the channel. Consequently, molecular docking is not applicable and QSAR gains increasingly interest for evaluating potent hERG inhibitors and identify toxicophore features. In the last decade, a wide range of QSAR methods have been applied to derive models for hERG blockers. 2D-QSAR and 3D-QSAR as well as pharmacophore (or toxicophore) analyses have been published. Table 9-3 summarizes the main features of some of these models.

Table 9-3. Examples of (Q)SAR for the prediction of hERG blockade

Method	Activity	N	r^2	q^2	References
Decision tree	TdP-causing activity (+/−).	264	–	–	[103]
Binary QSAR	IC_{50}	240	–	–	[104]
HQSAR	pIC_{50}	882	0.52	0.35	[105]
MLR	pIC_{50}	19	0.87	0.81	[83]
3D pharmacophore	pIC_{50}	322	0.76	0.72	[105]
Support vector regression	pIC_{50}	90	0.81	0.85	[106]

As an example of the type of study performed, a small set of 11 antipsychotics with IC_{50} values (for inhibition of hERG) have been used to generate a 3D-QSAR model with an r^2 of 0.77. Extending the approach to 22 molecules confirms the previous results and led to an r^2 of 0.99. From this model, four hydrophobic features and one ionizable feature have been identified as key elements for antipsychotics to block hERG [82]. In 2004 a simple 2D-QSAR with good statistical parameters was derived using the toxic potencies of 19 miscellaneous drugs. The model includes two significant parameters: log P corrected for ionization at pH 7.4 (log $D_{7.4}$) and the maximum diameter of molecules (Dmax). Many drugs in the training set possessed a positive charge at the physiological pH (7.4). Thus, the positive correlation obtained for the two parameters supports the roles of the steric property and the positive charge for hERG blockade [83].

Therefore, QSAR models have highlighted some structural features characterizing hERG blockers. Moreover, in silico methods allow for the discrimination of non-blockers from blockers. Thus computational tools, in particular QSAR, have found a place in pharmaceutical company for supporting the safety assessment of NCEs.

9.6.2. Example 2: Application of QSAR to Predict Blood–Brain Barrier Permeation

Targeting the brain represents an important challenge in pharmaceutical research [40]. The central nervous system (CNS) is the site of disruption in biochemical pathways, which leads to numerous brain disorders [41, 42]. Furthermore, some adverse affects can arise from cerebral activity of drugs developed for peripheral targets, as observed with the sedative effect by the first generation of histamine H1 antagonists [43, 44]. The amount of drug transferred into the brain is mainly dependent on their passive permeation through the blood–brain barrier (BBB), its ability to be transported by uptake or efflux systems as well as their degree of binding to plasma proteins. The major element limiting the brain uptake of a compound is the BBB formed by the endothelial cells of the brain–blood capillaries [84, 85].

Several in silico models based on the logarithm of the blood–brain ratio (log BB) and 2D (or 3D) physicochemical descriptors have been developed [45, 86–89] (Chapter 10). Generally, these models assume a purely passive BBB transfer. The models based on global physicochemical and topological descriptors (e.g., log P, hydrogen-bonding capacities, molecular weight, polar surface area) have shown that uptake into the brain is reduced by molecular polarity and increased by hydrophobicity [30, 90, 91]. In addition to whole molecule properties, VolSurf models derived from molecular interaction field (MIFs) descriptors have been applied. These have the advantage of providing an interpretation of molecular interactions from their 3D aspects [87, 88]. However, these VolSurf models have some limitations, either for virtual screening or for molecular structure optimization. In effect, the semi-quantitative model allows only for a classification in permeant and non-permeant compounds (BBB+ and BBB–) [87]. A quantitative VolSurf-based model [88] has been developed; however, in this model, the molecular interaction forces are described only in terms of polarity and hydrophobicity. This leads to little understanding of the effects of hydrogen-bonding acceptor and donor capacities and the effect of polarizability.

The 3D linear solvation energy (3D solvatochromic) method described in Section 9.4.3 has been applied to investigate the relationship between BBB permeation (measured in the rat by the permeability–surface area product (PS)) and the 3D MIFs descriptors [73]. The model is characterized by a cross-validated coefficient of determination (q^2) of 0.79. The plot of experimental versus predicted values is shown in Figure 9-5. The model can discriminate between the passive brain uptake of seven external compounds for which brain uptake values have been measured by in situ brain perfusion. Indeed, for this test set of seven compounds, the Spearman coefficient correlation is 1. One advantage of the 3D solvatochromic approach is that it takes into account the 3D distribution of MIFs, especially using the INTEGY moment (I) descriptor which measures the degree of localization–delocalization from any given MIFs.

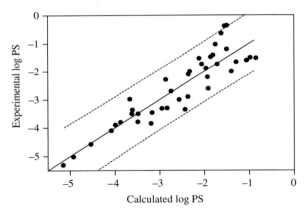

Figure 9-5. Logarithm of the experimental permeability–surface area product (PS) versus log PS calculated with a 3D solvatochromic (3D linear solvation energy) model. The 3D PLS model includes 41 compounds and two latent variables derived from 33 descriptors. The model is characterized by the following statistical parameters: $q^2 = 0.79$, $r^2 = 0.82$, $s = 0.44$, $F = 185$ [73]

9.6.3.   Example 3: Application of QSAR to Predict COX-2 Inhibition

In addition to optimizing the ADMET properties of drug molecules, QSAR has always been used to understand, model, and predict therapeutic effects. As an example, non-steroidal anti-inflammatory drugs (NSAIDs) provide relief from symptoms of inflammation and pain by inhibiting the cyclooxygenases (COX) enzymes COX-1 and COX-2. Classical NSAIDs include aspirin, ibuprofen, and diclofenac and are characterized by having a carboxylic acid group. The COX-1 and COX-2 enzymes are involved in the prostanoid synthesis pathway; however, expression of COX-1 is constitutive whereas expression of COX-2 is inducible. Clinical use of COX-1 inhibitors is associated with peptic ulceration; therefore a strategy consisted of developing NSAIDs with selectivity for COX-2. Nevertheless, COX-2 inhibitors are associated with side effects since they are known to increase cardiovascular risk. For example, in 2004, Rofecoxib was withdrawn from the world market due to its cardiovascular toxicity [92]. Besides the anti-inflammatory benefits, COX-2 inhibitors are also investigated for the treatment of colon cancer [93], neurodegenerative diseases such as Parkinson's [94] and Alzheimer's [95].

The crystal structures of many COX-2 enzymes have been reported, including crystal structures with bound inhibitors [96, 97]. The 3D structures obtained have provided important details about the binding mode of NSAIDs in COX-2. Since the COX-2 structure is available, molecular docking appears to be a suitable tool for in silico investigations of COX inhibitors. Molecular docking is a computational method which offers the analysis of the binding mode of the ligand into the target. A scoring function allows ranking of the ligands according to their predicted affinity [98, 99]. Nevertheless, inhibition (or activity) of a compound is not solely dependent on its binding mode and its affinity for the target; as a result, QSAR method remains an interesting and complementary strategy for predicting inhibition (or activity) and structural optimization.

The roles of hydrophobic, electronic, and steric descriptors have been investigated by deriving 2D-QSAR [100], 3D-QSAR CoMFA [101], and pharmacophore analysis for developing compounds with significant COX-2 inhibition. Multiple linear regression analysis models for diverse COX-2 inhibitory potency, such as those for terphenyls, cyclopentenes, and pyrazoles derivatives [100], have shown some significant contribution of calculated log P (ClogP), supporting the hypothesis of the contribution of hydrophobic interactions. For indomethacin analogs, a quadratic relationship (9-9) between the inhibition of COX-2 (IC_{50}) and ClogP shows an optimal ClogP value of 5.69:

$$\log IC_{50} = -10.2 \text{ C } \log P - 0.90 \text{ ClogP}^2 - 21.6$$
$$n = 7; \ r^2 = 0.99 \ ; q^2 = 0.94 \ ; s = 0.089$$

(9-9)

CoMFA and CoMSIA have been applied to model the activities of a series of 25 acanthonic acid derivatives [101]. Both models are characterized by satisfactory q^2 values (0.733 and 0.847, respectively). The steric and electrostatic and the hydrogen-bonding donor group fields have a positive contribution to the inhibitory potency, whereas the contribution of hydrogen-bonding group fields depends on the substituent positions.

The models described above are in accordance with the COX-2 crystal structure. The COX-2 binding site is formed by a long hydrophobic cavity, which contains three amino acids important for the binding of NSAID. Indeed, the carboxylic acid group of NSAIDs can interact with the guanidinium group of ARG120 [102]. Other binding modes involve a hydrogen bond between the acid carboxylic group and TYR385 and SER530 (e.g., diclofenac).

9.7.   CONCLUSION AND PERSPECTIVES

Since its conception in the 1960s, a keen interest for QSAR has been observed in the drug discovery area to enable the design of safe and potent drug candidates. During drug discovery and development phases, pharmacodynamic and pharmokinetic profiles of molecules can be derived using QSAR models. These in silico evaluations consist of the prediction of diverse properties (e.g., physicochemical, ADME) and activities to assist in the optimization and the prioritization of drug candidates. Numerous public, commercial, or corporate in silico tools including (Q)SAR models, decision trees, and molecular docking have been proposed to achieve these aims.

REFERENCES

1. Selassie CD (2003) History of quantitative structure-activity relationships. In: Abraham DJ (ed) Burger's medicinal chemistry and drug discovery, 6th edn. Wiley, New York
2. Drayer JI, Burns JP (1995) From discovery to market: the development of pharmaceuticals. In: Wolff ME (ed) Burger's medicinal chemistry and drug discovery, 5th edn. Willey, New York
3. Lesko LJ, Rowland M, Peck CC et al. (2000) Optimizing the science of drug development: Opportunities for better candidate selection and accelerated evaluation in humans. Eur J Pharm Sci 10:9–14
4. DiMasi JA (1995) Success rates for new drugs entering clinical testing in the United States. Clin Pharmacol Ther 58:1–14
5. Kennedy T (1997) Managing the drug discovery/development interface. Drug Disc Today 2: 436–444
6. Wenlock MC, Austin RP, Barton P et al. (2003) A comparison of physiochemical property profiles of development and marketed oral drugs. J Med Chem 46:1250–1256
7. Lin JH, Lu AYH (1997) Role of pharmacokinetics and metabolism in drug discovery and development. Pharmacol Rev 49:403–449
8. Cruciani G, Crivori P, Carrupt P-A et al. (2000) Molecular fields in quantitative structure permeation relationships: the VolSurf approach. J Mol Struct: THEOCHEM 503:17–30
9. Lombardo F, Gifford E, Shalaeva MY (2003) In silico ADME prediction: Data, models, facts and myths. Mini Rev Med Chem 8:861–875
10. Hansch C, Leo A, Mekapati SB et al. (2004) QSAR and ADME. Bioorg Med Chem 12:3391–3400
11. Lipinski CA, Lombardo F, Dominy BW et al. (1997) Experimental and computational approaches to estimate solubility and permeability in drug discovery and development settings. Adv Drug Del Rev 23:3–25
12. Yamashita F, Hashida M (2003) Mechanistic and empirical modeling of skin permeation of drugs. Adv Drug Del Rev 55:1185–1199
13. Wang Z, Yan A, Yuan Q et al. (2008) Explorations into modeling human oral bioavailability. Eur J Med Chem 43:2442–2452
14. Hansch C, Steward AR (1964) The use of substituent constants in the analysis of the structure–activity relationship in penicillin derivatives. J Med Chem 7:691–694
15. Hansch C, Maloney PP, Fujita T et al. (1962) Correlation of biological activity of phenoxyacetic acids with Hammett substituent constants and partition coefficients. Nature (London, UK) 194:178–180
16. Hansch C, Lien EJ (1968) An analysis of the structure–activity relationship in the adrenergic blocking activity of the βhaloalkylamines. Biochem Pharmacol 17:709–720

17. Hansch C, Leo A, Nikaitani D (1972) Additive-constitutive character of partition coefficients. J Org Chem 37:3090–3092
18. Hammett LP (1970) Physical organic chemistry: Reaction rates, equilibria and mechanism. McGraw-Hill, New York
19. Kubinyi H (1993) QSAR: Hansch analysis and related approaches. In: Mannhold R, Krogsgaard L, Timmerman H (eds) Methods and principles in medicinal chemistry, vol 1. VCH Publishers, New York
20. Free SM, Wilson JW (1964) A mathematical contribution to structure–activity studies. J Med Chem 7:395–399
21. Sciabola S, Stanton RV, Wittkopp S et al. (2008) Predicting kinase selectivity profiles using Free-Wilson QSAR analysis. J Chem Inf Model 48:1851–1867
22. Fujita T, Ban T (1971) Structure-activity relation. 3. Structure–activity study of phenethylamines as substrates of biosynthetic enzymes of sympathetic transmitters. J Med Chem 14:148–152
23. Taft RW, Abboud J-LM, Kamlet MJ et al. (1985) Linear solvation energy relations. J Sol Chem 14:153–186
24. Kamlet MJ, Abboud JLM, Abraham MH et al. (1983) Linear solvation energy relationships. 23. A comprehensive collection of the solvatochromic parameters, π^*, α, and β, and some methods for simplifying the generalized solvatochromic equation. J Org Chem 48:2877–2887
25. Kamlet MJ, Doherty RM, Abboud JLM et al. (1986) Linear solvation energy relationships: 36. Molecular properties governing solubilities of organic nonelectrolytes in water. J Pharm Sci 75:338–349
26. Murray JS, Politzer P, Famini GR (1998) Theoretical alternatives to linear solvation energy relationships. J Mol Struct: THEOCHEM 454:299–306
27. Famini GR, Penski CA, Wilson LY (1992) Using theoretical descriptors in quantitative structure activity relationships: Some physicochemical properties. J Phys Org Chem 5:395–408
28. Kamlet MJ, Doherty RM, Fiserova-Bergerova V et al. (1987) Solubility properties in biological media 9: Prediction of solubility and partition of organic nonelectrolytes in blood and tissues from solvatochromic parameters. J Pharm Sci 76:14–17
29. Abraham MH, Martins F (2004) Human skin permeation and partition: General linear free-energy relationship analyses. J Pharm Sci 93:1508–1523
30. Platts JA, Abraham MH, Zhao YH et al. (2001) Correlation and prediction of a large blood–brain distribution data set – an LFER study. Eur J Med Chem 36:719–730
31. Sangster J (1997) Octanol-water partition coefficients: Fundamentals and physical chemistry. Chichester, England.
32. Mannhold R, Poda GI, Ostermann C et al. (2009) Calculation of molecular lipophilicity: State-of-the-art and comparison of log P methods on more than 96,000 compounds. J Pharm Sci 98: 861–893
33. Lee CK, Uchida T, Kitagawa K et al. (1994) Skin permeability of various drugs with different lipophilicity. J Pharm Sci 83:562–565
34. Baláž Š (2000) Lipophilicity in trans-bilayer transport and subcellular pharmacokinetics. Persp Drug Disc Des 19:157–177
35. Testa B, Crivori P, Reist M et al. (2000) The influence of lipophilicity on the pharmacokinetic behavior of drugs: Concepts and examples. Persp Drug Disc Des 19:179–211
36. Efremov RG, Chugunov AO, Pyrkov TV et al. (2007) Molecular lipophilicity in protein modeling and drug design. Curr Med Chem 14:393–415
37. Testa B, Caron G, Crivori P et al. (2000) Lipophilicity and related molecular properties as determinants of pharmacokinetic behaviour. Chimia 54:672–677
38. Stella C, Galland A, Liu X et al. (2005) Novel RPLC stationary phases for lipophilicity measurement: solvatochromic analysis of retention mechanisms for neutral and basic compounds. J Sep Sci 28:2350–2362

39. Lombardo F, Shalaeva MY, Tupper KA et al. (2000) ElogPoct: A tool for lipophilicity determination in drug discovery. J Med Chem 43:2922–2928
40. George A (1999) The design and molecular modeling of CNS drugs. Curr Opin Drug Disc Dev 2:286–292
41. Youdim MBH, Buccafusco JJ (2005) CNS Targets for multi-functional drugs in the treatment of Alzheimer's and Parkinson's diseases. J Neural Transm 112:519–537
42. Van der Schyf CJ, Geldenhuys J, Youdim MBH (2006) Multifunctional drugs with different CNS targets for neuropsychiatric disorders. J Neurochem 99:1033–1048
43. Quach TT, Duchemin AM, Rose C et al. (1979) In vivo occupation of cerebral histamine H1-receptors evaluated with 3H-mepyramine may predict sedative properties of psychotropic drugs. Eur J Pharmacol 60:391–392
44. Bousquet J, Campbell AM, Canonica GW (1996) H1-receptors antagonists: Structure and classification. In: Simons FER, Dekkers M (eds) Histamine and H1-receptor antagonists in allergic diseases. Marcel Dekker, New York
45. Abraham MH (2004) The factors that influence permeation across the blood–brain barrier. Eur J Med Chem 39:235–240
46. Wiener H (1947) Structural determination of paraffin boiling points. J Am Chem Soc 69:17–20
47. Randic M (1975) Characterization of molecular branching. J Am Chem Soc 97:6609–6615
48. Balaban AT (1998) Topological and stereochemical molecular descriptors for databases useful in QSAR, similarity/dissimilarity and drug design. SAR QSAR Environ Res 8:1–21
49. Hall LH, Kier LB (1995) Electrotopological state indices for atom types: a novel combination of electronic, topological, and valence state information. J Chem Inf Comp Sci 35:1039–1045
50. Mezey PG (1992) Shape-similarity measures for molecular bodies: A 3D topological approach to quantitative shape-activity relations. J Chem Inf Comp Sci 32:650–656
51. Camenisch G, Folkers G, van de Waterbeemd H (1998) Shapes of membrane permeability-lipophilicity curves: Extension of theoretical models with an aqueous pore pathway. Eur J Pharm Sci 6:321–329
52. Hopfinger AJ (1980) A QSAR investigation of dihydrofolate reductase inhibition by Baker triazines based upon molecular shape analysis. J Am Chem Soc 102:7196–7206
53. Seri-Levy A, Salter R, West S et al. (1994) Shape similarity as a single independent variable in QSAR. Eur J Med Chem 29:687–694
54. Bondi A (1964) van der Waals Volumes and Radii. J Phys Chem 68:441–451
55. Connolly ML (1985) Computation of molecular volume. J Am Chem Soc 107:1118–1124
56. Arteca GA (1996) Molecular shape descriptors. In: Boyd DB, Lipkowitz KB (eds) Reviews in computational chemistry, vol 9. Wiley-VCH, New York
57. Verloop A (1987) The STERIMOL approach to drug design. Marcel Dekker, New York
58. Cruciani G, Pastor M, Guba W (2000) VolSurf: A new tool for the pharmacokinetic optimization of lead compounds. Eur J Pharm Sci 11:S29–S39
59. Andrews CW, Bennett L, Yu L (2000) Predicting human oral bioavailability of a compound: development of a novel quantitative structure–bioavailability relationship. Pharm Res 17:639–644
60. Veber DF, Johnson SR, Cheng H-Y et al. (2002) Molecular properties that influence the oral bioavailability of drug candidates. J Med Chem 45:2615–2623
61. Rekker RF, Mannhold R (1992) Calculation of drug lipophilicity. VCH, Weinheim.
62. Hansch C, Leo A, Nikaitani D (1972) Additive-constitutive character of partition coefficients. The Journal of Organic Chemistry 37:3090–3092
63. Austin RP, Davis AM, Manners CN (1995) Partitioning of ionizing molecules between aqueous buffers and phospholipid vesicles. J Pharm Sci 84:1180–1183
64. Scherrer RA, Howard SM (1977) Use of distribution coefficients in quantitative structure–activity relations. J Med Chem 20:53–58

65. Carrupt P-A, Testa B, Gaillard P (1997) Computational approaches to lipophilicity: Methods and applications. In: Boyd DB, Lipkowitz KB (eds) Reviews in computational chemistry, vol 11. Wiley-VCH, New York

66. Cramer RD, Wendt B (2007) Pushing the boundaries of 3D-QSAR. J Comp-Aided Mol Des 21:23–32

67. Goodford PJ (1985) A computational procedure for determining energetically favorable binding sites on biologically important macromolecules. J Med Chem 28:849–857

68. Cocchi M, Johansson E (1993) Amino acids characterization by GRID and multivariate data analysis. Quant Struct-Act Relat 12:1–8

69. Davis AM, Gensmantel NP, Johansson E et al. (1994) The use of the GRID program in the 3-D QSAR analysis of a series of calcium-channel agonists. J Med Chem 37:963–972

70. Pastor M, Cruciani G (1995) A novel strategy for improving ligand selectivity in receptor-based drug design. J Med Chem 38:4637–4647

71. Mannhold R, Berellini G, Carosati E et al. (2006) Use of MIF-based VolSurf descriptors in physicochemical and pharmacokinetic studies. In: Cruciani G, Mannhold R, Kubinyi H et al. (eds) Molecular interaction fields: Applications in drug discovery and ADME prediction. Wiley, Weinheim

72. Cianchetta G, Mannhold R, Cruciani G et al. (2004) Chemometric studies on the bactericidal activity of quinolones via an extended VolSurf approach. J Med Chem 47:3193–3201

73. Bajot F (2006) 3D solvatochromic models to derive pharmacokinetic in silico profiles of new chemical entities. Ph.D. Thesis, University of Geneva

74. Gaillard P, Carrupt P-A, Testa B et al. (1994) Molecular lipophilicity potential, a tool in 3D-QSAR. Method and applications. J Comp-Aided Mol Des 8:83–96

75. Rey S, Caron G, Ermondi G et al. (2001) Development of molecular hydrogen-bonding potentials (MHBPs) and their application to structure-permeation relations. J Mol Graphics Model 19:521–535

76. Weaver S, Gleeson MP (2008) The importance of the domain of applicability in QSAR modeling. J Mol Graphics Model 26:1315–1326

77. Netzeva TI, Worth AP, Aldenberg T et al. (2005) Current status of methods for defining the applicability domain of (quantitative) structure–activity relationships. ATLA 33:155–173

78. Schultz TW, Hewitt M, Netzeva TI et al. (2007) Assessing applicability domains of toxicological QSARs: Definition, confidence in predicted values, and the role of mechanisms of action. QSAR Comb Sci 26:238–254

79. Hoffmann P, Warner B (2006) Are hERG channel inhibition and QT interval prolongation all there is in drug-induced torsadogenesis? A review of emerging trends. J Pharmacol Toxicol Meth 53:87–105

80. Rangno R (1997) Terfenadine therapy: can we justify the risk? Can Med Assoc J 157:37–38

81. Jamieson C, Moir EM, Rankovic Z et al. (2006) Medicinal chemistry of hERG optimizations: Highlights and hang-ups. J Med Chem 49:5029–5046

82. Ekins S, Crumb WJ, Sarazan RD et al. (2002) Three-dimensional quantitative structure–activity relationship for inhibition of human ether-a-go-go-related gene potassium channel. J Pharmacol Exp Ther 301:427–434

83. Aptula AO, Cronin MTD (2004) Prediction of hERG K+ blocking potency: Application of structural knowledge. SAR QSAR Environ Res 15:399–411

84. Bradbury MWB (1984) The structure and function of the blood–brain barrier. Fed Proc 43:186–190

85. Bodor N, Brewster ME (1983) Problems of delivery of drugs to the brain. Pharmacol Ther 19:337–386

86. Rose K, Hall LH, Kier LB (2002) Modeling blood–brain barrier partitioning using the electro-topological state. J Chem Inf Comp Sci 42:651–666

87. Crivori P, Cruciani G, Carrupt PA et al. (2000) Predicting blood–brain barrier permeation from three-dimensional molecular structure. J Med Chem 43:2204–2216

88. Ooms F, Weber P, Carrupt P-A et al. (2002) A simple model to predict blood–brain barrier permeation from 3D molecular fields. Biochimi Biophys Acta – Mol Basis Disease 1587:118–125

89. Gerebtzoff G, Seelig A (2006) In silico prediction of blood–brain barrier permeation using the calculated molecular cross-sectional area as main parameter. J Chem Inf Model 46:2638–2650

90. Young RC, Mitchell RC, Brown TH et al. (1988) Development of a new physicochemical model for brain penetration and its application to the design of centrally acting H2 receptor histamine antagonists. J Med Chem 31:656–671

91. Clark DE (2003) In silico prediction of blood–brain barrier permeation. Drug Disc Today 8: 927–933

92. Reddy RN, Mutyala R, Aparoy P et al. (2007) Computer aided drug design approaches to develop cyclooxygenase based novel anti-inflammatory and anti-cancer drugs. Curr Pharm Des 13: 3505–3517

93. Fujimura T, Ohta T, Oyama K et al. (2007) Cyclooxygenase-2 (COX-2) in carcinogenesis and selective COX-2 inhibitors for chemoprevention in gastrointestinal cancers. J Gastrointest Canc 38:78–82

94. Esposito E, Di Matteo V, Benigno A et al. (2007) Non-steroidal anti-inflammatory drugs in Parkinson's disease. Exper Neurol 205:295–312

95. Hoozemans JJ, Rozemuller JM, van Haastert ES et al. (2008) Cyclooxygenase-1 and -2 in the different stages of Alzheimer's disease pathology. Curr Pharm Des 14:1419–1427

96. Kurumbail RG, Stevens AM, Gierse JK et al. (1996) Structural basis for selective inhibition of cyclooxygenase-2 by anti-inflammatory agents. Nature 384:644

97. Picot D, Loll PJ, Garavito RM (1994) The X-ray crystal structure of the membrane protein prostaglandin H2 synthase-1. Nature 367:243–249

98. Marshall GR, Taylor CM (2007) Introduction to computer-assisted drug design – Overview and perspective for the future. In: John BT, David JT (eds) Comprehensive medicinal chemistry II, vol 4. Elsevier, Oxford

99. Good A, John BT, David JT (2007) Virtual screening. In: John BT, David JT (eds) Comprehensive medicinal chemistry II, vol 4. Elsevier, Oxford

100. Garg R, Kurup A, Mekapati SB et al. (2003) Cyclooxygenase (COX) inhibitors: A comparative QSAR study. Chem Rev 103:703–732

101. Lee K-O, Park H-J, Kim Y-H et al. (2004) CoMFA and CoMSIA 3D QSAR studies on pimarane cyclooxygenase-2 (COX-2) inhibitors. Arch Pharm Res 27:467–470

102. Selinsky BS, Gupta K, Sharkey CT et al. (2001) Structural analysis of NSAID binding by prostaglandin H2 synthase: Time-dependent and time-independent inhibitors elicit identical enzyme conformations. Biochem 40:5172–5180

103. Gepp MM, Hutter MC (2006) Determination of hERG channel blockers using a decision tree. Bioorg Med Chem 14:5325–5332

104. Thai K-M, Ecker GF (2008) A binary QSAR model for classification of hERG potassium channel blockers. Bioorg Med Chem 16:4107–4119

105. Cianchetta G, Li Y, Kang J et al. (2005) Predictive models for hERG potassium channel blockers. Bioorg Med Chem Lett 15:3637–3642

106. Song M, Clark M (2006) Development and evaluation of an in silico model for hERG binding. J Chem Inf Model 46:392–400

CHAPTER 10

IN SILICO APPROACHES FOR PREDICTING ADME PROPERTIES

JUDITH C. MADDEN

*School of Pharmacy and Chemistry, Liverpool John Moores University, Liverpool, L3 3AF, England,
e-mail: j.madden@ljmu.ac.uk*

Abstract: A drug requires a suitable pharmacokinetic profile to be efficacious in vivo in humans. The relevant pharmacokinetic properties include the absorption, distribution, metabolism, and excretion (ADME) profile of the drug. This chapter provides an overview of the definition and meaning of key ADME properties, recent models developed to predict these properties, and a guide as to how to select the most appropriate model(s) for a given query. Many tools using the state-of-the-art in silico methodology are now available to users, and it is anticipated that the continual evolution of these tools will provide greater ability to predict ADME properties in the future. However, caution must be exercised in applying these tools as data are generally available only for "successful" drugs, i.e., those that reach the marketplace, and little supplementary information, such as that for drugs that have a poor pharmacokinetic profile, is available. The possibilities of using these methods and possible integration into toxicity prediction are explored.

Keywords: ADME, In silico methods, Biokinetics

10.1.   INTRODUCTION

One of the recent success stories in modern drug development has been the incorporation of in silico methods for the prediction of ADME properties into the design process (see Chapter 9). Significant savings have been made in time, cost, and animal use because rapid identification and rejection of pharmacokinetically unsuitable drug candidates means non-viable leads are not progressed from an early stage. In the period between 1991 and 2000, late stage candidate attrition due to pharmacokinetic reasons was shown to be reduced by approximately 30% [1]. There is also a growing interest in the application of in silico prediction of ADME properties to the area of toxicology, to improve accuracy in predicting adverse effects of a wide range of compounds (Chapter 11). The rationale for this is clear when one considers how any xenobiotic produces an effect within an organism as described below.

In order for a compound to be able to elicit a biological effect, be that a required therapeutic effect or an unwanted toxic effect, there are two key determinants:

T. Puzyn et al. (eds.), Recent Advances in QSAR Studies, 283–304.
DOI 10.1007/978-1-4020-9783-6_10, © Springer Science+Business Media B.V. 2010

(i) the intrinsic activity of the compound and (ii) its potential to reach the site of action in sufficient concentration for the requisite time period. Predicting intrinsic activity of compounds in drug design and toxicology is dealt with elsewhere in this volume (Chapters 4, 5, and 11). This current chapter is devoted to the consideration of the factors which determine whether or not a compound is likely to reach a specific site of action and how long it is likely to persist at that specific site and in the body as a whole.

The terminology used to describe these processes can be context dependent as highlighted by d'Yvoire et al. [2]; therefore a preliminary definition of the terms as used in this chapter is provided. Historically the absorption, distribution, metabolism, and excretion (ADME) properties of compounds were studied in relation to drug development. Hence, in general where the study of the movement (i.e., kinetics) of a compound within the body relates to a desirable effect of a therapeutic substance it is generally referred to as a pharmacokinetic (PK) property. As this appeared to restrict the definition to drugs (at therapeutic doses), the movement of toxic substances responsible for deleterious effects is now generally termed toxicokinetics (TK). The distinction between the terms pharmacokinetics and toxicokinetics, however, is not absolute. For example, where a drug produces a therapeutic effect this would be termed as pharmacokinetics; however, where the same drug produces side-effects or an excess therapeutic effect the drug's movement may be referred to as toxicokinetics. This has led to the general term biokinetics (BK), being proposed as a more inclusive term [2]. However, irrespective of terminology, in PK, TK, or BK studies, the fundamental properties of any compound of interest, i.e., its absorption, distribution, metabolism, and excretion (ADME) properties, are the same. Herein the term ADME will be used in discussion of the movement of all xenobiotics within the body.

ADME properties are arguably the most important consideration in determining the true potential of any compound to elicit a biological effect, desirable or undesirable, within the body. In this chapter, an overview of the definition and meaning of key ADME properties, recent models developed to predict these properties, and a guide as to how to select the most appropriate model(s) for a given query are presented.

10.1.1.   Overview of Key ADME Properties

The first requirement for the interaction of a xenobiotic with an organism is the uptake of the compound into the body (except for direct acting agents such as topical irritants). Absorption processes govern the transfer of compounds from the external to the internal environment. Uptake is dependent upon the route of exposure to the compound; there are numerous ways in which xenobiotics may enter the body. Absorption occurs across the gastrointestinal tract for food additives, for toxicants persisting in the food chain or water supply and for compounds leaching from food packaging. Absorption via the lungs or nasal mucosa is important for environmental contaminants present in the general atmosphere or in certain work places. Potential dermal absorption must be considered for cosmetics, personal care products, hair and clothing dyes in addition to general environmental contaminants. In vitro skin permeability measurements can be used to determine uptake across the dermal barrier. In therapeutics, drug formulation science has exploited every potential route of administering a compound into the body.

The percentage of available compound that is absorbed (% Abs) provides a preliminary measure of internal exposure from any route. For oral ingestion a common measure is termed the percentage human intestinal absorption (% HIA). However, absorption alone does not determine systemic availability. Acting in opposition to absorption processes are local, or first-pass, metabolism and active efflux processes (discussed below). Hence bioavailability (F) is often considered a more useful term as this refers to the percentage of available compound that appears in the systemic circulation. F is directly proportional to absorption and inversely proportional to local, or first pass, metabolism and active efflux.

Once a compound has successfully entered the systemic circulation, the next stage is distribution to other sites within the body. Distribution is a crucial consideration as drugs need to reach their intended site of action, but ideally should not distribute to where they may cause adverse side-effects. Toxicants may cause more severe effects in certain organs (such as the brain) than in others. Distribution and uptake into storage sites, such as adipose, has a significant impact on the time for which xenobiotics persist in the body. Distribution generally occurs via the blood stream, although the lymphatic system is relevant for some compounds. Overall the tendency of a compound to move out of the blood and into tissues is given by the apparent volume of distribution (V_d). This is a hypothetical volume into which a compound distributes and is determined from Eq. (10-1):

$$V_d = dose/C_0. \qquad (10\text{-}1)$$

where C_0 is the initial concentration measured in blood.

If a compound has a high tendency to move out of blood, the resulting concentration in blood is low and V_d is very large. Conversely, if it has a tendency to remain in blood C_0 will be high and V_d low. The tendency for a compound to remain in blood or move to other compartments is governed by its ability to pass through membranes and the relative affinity for tissue and blood proteins; hence the percentage of plasma protein binding or fraction bound (% PPB or f_b) and tissue binding are key determinants in distribution. As it is free (or unbound) drug that binds to targets the unbound volume of distribution (V_{du}) is often considered, this is shown in Eq. (10-2):

$$V_{du} = V_d/f_u \qquad (10\text{-}2)$$

where f_u is the fraction unbound (i.e. $1 - f_b$).

The more widely a compound is distributed throughout the body the more sites are available to elicit potentially toxic effects and the longer the compound will persist in the body. Individual tissue compositions can result in differing affinity for xenobiotics. Tissue:blood partition coefficients (PCs) indicate the likelihood of a compound being taken up by a specific tissue. One of the most important tissue: blood PCs is that for the blood–brain barrier (BBB) as the central nervous system is associated with some of the most significant toxic effects. The ability of a compound to cross the placenta is also of great concern. The properties used to indicate transfer across the placenta are the placental transfer index (TI) or the clearance index (CI).

Uptake into tissues is also influenced by influx and efflux transporters, which have been identified in many tissues. Uptake transporters include organic anion transporting polypeptides (OATPs), organic cation transporters (OCTs), organic anion transporters (OATs), and the organic cation/carnitine transporters (OCTNs).

Efflux transporters act in opposition to uptake processes. In particular, their presence in the gastrointestinal tract, blood–brain barrier and the placenta provides an important protective effect. ATP binding cassette (ABC) proteins have received much attention because of their recognized role in multi-drug resistance, i.e., their expression in tumor cells, resulting in active efflux of therapeutic agents, is known to be responsible for resistance to drug treatment. Examples of these include P-glycoprotein (P-gp), multi-drug resistance-associated protein (MRP2), and breast cancer resistance protein (BCRP). Phase 0 disposition refers to the ability of a xenobiotic to enter a cell where it may elicit a response. Consequently, the presence of influx and efflux transporters modulates biological response by increasing or decreasing cellular entry [3]. The role of transporters is linked with that of the metabolic processes.

Metabolism is the process by which the body converts xenobiotics usually into a less toxic, more polar form that can be readily excreted. However, in some cases metabolism may be necessary to convert an inactive drug into its active form (i.e., for pro-drugs) or it may lead to the formation of toxic metabolites. Phase I metabolism involves functionalization reactions where a polar group is added to, or exposed on, the molecule. These include reactions such as oxidation of nitrogen or sulfur groups, aliphatic or aromatic hydroxylation, de-amination and de-alkylation. The cytochrome (CYP) P450 superfamily of enzymes is responsible for the catalysis of many of these reactions. The isoforms CYP3A4, CYP2D6, and CYP2C9 are responsible for the metabolism of the vast majority of drugs. Whilst CYP1A1/2, CYP2A6, CYP2B1, and CYP2E1 play little role in drug metabolism, they do catalyze the activation of certain pro-carcinogenic environmental pollutants into their carcinogenic form and are therefore of toxicological importance [4]. Phase II reactions may be consecutive to, or independent of, phase I reactions and include the conjugation of a polar moiety to the compound (e.g., glucuronidation, sulfation, or acetylation) enabling renal or biliary excretion of the polar metabolite. Metabolizing enzymes have been found in all tissues of the body but are predominant in the liver, kidney, and intestine. First-pass metabolism is the process by which compounds that are absorbed into the gut travel via the hepatic portal vein to the liver and are metabolized before they reach the systemic circulation. Enterohepatic recycling, whereby compounds are excreted into bile and hence are returned to the gastrointestinal tract for reabsorption may lead to reappearance of xenobiotics in the blood stream and prolongation of effect. Gibbs et al. [5] discuss the role of skin metabolism in modulating activity of compounds presented to the body via the dermal route.

In terms of metabolism, there are three specific factors of importance:

- the nature of the metabolite (i.e., is it more or less active than the parent?);
- the extent to which it is formed (i.e., does it represent a major or minor metabolite?); and
- the rate at which it is formed (i.e., how much will be present in the body over time?).

Slow rates of metabolism can be associated with persistence and bioaccumulation of xenobiotics within the body, potentially leading to prolonged activity or toxicity.

Phase III disposition refers to the exit of metabolites from cells, a process which again can be modulated by efflux transporters. Szakacs et al. [3] refer to the concerted interaction of metabolizing enzymes and efflux transporters as an effective chemoimmune system by which the body may be protected from adverse effects of xenobiotics.

Excretion processes are those by which compounds are ultimately removed from the body. A major route of elimination is renal excretion, but excretion into sweat, feces, and expired air are also possible routes. The percentage of urinary excretion (% exc), usually refers to direct renal elimination of unchanged compound. Excretion into breast milk raises specific concerns of potential toxicity to newborns, particularly as this may be the sole food source for the infant; hence their exposure is relatively high. Milk:plasma ratios (m:p) are useful in determining relative concentrations in breast milk.

The rate at which a compound is eliminated from the body is referred to as the rate of clearance (Cl). This is defined as the volume of blood completely cleared of compound in a given time. Total clearance is that by any route but renal excretion and hepatic metabolic routes predominate.

Half-life (t $\frac{1}{2}$) is the time taken for the amount of compound in the body to fall by half. It is arguably the most important property as it dictates for how long the compound persists in the body and therefore the timescale over which it may elicit therapeutic or toxic effects. Half-life is determined according to Eq. (10-3) given below:

$$t^1/_2 = 0.693V_d/Cl \qquad (10\text{-}3)$$

By definition, half-life is governed by the extent to which the compound distributes throughout the body (V_d) and the rate at which it is cleared (Cl).

In determining biological effect, it is often desirable to relate activity to the concentration–time profile of a xenobiotic within a given tissue. For this physiologically based pharmacokinetic models (PBPK) may be used. In such models the biological system is represented as a series of organs, about which information, such as volumes and blood flows are known. These data are combined with compound-specific parameters (such as tissue partitioning, clearance) enabling the full time-course of the drug to be predicted in individual tissues. PBPK modeling and its use is discussed by d'Yvoire et al. [2].

From the above descriptions, it is clear that ADME plays a key role in determining the extent of overall effect any compound has on the body. Hence prediction of these properties has gained widespread interest in the quest to predict, more accurately, both therapeutic and toxic effects of xenobiotics. Table 10-1 provides a summary of key ADME properties, along with reviews/example models for their prediction [6–39]. Several of these models are discussed in more detail below.

Table 10-1. Summary of key ADME properties and references for reviews/example models for their prediction

Properties	Definition	References
% Abs; % HIA	Percentage of available compound that is absorbed across a barrier; percentage that is absorbed across the human gastrointestinal tract	[6–9]
Skin permeability (K_p)	Permeability of a solute through skin, determined by flux measurements	[10, 11]
F	Bioavailability – fraction of dose that enters the systemic circulation	[9, 12, 13]
% PPB; f_b; f_u	Percentage of compound bound to plasma proteins; fraction bound to plasma proteins; fraction unbound (i.e., free fraction)	[14, 15]
V_d; V_{du}	Apparent volume of distribution, i.e., the hypothetical volume into which a drug distributes; V_{du} is the volume of distribution for the unbound fraction of drug	[16–20]
Tissue:blood PCs	The ratio of concentrations between blood and tissue	[21, 22]
BBB partitioning	Blood–brain barrier partitioning, frequently expressed as ratio of concentrations between brain and blood (serum/plasma) or expressed in binary format to indicate likely or not likely to enter brain	[9, 23–26]
CI; TI	Clearance index; transfer index for placental transfer of compounds usually expressed as a ratio using antipyrine as a marker	[27]
Transporter substrates/non-substrates/inhibitors	Relates to the affinity of compounds for a wide range of transporters, (several of which are defined in the text)	[28–30]
% exc	The percentage of compound excreted unchanged in urine	[31]
m:p	The ratio of concentration between breast milk and plasma	[32]
Metabolism	The process by which xenobiotics are converted to an alternative compound (usually one which can be more readily excreted). Of significance is the nature of the metabolite, the enzyme responsible for the catalysis of the process and the rate at which it occurs	[33–36]
Cl; Cl_{tot}; Cl_h; Cl_r	Clearance, i.e., the volume of blood from which a compound is completely removed in a given time; clearance by all routes; clearance by hepatic route (i.e., metabolism); clearance by renal route (i.e., urinary excretion)	[37]
$t^{1/2}$	Half-life, i.e., the time taken for the concentration of a compound in the body to fall by half	[38, 39]

10.1.2. Data for Generation of in Silico Models

As with all areas of model development, as more data become available, greater opportunity arises to produce more accurate and robust models. In terms of predicting ADME properties, the majority of data has been generated for pharmaceutical products. Whilst, undoubtedly, hundreds of thousands of compounds have been screened in drug development projects, unfortunately the majority of these data are proprietary and therefore not publicly available for modeling (a similar situation is described with respect to toxicity data in Chapter 11). This means that the publicly available ADME data, and hence models, tend to be skewed toward that minority of candidate compounds which exist in pharmacokinetically viable space. This is because all commercially available drugs will have acceptable, although probably not ideal, pharmacokinetic properties. When generating models it is better to have unbiased data sets with uniform coverage of the parameter space.

An additional problem of the bias toward commercially available drugs is that efforts to predict toxicokinetics of, for example, environmental pollutants are severely hampered by the paucity of accessible data. Models generated from pharmaceutical data may not be suitable to predict the ADME properties for a wide range of organic compounds, such as industrial chemicals and pesticides. Consideration of the applicability domain of a model is critical in terms of selecting the most appropriate model for a given query. Many publications are available for a detailed discussion of applicability domain, e.g., Netzeva et al. [40] and therefore this will not be considered further here.

With the increased interest in generating predictive ADME models, there has been a corresponding increase in the number of relevant data sets published in recent years. Data mining, i.e., collating and structuring data, from either in silico repositories or literature publications can provide valuable information. Table 10-2 provides a list of potentially useful data sets and the references [10, 11, 13, 15, 18, 21, 22, 24–27, 30, 32, 36, 37, 41–50] from which the full data are available.

Table 10-2. Examples of resources for ADME data

Properties	Information available	Reference
Human intestinal absorption	Data for 648 chemicals	[41]
Human oral bioavailability	Data for 768 compounds	[42]
Human oral bioavailability	Data for 302 drugs	[13]
Skin permeability	K_p data for 124 compounds	[10]
Skin permeability	K_p data for 101 chemicals	[11]
Protein binding data	Percentage bound to human plasma protein for 1008 compounds	[15]
Volume of distribution	Data for 199 drugs in humans	[18]
Tissue:air partitioning	Data for 131 compounds partitioning into human blood, fat, brain, liver, muscle, and kidney (incomplete data for certain tissues)	[22]

Table 10-2. (continued)

Properties	Information available	Reference
Tissue:blood partitioning	Data for 46 compounds partitioning into kidney, brain, muscle, lung, liver, heart, and fat (incomplete data for certain tissues)	[20]
Air:brain partitioning	Human and rat air–brain partition coefficients for 81 compounds	[21]
Blood–brain partitioning	Blood/plasma/serum/brain partitioning data for 207 drugs in rat	[43]
Blood–brain partitioning	Log blood–brain barrier partitioning values for 328 compounds	[24]
Blood–brain barrier penetration	Binary data for 415 compounds (classified as blood–brain barrier penetrating or non-penetrating)	[25]
Blood–brain barrier penetration	Binary data for 1593 compounds (classified as blood–brain barrier crossing or non-crossing)	[26]
Placental transfer	Placental clearance index values for 86 compounds and transfer index values for 58 compounds	[27]
Clearance	Data for total clearance in human for 503 compounds	[37]
Metabolic pathways	Catalogue of all known bioactivation pathways of functional groups or structural motifs commonly used in drug design using 464 reference sources	[44]
CYP metabolism	List of 147 drugs with the CYP isoform predominantly responsible for their metabolism (CYP3A4, CYP2D6, and CYP2C9)	[36]
Clearance; plasma protein binding; volume of distribution	Total clearance, renal clearance, plasma protein binding, and volume of distribution data for 62 drugs in humans	[45]
Milk:plasma partitioning	Concentration ratio data for 123 drugs	[32]
Transporter data	117 substrates and 142 inhibitors of P-gp; 54 substrates and 21 inhibitors of MRP2; 41 substrates and 38 inhibitors of BCRP	[46]
PgP data substrates and non-substrates	Binary classification of 203 compounds as P-gp substrates (+) or non-substrates (−)	[30]
% urinary excretion; % plasma protein binding; clearance; volume of distribution; half-life, time to peak concentration, peak concentration	A compilation of ADME data for approximately 320 drugs (incomplete data for some drugs)	[47]
Half-life; therapeutic, toxic, and fatal blood concentrations	Data for over 500 drugs (incomplete data for some drugs)	[48]
Volume of Distribution; % plasma protein binding; % HSA binding (from HPLC retention data)	Data for 179 drugs (incomplete data for percentage plasma protein binding for some drugs)	[49]
Toxicogenomics micorarray data	Gives literature references for data on 36 compounds	[50]

10.2. MODELS FOR THE PREDICTION OF ADME PROPERTIES

For the ADME properties defined in Section 10.1.1 above, a range of in silico models have been developed. Table 10-1, along with a summary of key ADME properties provides references to example models, or broader reviews of models, available for each property. It is not possible here to provide a detailed review of all the models or modeling approaches available. Therefore, the discussion below will provide an overview of selected models and approaches. The reader is referred to the given reviews for further information.

Some of the simplest models in predictive ADME are those referred to as "rules of thumb." The most widely recognized of these is Lipinski et al's "rule of 5" [6]. This was devised to provide a screening tool for compounds that were likely to show absorption problems, i.e., poor absorption is more likely if

- molecular weight > 500:
- sum of OH and NH hydrogen bond donors > 5:
- sum of O and N hydrogen bond donors > 10:
- C log P > 5:

This type of tool found ready acceptance amongst users because of its simplicity and ready interpretability. It has led to an increasing number of rules of thumb being devised for other endpoints. In 2002 Veber et al. [12] proposed a model for predicting good bioavailability, i.e., good bioavailability is more likely for compounds with:

- $\leq$ 10 rotatable bonds;
- polar surface area $\leq$ 140 A^2; or
- sum of hydrogen bond donors and acceptors $\leq$ 12.

Norinder and Haberlein [23] proposed two rules of thumb for determining whether or not a compound is likely to cross the blood–brain barrier. These are:

- if number of N+O atoms is $\leq$ 5, then it is likely to enter brain;
- if log Kow – (N+O) is positive, then log BBB partition coefficient is positive.

Developing the theme of rapid screening for large corporate libraries, Lobell et al. [7] devised a traffic light system for "hit-selection." In their approach five "traffic lights" are calculated for ADMET properties relevant to absorption through the gastrointestinal tract. The requirements for a compound to be well absorbed are that it is reasonably soluble, not too polar, lipophilic, large, or flexible. These factors are determined by the solubility, polar surface area, C log P, molecular weight, and number of rotatable bonds. These properties are all readily calculable and the values for individual compounds are combined to give a traffic light value. From this the most promising candidates can be selected.

More recently in 2008, Gleeson [18] reported a series of ADMET rules of thumb for solubility, permeability, bioavailability, volume of distribution, plasma protein binding, CNS penetration, brain tissue binding, P-gp efflux, hERG inhibition, and inhibition of cytochromes CYP1A2/2C9/2C1/2D6/3A4. The influence of changing molecular weight and log P may have on these individual ADMET properties was demonstrated, providing a key as to how each of these may be optimized in drug development.

Simple structural information can also be useful in predicting other ADME properties. For example, the route of drug metabolism is determined by the presence of specific functional groups. Identification of such groups, hence deduction of likely routes of metabolism allows prediction of metabolic pathways important for identifying metabolites and potential drug–drug interactions. Manga et al. [36] showed the utility of using a recursive partitioning method (formal inference-based modeling) to predict the dominant form of P450 enzyme responsible for the metabolism of drugs. The model made use of descriptors for molecular weight, acidity, hydrogen bonding strength, molecular dimensions and log P. The model correctly identified which was the predominant enzyme responsible for metabolism for 94% of 96 compounds in the training set and 68% of 51 compounds in the test set.

Quantitative structure–activity relationship (QSAR) modeling has also proved useful in predicting ADME properties, although their use is generally restricted to smaller more homogenous data sets. QSAR models have been developed for many of the individual component processes in ADME.

Of the absorption processes, human intestinal absorption has received the greatest attention. Hou et al. [9] provide a review of 23 models for this endpoint including the use of multiple linear regression, non-linear regression, and partial least squares (in addition to neural network, support vector machine (SVM) and other analyses). Dermal absorption is of importance for both pharmaceuticals and environmental pollutants. Lian et al. [10] performed a comparative analysis of seven QSAR models to predict skin permeability, but concluded that more mechanistic studies were needed to improve predictions for this property.

General and specific distribution processes have also been successfully modelled using QSAR. In general terms, a global indication of distribution within the body is indicated by the apparent volume of distribution. Models for this global property have been developed by Ghafourian et al. [16] and Lombardo et al. [17]. Volume of distribution is dependent on the extent to which a compound binds to both plasma and tissue proteins. Colmenarejo [14] reviewed models available to predict binding to plasma proteins in addition to proposing new models to predict binding.

At a more local level, distribution into individual tissues is important in determining whether or not a compound is likely to distribute to a given site where it may elicit a therapeutic or toxic effect. Several models have been developed to predict tissue:blood partition coefficients, such as that described by Zhang [20]. In this model differential distribution into kidney, brain, muscle, lung, liver, heart and fat (based on tissue composition) was determined using 46 compounds. Of the tissue distribution models, partitioning into the brain has been the most extensively studied tissue. Hou et al. [9] provide a detailed review of 28 models to predict blood:brain barrier partitioning. More recently Konovalov et al. [24] using 328 log BBB values, proposed a system to benchmark QSARs for this endpoint to enable better comparison of current and future models.

From a toxicological perspective, an area of increasing concern is the partitioning of drugs and toxicants into the placenta as the potential to elicit toxic effects in the developing fetus is an important consideration. QSAR models to predict placental transfer of xenobiotics have been developed by Hewitt et al. [27].

HQSAR is a technique by which fragments of molecules are arranged to form a molecular hologram, such that three-dimensional information is implicitly encoded from input 2D structures. This technique was applied by Moda et al. [13] to the prediction of human oral bioavailability, providing reasonable correlations for this multi-factorial endpoint.

Three-dimensional modeling, although computationally more expensive, has also found a role particularly in binding studies, relevant to metabolism and determining affinity for efflux transporters. As discussed previously, in terms of metabolism, there is the potential for drug–drug interactions to occur between compounds that are metabolized by the same enzyme. Three-dimensional modeling of the specific interactions between ligands and their receptors provides greater understanding of the processes involved, which leads to improvement in predictions and helps to screen out such potential interactions during the design process. Three-dimensional QSAR modeling, pharmacophore generation, and homology modeling have all been applied to elucidate the role of P450s in drug metabolism. The application of 3D modeling to this field has been reviewed by de Groot [51] (Chapter 4). In particular, de Groot [51] discusses ligand-based and enzyme structure-based models. References are provided for pharmacophore models developed for P450s as well as references detailing known crystal structures for 15 bacterial, 2 fungal, and 7 mammalian P450s.

Determining which compounds are likely to act as substrates, non-substrates, or inhibitors for transport proteins is also important in predicting the overall internal exposure as well as tissue-specific exposure of xenobiotics. Chang et al. [28] provide a review of 3D QSAR studies for a wide range of membrane transporters including P-gp. Other studies on P-gp demonstrate another modeling technique which is proving useful in ADME modeling, i.e., the support vector machine (SVM) approach. Using this technique, Huang et al. [29] developed a model capable of distinguishing P-gp substrates from non-substrates with an average accuracy of over 91%.

Scientific opinion remains divided on the utility of neural networks to predict ADME and other endpoints. On the one hand, the flexibility of the approach enables non-linear relationships to be modelled. On the other, models are deemed to be non-transparent and difficult to interpret. However, many examples are available for the application of neural networks to this area. In terms of ADME models for excretion, one area of particular concern is the ability of a compound to be excreted into breast milk and the risk this may pose to neonates. Agatonovic-Kustrin et al. [32] used neural network methodology to develop a model identifying molecular features associated with transfer of drugs into breast milk. Turner et al. [45] also gives examples of neural network models for the prediction of total clearance, renal clearance, volume of distribution and fraction bound.

Whereas much of the above discussion relates to the development of individual models, one technique which appears to be growing in popularity is the use of consensus models. Banik [52] argues that as all models are simulations of reality, and therefore not totally accurate, a combination of individual models into a single consensus model can improve accuracy. An overview of different types of consensus models is presented, along with examples of where this approach has been shown to improve prediction accuracy.

The aim of the above discussion was to present an indication of the range of in silico modeling approaches available and information on where these have been applied to specific endpoints. However, within the literature there are also several reviews which cover the general application of in silico techniques to ADME predictions, as well as including comment on the status of the science. There are several examples of other useful reviews. Ekins et al. [53] provide an extensive review of available models and data sets. Duffy [54] gives an overview of models available and discussed selection of the most appropriate models for a given query; Gola et al. [55] review recent trends in predictive ADMET and argue for greater acceptance of in silico predictions, highlighting the importance of generating this data alongside activity data. Chohan et al. [56] draw on 61 references to review the status of QSARs for metabolism. Payne [35] provides an extensive review of techniques to predict metabolism. Winkler [57] discusses the role of neural networks in developing ADMET models. Dearden [58] reviews progress in the area of in silico ADMET modeling and provides a vision for future development in this area.

10.3.   SOFTWARE DEVELOPMENTS

Increased awareness of the importance of ADME in modulating both therapeutic and toxic effects of xenobiotics has led to an increased demand for software to predict these properties. Software providers have responded to this demand and there now exists a comprehensive range of computer packages to predict ADME properties. Table 10-3 lists some of the software available for the prediction of ADME properties and provides a brief description of functionalities available within the programs.

Table 10-3. Software for the prediction of ADME properties and commercial databases

Software provider	Software package	Predicted ADME properties	Websites
Accelrys	Discovery Studio ADMET	Absorption, BBB penetration, plasma protein binding, CYP2D6 binding Physicochemical properties/toxicity	http://www.accelrys. com
Bioinformatics and Molecular Design Research Centre	PreADME	Physicochemical properties; permeation through MDCK and Caco-2 cells; BBB permeation; human intestinal absorption; skin permeability; plasma protein binding	http://www. bmdrc.org/ 04_product/ 01_preadme.asp

Table 10-3.   (continued)

Software provider	Software package	Predicted ADME properties	Websites
BioRad	Know-it-All	Bioavailability, BBB permeability, half-life, absorption, plasma protein binding, volume of distribution, rule of 5 violations Physicochemical properties	http://www.knowitall.com
Chemistry Software Store	SLIPPER	Physicochemical properties, absorption	http://www.timtec.net/software/slipper/introduction.htm
ChemSilico	CSBBB CSHIA CSPB Other modules	BBB partitioning Human intestinal absorption Plasma protein binding Physicochemical properties/toxicities	http://www.chemsilico.com
CompuDrug	MetabolExpert MexAlert Rule of 5 Other modules	Metabolic fate of compounds Likelihood of first-pass metabolism Calculates rule of five parameters Physicochemical properties/ toxicities	http://www.compudrug.com
Cyprotex	Cloe® PK	Simulates concentration time course in blood and major organs; predicts renal excretion, hepatic metabolism and absorption; integrates experimental data	http://www.cyprotex.com
Laboratory of Mathematical Chemistry, Bourgas University	TIMES	Metabolic pathways	http://oasis-lmc.org/?section=software&swid=4
Genego	MetaDrug	Platform for the prediction of drug metabolism and toxicity	http://www.genego.com/metadrug.php
Lhasa	Meteor	Metabolic fate of compounds	http://www.lhasalimited.org/
Molecular Discovery	Metasite Volsurf+	Metabolic transformations Absorption, solubility, protein binding, volume of distribution, metabolic stability, BBB permeability	http://www.moldiscovery.com/index.php
MultiCASE	META/METAPC MCASE ADME module	Metabolic transformations Oral bioavailability, protein binding, urinary excretion, extent of metabolism, volume of distribution	http://www.multicase.com/products/products.htm
PharmaAlgorithms	ADME Boxes	P-gp substrate specificity; absorption; bioavailability; plasma protein binding; volume of distribution Physicochemical properties	http://www.pharma-algorithms.com/

Table 10-3. (continued)

Software provider	Software package	Predicted ADME properties	Websites
QuantumLead	q-ADME	Half-life, absorption, Caco-2 permeability, volume of distribution, humans serum albumin binding Physicochemical properties/toxicity	http://www.q-lead. com/adme_pk
Schrodinger	Qik Prop	Caco-2 and MDCK cell permeability, BBB permeation, serum albumin binding Physicochemical properties	http://www. schrodinger.com
SimCYP	Simcyp®	Population-based ADME simulator allowing profiles to be predicted in virtual populations [63]	http://www.simcyp. com/
Simulations Plus	ADMET Predictor	Intestinal permeability, absorption, BBB permeation, volume of distribution, plasma protein binding Physicochemical properties	http://www. simulations-plus. com
	GastroPlus	Physiological models for different species; dosage form effects; 1, 2, and 3 compartment models; complete PBPK models	

Provider	Package	Commercial Databases	Websites
Sunset Molecular	Wombat-PK	Database containing >6500 clinical pharmacokinetic measurements for 1125 compounds (bioavailability, percentage excretion, percentage plasma protein bound; clearance; volume of distribution, half-life, BBB permeation; metabolizing enzymes)	http://www. sunsetmolecular. com/
University of Washington	Metabolism and transport drug-interaction database	Database for enzyme and transporter interactions	http://www. druginteractioninfo. org/

The capabilities of these packages are diverse and are suitable for answering a variety of different queries. Facilities to predict key ADME properties, e.g., absorption, blood–brain barrier partitioning, and percentage plasma protein binding are available in many programs and require only the compound's structure for input. Examples include ADME boxes, ADMET predictor, Know-it-All, etc. (refer to Table 10-3 for more examples). Other software, such as Cloe®, requires the input of measured data (such as percentage plasma protein binding) in order to develop more accurate and comprehensive physiologically-based pharmacokinetic (PBPK) models of internal exposure. Within populations, different responses to xenobiotics are anticipated due to age, sex, health status, and genetic predisposition of individuals.

Simcyp® is a population-based ADME simulator that combines information from in vitro systems, drug-specific physicochemical information, and demographic, physiological, and genetic information to simulate ADME processes for drug candidates within a population. This provides a more mechanistic interpretation of pharmacokinetic or toxicokinetic behavior enabling predictions for subgroups of the population. Rapid expansion in the number of software packages available and their increasing sophistication is one of the hallmarks of recent advances in in silico tools for ADME prediction.

10.4.   SELECTING THE MOST APPROPRIATE MODELING APPROACH

The information provided above demonstrates that a wide range of literature models and computational tools for predicting ADME properties is currently available and continuously expanding. The models range from high-throughput initial screening rules to detailed three-dimensional enzyme binding analyses, requiring a high level of computational power. As with all QSAR studies, the selection of the most appropriate model to use is dependent on the nature of the query, i.e., what level of detail is necessary to answer the question posed and which model is most readily interpretable and useful to the user.

Simple screening tools such as Lipinski's "rule of five" [6] or Gleeson's rules of thumb [8] have perennial appeal because of their simplicity. Such models relying on cut-offs for simple molecular features are readily interpreted by end users and easily acted upon. For example in drug design, knowing that drugs with a molecular weight above 500 may be associated with absorption problems, compounds can be designed with a maximum molecular weight of 450 Da, allowing for subsequent structural modification in the later optimization stages. Such approaches work well in certain circumstances, such as when a large diverse library of compounds is available and the aim is simply to begin to narrow down the number of potential drug candidates, or in priority setting where large numbers of compounds need to be considered. It must be noted, however, that these are simplistic models and inevitably there will be exceptions. The degree of inaccuracy acceptable is dependent on the purpose of the study. Whilst it may be acceptable to prioritize only those drugs with molecular weights below 500 Da in drug design, it would not be acceptable to presume that a toxicant present in food would not exhibit oral toxicity simply because its molecular weight indicates it may be associated with poor absorption characteristics.

Within a given series of compounds, a traditional QSAR approach can work well. Many QSAR models are available to choose from, but it is essential that the query compound falls within the applicability domain of the given model. Whilst more quantitative and detailed information may be available from these models, they tend to be of more limited applicability than the global screening methods. Global QSARs covering large numbers of diverse compounds have been developed, but there is a danger that reasonable global statistics may obfuscate poorer statistics for subgroups of compounds. This presents a real danger of misprediction if the query compound falls within this category [59]. It is advisable to ascertain the reliability of the model to predict the required properties within the region of chemical space under investigation. One way of addressing this issue is by the use of "trainable" models. In this approach commercial or proprietary models are continually updated or re-trained using new data as it becomes available. The ADME Boxes approach from PharmaAlgorithms is an example of commercial software with this functionality. The chemical space within proprietary databases is unlikely to be fully represented by the chemistry used to train commercial models. It is important therefore that this additional knowledge can be captured and utilized to improve predictions in the chemical space relevant to the user. The region of chemical space under investigation at any given time is not a constant entity. The importance of allowing models to evolve over time has been exemplified by the study of Rodgers et al. [60]. In this study the authors demonstrated that models that were updated with new information over a two year period were more predictive than static models that did not evolve over time and advocated the "autoupdating" of QSAR models.

Once the nature of the query becomes more precise, a more specific and informative modeling tool may be required. For example, if detailed information is required on whether or not a drug may interact with a specific enzyme and therefore promote drug–drug interactions, three-dimensional modeling tools may become necessary. Another consideration in the selection of the most appropriate model is the selection of the most appropriate endpoint. For example, models exist to predict half-life of compounds; however half-life itself is a composite parameter based on volume of distribution and clearance. It is difficult to appreciate the influence of each of these individually from a global model for half-life. What may be more appropriate is to use individual models for volume of distribution and clearance, and then consider how these factors, acting in concert, may influence overall half-life. A similar argument can be made for the prediction of bioavailability which a composite parameter based on absorption, metabolism, and affinity for transporters. Features dictating the extent of absorption may not be the same as those dictating extent of metabolism or transporter affinity, but it is the combination of all of these parameters that controls overall bioavailability.

Figure 10-1 summarizes the different levels of information provided by the various modeling approaches and how this information can be fed into overall predictions of ADME behavior.

10.5.   FUTURE DIRECTION

Overall, the application of in silico predictive methods to the area of ADME has shown much success in recent years. It is anticipated that this will be a subject of continual development in future not only in drug design applications but also in the area of predictive toxicology. This is because a better knowledge of the internal exposure of xenobiotics provides greater accuracy and understanding in the prediction of any biological effect. Being of relevance to both toxicology and drug development, affords the opportunity to address issues in predictive ADME from different perspectives and allows for a cross-fertilization of ideas. In drug design there are benefits in selecting more appropriate candidates to take forward

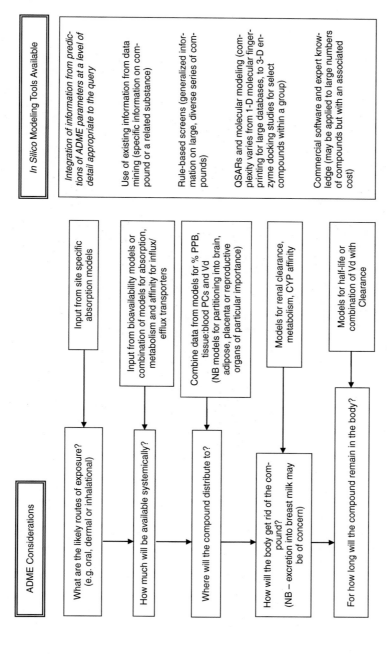

Figure 10-1. Flow diagram for incorporation of in silico ADME information into predictions for biological activity

(Chapter 9), whereas in toxicological risk assessment the science may be applied, for example, in priority setting for the testing of industrial chemicals (Chapter 11). Such data could be used to feed into integrated testing strategies and help determine the most appropriate testing protocols.

Previous, more naïve, concepts of biological systems have now been superseded by more mechanistic understanding. For example, uptake of a compound is not considered to be solely a passive absorption issue, but reliant on the orchestration of an entire chemoimmunity system of alternative absorption pathways, transporter affinities and integrated metabolic processes. The systems biology approach is providing more understanding of the interaction between xenobiotics and organisms as a whole [61]. As more is understood about the inter-relatedness of the biological pathways, the need to develop models for the individual processes becomes clear. The next step is the rational integration of all of these individual predictive models to enable the overall ADME behavior of a compound to be accurately predicted.

There are many opportunities for improvements in this field in the future. Predictions are currently limited due to a lack of high-quality experimental data; increasing the availability of this data will be crucial to model development. Within drug companies, an enormous amount of data has no doubt been generated on ADME properties of drug candidates; however, this is not publicly available. Information is also severely lacking in terms of ADME information for industrial chemicals and this presents a real challenge for the development of useful models for this type of compound. One potential way forward would be via the use of an "honest broker" as proposed by Dearden [58]. The broker could liaise with industry and enable proprietary data to be incorporated into the model development process in a secure environment. Increased availability of data would enhance the capacity to build more robust models. Current models for ADME properties tend to be skewed toward pharmacokinetically viable compounds as information is publicly available for these. Inclusion of information on the non-viable compounds would enable more balanced and robust models to be developed. A unified system for the collation and organization of available data would also be beneficial, but would require agreement on definitions and terminology.

In their discussion of why ADME models fail Stouch et al. [62] indicated that model transparency is a major factor. Model users need to be able to identify which compounds are used within the training set to ascertain if the model is suitable for their query compound (Chapters 5, 6, 12). Without this knowledge models can be criticized for performing poorly, but the reality is that the model may not have been suitable for the compound in question. Better relationships between model users and vendors could provide a mutually beneficial strategy in future. This approach has already seen success in the development of the DEREK for Windows software where a dynamic relationship exists between the software developers (Lhasa Ltd, Leeds) and the user group. Applying the same philosophy to ADME data may bring about corresponding improvements for this type of model. Similarly, extending the concept of trainable models, where in-house knowledge and expertise can be integrated into model development, is also a promising prospect.

Simultaneous development and integration of ADME predictions with the acquisition of knowledge on activity needs to become an intrinsic part of process in predicting the overall behavior of xenobiotics. This requires organizations to maximize collaboration and knowledge sharing between researchers in these areas.

10.6.   CONCLUSION

For more than 30 years, drug development has benefited from in silico approaches to predict the activity and toxicity of drugs. Although, in general, researchers were slower at realizing the potential of these techniques to be applied to ADME predictions, this situation is now very different. One advantage of this is that many of the methods for in silico prediction have been rigorously investigated and developed for predicting activity, and the knowledge gained within that field can now be applied to the development of models to predict ADME. Many tools, using the state-of-the-art in silico methodology are now available to users and it is anticipated that the continual evolution of these tools will provide greater ability to predict ADME properties in the future.

ACKNOWLEDGEMENT

The funding of the European Union 6th Framework CAESAR Specific Targeted Project (SSPI-022674-CAESAR) and OSIRIS Integrated Project (GOCE-037017-OSIRIS) is gratefully acknowledged.

REFERENCES

1. Kerns EH, Di L (2008) Drug-like properties: Concepts, structure design and methods. Elsevier, Burlington, USA
2. d'Yvoire MB, Prieto P, Blaauboer BJ et al. (2007) Physiologically-based kinetic modelling (PBK modelling): Meeting the 3Rs agenda. The report and recommendations of ECVAM workshop 63. ATLA 35:661–671
3. Szakacs G, Varadi A, Ozvegy-Laczka C, Sarkadi B (2008) The role of ABC transporters in drug absorption, distribution, metabolism, excretion and toxicity (ADME-Tox). DDT 13:379–393
4. Wilkinson GG (2001) Pharmacokinetics: The dynamics of drug absorption, distribution and elimination. In: Hardman J, Limbird E (eds) Goodman and Gilman's the pharmacological basis of therapeutics, McGraw-Hill, New York, pp 3–29
5. Gibbs S, van de Sandt JJM, Merk HF et al. (2007) Xenobiotic metabolism in human skin and 3-D human constructs: A review. Curr Drug Metab 8:758–772
6. Lipinski CA, Lombardo F, Dominy BW, Feeney PJ (1997) Experimental and computational approaches to estimate solubility and permeability in drug discovery and development settings. Adv Drug Deliv Rev 23:3–25
7. Lobell M, Hendrix M, Hinzen B, Keldenrich J (2006) *In Silico* ADMET traffic lights as a tool for the prioritization of HTS hits. Chem Med Chem 1:1229–1236
8. Gleeson MP (2008) Generation of a set of simple, interpretable ADMET rules of thumb. J Med Chem 51:817–834

9. Hou T, Wang J, Zhang W et al. (2006) Recent advances in computational prediction of drug absorption and permeability in drug discovery. Curr Med Chem 13:2653–2667

10. Lian G, Chen L, Han L (2008) An evaluation of mathematical models for predicting skin permeability. J Pharm Sci 97:584–598

11. Basak SC, Mills D Mumtaz MM (2007) A quantitative structure-activity relationship (QSAR) study of dermal absorption using theoretical molecular descriptors. SAR QSAR Env Res 18:45–55

12. Veber DF, Johnson SR, Cheng H-Y et al. (2002) Molecular properties that influence the oral bioavailability of drug candidates. J Med Chem 45:2615–2623

13. Moda TL, Monanari CA, Andricopulo AD (2007) Hologram QSAR model for the prediction of human oral bioavailability. Bioorg Med Chem 15:7738–7745

14. Colmenarejo G (2003) *In silico* prediction of drug-binding strengths to human serum albumin. Med Res Rev 23:275–301

15. Votano JR, Parham M, Hall ML et al. (2006) QSAR modeling of human serum protein binding with several modeling techniques utilizing structure-information representation. J Med Chem 49: 7169–7181

16. Ghafourian T, Barzegar-Jalali M, Hakimiha N et al. (2004) Quantitative structure-pharmacokinetic relationship modelling: apparent volume of distribution. J Pharm Pharmacol 56:339–350

17. Lombardo F, Obach RS, Shalaeva MY, Gao F (2004) Prediction of human volume of distribution values for neutral and basic drugs. 2. Extended data set and leave-class-out statistics. J Med Chem 47:1242–1250

18. Gleeson MP, Waters NJ, Paine SW et al. (2006) *In silico* human and rat Vss quantitative structure-activity relationship models. J Med Chem 49:1953–1963

19. Sui XF, Sun J, Wu X et al. (2008) Predicting the volume of distribution of drugs in humans. Curr Drug Metab 9:574–580

20. Zhang H (2005) A new approach for the tissue-blood partition coefficients of neutral and ionized compounds. J Chem Inf Model 45:121–127

21. Abraham MH, Ibrahim A, Acree WE Jr (2006) Air to brain, blood to brain and plasma to brain distribution of volatile organic compounds: linear free energy analyses. Eur J Med Chem 41: 494–502

22. Basak SC, Mills D, Gute BD (2006) Prediction of tissue–air partition coefficients – theoretical vs experimental methods. QSAR SAR Env Res 17:515–532

23. Norinder U, Haberlein M (2002) Computational approaches to the prediction of the blood-brain distribution. Adv Drug Deliv Rev 54:291–313

24. Konovalov DA, Coomans D, Deconinck E, Heyden YV (2007) Benchmarking of QSAR models for blood-brain barrier permeation. J Chem Inf Model 47:1648–1656

25. Li H, Yap CW, Ung CY et al. (2005) Effect of selection of molecular descriptors on the prediction of blood brain barrier penetrating and nonpenetrating agents by statistical learning methods. J Chem Inf Model 45:1376–1384

26. Zhao YH, Abraham MH, Ibrahim A et al. (2007) Predicting penetration across the blood-brain barrier from simple descriptors and fragmentation schemes. J Chem Inf Model 47:170–175

27. Hewitt M, Madden JC, Rowe PH, Cronin MTD (2007) Structure-based modelling in reproductive toxicology: (Q)SARs for the placental barrier. SAR QSAR Env Res 18:57–76

28. Chang C, Ray A, Swaan P (2005) *In Silico* strategies for modeling membrane transporter function. DDT 10:663–671

29. Huang J, Ma G, Muhammad I, Cheng Y (2007) Identifying P-glycoprotein substrates using a support vector machine optimised by a particle swarm. J Chem Inf Mod 47:1638–1647

30. Cabrera MA, Gonzalez I, Fernandez C et al. (2006) A topological substructural approach for the prediction of P-glycoprotein substrates. J Pharm Sci 95:589–606

31. Manga N, Duffy JC, Rowe PH, Cronin MTD (2003) A hierarchical QSAR model for urinary excretion of drugs in humans as a predictive tool for biotransformation. QSAR Comb Sci 22:263–273

32. Agatonovic-Kustrin S, Ling LH, Tham SY, Alany RG (2002) Molecular descriptors that influence the amount of drugs transfer into human breast milk. J Pharm Biomed Anal 29:103–119

33. Afzelius L, Arnby CH, Broo A et al. (2007) State-of-the-art tools for computational site of metabolism predictions: comparative analysis, mechanistical insights, and future applications. Drug Metab Rev 39:61–86

34. Madden JC, Cronin MTD (2006) Structure-based methods for the prediction of drug metabolism. Expert Opin Drug Metab Toxicol 2:545–557

35. Payne M (2004) Computer-based methods for the prediction of chemical metabolism and biotransformation within biological organisms. In: Cronin MTD, Livingstone DJ (eds) Predicting chemical toxicity and fate. CRC Press, Boca Raton

36. Manga N, Duffy JC, Rowe PH, Cronin MTD (2005) Structure-based methods for the prediction of the dominant P450 enzyme in human drug biotransformation: Consideration of CYP3A4, CYP2C9, CYP2D6. SAR QSAR Env Res 16:43–61

37. Yap CW, Li ZR, Chen YZ (2006) Quantitative structure-pharmacokinetic relationships for drug clearance by using statistical learning methods. J Mol Graph Mod 24:383–395

38. Quinones C, Caceres J, Stud M, Martinez A (2000) Prediction of drug half-life values of antihistamines based on the CODES/neural network model. Quant Struct Act Relat 19:448–454

39. Quinones-Torrelo C, Sagrado S, Villaneuva-Camanas RM, Medina-Hernandez MJ (2001) Retention pharmacokinetic and pharmacodynamic parameter relationships of antihistmaine drugs using biopartitioning micellar chromatography. J Chromatogr B 761:13–26

40. Netzeva TI, Worth AP, Aldenberg T et al. (2005) Current status of methods for defining the applicability domain of (quantitative) structure–activity relationships. The report and recommendations of ECVAM Workshop 52. ATLA 33:155–173

41. Hou T, Wang J, Zhang W, Xu X (2007) ADME evaluation in drug discovery. 7. Prediction of oral absorption by correlation and classification. J Chem Inf Model 47:208–218

42. Hou T, Wang J, Zhang W, Xu X (2007) ADME evaluation in drug discovery. 6. Can oral bioavailability in humans be effectively predicted by simple molecular property-based rules. J Chem Inf Model 47:460–463

43. Abraham MH, Ibrahim A, Zhao Y, Acree WE Jr (2006) A database for partition of volatile organic compounds and drugs from blood/plasma/serum to brain and an LFER analysis of the data. J Pharm Sci 95:2091–2100

44. Kalgutkar AS, Gardner I, Obach S et al. (2005) A comprehensive listing of bioactivation pathways of organic functional groups. Curr Drug Metab 6:161–225

45. Turner JV, Maddalena DJ, Cutler DJ (2004) Pharmacokinetic parameter prediction from drug structure using artificial neural networks. Int J Pharmac 270:209–219

46. Takano M, Yumoto R, Murakami T (2006) Expression and function of efflux transporters in the intestine. Pharmacol Ther 109:137–161

47. Thummel KE, Shen GG (2001) Design and optimization of dosage regimens: Pharmacokinetic data. In: Hardman J, Limbird E (eds) Goodman and Gilman's the pharmacological basis of therapeutics. McGraw-Hill, New York, pp 1917–2023

48. Schultz M, Schmoldt A (1997) Therapeutic and toxic blood concentrations of more than 500 drugs. Pharmazie 52(12):895–911

49. Hollósy F, Valkó K, Hersey A et al. (2006) Estimation of volume of distribution in humans from high throughput HPLC-based measurements of human serum albumin binding and immobilised artificial membrane partitioning. J Med Chem 49:6958–6971

50. Ekins S (2006) Systems-ADME/Tox: Resources and network approaches. J Pharmacol Toxicol Meth 53:38–66

51. de Groot M (2006) Designing better drugs: Predicting cytochrome P450 metabolism. Drug Disc Today 11:601–606

52. Banik GM (2004) *In silico* ADME-Tox prediction: the more the merrier. Curr Drug Disc 4:31–34

53. Ekins S, Waller CL, Swann PW (2000) Progress in predicting human ADME parameters *in silico*. J Pharmacol Toxicol Meth 44:251–272

54. Duffy JC (2004) Prediction of pharmacokinetic parameters in drug design and toxicology. In: Cronin MTD, Livingstone DJ (eds) Predicting chemical toxicity and fate. CRC Press, Boca Raton

55. Gola J, Obrezanova O, Champness E, Segall M (2006) ADMET property prediction: The state of the art and current challenges. QSAR Comb Sci 25:1172–1180

56. Chohan KK, Paine SW, Water, NJ (2006) Quantitative structure activity relationships in drug metabolism. Curr Top Med Chem 6:1569–1578

57. Winkler DA (2004) Neural networks in ADME and toxicity prediction. Drugs Fut 29:1043–1057

58. Dearden JC (2007) *In silico* prediction of ADMET properties: How far have we come? Expert Opin Drug Metab Toxicol 3:635–639

59. Enoch SJ, Cronin MTD, Schultz TW, Madden JC (2008) An evaluation of global QSAR models for the prediction of the toxicity of phenols to Tetrahymena pyriformis. Chemosphere 71:1225–1232

60. Rodgers SL, Davis AM, van de Waterbeemd H (2007) Time-series QSAR analysis of human plasma protein binding data. QSAR Comb Sci 26:511–521

61. Ekins S, Andreyev S, Ryabov A et al. (2005) Computational prediction of human drug metabolism. Expert Opin Drug Metab Toxicol 1:303–324

62. Stouch TR, Kenyon JR, Johnson SR et al. (2003) *In Silico* ADME/Tox: Why models fail. J Comput-Aid Mol Des 17:83–92

63. Jamei M, Marciniak S, Feng K (2009) The Simcyp® population based ADME simulator. Expert Opin Drug Metab Toxicol 5:211–223

CHAPTER 11

PREDICTION OF HARMFUL HUMAN HEALTH EFFECTS OF CHEMICALS FROM STRUCTURE

MARK T.D. CRONIN

School of Pharmacy and Chemistry, Liverpool John Moores University, Liverpool L3 3AF, England,
e-mail: m.t.cronin@ljmu.ac.uk

Abstract: There is a great need to assess the harmful effects of chemicals to which man is exposed. Various in silico techniques including chemical grouping and category formation, as well as the use of (Q)SARs can be applied to predict the toxicity of chemicals for a number of toxicological effects. This chapter provides an overview of the state of the art of the prediction of the harmful effects of chemicals to human health. A variety of existing data can be used to obtain information; many such data are formalized into freely available and commercial databases. (Q)SARs can be developed (as illustrated with reference to skin sensitization) for local and global data sets. In addition, chemical grouping techniques can be applied on "similar" chemicals to allow for read-across predictions. Many "expert systems" are now available that incorporate these approaches. With these in silico approaches available, the techniques to apply them successfully have become essential. Integration of different in silico approaches with each other, as well as with other alternative approaches, e.g., in vitro and -omics through the development of integrated testing strategies, will assist in the more efficient prediction of the harmful health effects of chemicals

Keywords: In silico toxicology, (Q)SAR, Harmful health effects

11.1.   INTRODUCTION

Modern society demands a safe environment to live, work, and study in. Man can be exposed to chemicals through a number of routes, whether they be deliberate, accidental, or through occupation. Thus the hazards associated with exposure chemicals should be assessed, allowing for proper risk assessment and, if required, risk management [1].

Mankind is exposed to chemicals from various sources, from the food and drink we imbibe, use of pesticides in food production, prescription of pharmaceuticals, application of cosmetic and personal products, use of detergents and washing agents,

T. Puzyn et al. (eds.), Recent Advances in QSAR Studies, 305–325.
DOI 10.1007/978-1-4020-9783-6_11, © Springer Science+Business Media B.V. 2010

clothes, paints in addition to exposure (often in higher concentration) in occupational settings. In order to control risks to mankind, governments have developed legislative strategies to gain information to enable regulatory decisions to be made. Often these are different depending on the type of product, e.g., pharmaceutical, food additive, pesticide or cosmetic ingredient. Typically products which are taken orally, in particular pharmaceuticals, are the most closely regulated compounds thus requiring the most comprehensive dossiers of toxicological information.

Whilst the uses, quantities and routes of exposure to chemicals are different, the tests to obtain the toxicological information are broadly the same. Thus the testing protocols for acute toxicity, carcinogenicity, etc., will be the same regardless of the compound considered. In terms of replacing animal tests, this should mean that alternative methods should be transferable across regulatory applications.

There are various and well-recognized reasons for "updating toxicology" through the replacement and refinement of animal tests. The principles of the 3Rs, i.e., replacement, refinement, and reduction [2] are well established throughout the science. In silico approaches make up a significant contribution to the 3Rs in the toxicological assessment of chemicals. Aside from animal welfare, the reasons for the interest in in silico toxicology are quite obvious, namely the reduction in time and cost of product development and the ability to identify toxicity early on. These requirements to seek alternatives are supported in many regions of the world by legislation and government funding strategies. In the product development scenario, in silico assessment has the added advantage of being able to design compounds rationally not only to optimize beneficial activity, but also to reduce or eliminate unwanted side-effects. For existing compounds, i.e., those already produced and/or marketed, it is well recognized that there are insufficient toxicity data for the vast majority of them [3]. Indeed, safeguarding of man and the environment from harmful existing chemicals, for which limited data currently exist, is at the heart of the REACH legislation [4–6] (Chapter 13).

11.1.1.   Prediction of Harmful Effects to Man?

Before discussing the role of in silico alternatives in toxicological testing, an issue must first be considered with regard to the use of animals themselves as surrogates for man. As stated throughout this book, data are required for modeling purposes to develop (Q)SARs and related techniques. We are using these data to establish whether or not a compound will be toxic to man, whilst the experiment itself may have been performed in another species. This can be a controversial situation, and critics will argue that the animal tests themselves do not predict effects in man in some circumstances. Thus, the developer and user of in silico toxicology approaches must be aware not only of the shortcomings of the predictive methods themselves, but also of the data on which they are originally based.

11.1.2. Relevant Toxicity Endpoints Where QSAR Can Make a Significant Contribution

There is an expectation that animal testing for toxicity will, one day, be replaced by alternative methods. Amongst the alternatives "computational models" are usually referred to. It must be understood that we are a long way from replacing animals in all forms of toxicological testing. However, currently there are real possibilities

Table 11-1. Summary of major toxicological endpoints and availability of in silico models. The readers are referred to Table 11-3 for information on the individual expert systems

Toxicological endpoint	Availability of in silico models[a]
Skin sensitization	A number of expert systems are available for skin sensitization. There is a strong basis to grouping approaches based on protein binding. Local QSARs are also available
Skin and eye irritation and corrosion	There are a number of expert systems containing structural alerts relating to irritation. Expert systems for corrosivity exist
Mutagenicity	There are many approaches and models to predict mutagenicity. A significant number of structural alerts relating, mainly, to DNA binding are available in various systems. Also, quantitative expert systems and local QSAR models (e.g., for chemical classes) can be applied
Carcinogenicity	There are a significant number of approaches to predict carcinogenicity. A number of structural alerts relating, mainly, to genotoxic mechanisms are available. There are also other quantitative expert systems
Reproductive and developmental toxicity	It is acknowledged that there are relatively few (Q)SAR models for reproductive toxicity as an "endpoint," although limited (Q)SARs exist for individual effects, e.g., teratogenicity [42]. Information on a limited number of modes and mechanisms of action may assist in the grouping of similar chemicals. Models for some effects, e.g., endocrine disruption, are, however well developed and successful
Acute (mammalian) toxicity	A small number of expert systems and QSARs exist. This is considered a difficult area to make predictions in due to the nature of the test, i.e., many mechanisms, a short duration (therefore a steady-state is unlikely to be reached) and considerable variability in the data modelled
Chronic (mammalian) toxicity	There are few expert system or QSAR approaches to predict chronic toxicity. This is a very difficult endpoint to model by conventional QSAR methods and may be more applicable to grouping approaches
Cardio toxicity (this is non-regulatory endpoint, but has received much attention due to its importance for drug toxicity)	A large number of papers have been published recently on topics such as HERG inhibition (Chapter 9). Many of these provide potential models for screening molecules

[a]Please note availability of models does not imply a prediction will be correct, this is dependent on the robustness of the model, applicability domain, and other acceptance criteria.

for using in silico approaches for certain endpoints. The intention of Table 11-1 is to give some indication of where in silico approaches may provide some benefit as alternatives to the use of animals. Despite what may be stated in this table, the reader is reminded that all these methods and recommendations are subject to the usually caveats of using predictive approaches, i.e., the restrictions of the models and applicability domains, etc.

11.2.   IN SILICO TOOLS FOR TOXICITY PREDICTION

There are a number of approaches to predict the toxicity of single chemical substances from structure. These can vary from simple, i.e., retrieval of existing data from a database, through to modeling approaches based on chemical groupings, to complex multivariate techniques. The approaches described in this section illustrate the utility of many of the methods described elsewhere in this volume.

11.2.1.   Databases

Toxicity data are required in the in silico prediction of toxicity for a number of reasons, including

- Should an acceptable test result be already available, there may be no need to perform the test.
- Toxicity data are themselves required to populate categories or classes of molecules to allow for read-across.
- Toxicity data are required to formulate the individual QSARs and models.

There are numerous sources of toxicity data [7]. These come in a variety of shapes and sizes, of varying quality, and with different issues to be addressed. A recent overview of toxicological databases is given by Bassan and Worth [8], who have provided an invaluable resource for modelers and model users. Representative types of databases are summarized in Table 11-2 – it should be noted that this is only a selection of the databases available, many more are available in the literature or in freely available tools such as the OECD (Q)SAR Application Toolbox.

Many of the modern toxicological databases go much beyond simple data retrieval. Several (e.g., AMBIT, DSSTox) include possibilities to search the databases for "similar" compounds or for (sub-)fragments. This is an exceptionally useful feature, although great care must be taken in extrapolating information from so-called similar chemicals. These ideas lead naturally into the approaches based around grouping chemicals together, or category formation, described below and in Chapter 7.

Table 11-2. Summary of the types of databases available for human toxicity data and their potential for use in in silico toxicology

Type of database	Illustrative examples and references	Comments
Single literature data sets	Skin sensitization data from a single source for the guinea pig maximization test [13]	These are often high-quality data sets as they may have been measured in the same laboratory, and even by the same procedure and/or operators. They are likely to be small, but may be highly suitable for (local) modeling
Literature compilations	Skin sensitization databases of historical local lymph node assay data from various laboratories [14]; Carcinogenic Potency Database (CPDB) (http://potency.berkeley.edu/)	If available, these databases may provide a good starting point for modeling studies. Data are usually from a single protocol or test, but may vary in consistency as they are measured in different laboratories
Freely accessible compilations of toxicity information	ToxNet http://toxnet.nlm.nih. gov; AMBIT http://ambit.acad.bg/ambit/php/	Large databases covering many endpoints and effects. Most data will be of variable quality and may include non-standard endpoints
Refined (and hence potentially higher quality) databases	DSSTox http://www.epa.gov/ncct/ dsstox/index.html. Other databases are available from collaborative projects such as the European Union Framework Projects, e.g., the CAESAR project (http://www.caesar-project.eu/)	Well-curated databases in terms of checking chemical structure and information and data quality. These are freely available and make ideal starting points for modeling
Databases developed for regulatory purposes	European chemical Substances Information System (ESIS) http://ecb.jrc.ec.europa.eu/esis/	Collections of toxicological information and data relating to substances (usually industrial chemicals). Data quality is not necessarily checked on entry to the database
Commercial databases	Leadscope (including FDA) databases (http://www.leadscope.com/); VITIC database from Lhasa Ltd. (http://www.lhasalimited.org)	These are often large databases with many compounds, a wide variety of chemistry and a number of different endpoints. Modelers will need to ascertain the ownership of the data and information within them before reproducing the data in a published model

Table 11-2. (continued)

Type of database	Illustrative examples and references	Comments
Corporate and/or closed databases	Some information and/or data are available openly on corporate websites, e.g., of the pharmaceutical companies – these can be searched for material safety data sheets (MSDS). MSDS can also be found from other chemical suppliers (e.g., SigmaAldrich – http://www.sigmaaldrich.com) Other corporate databases are not openly available and are used as research tools. These may vary in size and quality and even whether they are in a computerized form or not.	There are many data in chemical supplier or pharmaceutical databases. They may be difficult to use since the entries will be for single substances. Details on the test procedure may not be available. For some substance, e.g., pharmaceuticals, a test result may not be available, the record may simply state the likely presence of a toxicity

Whilst toxicological data and information may be available from the types of resources listed in Table 11-2, not all will be suitable for use in in silico toxicology. To develop high-quality (Q)SARs, high quality, consistent, and reliable toxicity data are required [9]. Ideally this would suggest data obtained from a consistent and reliable protocol, performed to the same standards and undertaken in the same laboratory and even by the same technicians. Very few of the examples noted in Table 11-2 would conform to these standards. Therefore, the considerable issues of evaluating data quality must be borne in mind. In particular, the following considerations must be addressed:

- Is the chemical structure, its name(s), and other identifiers (e.g., CAS number, SMILES, INChI codes) correct? This is the most fundamental issue with the recording of data into databases and a remarkable source of errors. The author's own anecdotal experience suggests that there may be a remarkable variety of errors possible in recording data (cf. Dearden et al. [10]). Young et al. [11] document some of the errors in publicly and privately available databases.
- The toxicity data themselves must be evaluated for quality. Whilst complete accuracy can probably never be ensured, one would hope to obtain toxicity data from standardized and internationally recognized protocol, e.g., OECD, performed to Good Laboratory Practice (GLP) standards. Care must be taken to ensure methods are comparable. Klimisch et al. [12] provided a simple scoring system (1-4) for ecotoxicological data. Inevitably this system has been criticized in some quarters as being too simplistic, but in over 10 years it has not been bettered. It may be possible to adapt a similar scoring scheme for human endpoints.

Thus, when extracting data from databases, one should ensure the chemical structures are correct and the overall quality of the toxicity data being utilized. With regard to data quality, it must be remembered that one may be able to compile large numbers of individually high-quality data, the data set for (Q)SAR modeling may be of reduced quality if there is a mixture of endpoints and methodologies.

11.2.2.   QSARs

QSARs can be developed or used for any toxicological endpoint for which suitable and sufficient toxicity data are available. The full range of QSAR techniques as described in this book can be applied. The types of endpoints employed mean that, in general, either continuous data (e.g., toxicity potency data) or categoric data (e.g., qualitative toxicity data) are modelled. It is well beyond the scope of a single chapter to cover and describe the full selection of (Q)SARs available for all toxicological endpoints. Thus, this section will illustrate a number of types of (Q)SARs with examples from skin sensitization.

Skin sensitization is chosen as an illustrative example as it is one of the most important endpoints in terms of human exposure to chemicals. It is also one of the first endpoints to be triggered in the European Union's REACH legislation and hence alternatives are actively sought. It is, in parts, a well-understood toxicological event with established modes and mechanisms of action. It is supported by a number of high-quality databases, albeit measured in different protocols, such as data for the guinea pig maximization test [13] and the local lymph node assay [14, 15]. There are a number of excellent reviews in the use of (Q)SAR to predict skin sensitization, such as Patlewicz et al. [16, 17] and Roberts et al. [18]. Therefore, this section will only draw on the available literature in an attempt to illustrate the types of models that are possible – it is intended that the same modeling principles and processes could be applied across a number of human health and toxicological endpoints.

11.2.2.1.   Skin Sensitization Data for Modeling

As noted above, there are a reasonable number of skin sensitization data available for modeling. The immediate decision faced by the modeler is whether they wish to use the data in a qualitative or quantitative manner. For classification and labeling (e.g., for regulatory purposes) all that may be required is a categoric assignment of a chemical as a sensitizer or non-sensitizer. For risk assessment, a potency may be required to identify compounds with high or extreme sensitizing potential from those with only a lower potential for sensitization. Both qualitative and quantitative data are available from the local lymph node assay, and qualitative and semi-quantitative from the guinea pig maximization test. The decision on which data to utilize must be made partly in the context of what is required of the prediction, and also on the quantity and quality of the available data.

11.2.2.2.   SAR (Qualitative) Models for Skin Sensitization

There are number of approaches to identify compounds with the potential to elicit skin sensitization qualitatively. The simplest is to determine feature of a molecule that are likely to promote this endpoint. If identified, these features can be coded such that they can be identified in new molecules. For instance, in one of several early attempts to derive such fragments groups, in a naïve manner, compounds were grouped together and suggestions made for fragments likely to promote skin sensitization (and those not associated with it) [19]. These were later rationalized with a more mechanistic emphasis, cf. Payne and Walsh [20]. The latter rules were ultimately adapted and provided the source of part of the skin sensitization rulebase that entered the DEREK for Windows expert system [21]. As with all in silico models, this approach works best when it is supported by a mechanistic justification. Through a gradual process of evolution and development of the knowledge by a number of workers, these structural features have been associated with a total of six mechanisms of action with regard to organic mechanistic chemistry [22, 23]. The power of this approach is that the chemistry can be encoded computationally, for instance Enoch et al. [24] have written SMARTS strings to capture the chemistry associated with mechanisms of action.

The process of assigning a molecule to a categoric classification is also amenable to a number of QSAR techniques (and forms the basis of a number of expert systems). One of the simplest of these techniques is discriminant analysis; this was applied by Cronin and Basketter [13] to the guinea pig maximization database. Other techniques applied to model skin sensitization, with increasing levels of complexity include logistic regression analysis [25]; classification trees and random forests [26]; a back propagation neural network [27]; and support vector machines [28]. As these techniques become more non-linear (and hence more complex), many workers claim improved accuracy in their predictions. This may be, however, at the cost of transparency of the model and its portability to new situations as defined by the OECD Principles for the Validation of (Q)SARs (see Chapter 13).

11.2.2.3.   *QSAR Models for Skin Sensitization*

A number of QSAR models and approaches have been developed to predict the potency of skin sensitizers. The endpoint from the local lymph node assay can be converted to a value that is suitable for QSAR analysis, i.e., the concentration of each chemical necessary to stimulate a threefold increase in proliferation in draining (murine) lymph nodes compared to concurrent vehicle-treated controls (EC_3) [29]. There has been some success in relating these values within restricted groups of compounds. This approach relies first on identifying a "domain" and then developing a model within it. Two easy approaches to develop a domain are to group compounds together according to structural analogs or by those with comparable mechanisms of toxic action. The obvious drawback of this approach to developing "local QSARs" is the very restricted domains in which they operate.

As an example of a simple structural series, Basketter et al. [30] demonstrated that the potency in the local lymph node assay of nine bromoalkanes was related to a quadratic function with the logarithm of the octanol–water partition coefficient (log P) (see Eq. 11-1):

$$pEC3 = 1.61 \log P - 0.09(\log P)^2 - 7.4$$

$$n = 9; R = 0.97; s = 0.11; F = 50.0 \tag{11-1}$$

where n is the number of compounds, R is the correlation coefficient, s is the standard error of the estimate, and F is the Fisher statistic.

This relationship indicates that the reactivity of the bromoalkanes remains the same; the variation in potency is probably a function of transport and the ability to permeate the skin. The mechanism of action for these compounds has been identified as being the S_N2 reaction domain [18, 23].

In the example of the bromoalkanes, the local QSAR was developed on a structural basis with a mechanistic interpretation placed on it at a later date. Subsequent studies have attempted to form groups of compounds on the basis of a common mechanism of action. For instance, Patlewicz et al. [31, 32] brought together a group of compounds that are likely to act as Schiff's bases (e.g., aliphatic and aryl aldehydes). In this case, a two-parameter equation (11-2) was developed incorporating not only log P but also the Taft σ^* substituent constant (σ^*):

$$pEC3 = 0.17 + 0.30 \log P + 0.93\,\sigma^*$$

$$n = 14; R^2 = 0.87; s = 0.165; F = 37.7 \tag{11-2}$$

The equation can be interpreted in terms of transport to the active site (in this case the skin immuno-protein) and the relative reactivity of the Schiff base with the protein.

A recent extension to the development of local QSARs is the use of quantitative read-across. Enoch et al. [33] have demonstrated that for a given group of compounds which can be identified from their structure as belonging to a particular mechanistic domain, limited read-across may be possible. The approach taken by Enoch et al. [33] was to use a calculated "electrophilicity index" to account for relative reactivity. Within the group (or category) skin sensitization potential was "interpolated" on the basis of electrophilicity.

Thus, there are a number of approaches for developing local QSARs for skin sensitization potency. The advantages of the approaches described, in addition to the many others in the literature, are that they are clear, transparent, and simple to develop and use. They work successfully as they assume within the limited domain that the molecules act by a similar mechanism of action and hence rate limiting steps are reduced to one or two factors (i.e., transport and relative reactivity). Requiring relatively few data (especially quantitative read-across) means that predictions can be derived on very limited data sets. This obvious disadvantage of this strategy is the very limited domain of the models.

There have been a number of attempts to develop models on the basis of a larger data set with more chemical structures in it. Such models are sometimes termed "global QSARs" as opposed to the local models described above. The larger data sets of local lymph node assay data (such as Gerberick et al. [14]) are ideal for the development of models. For instance, Miller et al. [34] used a relatively simple algorithm to predict EC_3 for a total of 65 chemicals (with a 22 chemical test set); Estrada et al. [35] used the TOPSMODE approach on a similar data set. Both models (and others) are reviewed critically by Patlewicz et al. [16, 17] and the readers are referred to those publications prior to a consideration of using them. There are also a number of expert system approaches to predicting skin sensitization (in addition to DEREK for Windows noted previously) and the readers are referred to the next section for a discussion of those. There are, of course, clear advantages in using an appropriate global QSAR – they are much more broadly applicable and will be more widely utilized. There may be some disadvantages as well in the global QSAR approaches: complex and non-linear models are difficult to interpret. Such models also generalize differences in experimental methodology – a good example being the different vehicles used in the local lymph node assay. The most significant disadvantage is that a model is created across multiple mechanisms of action and hence runs the risk of spurious correlations.

11.2.2.4.   *General Comments of the Use of QSAR Models for Predicting Human Health Effects*

The development of (Q)SARs for skin sensitization described in Section 11.2.2 is, of course, only for illustration and should not be considered as a full overview of the state-of-the-art for this important endpoint. Whilst the development of (Q)SARs for other endpoints will require particular attention and individual consideration of the issues, there are some general comments that can be made from the illustration of skin sensitization:

- It is possible to develop similar (Q)SAR models for many, if not most, human health effects – providing suitable toxicity data exist for modeling.
- For most endpoints, e.g., carcinogenicity, reproductive toxicity, eye and skin irritation, both qualitative and quantitative models could be developed. The developer must decide what information he/she wants to build the model appropriately.
- Data quality must be understood and evaluated for the endpoint of interest.
- SARs and fragments associated with a particular endpoint are a valuable method to identify hazardous molecules.
- Local QSARs can be developed within strictly defined and restricted domains of structural analogs or individual mechanisms. Used carefully, they can provide accurate predictions of toxicity.
- Global QSARs are more broadly applicable, but may provide less accurate predictions due to their generalist nature.

- Attention should be paid to model quality and, even if not used for regulatory purposes, the OECD Principles for the Validation of (Q)SARs (Chapter 13).

11.2.3. Expert Systems

The user of in silico techniques to predict toxicity is provided with a variety of "off-the-shelf" software packages. These are commonly referred to as "expert systems" for toxicity prediction. It is the author's personal opinion that the usage of the term "expert system" in this context is very broad. In its truest sense, this term would refer a system that captures the knowledge of an expert. Some approaches do, in fact, exactly this, notably DEREK for Windows and OncoLogic. Other approaches can be considered to be a formalization of a QSAR. These extend the definition of expert systems for toxicity prediction, such that it is probably easiest to utilize the definition provided by Dearden et al. [36], namely

> An expert system for predicting toxicity is considered to be any formalized system, not necessarily computer-based, which enables a user to obtain rational predictions about the toxicity of chemicals. All expert systems for the prediction of chemical toxicity are built upon experimental data representing one or more toxic manifestations of chemicals in biological systems (the database), and/or rules derived from such data (the rulebase).

Whilst over a decade old, Dearden et al.'s terminology is still highly relevant. There are a number of reasons for the popularity of expert systems for toxicity prediction:

- These products can be obtained and implemented easily – some are developed and distributed on a commercial basis, a small number are made freely available.
- Generally they do not require the user to develop a particular model, or repeat the development of a model.
- Usually, the software is well documented and supported.
- Predictions can be made by a non-specialist toxicologist, chemist, or QSAR practitioner – however, it is recommended that all users of expert systems and their prediction have some expertise and training.
- Often they can be integrated into automated workflows, e.g., for discovery or lead optimization.
- Many come with significant other capabilities, i.e., searching databases and assessment of applicability domain.
- For regulatory assessment, a user in a business will know the prediction that a scientist in a governmental regulatory agency will obtain.

There are plenty of recent reviews in the area of expert systems for toxicity prediction, see Helma [37]; Mohan et al. [38]; Muster et al. [39]; Nigsch et al. [40] as well as the endpoint specific reviews of Bassan and Worth [8]; Gallegos-Saliner et al. [41]; Patlewicz et al. [17]; and Cronin and Worth [42]. A summary of the main commercial and freely available expert systems for predicting toxicity is provided in Table 11-3.

Table 11-3. A non-exhaustive list of expert systems to predict toxicity

Expert system (and distributor)	Description	Main (indicative) endpoints covered	Further information and/or reference
DEREK for Windows (Lhasa Ltd, Leeds, England)	A knowledge-based expert system created with knowledge of structure–toxicity relationships. The software contains over 500 structural alerts and supporting information	Carcinogenicity, mutagenicity, genotoxicity, skin sensitization, teratogenicity, irritancy, respiratory sensitization, hepatotoxicity, neurotoxicity, ocular toxicity	http://www.lhasalimited.org/index.php?cat=2&sub_cat=64
EPISuite (freely available from the United States Environmental Protection Agency)	A suite of programs that integrate a number of estimation models for the prediction of environmental and physical/chemical properties	Mainly environmental endpoints, although includes a dermal penetration algorithm (DERMWIN) which calculates skin permeability coefficient; dermally absorbed dose per event	http://www.epa.gov/opptintr/exposure/pubs/episuite.htm
HazardExpert Pro	A rule based program for the prediction of toxicity using structural fragments. The software incorporates a consideration of bioavailability via the calculation of $\log P$ and pK_a	Oncogenicity, mutagenicity, teratogenicity, membrane irritation, sensitization, immunotoxicity, and neurotoxicity	http://www.compudrug.com/
LAZAR	A fragment-based approach	Liver toxicity, mutagenicity, and carcinogenicity	http://www.in-silico.de/
MC4PC	A statistically based algorithm based on fragments encoding features promoting toxicity (toxicophores) and deactivating (toxicophobes)	Over 180 modules (i.e., individual endpoints) including those for the following endpoints: acute toxicity in mammals; adverse effects in humans; carcinogenicity, cytotoxicity, developmental toxicity, teratogenicity, genetic toxicity, skin and eye irritation, allergies	http://www.multicase.com/

Table 11-3. (continued)

Expert system (and distributor)	Description	Main (indicative) endpoints covered	Further information and/or reference
OncoLogic	Application of SAR analysis and incorporation of knowledge of the mechanisms of action and human epidemiological studies	Cancer	US EPA has recently purchased the right to the system and is currently updating the system for free distribution to the public (more details from Dr Yin-tak Woo; e-mail: woo.yintak@epa.gov)
OSIRIS Property Explorer	Freely available Internet software where predictions are made from structural fragments	Mutagenicity, tumorigenicity, irritation, and reproductive effects	http://www. organic-chemistry. org/prog/peo/
PASS	Predicts activity from the structure	A wide variety of endpoints relevant to toxicity, particularly those based on receptor-mediated effects	http://195.178.207.233 /PASS/index.html
TIMES-SS (Laboratory of Mathematical Chemistry, University "Prof. Assen Zlatarov", Bourgas, Bulgaria)	A hybrid expert system encoding structure–toxicity and structure–metabolism relationships	Skin sensitization, mutagenicity, chromosomal aberration and estrogen, and androgen receptor binding affinities of chemicals	http://oasis-lmc.org/? section= software&swid=10
TOPKAT	Statistical QSARs based on regression and discriminant analysis of 2D descriptors, utilizing large heterogeneous data sets	Rodent carcinogenicity, Ames mutagenicity, rat oral LD_{50}, rat chronic LOAEL, developmental toxicity potential, skin sensitization, rat maximum-tolerated dose, eye and skin irritancy, rat inhalation toxicity LC_{50}, and rat maximum-tolerated dose	http://accelrys.com/ products/ discovery-studio/ toxicology/

Table 11-3. (continued)

Expert system (and distributor)	Description	Main (indicative) endpoints covered	Further information and/or reference
ToxTree	Groups chemicals and predicts various types of toxicity based on decision-tree approaches	The Cramer classification scheme, BfR rules for predicting skin and eye irritation and corrosion, and the Benigni-Bossa rulebase for mutagenicity and carcinogenicity	http://ecb.jrc.ec.europa.eu/qsar/qsar-tools/index.php?c=TOXTREE
TerraQSAR-Skin	Neural network-based program to compute the skin irritation potential of organic chemicals for the rabbit (Draize test)	Skin irritation	http://www.terrabase-inc.com/

11.2.4.   Grouping Approaches

The grouping of chemicals together to form "categories" of similar substances, and hence with similar properties, is a simple yet immensely powerful technique for exploring toxicity data and making predictions. The specifics of these approaches are described in Chapter 7. Predictions can be made from the grouping of chemicals together through the careful application of read-across and other methods for inter-polation and extrapolation. The formation of predictions to allow for toxicological read-across is considered to be a simple, highly transparent, and mechanistically based method of in silico toxicity prediction. As such it is growing in acceptance for regulatory purposes.

The process of category formation and read-across is best described through the OECD Guidance [43]. Some recent examples of the development of categories to allow for read-across for human health effects include the work of Cunningham et al. [44] for carcinogenicity; Enoch et al. [33] for skin sensitization; Koleva et al. [45] for other endpoints including mutagenicity and acute (environmental) toxicity; and Enoch et al. [46] for teratogenicity.

The key to the successful formation of a category to allow for toxicological read-across is to identify "similar" chemicals. As described more fully by Enoch (Chapter 7), similarity can be considered in terms of

- Structural analogs, e.g., the same functional group with varying alkyl substituents implying the same mechanism of action;
- The same mechanism of action, as defined by structures without a common structural analog (e.g., protein/DNA binding, receptor–ligand interactions)
- Chemical similarity on the basis of an algorithm to assess it.

In general, the simpler the category formed, the easier it will be to apply. A number of pieces of software have been developed to allow for category formation. All will form categories around a substance. For example, the OECD (Q)SAR Application Toolbox does this on the basis of a large database (several hundred thousand structures) for which there are only a small number of toxicity data. The model user therefore needs to populate the category with data for successful read-across. Other software (e.g., Toxmatch) attempts to establish categories within known databases, allowing the user to make their own decisions regarding read-across [47]. The tools that are available to form categories are summarized in Table 11-4.

Table 11-4. A selection of the software tools available to form chemical categories to allow for read-across

Expert system (and distributor)	Description	Approach to chemical grouping	Further information and/or reference
AMBIT (developed by IdeaConsult Ltd, Sofia, Bulgaria)	Publicly available chemical databases and functional tools, including a tool for defining applicability domain of QSAR models. Contains over 450,000 records for individual chemicals. Searchable by chemical structure and fragment	Performs chemical grouping and assesses the applicability domain of a QSAR offering a variety of methods including: statistical approaches that rely on "descriptor space;" approaches based on mechanistic understanding; and approaches based on structural similarity	http://ambit.acad.bg
Analog Identification Method (AIM) (developed from the United States Environmental Protection Agency)	A freely available, web-based, computerized tool that identifies chemical analogs based on structure. The AIM database contains 31,031 potential analogs with publicly available toxicity data	Uses a chemical fragment-based approach with 645 individual chemical fragments to identify potential analogs	http://www.epa.gov/oppt/sf/tools/methods.htm#new
Leadscope (developed by Leadscope Inc.)	A data management and data mining tool; it is possible to import additional data sets and perform comparisons with existing databases containing toxicological information	Based on 27,000 chemical fingerprints. A number of statistical algorithms are also embedded to enable functionalities, such as clustering of chemicals and data, extraction of structural rules, development of QSAR models as well as development of chemical categories	http://www.leadscope.com

Table 11-4. (continued)

Expert system (and distributor)	Description	Approach to chemical grouping	Further information and/or reference
OECD (Q)SAR Application Toolbox (developed by the Organization for Economic Co-operation and Development)	A freely available, web-based, tool provides a workflow for category formation. The toolbox contains databases with results from experimental studies; a library of QSAR models; tools to estimate missing experimental values by read-across, and trend analysis	Categories are formed on the basis of structural analogs, mechanisms or modes of action, i.e., protein or DNA binding	http://www.oecd.org – search for "QSAR" at the OECD web-site
ToxMatch (developed by Ideaconsult Ltd (Sofia, Bulgaria) under the terms of a European Commission Joint Research Centre contract)	Freely available software which includes the ability to compare data sets based on various structural and descriptor-based similarity indexes as well as the means to calculate pair-wise similarity between compounds or aggregated similarity of a compound to a set	Several chemical similarity indexes facilitate the grouping of chemicals	http://ecb.jrc.ec. europa.eu/qsar/ qsar-tools/index. php?c= TOXMATCH

11.3.   THE FUTURE OF IN SILICO TOXICITY PREDICTION

The (immediate) future brings with it is a possibility for a shift in thinking in the toxicological assessment of chemicals, if not a complete re-invention of the science. For many toxicological effects and regulatory endpoints, it is very unlikely that a single "alternative" test – be it -omics, in vitro, or in silico will be sufficient to replace an existing in vivo test. In particular, it is the view of the author that no single in silico assay will be sufficient to replace a toxicity assay, especially for complex human health effects. Therefore, the key to replacing animal tests for toxicity will be to integrate methods, technologies, and approaches.

11.3.1.   Consensus (Q)SAR Models

There are well-established methods to combine (Q)SARs to form "consensus" models. For instance, Gramatica et al. [48] describe the compilation of in silico predictions from the same or similar techniques. Often these are regression-based

QSAR models which are combined together (Chapters 5 and 12). Typically these methods work most efficiently with large, heterogeneous data sets of activities to model. In order to obtain a variety of "equivalent" models, they require a large group of physico-chemical descriptors and/or properties and a method to select them – such as a genetic algorithm. The idea is to create a pool of models from which either the best or most diverse QSARs are selected. The predictions are then either averaged or weighted by some method. This "consensus" approach will normally perform better than a single QSAR, although some workers have noted that the improvement in statistical fit is at the cost of increased complexity of the model [49].

A more comprehensive method to form a consensus is the compilation of in silico predictions from the different techniques. A good example of this is the prediction of carcinogenicity reported by workers from the United States Food and Drug Administration (FDA) [50]. QSAR models were based upon a weight-of-evidence paradigm. Identical training data sets were configured for four QSAR programs and QSAR models, namely MC4PC, MDL-QSAR, BioEpisteme, and Leadscope. Models were constructed for the male/female/composite rat and mouse and composite rodent endpoints. The predictions from the models were adjusted to favor high specificity. A number of important findings were determined including complementary predictions of carcinogenesis from individual models; consensus predictions for two programs; and the ability to achieve better performance and better confidence predictions with the weight of evidence approach. Overall, the use of four QSAR programs was able to predict carcinogenicity with high specificity (85%).

Other consensus approaches to predicting toxicity include that of Votano [51]. In addition Abshear et al. [52] made attempt to predict mutagenicity from three expert systems (HazardExpert; CSGenoTox; and EqubitsMutagen). These predictions were combined in the "KnowItAll®" software (from Bio-Rad Laboratories, Inc.) with good success.

Technological platforms also now allow for combination of various methods. In an interesting approach, four commercial developers of software, Leadscope Inc., Lhasa Ltd, MultiCASE Inc., and Molecular Networks GMBH have launched a service termed InSilicoFirst, with more information being available from http://insilicofirst.com/. This allows, via a single platform, access to integrated tools for the prediction of environmental toxicity. This type of approach of integrating well-established and recognized models could provide a great benefit if also extended to human health effects.

11.3.2.   Integrated Testing Strategies (ITS)

The concept of integrated (formerly intelligent) testing strategies (ITS) for toxicity testing has been developed to allow for the rational and intelligent integration of data from alternative test methods. A key initial stage to the use of ITS is the use of in silico data. This can include existing data, i.e., from databases, read-across, and (Q)SAR models. Further information on integrated testing strategies can be obtained from Grindon, Combes et al. [53–60], and from the guidance provided by the European Chemicals Agency [61]. Techniques to combine data together, or make a decision on the basis of limited (test or non-test) data, using weight of evidence ideas are required to make this powerful technology and practical reality.

11.4. CONCLUSIONS

We live in exciting times for computational toxicology. A number of issues, commercial, regulatory, ethical, and technological have brought this science to the fore – allowing much greater possibilities for the in silico assessment of toxicity. A large number of techniques and software and much guidance are available for the prediction of the harmful effects of chemicals to human health. The user of these models is now able to predict toxicity with increasing confidence and be able to demonstrate the validity of the prediction.

ACKNOWLEDGEMENT

This project was sponsored by Defra through the Sustainable Arable Link Programme. The funding of the European Union 6th Framework OSIRIS Integrated Project (GOCE-037017-OSIRIS) is also gratefully acknowledged.

REFERENCES

1. van Leeuwen CJ, Vermeire T (eds) (2007) Risk assessment of chemicals: An introduction, 2nd edn. Springer, Dordrecht, The Netherlands
2. Hester RE, Harrison RM (eds) (2006) Alternatives to animal testing. RSC Publishing, Cambridge
3. van Leeuwen CJ, Bro-Rasmussen F, Feijtel TCJ et al. (1996) Risk assessment and management of new and existing chemicals. Environ Toxicol Pharmacol 2:243–299
4. European Union (2007) Corrigendum to Regulation (EC) No 1907/2006 of the European Parliament and of the Council of 18 December 2006 concerning the Registration, Evaluation, Authorization and Restriction of Chemicals (REACH), establishing a European Chemicals Agency, amending Directive 1999/45/EC and repealing Council Regulation (EEC) No 793/93 and Commission Regulation (EC) No 1488/94 as well as Council Directive 76/769/EEC and Commission Directives 91/155/EEC, 93/67/EEC, 93/105/EC and 2000/21/EC (OJ L396, 30.12.2006). Off J Eur Union L 136:50
5. Schaafsma G, Kroese ED, Tielemans, ELJP et al. (2009) REACH, non-testing approaches and the urgent need for a change in mind set. Reg Tox Pharm 53:70–80
6. van Leeuwen CJ, Hansen BG, de Bruijn JHM (2007) Management of industrial chemicals in the European Union (REACH). In: Van Leeuwen CJ and Vermeire TG (eds), *Risk Assessment of Chemicals An Introduction*. 2nd edn, Springer Publishers, Dordrecht, The Netherlands, pp. 511–551
7. Cronin MTD (2005) Toxicological information for use in predictive modeling: quality, sources and databases. In Helma C (ed) Predictive toxicology. Taylor and Francis, Boca Raton, FL, pp. 93–133
8. Bassan A, Worth AP (2008) The integrated use of models for the properties and effects of chemicals by means of a structured workflow. QSAR Comb Sci 27:6–20
9. Cronin MTD, Schultz TW (2003) Pitfalls in QSAR. J Mol Struct (Theochem) 622:39–51
10. Dearden JC, Cronin MTD, Kaiser KLE (2009) How not to develop a quantitative structure–activity or structure–property relationship (QSAR/QSPR). SAR QSAR Environ Res 20:241–266
11. Young D, Martin T, Venkatapathy R et al. (2008) Are the chemical structures in your QSAR correct? QSAR Comb Sci 27:1337–1345

12. Klimisch HJ, Andreae E, Tillmann U (1997) A systematic approach for evaluating the quality of experimental and ecotoxicological data. Reg Tox Pharm 25:1–5
13. Cronin MTD, Basketter DA (1994) A multivariate QSAR analysis of a skin sensitization database. SAR QSAR Environ Res 2:159–179
14. Gerberick GF, Ryan CA, Kern PS et al. (2005) Compilation of historical local lymph node data for evaluation of skin sensitization alternative methods. Dermatitis 16:157–202
15. Roberts DW, Patlewicz G, Dimitrov SD et al. (2007) TIMES-SS—A mechanistic evaluation of an external validation study using reaction chemistry principles. Chem Res Toxicol 20:1321–1330
16. Patlewicz G, Aptula AO, Uriarte E et al. (2007) An evaluation of selected global (Q)SARs/expert systems for the prediction of skin sensitisation potential. SAR QSAR Environ Res 18:515–541
17. Patlewicz G, Aptula AO, Roberts DW et al. (2008) A minireview of available skin sensitization (Q)SARs/expert systems. QSAR Comb Sci 27:60–76
18. Roberts DW, Aptula AO, Cronin MTD et al. (2007) Global (Q)SARs for skin sensitisation – assessment against OECD principles. SAR QSAR Environ Res 18:343–365
19. Cronin MTD, Basketter DA (1993). A QSAR evaluation of an existing contact allergy database. In: Wermuth CG (ed) Trends in QSAR and molecular modelling 92. Escom, Leiden, pp. 297–298
20. Payne MP, Walsh PT (1994) Structure–activity relationships for skin sensitization potential: Development of structural alerts for use in knowledge-based toxicity prediction systems. J Chem Inf Comput Sci 34:154–161
21. Barratt MD, Basketter DA, Chamberlain M et al. (1994) An expert system rulebase for identifying contact allergens. Toxicol In Vitro 8:1053–1060
22. Aptula AO, Roberts DW (2006) Mechanistic applicability domains for nonanimal-based prediction of toxicological end points: General principles and application to reactive toxicity. Chem Res Toxicol 19:1097–1105
23. Roberts DW, Aptula AO, Patlewicz G (2007) Electrophilic chemistry related to skin sensitization. Reaction mechanistic applicability domain classification for a published data set of 106 chemicals tested in the mouse local lymph node assay. Chem Res Toxicol 20:44–60
24. Enoch SJ, Madden JC, Cronin MTD (2008) Identification of mechanisms of toxic action for skin sensitisation using a SMARTS pattern based approach. SAR QSAR Environ Res 19:555–578
25. Fedorowicz A, Singh H, Soderholm S et al. (2005) Structure–activity models for contact sensitization. Chem Res Toxicol 18:954–969
26. Li S, Adam Fedorowicz A, Singh H et al. (2005) Application of the random forest method in studies of local lymph node assay based skin sensitization data. J Chem Inf Model 45:952–964
27. Devillers J (2000) A neural network SAR model for allergic contact dermatitis. Toxicol Mech Meth 10:181–193
28. Ren Y, Liu H, Xue C et al. (2006) Classification study of skin sensitizers based on support vector machine and linear discriminant analysis. Anal Chim Acta 572:272–282
29. Basketter D, Darlenski R, Fluhr JW (2008) Skin irritation and sensitization: mechanisms and new approaches for risk assessment. 2. Skin sensitization. Skin Pharmacol Physiol 21:191–202
30. Basketter DA, Roberts DW, Cronin M et al. (1992) The value of the local lymph node assay in quantitative structure activity investigations. Contact Derm 27:137–142
31. Patlewicz G, Roberts DW, Walker JD (2003) QSARs for the skin sensitization potential of aldehydes and related compounds. QSAR Comb Sci 22:196–203
32. Patlewicz GY, Basketter DA, Pease CKS et al. (2004) Further evaluation of quantitative structure–activity relationship models for the prediction of the skin sensitization potency of selected fragrance allergens. Contact Derm 50:91–97
33. Enoch SJ, Cronin MTD, Schultz TW et al. (2008) Quantitative and mechanistic read across for predicting the skin sensitization potential of alkenes acting via Michael addition. Chem Res Toxicol 21:513–520

34. Miller MD, Yourtee DM, Glaros AG et al. (2005) Quantum mechanical structure–activity relationship analyses for skin sensitization. J Chem Inf Model 45:924–929

35. Estrada E, Patlewicz G, Chamberlain M et al. (2003) Computer-aided knowledge generation for understanding skin sensitization mechanisms: The TOPS-MODE approach. Chem Res Toxicol 16:1226–1235

36. Dearden JC, Barratt MD, Benigni R et al. (1997) The development and validation of expert systems for predicting toxicity. Altern Lab Anim 25:223–252

37. Helma C (ed) (2005) Predictive toxicology. CRC Press, Boca Raton, FL

38. Mohan CG, Gandhi T, Garg D (2007) Computer-assisted methods in chemical toxicity prediction. Mini-Rev Med Chem 7:499–507

39. Muster W, Breidenbach A, Fischer H et al. (2008) Computational toxicology in drug development. Drug Disc Today 13:303–310

40. Nigsch F, Macaluso NJM, Mitchell JBO et al. (2009) Computational toxicology: an overview of the sources of data and of modelling methods. Exp Opin Drug Metab Toxicol 5:1–14

41. Gallegos-Saliner A, Patlewicz G, Worth AP (2008) A review of (Q)SAR models for skin and eye irritation and corrosion. QSAR Comb Sci 27:49–59

42. Cronin MTD, Worth AP (2008) (Q)SARs for predicting effects relating to reproductive toxicity. QSAR Comb Sci 27:91–100

43. Organization for Economic Cooperation and Development (OECD) (2007) Guidance on grouping of chemicals. OECD Environment Health and Safety Publications Series on Testing and Assessment. No. 80. OECD, Paris, France. ENV/JM/MONO(2007)28 available from www.oecd.org.

44. Cunningham AR, Moss ST, Lype SA et al. (2008) Structure–activity relationship analysis of rat mammary carcinogens. Chem Res Toxicol 21:1970–1982

45. Koleva YK, Madden JC, Cronin MTD (2008) Formation of categories from structure–activity relationships to allow read-across for risk assessment: toxicity of an unsaturated carbonyl compounds. Chem Res Toxicol 21:2300–2312

46. Enoch SJ, Cronin MTD, Hewitt M (2009) Formation of structural categories to allow for read-across for teratogenicity. QSAR Comb Sci 28:696–708

47. Patlewicz G, Jeliazkova N, Gallegos-Saliner A et al. (2008) Toxmatch – A new software tool to aid in the development and evaluation of chemically similar groups. SAR QSAR Environ Res 19: 397–412

48. Gramatica P, Pilutti P, Papa E et al. (2004) Validated QSAR prediction of OH tropospheric degradation of VOCs: splitting into training-test sets and consensus modelling. J Chem Inf Comput Sci 44:1794–1802

49. Hewitt M, Cronin MTD, Madden JC et al. (2007) Consensus QSAR Models: Do the benefits outweigh the complexity? J Chem Inf Model 47:1460–1468

50. Matthews EJ, Kruhlak NL, Benz RD et al. (2008) Combined use of MC4PC, MDL-QSAR, BioEpisteme, Leadscope PDM, and Derek for Windows software to achieve high-performance, high-confidence, mode of action-based predictions of chemical carcinogenesis in rodents. Toxicol Mech Meth 18:189–206

51. Votano JR, Parham M, Hall LH et al. (2004) Three new consensus QSAR models for the prediction of Ames genotoxicity. Mutagenesis 19:365–377

52. Abshear T, Banik GM, D'Souza ML et al. (2006) A model validation and consensus building environment. SAR QSAR Environ Res 17:311–321

53. Grindon C, Combes R, Cronin MTD et al. (2006) Integrated testing strategies for use in the EU REACH system. Altern Lab Anim 34:407–427

54. Combes R, Grindon C, Cronin MTD et al. (2008) Integrated decision-tree testing strategies for acute systemic toxicity and toxicokinetics with respect to the requirements of the EU REACH legislation. Altern Lab Anim 36:45–63

55. Combes R, Grindon C, Cronin MTD et al. (2008) An integrated decision-tree testing strategy for eye irritation with respect to the requirements of the EU REACH legislation. Altern Lab Anim 36:81–92

56. Combes R, Grindon C, Cronin MTD et al. (2008) Integrated decision-tree testing strategies for developmental and reproductive toxicity with respect to the requirements of the EU REACH legislation. Altern Lab Anim 36:65–80

57. Combes R, Grindon C, Cronin MTD et al. (2008) An integrated decision-tree testing strategy for repeat dose toxicity with respect to the requirements of the EU REACH legislation. Altern Lab Anim 36:93–101

58. Grindon C, Combes R, Cronin MTD et al. (2007) Integrated decision-tree testing strategies for skin sensitisation with respect to the requirements of the EU REACH Legislation. Altern Lab Anim 35:683–697

59. Grindon C, Combes R, Cronin MTD et al. (2007) Integrated decision-tree testing strategies for skin corrosion and irritation with respect to the requirements of the EU REACH Legislation. Altern Lab Anim 35:673–682

60. Combes R, Grindon C, Cronin MTD et al. (2007) Proposed integrated decision-tree testing strategies for mutagenicity and carcinogenicity in relation to the EU REACH legislation. Altern Lab Anim 35:267–287

61. European Chemicals Agency (2008) Guidance on information requirements and chemical safety assessment, EChA, Helsinki. Available from http://guidance.echa.europa.eu/docs/guidance_document/information_requirements_en.htm?time=1238407373

CHAPTER 12

CHEMOMETRIC METHODS AND THEORETICAL MOLECULAR DESCRIPTORS IN PREDICTIVE QSAR MODELING OF THE ENVIRONMENTAL BEHAVIOR OF ORGANIC POLLUTANTS

PAOLA GRAMATICA

QSAR Research Unit in Environmental Chemistry and Ecotoxicology,
Department of Structural and Functional Biology, University of Insubria, Varese, Italy,
e-mail: paola.gramatica@uninsubria.it; http://www.qsar.it

Abstract: This chapter surveys the QSAR modeling approaches (developed by the author's research group) for the validated prediction of environmental properties of organic pollutants. Various chemometric methods, based on different theoretical molecular descriptors, have been applied: explorative techniques (such as PCA for ranking, SOM for similarity analysis), modeling approaches by multiple-linear regression (MLR, in particular OLS), and classification methods (mainly k-NN, CART, CP-ANN). The focus of this review is on the main topics of environmental chemistry and ecotoxicology, related to the physico-chemical properties, the reactivity, and biological activity of chemicals of high environmental concern. Thus, the review deals with atmospheric degradation reactions of VOCs by tropospheric oxidants, persistence and long-range transport of POPs, sorption behavior of pesticides (K_{oc} and leaching), bioconcentration, toxicity (acute aquatic toxicity, mutagenicity of PAHs, estrogen binding activity for endocrine disruptors compounds (EDCs)), and finally persistent bioaccumulative and toxic (PBT) behavior for the screening and prioritization of organic pollutants. Common to all the proposed models is the attention paid to model validation for predictive ability (not only internal, but also external for chemicals not participating in the model development) and checking of the chemical domain of applicability. Adherence to such a policy, requested also by the OECD principles, ensures the production of reliable predicted data, useful also in the new European regulation of chemicals, REACH.

Keywords: QSAR, Chemometric methods, Theoretical molecular descriptors, MLR, Classification, Environmental pollutants, Ranking

12.1.   INTRODUCTION

The QSAR world has undergone profound changes since the pioneering work of Corwin Hansch, considered the founder of modern QSAR modeling [1, 2]. The main change is reflected in the growth of a parallel and quite different conceptual

327

T. Puzyn et al. (eds.), Recent Advances in QSAR Studies, 327–366.
DOI 10.1007/978-1-4020-9783-6_12, © Springer Science+Business Media B.V. 2010

approach to the modeling of the relationships among a chemical's structure and its activity/properties.

In the Hansch approach, still applied widely and followed by many QSAR modelers (for instance, [3–5]), molecular structure is represented by only a few molecular descriptors (typically log K_{ow},[1] Hammett constants, HOMO/LUMO, some steric parameters) selected personally by the modeler and inserted in the QSAR equation to model a studied endpoint. Alternatively, in a different approach chemical structure is represented, in the first preliminary step, by a large number of theoretical molecular descriptors which are then, in a second step, selected by different chemometric methods as the best correlated with the response and, finally, included in the QSAR model (the algorithm), the fundamental aim being the optimization of model performance for prediction.

According to the Hansch approach, descriptor selection is guided by the modeler's conviction to have *a priori* knowledge of the mechanism of the studied activity/property. The modeler's presumption is to assign mechanistic meaning to any used molecular descriptor selected by the modeler from among a limited pool of potential modeling variables. These descriptors are normally well known and used repeatedly (for instance, log K_{ow} is a universal parameter mimicking cell membrane permeation, thus it is used in models for toxicity, but it is also related to various partition coefficients such as bioconcentration/bioaccumulation, soil sorption coefficient; HOMO/LUMO energies are often selected for modeling chemical reactivity, etc.).

On the other hand, the "statistical" approach, an approach parallel to the previous so-called "mechanistic" one, is based on the fundamental conviction that the QSAR modeler should not influence, *a priori* and personally, the descriptor selection through mechanistic assumptions. Instead they should apply unbiased mathematical tools to select, from a wide pool of input descriptors, those descriptors most correlated to the studied response. The number and typology of the available input descriptors must be as wide and different as possible in order to guarantee the possibility of representing any aspect of the molecular structure. Different descriptors are different ways or perspectives to view a molecule. Descriptor selection should be performed by applying mathematical approaches to maximize, as an optimization parameter, the predictive power of the QSAR model, as the real utility of any model considered is its predictivity.

The first aim of any modeler should be the validation for predictive purposes of the QSAR model, for both the mechanistic and statistical approaches; in fact, a QSAR model must, first of all, be a real model, robust and predictive, to be considered a reliable model; only a stable and predictive model can be usefully interpreted for its mechanistic meaning, even so this is not always easy or feasible [6]. However, this is a second step in the statistical QSAR modeling.

---

[1] The symbol refers to the same property as log P (namely to the n-octanol/water partition coefficient). However, in many environmental studies this partition coefficient is abbreviated by "log K_{ow}" to be consistent with the other environmentally relevant coefficients, e.g., n-octanol/air partition coefficient (K_{oa}), air/water partition coefficient (K_{aw}).

QSAR model validation has been recognized by specific OECD expert groups as a crucial and urgent requirement in recent years, and this has led to the development, for regulatory purposes, of the "OECD principles for the validation of (Q)SAR models" (http://www.oecd.org/document/23/0,3343,fr_2649_34365_33957015_1_1_1_1,00.html).

The need for this important action was mainly due to the recent new chemicals policy of the European Commission (REACH: *R*egistration, *E*valuation, *A*uthorization and restriction of *Ch*emicals) (http://europa.eu.int/comm/environment/chemicals/reach.htm) that explicitly states the need to use (Q)SAR models to reduce experimental testing (including animal testing). Obviously, to meet the requirements of the REACH legislation (see also Chapter 13) it is essential to use (Q)SAR models that produce reliable estimates, i.e., validated (Q)SAR models. Thus, reliable QSAR model must be associated with the following information: (1) a defined endpoint; (2) an unambiguous algorithm; (3) a defined domain of applicability; (4) appropriate measures of goodness-of-fit, robustness and predictivity; (5) a mechanistic interpretation, if possible.

Some crucial points of the statistical approach of QSAR modeling, applied by the author's group, are put into context, according to the guidelines of the OECD principles, which are the chemometric approach steps.

12.2. A DEFINED ENDPOINT (OECD PRINCIPLE 1)

The most common regulatory endpoints, associated with OECD test guidelines, are related to (a) physico-chemical properties (such as melting and boiling points, vapor pressure, K_{ow}, K_{oc}, water solubility); (b) environmental fate (such as biodegradation, hydrolysis, atmospheric oxidation, bioaccumulation); (c) human health (acute oral, acute inhalation, acute dermal, skin irritation, eye irritation, skin sensitization, genotoxicity, reproductive and developmental toxicity, carcinogenicity, specific organ toxicity (e.g., hepatotoxicity, cardiotoxicity)); and (d) ecological effects (acute fish, acute daphnid, alga, long-term aquatic, and terrestrial toxicity) of chemicals.

The various experimental endpoints that have been modelled by the QSAR Research Unit of Insubria University are described in the following sections, after the discussion on the main methodological topics. A distinction will be made between single endpoints and cumulative endpoints, which take into account a contemporaneous contribution of different properties or activities.

12.3. AN UNAMBIGUOUS ALGORITHM (OECD PRINCIPLE 2)

The algorithms used in (Q)SAR modeling should be described thoroughly so that the user will understand exactly how the estimated value was produced and can reproduce exactly the calculations also for new chemicals, if desired.

When the studied endpoint needs to be modelled using more than one descriptor (selected by different approaches) multivariate techniques are applied. As there can be multiple steps in estimating the endpoint of a chemical, it is important that the nature of the used algorithms be unambiguous, as required by OECD Principle 2.

12.3.1. Chemometric Methods

12.3.1.1. Regression Models

Regression analysis is the use of statistical methods for modeling a dependent variable Y, a quantitative measures of response (e.g., boiling point, LD_{50}), in terms of predictors X (independent variables or molecular descriptors).

There are many different multivariate methods for regression analysis, more or less widely applied in QSAR studies: multiple linear regression (MLR), principal component regression (PCR), partial least squares (PLS), artificial neural networks (ANNs), fuzzy clustering and regression are among more commonly used approaches for regression modeling.

Although all QSAR models (linear and not linear) are based on algorithms, the most common regression method, which describes models by completely transparent and easily reproducible mathematical equations, is multiple linear regression (MLR), in particular ordinary least squares (OLS) method. This method has been applied by the author in her QSAR studies; to cite some most recent papers, see [7–28] and Chapter 6. Some of these models are commented on in the following paragraphs.

The correlation of the variables in the modeling must be controlled carefully (for instance, by applying the QUIK rule [29]) and the problem of possible overfitting [30], common also to other modeling methods, must also be checked by statistical validation methods to verify robustness and predictivity. The selection of descriptors in MLR can be performed either *a priori* by the model developer on a mechanistic basis or by evolutionary techniques such as genetic algorithms. In this second approach, the model's developer should try to interpret mechanistically the descriptors selected, but only after model development and statistical validation for predictivity.

12.3.1.2. Classification Models

Another common problem in QSAR analysis is prediction of the group membership from molecular descriptors. In the simplest case, chemicals are categorized into one, two, or more groups depending on their activity, indicated by the same value of a categorical variable: active/inactive or, for instance, toxic/non-toxic.

Classification models are quantitative models based on relationships between independent variables X (in this case molecular descriptors) and a categorical response variable of integer numerical values, each representing the class of the corresponding sample.

The term "quantitative" is referred to the numerical value of the variables necessary to classify the chemicals in the qualitative classes (a categorical response) and it specifies the quantitative meaning of a QSAR-based classification process.

Such classification, also called supervised pattern recognition, is the assignment, on the basis of a classification rule, of chemicals to one of the classes defined *a priori* (or of groups of chemicals in the training set). Thus, the goal of a classification method is to develop a classification rule (by selecting the predictor variables) based on a training set of chemicals of known classes so that the rule can be applied to a test set of compounds of unknown classes. A wide range of classification methods exists, including discriminant analysis (DA; linear quadratic, and regularized DA), soft independent modeling of class analogy (SIMCA), k-nearest neighbors (k-NN), classification and regression tree (CART), artificial neural network, support vector machine, etc.

The QSAR Research Unit of Insubria University has developed some satisfactory, validated, and usable classification models (for instance, among the more recent [16, 31–35]) by applying different classification methods, mainly classification and regression tree (CART) [36, 37], k-nearest neighbor (k-NN) [38], and artificial neural networks (in particular, Kohonen maps or self-organizing maps (SOM) [39–41]).

CART is a non-parametric unbiased classification strategy to classify chemicals with automatic stepwise variable selection. As the final output, CART displays a binary, immediately applicable, classification tree; each non-terminal node corresponds to a discriminant variable (with the threshold value of that molecular descriptor) and each terminal node corresponds to a single class. To classify a chemical, at each binary node, the tree branch, matching the values of the chemical on the corresponding splitting descriptor, must be followed.

The k-NN method is a non-parametric unbiased classification method that searches for the k-nearest neighbors of each chemical in a data set. The compound under study is classified by considering the majority of classes to which the kth nearest chemicals belong. k-NN is applied to autoscaled data with *a priori* probability proportional to the size of the classes; the predictive power of the model is checked for k nearest neighbors between 1 and 10.

Counter-propagation artificial neural networks (CP-ANNs), particularly Kohonen maps, are supervised classification methods. Input variables (molecular descriptors) calculated for the studied chemicals provide the input for the net or the Kohonen layer. The architecture of the net is constituted by $N \times N \times p$, where p is the number of input variables and each p-dimensional vector is a neuron (N). Thus, the neurons are vectors of weights, corresponding to the input variables. During the learning, n chemicals are presented to the net – one at a time – a fixed number of times (epochs); each chemical is then assigned to the cell for which the distance between the chemical vector and the neuron is minimum. The target values (i.e., the classes to be modelled) are given to the output layer (the top-map: a two-dimensional plane of response), which has the same topological arrangement of neurons as the Kohonen layer. The position of the chemicals is projected to the output layer and the weights are corrected in such a way that they fit the output values (classes) of corresponding chemicals. The Kohonen-ANN automatically adapts itself in such a way that similar input objects are associated with topologically close neurons in the top-map. The chemical similarity decreases with increasing of the topological distance.

The trained network can be used for predictions; a new object in the Kohonen layer will lie on the neuron with the most similar weights. This position is then projected to the top-map, which provides a predicted output value. It is important to remember that the Kohonen top-map has toroid geometry; each neuron has the same number of neighbors, including the neurons on the borders of the top-map.

According to the OECD principles, for a QSAR model to be acceptable for use to make regulatory decisions it must be clearly defined, easily and continuously applicable in such a way that the calculations for the prediction of the endpoint can be reproduced by everyone, and applicable to new chemicals. The unambiguous algorithm is characterized not only by the mathematical method of calculation used, but also by the specific molecular descriptors required in the model mathematical equation. Thus, the exact procedure used to calculate the descriptors, including compound pre-treatment (e.g., energy minimization, partial charge calculation), the software employed, and the variable selection method for QSAR model development should be considered integrative parts of the overall definition of an unambiguous algorithm.

12.3.2.   Theoretical Molecular Descriptors

It has become quite common to use a wide set of molecular descriptors of different kinds (experimental and/or theoretical) that are able to capture all the structural aspects of a chemical to translate the molecular structure into numbers. The various descriptors are different ways or perspectives to view a molecule, taking into account the various features of its chemical structure, not only one-dimensional (e.g., the simple counts of atoms and groups), but also two-dimensional from a topological graph or three-dimensional from a minimum energy conformation. Livingstone has published a survey of these approaches [42]. Much of the software calculates broad sets of different theoretical descriptors, from SMILES, 2D-graphs to 3D-x,y,z-coordinates. Some of the frequently used descriptor calculation software includes ADAPT [43], OASIS [44], CODESSA [45], DRAGON [46], and MolConnZ [47]. It has been estimated that more than 3000 molecular descriptors are now available, and most of them have been summarized and explained [48–50]. The great advantage of theoretical descriptors is that they can be calculated homogeneously by a defined software for all chemicals, even those not yet synthesized, the only need being a hypothesized chemical structure. This peculiarity explains their wide and successful use in QSAR modeling. The DRAGON software has always been used in models developed by the author's group. In the version more frequently used by the author (5.4), 1664 molecular descriptors of the following different typologies were calculated: (a) 0D-48 constitutional (atom and group counts); (b) 1D-14 charge descriptors; (c) 1D-29 molecular properties; (d) 2D-119 topological; (e) 2D-47 walk and path counts, (f) 2D-33 connectivity index; (g) 2D-47 information index; (h) 2D-96 various auto-correlations from the molecular graph; (i) 2D-107 edge adjacency indices; (j) 2D-64 descriptors of Burden (BCUTs eigenvalues); (k) 2D-21 topological charge indices; (l) 2D-44 eigenvalue-based indices; (m) 3D-41 Randic molecular profiles; (n) 3D-74 geometrical descriptors; (o) 3D-150 radial distribution function; (p) 3D-160 Morse; (q) 3D-99 weighted holistic invariant molecular descriptors (WHIMs) [51–53]; (r) 3D-197 geometry, topology and atom-weights assembly (GETAWAY) descriptors [54, 55]; (s) 154 functional groups; (t) 120 atom-centered fragments. The list and meaning of the molecular descriptors are provided by the DRAGON package and the calculation procedure is explained in detail, with related literature references, in the Handbook of Molecular Descriptors from Todeschini and Consonni [50] and in Chapter 3. The DRAGON software is continuously implemented with new descriptors.

12.3.3. Variable Selection and Reduction. The Genetic Algorithm Strategy for Variable Selection

The existence of a huge number of different molecular descriptors, experimental or theoretical, to describe chemical structure is a great resource as it allows QSAR modelers (particularly those working with the statistical approach) to have different X-variables available that take into account each structural feature in various ways. In principle, all the different possible combinations of the X-variables should be investigated to find the most predictive QSAR model. However, this can be quite taxing, mainly for reasons of time.

Sometimes molecular descriptors, which are only different views of the same molecular aspect, are highly correlated. Thus, when dealing with a large number of highly correlated descriptors, variable selection is necessary to find a simple and predictive QSAR model, which must be based on the minimum number of descriptors, and the least correlated, as possible. First, objective selection is applied using only independent variables (X): descriptors to discard are identified by tests of identical values and pairwise correlations, looking for descriptors less correlated to one another.

Secondly, modeling variable selection methods, which additionally use dependent variable values (Y), are applied to this pre-reduced set of descriptors to further reduce it to the true modeling set, not only in fitting but, most importantly, in prediction. Such selection is performed by alternative variable selection methods.

Several strategies for variable subset selection have been applied in QSAR (stepwise regressions, forward selection, backward elimination, simulated annealing, evolutionary and genetic algorithms, among those most widely applied). A comparison of these methods [56] has demonstrated the advantages, and the success, of genetic algorithms (GAs) as a variable selection procedure for QSAR studies.

GAs are a particular kind of evolutionary algorithms (EAs), shown to be able to solve complex optimization problems in a number of fields, including chemistry [57–59]. The natural principles of the evolution of species in the biological world are applied, i.e., the assumption that conditions leading to better results will prevail over poorer ones, and that improvement can be obtained by different kinds of recombination of independent variables, i.e., reproduction, mutation, and crossover. The goodness-of-fit of the selected solution is measured by a function that has to be optimized.

Genetic algorithms, first proposed as a strategy for variable subset selection in multivariate analysis by Leardi et al. [60] and applied to QSAR modeling by Rogers and Hopfinger [61], are a very effective tool with many merits compared to other methods. GAs are now widely and successfully applied in QSAR approaches, where there is quite a number of molecular descriptors, in various modified versions, depending on the way of performing reproduction, crossover, mutation, etc. [62–66].

In variable selection for QSAR studies, a bit equal to 1 denotes a variable (molecular descriptor) present in the regression model or equal to 0 if excluded. A population, constituted by a number of 0/1 bit strings (each of length equal to the total number of variables in the model), is evolved following genetic algorithm rules, maximizing the predictive power of the models (verified by the explained variance in prediction, Q^2_{cv} or by the root mean squared error of prediction, RMSEcv). Only models producing the highest predictive power are finally retained and further analyzed with additional validation techniques.

Whereas EAs search for the global optimum and end up with only one or very few results [64, 65, 67], GAs simultaneously create many different results of comparable quality in larger populations of models with more or less the same predictive power. Within a given population the selected models can differ in the number and kind of variables. Similar descriptors, which are able to capture some specific aspects of chemical structure, can be selected by GA in alternative combinations for modeling the response. Thus, similarly performing models can be considered as different perspectives to arrive at essentially the same conclusion. Owing to this, the GA-based approach has no single "best" set of descriptors related to the Y-dependent variable; there is a population of good models of similar performance that could be also combined in consensus modeling approaches [18, 19] to obtain averaged predictions.

Different rules can be adopted to select the final preferred "best" models. In the author's researches the QUIK (Q under influence of K) rule [29] is always applied as the first filter to avoid multi-collinearity in model descriptors without prediction power or with "apparent" prediction power (chance correlation). According to this rule, only models with a K multivariate correlation calculated on the X+Y block, at least 5% greater than the K correlation of the X-block, are considered statistically significant and checked for predictivity (both internally by different cross-validations and externally on chemicals which do not participate in model development).

Another important parameter that must be considered is the root mean squared error (RMSE) that summarizes the overall error of the model; it is calculated as the root square of the sum of squared errors in calculation (RMSE) or prediction (RMSEcv and RMSEp) divided by their total number. The best model has the smallest RMSE and very similar RMSE values for training and external prediction chemicals, highlighting the model's generalizability [68].

12.4.   APPLICABILITY DOMAIN (OECD PRINCIPLE 3)

The third OECD Principle takes into consideration another crucial problem: the definition of the applicability domain (AD) of a QSAR model. Even a robust, significant, and validated QSAR model cannot be expected to reliably predict the property modelled for the entire universe of chemicals. In fact, only predictions for chemicals falling within the domain of the developed model can be considered reliable and not model extrapolations. This topic was dealt with at a recent workshop where several different approaches for linear and non-linear models were proposed [69], in relation to different model types.

The AD is a theoretical spatial region defined by the model descriptors and the response modelled, and is thus defined by the nature of the chemicals in the training set, represented in each model by specific molecular descriptors. To clarify recent doubts [70], it is important to note that each QSAR model has its own specific AD based on the training set chemicals, not just on the kind of included chemicals but also on the values of the specific descriptors used in the model itself; such descriptors are dependent on the type of the training chemicals.

As was explained above, a population of MLR models of similar good quality, developed by variable selection performed with a genetic algorithm [66] can include a 100 different models developed on the same training set but based on different descriptors: even if developed on the same chemicals, the AD for new chemicals can differ from model to model, depending on the specific descriptors. Through the leverage approach [71] (shown below) it is possible to verify whether a new chemical will lie within the model domain (in this case predicted data can be considered as interpolated and with reduced uncertainty, at least similar to that of training chemicals, thus more reliable) or outside the domain (thus, predicted data are extrapolated by the model and must be considered of increased uncertainty, thus less reliable). If it is outside the model domain a warning must be given. Leverage is used as a quantitative measure of the model applicability domain and is suitable for evaluating the degree of extrapolation, which represents a sort of compound "distance" from the model experimental space (the structural centroid of the training set). It is a measure of the influence a particular chemical's structure has on the model: chemicals close to the centroid are less influential in model building than extreme points. A compound with high leverage in a QSAR model would reinforce the model if the compound is in the training set, but such a compound in the test set could have unreliable predicted data, the result of substantial extrapolation of the model.

The prediction should be considered unreliable for compounds in the test set with high-leverage values ($h > h^*$, the critical value being $h^* = 3p'/n$, where p' is the number of model variables plus one and n is the number of the objects used to calculate the model). When the leverage value of a compound is lower than the critical value, the probability of accordance between predicted and actual values is as high as that for the training set chemicals. Conversely, a high-leverage chemical is structurally distant from the training chemicals, thus it can be considered outside the AD of the model. To visualize the AD of a QSAR model, the plot of standardized cross-validated residuals (R) vs. leverage (Hat diagonal) values (h) (the Williams plot) can be used for an immediate and simple graphical detection of both the response outliers (i.e., compounds with cross-validated standardized residuals greater than three standard deviation units, $> 3\sigma$) and structurally influential chemicals in a model ($h > h^*$).

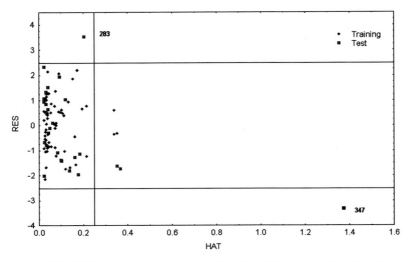

Figure 12-1. Williams plot for an externally validated model for the toxicity to *Pimephales promelas* of polar narcotics. Cut-off value: 2.5 h* (with copyright permission from [26])

It is important to note that the AD of a model cannot be verified by studying only a few chemicals, as in such cases [72] it is impossible to obtain conclusions that can be generalized on the applicability of the model itself.

Figure 12-1 shows the Williams plot of a model for compounds that act as polar narcotics to *Pimephales promelas* [26]; as an example, here the toxicity of chemical no. 347 is incorrectly predicted (>3σ) and it is also a test chemical completely outside the AD of the model, as defined by the Hat vertical line (high h leverage value), thus it is both a response outlier and a high-leverage chemical. Two other chemicals (squares at 0.35 h) slightly exceed the critical hat value (vertical line) but are close to three chemicals of the training set (rhombus), slightly influential in the model development: the predictions for these test chemicals can be considered as reliable as those of the training chemicals. The toxicity of chemical no. 283 is incorrectly predicted (>3σ), but in this case it belongs to the model AD, being within the cut-off value of Hat. This erroneous prediction could probably be attributed to error or variability in the experimental data rather than to molecular structure or model.

12.5.   MODEL VALIDATION FOR PREDICTIVITY (OECD PRINCIPLE 4)

Model validation must always be used to avoid the possibility of "overfitted" models, i.e., models where too many variables, useful only for fitting the training data, have been selected, and to avoid the selection of variables randomly correlated (by chance) with the dependent response. Particular care must be taken against overfitting [30], thus subsets with the fewest variables are favored, as the chance of finding "apparently acceptable" models increases with increasing X-variables. The proportion of random variables selected by chance correlation could also increase [73]. The ratio of chemicals to variables should always be higher than five for a small data set, but the number of descriptors must be the lowest as possible for bigger data sets too (according to the Ockham's Razor: "avoid complexity if not necessary").

Therefore, a set of models of similar performance, verified by leave-one-out model validation, need to be further validated by leave-more-out cross-validation or bootstrap [74, 75]. This is done to avoid overestimation of the model's predictive power by Q^2_{LOO} [76, 77] and to verify the stability of model predictivity (robustness). Response permutation testing (Y scrambling) [6] or other resampling techniques are also applied for excluding that the developed model is based on descriptors that could be related to the response only by chance. Finally, for the most stringent evaluation of model applicability for prediction of new chemicals, external validation (verified by Q^2_{EXT} or R^2_{EXT}) of all models is recommended as the last step after model development, and for the assessment of true predictive ability [6, 10, 78].

The preferred model will be that with the highest prediction parameter values and the most balanced results between the cross-validation parameters on the training chemicals (Q^2_{cv}, Q^2_{LMO}, Q^2_{BOOT}), verified during descriptor selection, and the predictive power (Q^2_{EXT} or R^2_{EXT}), verified later on the external prediction chemicals.

The limiting problem for efficient external validation of a QSAR model is, obviously, data availability. Given the availability of a sufficiently large number (never less than five or 20% of training set) of really new and reliable experimental data, the best proof of an already developed model accuracy is to test model performance on these additional data, at the same time checking the chemical AD. However, it is usually difficult to have data available for new experimentally tested compounds (in useful quantity and quality) for external validation purposes, thus, in the absence of additional data, external validation by *a priori* splitting the available data can be usefully applied to define the actual predictive power of the model more precisely.

12.5.1. Splitting of the Data Set for the Construction of an External Prediction Set

In the absence of new additional data, we assume that there is less data than is actually available; this is the reason for splitting the data in a reasonable way (commented on below) into a training set and a prediction set of "momentarily forgotten chemicals."

Thus, before model development, the available input data set can be split adequately by different procedures into the training set (for model development) and the prediction set (never used for variable selection and model development, but used exclusively once for model predictive assessment, performed only after model development). At this point the underlying goal is to ensure that both the training and prediction sets separately span the whole descriptor space occupied by the entire data set, and that the chemical domain in the two data sets is not too dissimilar [77, 79–81] as it is impossible for a model to be applied outside its chemical domain and obtain reliable predictions. The composition of the training and prediction sets is of crucial importance. The best splitting must guarantee that the training and prediction sets are scattered over the whole area occupied by representative points in the descriptor space (representativity), and that the training set is distributed over an area occupied by representative points for the whole data set (diversity). The more widely applied splitting methodologies are based on structural similarity analysis (for instance, Kennard Stone, duplex, D-optimal distance [11–13, 17, 18, 20, 21, 81, 82], self-organizing map (SOM) or Kohonen-map ANN [17, 18, 20, 21, 26, 27, 35, 39, 41, 80]. Alternatively, to split the available data without any bias for structure, random selection through activity sampling can be applied. Random splitting is highly useful if applied iteratively in splitting for CV internal validation and can be considered quite similar to real-life situations, but it can give very variable results when applied in this external validation, depending greatly on set dimension and representativity [80, 83, 84]. In addition, in this last case there is a greater probability of selecting chemicals outside the model structural AD in the prediction set; thus, the predictions for these chemicals could be unreliable, simply as they are extrapolated by the model.

12.5.2.   Internal and External Validation

External validation should be applied to any proposed QSAR model to determine both its generalizability for new chemicals that, obviously, must belong to the model AD and the "realistic" predictive power of the model [6, 83–85]. The model must be tested on a sufficiently large number of chemicals not used during its development, at least 20% of the complete data set is recommended, but the most stable models (of easily modelled endpoints) can also be checked on a prediction set larger than the training set [19, 85]; this will avoid "supposed" external validation based on too few chemicals [72]. In fact, it has been demonstrated that if the test set consists only of a small number of compounds, there is increased possibility of chance correlation between the predicted and observed response of the compounds [79].

  It is not unusual for models with high internal predictivity, verified by internal validation methods (LOO, LMO, Bootstrap), but externally less predictive or even absolutely unpredictive, to be present in populations of models developed using evolutionary techniques to select the descriptors. The statistical approach to QSAR modeling always carefully checks this possibility by externally validating any model, stable in cross-validation, before its proposal. In fact, cross-validation is necessary but is not a sufficient validation approach for really predictive models [6, 77–79]. In relation to this crucial point of QSAR model validation, there is a wide debate and discordant opinions in the QSAR community concerning the different outcomes of internal and external validation on QSAR models. A mini-review dealing with this problem has been recently published by the author [84], where an examination is made of the OECD Principles 2, 3, and 4, and particular attention has been paid to the differences in internal and external validation. The theoretical constructs are illustrated with examples taken from both the literature and personal experience, derived also from a recent report for the European Centre for Validation of Alternative Methods (ECVAM) on "Evaluation of different statistical approaches to the validation of Quantitative Structure–Activity Relationships" [83].

Since GAs simultaneously create many different, similarly acceptable models in a population, the user can choose the "best model" according to need: the possibility of having reliable predictions for some chemicals rather than others, the interpretability of the selected molecular descriptors, the presence of different outliers, etc.

In the statistical approach the best model is selected by maximizing all the CV internal validation parameters, by applying CV in the proper way and step. Then, only the good models ($Q^2_{LOO}>0.7$), stable and internal predictive (with similar values of all the different CV-Q^2), are subjected to external validation on the *a priori* split prediction set.

In our works we always select, from among the best externally predictive models, those with the smallest number of response outliers and structurally influential chemicals, especially those in the prediction set.

12.5.3. Validation of Classification Models

To assess the predictive ability of classification models, the percentage of misclassified chemicals, as error rate (ER%) and error rate in prediction (ER_{cv}%), are calculated by the leave-one-out method (where each chemical is taken out of the training set once and predicted by the model). Comparison with the no-model error rate (NoMER) is used to evaluate model performance. NoMER represents the object distribution in the defined classes before applying any classification method, and is calculated as an error rate by considering all the objects as misclassified into the greatest class. This provides a reference classification parameter to evaluate the actual efficiency of a classifier: the greater the difference between NoMER and the actual ER, the better the model performance.

The outputs of a classification model are the class assignments and the misclassification matrix, which shows how well the classes are separated. The goodness of the classification models is also assessed by the following parameters: accuracy or concordance (the proportion of correctly classified chemicals), sensitivity (the proportion of active chemicals predicted to be active), specificity (the proportion of non-active chemicals predicted to be non-active), false negatives (the proportion of active chemicals falsely predicted as non-active) and false positives (the proportion of non-active chemicals falsely predicted as active). Depending on the intended application of the predictive tool, the classification model can be optimized in either direction. In drug design the objective is to obtain a high specificity as a false positive prediction could result in the loss of a valuable candidate. In the regulatory environment, for safety assessment and consumer protection, the precautionary principle must be applied, so an optimization of sensitivity would be desirable, as every false negative compound could result in a lack of protection and consequently pose a risk for the user.

12.6. MOLECULAR DESCRIPTOR INTERPRETATION, IF POSSIBLE (OECD PRINCIPLE 5)

Regarding the interpretability of the descriptors it is important to take into account that the response modelled is frequently the result of a series of complex biological or physico-chemical mechanisms, thus it is very difficult and reductionist to ascribe too much importance to the mechanistic meaning of the molecular descriptors used in a QSAR model. Moreover, it must also be highlighted that in multivariate models such as MLR models, even though the interpretation of the singular molecular descriptor can certainly be useful, it is only the combination of the selected set of descriptors that is able to model the studied endpoint. If the main aim of QSAR modeling is to fill the gaps in available data, the modeler's attention should be focused on model quality. In relation to this point, Livingstone, in an interesting perspective paper [42] states: "The need for interpretability depends on the application, since a validated mathematical model relating a target property to chemical features may, in some cases, be all that is necessary, though it is obviously desirable to attempt some explanation of the 'mechanism' in chemical terms, but it is often not necessary, per se." Zefirov and Palyulin [78] took the same position, differentiating predictive QSARs, where attention essentially concerns the best prediction quality, from descriptive QSARs where the major attention is paid to descriptor interpretability.

The author's approach to QSAR modeling will be illustrated in the following sections of this chapter through the modeling of environmental endpoints. The approach starts with a statistical validation for predictivity and continues on through further interpretation for the mechanistic meaning of the selected descriptors, but only if possible, as set down by the fifth OECD principle [6]. Therefore, the application domain of this approach (the "statistical approach") is mainly related to the production of predicted data (predictive QSAR), strongly verified for their reliability; such data can be more usefully applied to screen and rank chemicals providing priority lists.

12.7. ENVIRONMENTAL SINGLE ENDPOINTS

12.7.1. Physico-chemical Properties

Organic chemicals now need to be characterized by many parameters, either because of the registration policy required to chemical industries (see for example, the new European REACH policy) or for an understanding of the environmental behavior of chemicals present as pollutants in various compartments. Unfortunately there is an enormous lack of knowledge for many important endpoints, such as various physico-chemical properties (for instance, melting point, boiling point, aqueous solubility, volatility, hydrophobicity, various partition coefficients), environmental reactivity and derived persistence, toxicity, mutagenicity. This lack of knowledge calls for a predictive approach to the assessment of chemicals, such as by QSAR modeling.

A set of various physico-chemical properties for important classes of chemicals present in the environment, pollutant compounds such as PAHs [86] haloaromatics [87], PCBs [88], chemicals of EEC Priority List 1 [89] have been modelled using the weighted holistic invariant molecular (WHIM) descriptors [51–53, 90, 91]. WHIM descriptors are theoretical three-dimensional molecular indices that contain information, in terms of size, shape, symmetry, and atom distribution, on the whole molecular structure. These indices are calculated from the (x, y, z) coordinates of a molecule within different weighting schemes by principal component analysis and represent a very general approach to describe molecules in a unitary conceptual framework, independent from the molecular alignment. Their meaning is defined by the same mathematical properties of the algorithm used for their calculation, and their application in QSAR modeling was very successful. A recent paper [92] again highlighted that, contrary to erroneous statements in the literature [93, 94], one set of WHIM descriptors, the k descriptors, are very useful in discriminating the shape of chemicals and can thus be used to study structural similarity.

Since then other physico-chemical properties have been modelled successfully by combining different kinds of theoretical molecular descriptors (mono-dimensional, bi-dimensional, and three-dimensional) calculated by the DRAGON software [46]: the basic physico-chemical properties of organic solvents [95], esters [15] and brominated flame retardants, mainly polybromodiphenyl ethers (PBDE) [24], the soil sorption coefficient (K_{oc}) for pesticides [19, 96] (discussed below in Section 12.7.1.1).

A general classification of 152 organic solvents has been proposed [95] by applying the k-nearest neighbor method and counter propagation artificial neural networks (CP-ANN), in particular Kohonen-maps. A good separation for five classes was obtained by the net architecture ($20 \times 20 \times 4$, 200 iterations), based on simple molecular descriptors (unsaturation index – UI, hydrophilicity factor – Hy, average atomic composition – AAC, and the number of nitrogen atoms in the molecular structure – nN). The performances were very satisfactory: ER (%)=4.4 and ER_{cv} (%)=11.4 (to be compared with the error rate without the model NoMER (%)=69.5.)

12.7.1.1. Soil Sorption of Pesticides

Sorption processes play a major role in determining the environmental fate, distribution, and persistence of chemicals. An important parameter when studying soil mobility and environmental distribution of chemicals is the soil sorption coefficient, expressed as the ratio between chemical concentration in soil and in water, normalized to organic carbon (K_{oc}).

Many QSAR papers on soil sorption coefficient prediction have been published and reviewed by some authors [85, 96–104].

The proposed models were mainly based on the correlation with octanol/water partition coefficients (K_{ow}) and water solubility (S_w), others on theoretical molecular structure descriptors. A recent paper by the author dealt with log K_{oc} of a heterogeneous set of 643 organic non-ionic compounds [19]; the response range

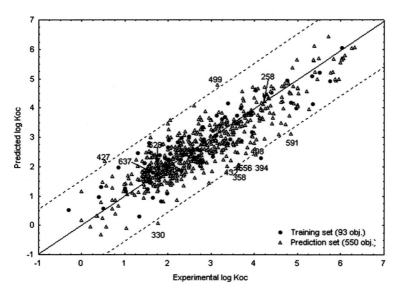

Figure 12-2. Plot of experimental vs. predicted log K_{OC} for the Eq. (12-1). The values for the training and prediction set chemicals are labeled differently, the outliers are numbered. The *dotted lines* indicate the 3σ interval (with copyright permission from [19])

was more than six log units, and prediction was made by a statistically validated QSAR modeling approach based on MLR and theoretical molecular descriptors, selected by GA from DRAGON (see Eq. 12-1). The high generalizability of one of the proposed models (scatter plot in Figure 12-2) was verified on external chemicals, performed by adequately splitting, by SOM and also randomly, the available set of experimental data into a very reduced representative training set (even less than 15% of the original data set) for model development and a large prediction set (more than 85% of the original data) used only for model performance inspection.

$$\log K_{oc} = -2.19(\pm 0.30) + 2.10(\pm 0.14)VED1 - 0.34(\pm 0.04)nHAcc - 0.31$$
$$(\pm 0.05)MAXDP - 0.33(\pm 0.12)CIC0$$

$$n(\text{training}) = 93\ R^2 = 0.82\ Q_{cv}^2 = 0.80\ Q_{BOOT}^2 = 0.79\ RMSE$$
$$= 0.523\ RMSEp_{LOO} = 0.523$$

$$n(\text{prediction set}) = 550\ Q_{EXT}^2 = 0.78\ RMSEp_{EXT} = 0.560 \qquad (12\text{-}1)$$

The proposed models have good stability, robustness, and predictivity when verified by internal validation (cross-validation by LOO and Bootstrap) and also by external validation on a much greater data set. The stability of RMSE/RMSEp for both the training and prediction sets is further proof of model predictivity.

The chemical applicability domain is verified by the Williams graph: nine out-liers for response and three structurally influential chemicals have been highlighted (numbered in Figure 12-2).

The selected molecular descriptors have a clear mechanistic meaning; they are related to both the molecular size of the chemical and its electronic features relevant to soil partitioning, as well as to the chemical's ability to form hydrogen bonds with water. A combination of different models from the GA-model population also allowed the proposal of predictions obtained by the better consensus model that, compared with published models and EPISuite predictions [105], are always among the best. The proposed models fulfill the fundamental points set down by OECD principles for the regulatory acceptability of a QSAR and could be reliably used as scientifically valid models in the REACH program.

The application of a single and general QSAR model, based on theoretical molecular descriptors for a large set of heterogeneous compounds, could be very useful for the screening of big data sets and for designing new chemicals, environmentally friendly as safer alternatives to dangerous chemicals.

12.7.2. Tropospheric Reactivity of Volatile Organic Compounds with Oxidants

The troposphere is the principal recipient of volatile organic compounds (VOCs) of both anthropogenic and biogenic origin. An indirect measure of the persistence of organic compounds in the atmosphere, and therefore a necessary parameter in environmental exposure assessment, is the rate at which these compounds react. The tropospheric lifetime of most organic chemicals, deriving from terrestrial emissions, is controlled by their degradation reaction with the OH radical and ozone during the daytime and NO_3 radicals at night.

In recent years, several QSAR/QSPR models predicting oxidation rate constants with tropospheric oxidants have been published and the different approaches to molecular description and the adopted methodology have been compared [13, 14, 18, 23, 106–117].

The most used method, implemented in AOPWIN of EPISUITE [118] for estimating tropospheric degradation by hydroxyl radicals is Atkinson's fragment contribution method [107]. New general MLR models of the OH radical reaction rate for a wide and heterogeneous data set of 460 volatile organic compounds (VOCs) were developed by the author's group [18]. The special feature of these models, in comparison to others, is the selection of theoretical molecular descriptors by a genetic algorithm as a variable subset selection procedure, their applicability to heterogeneous chemicals, and their validation for predictive purposes by both internal and external validation. External validation was performed by splitting the original data set by two different methods: the statistical experimental design pro-cedure (D-optimal distance) and the Kohonen self-organizing map (SOM); this was performed to verify the impact that the structural heterogeneity (in chemicals' split into training and prediction sets) has on model performance. The consequences on the model predictivity are also compared. D-optimal design, where the most dissimilar chemicals are always selected for the training set, leads to models with better predictive performance than models developed on the training set selected by SOM. The chemical applicability domain of the models and the reliability of the predictions are always verified by the leverage approach. The best proposed predictive model is based on four molecular descriptors and has the following equation (12-2):

$$\log\ k(OH) = 5.15(\pm 0.35) - 0.66(\pm 0.03)HOMO + 0.33(\pm 0.03)nX - 0.37$$
$$(\pm 0.04)CIC0 + (\pm 0.02)0.13\ nCaH$$

$$n(training) = 234\ R^2 = 0.83\ Q^2 = 0.82\ Q^2LMO(50\%) = 0.81\ RMSE = 0.473$$

$$n(test) = 226\ Q^2_{EXT} = 0.81\ RMSEp = 0.484\ K_{xx} = 33.8\%\ K_{xy} = 44.6\%\quad (12\text{-}2)$$

It is evident from the statistical parameters that the proposed model has good stability, robustness, and predictivity verified by internal (cross-validation by LOO and LMO) and also external validation. The influential chemicals are mainly the highly fluorinated chemicals, which have a strong structural peculiarity that the model is not able to capture. In Figure 12-3 the experimental values vs. those predicted by Eq. (12-2) are plotted.

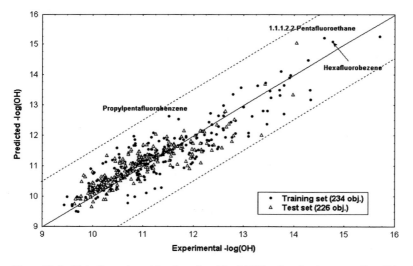

Figure 12-3. Plot of experimental and predicted log k(OH) values for the externally validated model by experimental design splitting. The training and test set chemicals are labeled differently, the outliers and influential chemicals are highlighted. The *dotted lines* indicate the 3σ interval (with copyright permission from [18])

The availability in the GA population of several possible models, similarly reliable for response prediction, also allowed the proposal of a consensus model which provides better predicted data than the majority of individual models, taking into account the more unique aspects of a particular structure.

While good models for OH rate constants are proposed in the literature for various chemical classes [107, 110–113, 115, 117], the modeling of reactivity with NO_3 radicals is more problematic. Most published QSAR models were obtained from separate training sets for aliphatic and aromatic compounds and the rate constants of aliphatic chemicals with NO_3 radicals were successfully predicted [106, 108, 109]; however, the models for aromatic compounds do not appear to be so satisfactory, often being only local models built on very small training sets and, consequently, without any reasonable applicability for data prediction.

New general QSAR models for predicting oxidation rate constants (kNO_3) for heterogeneous sets containing both aliphatic and aromatic compounds, based on few theoretical molecular descriptors (for instance, HOMO, number of aromatic rings, and an autocorrelation descriptor, MATS1m), were recently developed by the author's group [13, 23]. The models have high predictivity even on external chemicals, obtained by splitting the available data using different methods. The possibility of having molecular descriptors available for all chemicals (even those not yet synthesized), the good prediction performance of models applicable to a wide variety of aromatic and aliphatic chemicals, and the possibility of verifying the chemical domain of applicability by the leverage approach makes these useful models for producing reliable estimated NO_3 radical rate constants, when experimental parameters are not available.

The author has also proposed a predictive QSAR model of reaction rate with ozone for 125 heterogeneous chemicals [14]. The model, based on molecular descriptors, always selected by GA (HOMO–LUMO gap plus four molecular descriptors from DRAGON), has good predictive performance, also verified by statistical external validation on 42 chemicals not used for model development ($Q^2_{EXT}=0.904$, average RMS$=0.77$ log units). This model appears more predictive than the model previously proposed by Pompe and Veber [114], a six-parameter MLR model developed on 116 heterogeneous chemicals and based on molecular descriptors, calculated by the CODESSA software, selected by a stepwise selection procedure. The predictive performance of this model was verified only internally by cross-validation with 10 groups of validation ($Q^2=0.83$) and had an average RMS of 0.99 log units.

12.7.3. Biological Endpoints

12.7.3.1. *Bioconcentration Factor*

The bioconcentration factor (BCF) is an important parameter in environmental assessment as it is an estimate of the tendency of a chemical to concentrate and, consequently, to accumulate in an organism. The most common QSAR method, and the oldest, for estimating chemical bioconcentration is to establish correlations between

BCF and chemical hydrophobicity using K_{ow}, i.e., the n-octanol/water partition coefficient. A comparative study of BCF models based on log K_{ow} was performed by Devillers et al. [119]. Different models for BCF using theoretical molecular descriptors have been developed, among others: [120–124] and also by the author's group [8, 9, 27], with particular attention, as usual, to the external predictivity and the chemical applicability domain.

An example is the model reported by the following equation (12-3):

$$\log BCF = -0.74(\pm 0.35) + 2.55(\pm 0.13)^V I^M_{D,deg} - 1.09(\pm 0.11)HIC$$
$$-0.42(\pm 0.03)nHAcc - 1.22(\pm 0.17)GATS1e - 1.55(\pm 0.34)MATS1p$$

$$n_{(training)} = 179 \ R^2 = 0.81 \ Q^2_{LOO} = 0.79 \ Q^2_{BOOT} = 0.79$$
$$RMSE_{(train \ set)} = 0.56 \ RMSE_{(cross-val. \ set)} = 0.58$$

$$n_{(prediction)} = 59 \ Q^2_{EXT} = 0.87 \ RMSE_{(prediction \ set)} = 0.57 \qquad (12\text{-}3)$$

12.7.3.2.   Toxicity

Acute aquatic toxicity. The European Union's so-called "List 1" of priority chemicals dangerous for the aquatic environment (more than 100 heterogeneous chemicals) was modelled for ecotoxicological endpoints (aquatic toxicity on bacteria, algae, *Daphnia*, fish, mammals) [89] by different theoretical descriptors, mainly WHIM. In addition, WHIM descriptors were also satisfactory in the modeling of a more reduced set of toxicity data on *Daphnia* (49 compounds including amines, chlorobenzenes, organotin and organophosphorous pesticides) [125].

An innovative strategy for the selection of compounds with a similar toxicological mode of action was proposed as a key problem in the study of chemical mixtures (PREDICT European Research Project) [126]. A complete representation of chemical structures for phenylureas and triazines by different molecular descriptors (1D-structural, 2D-topological, 3D-WHIM) allowed a preliminary exploration of structural similarity based on principal components analysis (PCA), multidimensional scaling (MDS), and hierarchical cluster. The use of a genetic algorithm to select the most relevant molecular descriptors in modeling toxicity data makes it possible both to develop good predictive toxicity models and select the most similar phenylureas and triazines. The way of doing this is to apply chemometric approaches based only on molecular similarity related to toxicological mode of action.

The Duluth data set of toxicity data to *P. promelas* was recently studied by the author group [26] and new statistically validated MLR models were developed to predict the aquatic toxicity of chemicals classified according to their mode of action (MOA). Also, a unique general model for direct toxicity prediction (DTP model) was developed to propose a predictive tool with a wide applicability domain, applicable independently of *a priori* knowledge of the MOA of chemicals.

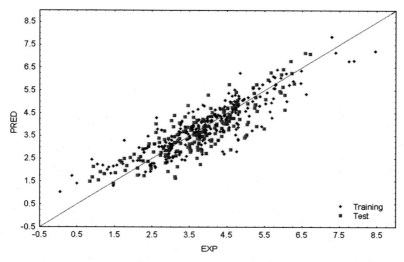

Figure 12-4. Plot of experimental and predicted toxicity values (*Pimephales promelas*) of the externally validated general-DTP log P-free model developed on a training set of 249 compounds (with copyright permission from [26])

The externally validated general-DTP log P-free model, reported below (Eq. 12-4) with statistical parameters, was developed on a training set of 249 compounds and applied for the prediction of the toxicity of 200 external chemicals, obtained by splitting the data by SOM (scatter plot in Figure 12-4):

$$\log(1/LC_{50})_{96h} = -2.54 + 0.91WA + 6.2Mv + 0.21nCb^-$$
$$+0.08H - 046 - 0.19MAXDP - 0.33nN$$

$$n_{training} = 249 \ R^2 = 0.79 \ Q^2_{LOO} = 0.78 \ Q^2_{BOOT} = 0.78 \ RMSE = 0.595$$

$$n_{test} = 200 \ Q^2_{EXT} = 0.71 \ RMSEcv = 0.613 \ RMSEp = 0.64 \qquad (12\text{-}4)$$

Chronic toxicity: mutagenicity. The potential for mutagenicity of chemicals of environmental concern, such as aromatic amines and PAHs, is of high relevance; many QSAR models, based on the mechanistic approach, have been published on this topic and reviewed by Benigni [5, 127].

With regard to this important topic, our group has published useful MLR models, always verified for their external predictivity on new chemicals, for the Ames test results on amines [12] and nitro-PAHs [20]. Externally validated classification models, by k-NN and CART, were also developed for the mutagenicity of benzocyclopentaphenanthrenes and chrysenes, determined by the Ames test [128], and PAH mutagenicity, determined on human B-lymphoblastoid [35].

Endocrine Disruption. A large number of environmental chemicals, known as endocrine disruptor chemicals (EDCs), are suspected of disrupting endocrine functions by mimicking or antagonizing natural hormones. Such chemicals may pose a serious threat to the health of humans and wildlife; they are thought to act through a variety of mechanisms, mainly estrogen receptor-mediated mechanisms of toxicity. Under the new European legislation REACH (http://europa.eu.int/comm/environment/chemicals/reach.htm) EDCs will require an authorization to be produced and used, if safer alternative are not available. However, it is practically impossible to perform thorough toxicological tests on all potential xenoestrogens, thus QSAR modeling has been applied by many other authors in these last years [129–142] providing promising methods for the estimation of a compound's estrogenic activity.

QSAR models of the estrogen receptor binding affinity of a large data set of heterogeneous chemicals have been built also in our laboratory using theoretical molecular descriptors [21, 33] giving full consideration, during model construction and assessment, to the new OECD principles for the regulatory acceptance of QSARs. A data set of 128 NCTR compounds (EDKB, http://edkb.fda.gov/databasedoor.html) including several different chemical categories, such as steroidal estrogens, synthetic estrogens, antiestrogens, phytoestrogens, other miscellaneous steroids, alkylphenols, diphenyl derivatives, organochlorines, pesticides, alkylhydroxybenzoate preservatives (parabens), phthalates, and a number of other miscellaneous chemicals, was studied. An unambiguous multiple linear regression (MLR) algorithm was used to build the models by selecting the modeling descriptors by a genetic algorithm. (Table 12-1 presents the statistical parameters of the best-selected model.) The predictive ability of the model was validated, as usually, by both internal and external validation, and the applicability domain was checked by the leverage approach to verify prediction reliability.

Twenty-one chemicals of the Kuiper data set [143] were used for external validation, with the following highly satisfying results: $R^2_{pred}=0.778$, $Q^2_{EXT}=0.754$, RMSE of prediction of 0.559 (Figure 12-5).

The results of several validation paths using different splitting methods performed in parallel (D-optimal design, SOM, random on activity sampling) give additional proof that the proposed QSAR model is robust and satisfactory (R^2_{pred} range: 0.761–0.807), thus providing a feasible and practical tool for the rapid screening of the estrogen activity of organic compounds, supposed endocrine disruptors chemicals.

On the same topic, satisfactory predictive models for the EDC classification based on different classification methods have been developed and recently proposed [33]. In this study, QSAR models were developed to quickly and effectively identify possible estrogen-like chemicals based on 232 structurally diverse chemicals from the NCTR database (training set) by using several non-linear classification methodologies (least square support vector machine (LS-SVM), counter propagation artificial neural network (CP-ANN), and k-nearest neighbor (kNN)) based on molecular structural descriptors. The models were validated externally with 87 chemicals (prediction set) not included in the training set. All three methods gave

Table 12-1. The MLR model between the structural descriptor and the log RBA of estrogens

Variable	Full name of variable	Reg.coeff.	Err.coeff.	Std.coeff.
Intercept		15.83	2.20	0.00
X2A	Average connectivity index chi-2	−43.75	5.28	−0.49
TIC1	Total information index (neighborhood symmetry first-order)	0.04	0.00	0.89
EEig02d	Eigenvalue 2 from edge adjacency matrix weighted dipole moments	−2.67	0.31	−0.56
JGI10	Mean topological charge index of order 10	79.92	10.85	0.32
SPH	Spherosity index	2.60	0.56	0.24
E1u	The first component accessibility directional WHIM index/unweighted	−7.12	1.57	−0.25
RTm+	R maximal index weighed by atomic masses	4.78	0.74	0.28
nArOR	The number of aromatic ether groups	−1.25	0.15	−0.39

Model parameters: $n=128$, $R^2=0.824$, $R^2_{adj}=0.812$, $Q^2_{LOO}=0.793$, $Q^2_{BOOT}=0.780$, RMSEcv=0.7484, RMSEp=0.8105, $K_x=35.13$, $K_{xy}=37.89$, and $s=0.7762$.

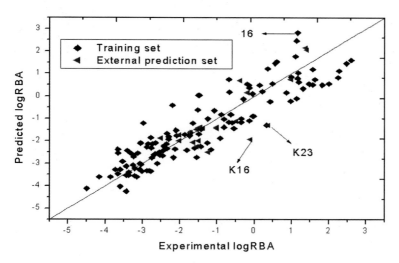

Figure 12-5. Predicted Log RBA values vs. experimental values for the original data set of estrogens (NCTR data set) and external prediction set (Kuiper's data set) (with copyright permission from [21])

satisfactory prediction results both for training and prediction sets; the most accurate model was obtained by the LS-SVM approach. The highly important feature of all these models is their low false negative percentage, useful in a precautionary approach. Our models were also applied to about 58,000 discrete organic chemicals from US-EPA; about 76% were predicted, by each model, not to bind to an estrogen receptor.

The obtained results indicate that the proposed QSAR models are robust, widely applicable, and could provide a feasible and practical tool for the rapid screening of potential estrogens. It is very useful information to prioritize chemicals for more expensive assays. In fact, the common 40,300 negative compounds could be excluded from the potential estrogens without experiments and a high accuracy (low false negative value).

A review on the applications of machine learning algorithms in the modeling of estrogen-like chemicals has been recently published [144].

12.8. MODELING MORE THAN A SINGLE ENDPOINT

12.8.1. PC Scores as New Endpoints: Ranking Indexes

The environment is a highly complex system in which many parameters are of contemporaneous relevance: the understanding, rationalization, and interpretation of their covariance are the principal pursuit of any environmental researcher. Indeed, environmental chemistry deals with the behavior of chemicals in the environment, behavior which is regulated by many different variables such as physico-chemical properties, chemical reactivity, biological activity.

The application of explorative methods of multivariate analysis to various topics of environmental concern allows a combined view that generates ordination and grouping of the studied chemicals, in addition to the discovering of variable relationships. Any problem related to chemical behavior in the environment can be analyzed by multivariate explorative techniques, the outcome being to obtain chemical screening and ranking according to the studied properties, reactivities, or activities and, finally, the proposal of an index.

This was the starting point, and also the central core, of most of the author 15-year research of QSAR modeling at Insubria University.

The significant combination of variables from multivariate analysis can be used as a score value (a cumulative index), and modelled as a new endpoint by the QSAR approach to exploit already available information concerning chemical behavior, and to propose models able to predict such behavior for chemicals for which the same information is not yet known, or even for new chemicals before their synthesis. In fact, our QSAR approach, both for modeling quantitative response by regression methods and qualitative response by classification methods, is based on theoretical molecular descriptors that can be calculated for any drawn chemicals starting from the atomic coordinates, thus without the knowledge of any experimental parameter.

12.8.2. Multivariate Explorative Methods

The principal aim of any explorative technique is to capture the information available in any multivariate context and condense it into a more easily interpretable view (a score value or a graph). Thus, from these exploratory tools a more focused investigation can be made into chemicals of higher concern, directing the next investigative steps or suggesting others. Some of the more commonly used exploratory techniques are commented on here and applied in environmental chemistry and ecotoxicology.

12.8.2.1. Principal Component Analysis

Probably the most widely known and used explorative multivariate method is principal component analysis (PCA) [145, 146] (Chapter 6). In PCA, linear combinations of the studied variables are created, and these combinations explain, to the greatest possible degree, the variation in the original data. The first principal component (PC1) accounts for the maximum amount of possible data variance in a single variable, while subsequent PCs account for successively smaller quantities of the original variance. Principal components are derived in such a way that they are orthogonal. Indeed, it is good practice, especially when the original variables have different ranges of scales, to derive the principal components from the standardized data (mean of 0 and standard deviation of 1), i.e., via the correlation matrix. In this way all the variables are treated as if they are of equal importance, regardless of their scale of measurement. To be useful, it is desirable that the first two PCs account for a substantial proportion of the variance in the original data, thus they can be considered sufficiently representative of the main information included in the data, while the remaining PCs condense irrelevant information or even experimental noise. It is quite common for a PCA to be represented by a score plot, loading plot, or biplot, defined as the joint representation of the rows and columns of a data matrix; points (scores) represent the chemicals and vectors or lines represent the variables (loadings). The lengths of the vectors indicate the information associated with the variable, while the cosine of the angle between the vectors reflects their correlation. In our environmental chemistry studies, PCA has been widely used for screening and ranking purposes in many contexts: (a) tropospheric degradability of volatile organic compounds (VOCs) [11, 17, 106]; (b) mobility in the atmosphere or long-range transport of persistent organic pollutants (POPs) [16, 31, 147]; (c) environmental partitioning tendency of pesticides [7, 32]; (d) POP and PBT screening [10, 24, 34, 147–149].

In addition, this multivariate approach was adopted to study aquatic toxicity of EU-priority listed chemicals on different endpoints [150] and esters [25], the endocrine disrupting activity based on three different endpoints [33] and the abiotic oxidation of phenols in an aqueous environment [9].

12.8.2.2. QSAR Modeling of Ranking Indexes

Tropospheric Persistence/Degradability of Volatile Organic Compounds (VOCs). Studies has been made of the screening/ranking of volatile organic chemicals according to their tendency to degrade in the troposphere. Indeed, as the atmospheric persistence of a chemical is mainly dependent on the degradation rates of its reaction with oxidants, the contemporaneous variation and influence of the rate constants for their degradation by OH, NO_3 radicals, and ozone (kOH, kNO_3, and kO_3), in determining the inherent tendency to degradability, was explored by principal component analysis (PCA).

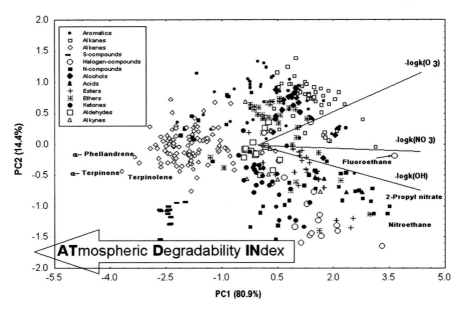

Figure 12-6. Score plot and loading plot of the two principal component analysis of three rate constants (kOH, kNO3, kO3) for 399 chemicals (labeled according to chemical classes). ATDIN: ATmospheric Degradability INdex. Cumulative explained variance: 95.3%. Explained Variance of PC1 (ATDINdex)=80.9% (with copyright permission from [17])

In a preliminary study, the experimental data allowed the ranking of a set of 65 heterogeneous VOCs, for which all the degradation rate constants were known; an atmospheric persistence index (ATPIN) had been defined and modelled by theoretical molecular descriptors [11]. Later, the application of our MLR models, developed for each studied degradation rate constant (kNO3, kO3, and kOH) [13, 14, 18], allowed a similar PC analysis (Figure 12-6) of a much larger set of 399 chemicals.

This new more informative index (PC1 score of Figure 12-6, 80.9% of explained variance, newly defined ATDIN – atmospheric degradability index), based on a wider set of more structurally heterogeneous chemicals, was also satisfactorily modelled by MLR based on theoretical molecular descriptors and externally validated (Q^2 0.94; Q^2_{EXT} 0.92) (scatter plot in Figure 12-7) [17].

Mobility in Atmosphere and Long-Range Transport of Persistent Organic Pollutants (POPs). The intrinsic tendency of compounds toward global mobility in the atmosphere has been studied, since it is a necessary property for the evaluation of the long-range transport (LRT) of POPs [16, 31]. As the mobility potential of a chemical depends on the various physico-chemical properties of a compound, principal component analysis was used to explore the contemporaneous variation and influence of all the properties selected as being the most relevant to LRT potential (such as vapor pressure, water solubility, boiling point, melting point, temperature of condensation, various partition coefficients among different compartments;

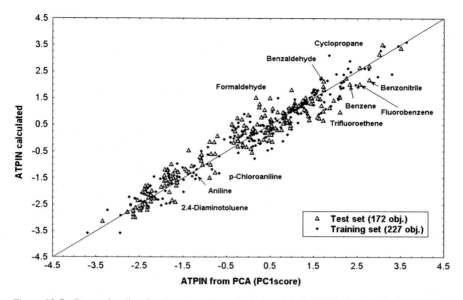

Figure 12-7. Regression line for the externally validated model of ATPIN (ATmospheric Persistence Index: the opposite of ATDIN). The training and test set chemicals are differently highlighted, the outliers and influential chemicals are named (with copyright permission from [17])

for instance, Henry's law constant, octanol/water partition coefficient, soil sorption coefficient, octanol/air partition coefficient).

A simple interpretation of the obtained PC1 is as a scoring function of intrinsic tendency toward global mobility. We have proposed this PC1 scoring as the ranking score for the 82 possible POPs in four *a priori* classes: high, relatively high, relatively low, and low mobility.

These classes have been successfully modelled by the CART method, based on four theoretical molecular descriptors (two Kier and Hall connectivity indexes, molecular weight, and sum of electronegativities) with only 6% of errors in cross-validation. The main aim was to develop a simple and rapid framework to screen, rank, and classify also new organic chemicals according to their intrinsic global mobility tendency, just from the knowledge of their chemical structure.

An analogous approach was previously applied to a subset of 52 POPs to define a long-range transport (LRT) index derived from the PC1 score, on the basis of physico-chemical properties and additionally taking into account atmospheric half-life data [147].

Environmental partitioning tendency of pesticides. The partitioning of pesticides into different environmental compartments depends mainly on the physico-chemical properties of the studied chemical, such as the organic carbon partition coefficient (K_{oc}), the n-octanol/water partition coefficient (K_{ow}), water solubility (S_w), vapor pressure (V_p), and Henry's law constant (H). To rank and classify the 54 studied pesticides, belonging to various chemical categories, according to their distribution

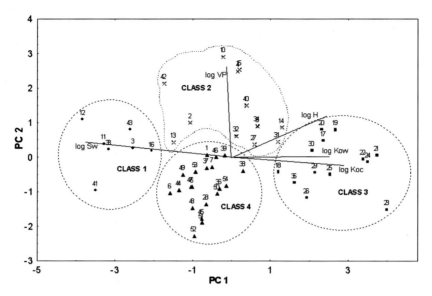

Figure 12-8. Score plot and loading plot of the two first principal components of PCA of five physico-chemical properties (K_{oc}, K_{ow}, S_w, V_p, and Henry's law constant) for 54 pesticides. Cumulative explained variance: 94.6%; explained variance of PC1: 70.1% (with copyright permission from [32])

tendency in various media, we applied [32] a combination of two multivariate approaches: principal component analysis (Figure 12-8) for ranking and hierarchical cluster analysis for the definition of the four *a priori* classes, according to their environmental behavior (1. soluble, 2. volatile, 3. sorbed, and 4. non-volatile/medium class) (*circles* in Figure 12-8).

The pesticides were finally assigned to the defined four classes by different classification methods (CART, k-NN, RDA) using theoretical molecular descriptors (for example, the CART tree is reported in Figure 12-9). Two of the selected molecular descriptors are quite easily interpretable, in particular (a) MW encodes information on molecule dimension; it is well known that big molecules have the greatest tendency to bind, by van der Waals forces, to the organic component of the soil, becoming the most sorbed in organic soils but the least soluble in water (Class 3) and (b) the possibility of a chemical to link by hydrogen bonds to water molecules (encoded in the molecular descriptor nHDon) results in the higher solubility of the Class 1 pesticides; furthermore, chemicals with fewer intramolecular hydrogen bonds are the most volatile (Class 2). The last topological descriptor J, that discriminates Class 4 of the medium-behavior pesticides, is not easily interpretable.

A wider, heterogeneous, and quite representative data set of pesticides of different chemical classes (acetanilides, carbamates, dinitroanilines, organochlorides, organophosphates, phenylureas, triazines, triazoles), already studied for their K_{oc}

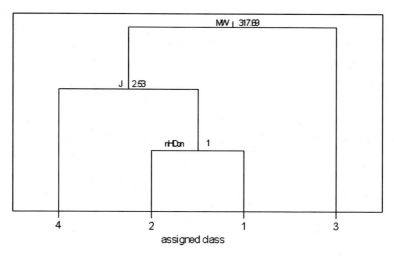

Figure 12-9. Classification tree by classification and regression tree (CART) of mobility classes for 54 pesticides. Error rate (ER) 11.11%; ER in prediction: 18.53%; NoMER: 62.96% (with copyright permission from [32])

modeling [96] has also undergone PC analysis of various environmental partitioning properties (solubility, volatility, partition coefficients, etc.) to study leaching tendency [7]. The resultant macrovariables, PC1 and PC2 scores, called the leaching index (LIN) and volatility index (VIN), have been proposed as cumulative environmental partitioning indexes in different media. These two indexes were modelled by theoretical molecular descriptors with satisfactory predictive power (Q^2 leave-30%-out=0.85 for LIN). Such an approach allows a rapid pre-determination and the screening of the environmental distribution of pesticides, starting only from the molecular structure of the pesticide without any *a priori* knowledge of the physico-chemical properties.

The proposed index LIN was used in a comparative analysis with GUS and LEACH index for highlighting the pesticides most dangerous to the aquatic compartment among those widely used in Uzbekistan, in the Amu-Darya river basin [151].

POPs and PBTs. QSAR approaches, based on molecular structure for the prioritization of chemicals for persistence, particularly persistent organic pollutants (POPs) screening and ranking method for global half-life, have recently been proposed [10, 24, 148, 149].

Persistence in the environment is an important criterion in prioritizing hazardous chemicals and in identifying new persistent organic pollutants (POPs). Degradation half-life in various compartments is among the more commonly used criteria for studying environmental persistence, but the limited availability of experimental data or reliable estimates is a serious problem. Available half-life data for degradation in air, water, sediment, and soil, for a set of 250 organic POP-type chemicals, have

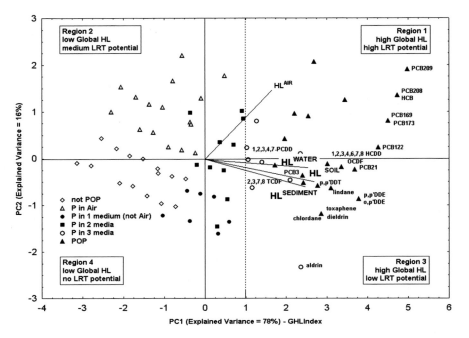

Figure 12-10. Principal component analysis on half-life data for 250 organic compounds in the various compartments (air, water, sediment, and soil) (PC1–PC2: explained variance=94%). P=persistent (with copyright permission from [10])

been combined in a multivariate approach by principal component analysis. This PCA distributes the studied compounds according to their cumulative, or global, half-life and relative persistence in different media, to obtain a ranking of the studied organic pollutants according to their relative overall half-life.

The biplot relative to the first and second components is reported in Figure 12-10, where the chemicals (points or scores) are distributed according to their environmental persistence, represented by the linear combination of their half lives in the four selected media (loadings shown as lines). The cumulative explained variance of the first two PCs is 94%, and the PC1 alone provides the largest part, 78%, of the total information. The loading lines show the importance of each variable in the first two PCs.

It is interesting to note that all the half-life values (lines) are oriented in the same direction along the first principal component, thus PC1, derived from a linear combination of half-life in different media, is a new macro-variable condensing chemical tendency to environmental persistence. PC1 ranks the compounds according to their cumulative half-life and discriminates between them with regard to persistence; chemicals with high half-life values in all the media (highlighted in the PCA graph) are located to the right of the plot, in the zone of global higher persistence (very persistent chemicals anywhere); chemicals with a lower global half-life fall to the left of the graph, not being persistent in any medium (labeled in Figure 12-10) or persistent in only one medium; chemicals persistent in 2 or 3 media are located in the intermediate zone of Figure 12-10.

PC2, although less informative (E.V. 16%), is also interesting; it separates the compounds more persistent in air (upper parts in Figure 12-10, regions 1 and 2), i.e., those with higher LRT potential from those more persistent in water, soil, and sediment (lower parts in Figure 12-10, regions 3 and 4).

A deeper analysis of the distribution of the studied chemicals gives some interesting results and confirms experimental evidence: to the right, among the very persistent chemicals in all the compartments (*full triangles* in Figure 12-10), we find most of the compounds recognized as POPs by the Stockholm Convention [152]. Highly chlorinated PCBs and hexachlorobenzene are among the most persistent compounds in our reference scenario. All these compounds are grouped in Region 1 owing to their global high persistence, especially in air. The less chlorinated PCBs (PCB-3 and PCB 21) fall in the zone of very persistent chemicals, but not in the upper part of Region 1, due to their lower persistence in air compared with highly chlorinated congeners. p,p'-DDT, p,p'-DDE, o,p'-DDE, highly chlorinated dioxins and dioxin-like compounds, as well as pesticides toxaphene, lindane, chlordane, dieldrin, and aldrin fall in Region 3 (highly persistent chemicals mainly in compartments different from air).

A global half-life index (GHLI) obtained from existing knowledge of generalized chemical persistence over a wide scenario of 250 chemicals, which reliability was verified through comparison with multimedia model results and empirical evidence, was proposed from this PC analysis [10]. This global index, the PC1 score, was then modelled as a cumulative endpoint using a QSAR approach based on theoretical molecular descriptors; a simple and robust regression model externally validated for its predictive ability [6, 84] has been derived. The original set of available data was first randomly split into training and prediction sets; 50% of the compounds were put into the prediction set (125 compounds) while the other 50% was used to build the QSPR model by MLR. Given below (Eq. 12-5) is the best QSPR model, selected by statistical approaches and its statistical parameters (Figure 12-11 shows the plot of GHLI values from PCA vs. predicted GHLI values):

$$\text{GHL Index} = -3.12(\pm 0.77) + 0.33(\pm 4.5E - 2)X0v + 5.1(\pm 0.99)Mv - 0.32$$
$$(\pm 6.13E - 2)MAXDP - 0.61(\pm 0.10)nHDon - 0.5(\pm 1.15)CIC0$$
$$-0.61(\pm 0.13)O - 060$$

$$n_{training} = 125 \ R^2 = 0.85 \ Q^2_{LOO} = 0.83 \ Q^2_{BOOT} = 0.83$$
$$RMSE = 0.76 \ RMSEcv = 0.70;$$

$$n_{prediction} = 125 \ R^2_{EXT} = 0.79 \ RMSEp = 0.78 \tag{12-5}$$

This model presents good internal and external predictive power, a result that must be highlighted as proof of model robustness and real external predictivity.

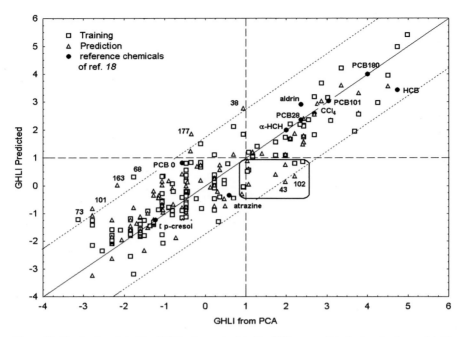

Figure 12-11. Scatter plot of the GHLI values calculated by PCA vs. predicted values by the model. The GHLI values for the training and prediction set chemicals are labeled differently. The *diagonal dotted lines* indicate the 2.5σ interval and response outliers are numbered. *Vertical and horizontal dotted lines* identify the cut-off value of GHLI=1 for high-persistent chemicals (with copyright permission from [10])

The only really dangerous zone in the proposed model is the underestimation zone (*circled* in Figure 12-11).

The application of this model, using only a few structural descriptors, could allow a fast preliminary identification and prioritization of not yet known POPs, just from the knowledge of their molecular structure. The proposed multivariate approach is particularly useful not only to screen and to make an early prioritization of environmental persistence for pollutants already on the market, but also for compounds not yet synthesized, which could represent safer alternative and replacement solutions for recognized POPs. No method other than QSAR is applicable to detect the potential persistence of new compounds.

Similarly, highly predictive classification models, based on k-NN, CART, and CP-ANN, have been developed and can be usefully applied for POP pre-screening. The *a priori* classes have been defined by applying hierarchical cluster analysis to the half-life data [34].

An approach analogous to GHLI has been successfully applied to the PCA-combination of data obtained from the above cumulative half-lives for persistence GHLI, bioconcentration data of fish, and acute toxicity data of *P. promelas* in order to propose, and then model by QSPR approach, a combined index of PBT behavior

[24, 148, 149]. A simple model, based on easy calculable molecular descriptors and with high external predictivity ($Q^2_{EXT} > 0.8$), has been developed and will be published. This PBT index can be applied also to chemicals without any experimental data and even to not yet synthesized compounds.

These QSAR-based tools, validated for their predictivity on new chemicals, could help in highlighting the POP and PBT behavior also of chemicals not yet synthesized, and could be usefully applied for the new European Regulation REACH, which requires most demanding authorization steps for PBTs and the design of safer alternatives. The results of our predictions were comparable with those from the US-EPA PBT profiler (http://www.epa.gov/pbt/tools/toolbox.htm).

12.9. CONCLUSIONS

A statistical approach to QSAR modeling, based on heterogeneous theoretical molecular descriptors and chemometric methods and developed with the fundamental aim of predictive applications, has been introduced and discussed in this review. Several applications to environmentally relevant topics related to organic pollutants, performed by the Insubria QSAR Research Unit in last 15 years, have been presented. Different endpoints related to physico-chemical properties, persistence, bioaccumulation, and toxicity have been modelled, not only singularly, but also as combined endpoints, obtained by multivariate analysis; the approach is innovative and highly useful for ranking and prioritizing purposes. All the proposed models characteristically check the predictive performance and applicability domain of the chemicals, even new chemicals that never participated in the model development. The fulfillment of the "OECD principles for QSAR validation" is a guarantee for the reliability of the predicted data obtained by our models and their possible applicability in the context of REACH.

ACKNOWLEDGEMENT

Many thanks to my collaborators who participated in the research, reviewed here, carried out over the past 15 years, particularly Ester Papa and Pamela Pilutti. Thanks are also due to Roberto Todeschini who was my teacher of chemometric QSAR.

REFERENCES

1. Hansch C, Fujita T (1964) *p*-s-p analysis. A method for the correlation of biological activity and chemical structure. J Am Chem Soc 86:1616–1626
2. Hansch C, Leo A (1995) Exploring QSAR: Fundamentals and applications in chemistry and biology. American Chemical Society, Washington, DC 490–496
3. Schultz TW, Cronin MTD, Netzeva TI et al. (2002) Structure–toxicity relationships for aliphatic chemicals evaluated with Tetrahymena pyriformis. Chem Res Toxicol 15:1602–1609
4. Veith GD, Mekenyan O (1993) A QSAR approach for estimating the aquatic toxicity of soft electrophiles (QSAR for soft electrophiles). Quant Struct -Act Relat 12:349–356
5. Benigni R (2005) Structure–activity relationship studies of chemical mutagens and carcinogens: Mechanistic investigations and prediction approaches. Chem Rev 105:1767–1800

6. Tropsha A, Gramatica P, Gombar VK (2003) The importance of being earnest: Validation is the absolute essential for successful application and interpretation of QSPR models. QSAR Comb Sci 22:69–77

7. Gramatica P, Di Guardo A (2002) Screening of pesticides for environmental partitioning tendency. Chemosphere 47:947–956

8. Gramatica P, Papa E (2003) QSAR modeling of bioconcentration factor by theoretical molecular descriptors. QSAR Comb Sci 22:374–385

9. Gramatica P, Papa E (2005) An update of the BCF QSAR model based on theoretical molecular descriptors. QSAR Comb Sci 24:953–960

10. Gramatica P, Papa E (2007) Screening and ranking of POPs for global half-life: QSAR approaches for prioritization based on molecular structure. Environ Sci Technol 41:2833–2839

11. Gramatica P, Pilutti P, Papa E (2002) Ranking of volatile organic compounds for tropospheric degradability by oxidants: A QSPR approach. SAR QSAR Environ Res 13:743–753

12. Gramatica P, Consonni V, Pavan M (2003) Prediction of aromatic amines mutagenicity from theoretical molecular descriptors. SAR QSAR Environ Res 14:237–250

13. Gramatica P, Pilutti P, Papa E (2003) Predicting the NO3 radical tropospheric degradability of organic pollutants by theoretical molecular descriptors. Atmos Environ 37:3115–3124

14. Gramatica P, Pilutti P, Papa E (2003) QSAR prediction of ozone tropospheric degradation. QSAR Comb Sci 22:364–373

15. Gramatica P, Battaini F, Papa E (2004) QSAR prediction of physico-chemical properties of esters. Fresenius Environ Bull 13:1258–1262

16. Gramatica P, Papa E, Pozzi S (2004) Prediction of POP environmental persistence and long range transport by QSAR and chemometric approaches. Fresenius Environ Bull 13:1204–1209

17. Gramatica P, Pilutti P, Papa E (2004) A tool for the assessment of VOC degradability by tropospheric oxidants starting from chemical structure. Atmos Environ 38:6167–6175

18. Gramatica P, Pilutti P, Papa E (2004) Validated QSAR prediction of OH tropospheric degradation of VOCs: Splitting into training-test sets and consensus modeling. J Chem Inf Comput Sci 44:1794–1802

19. Gramatica P, Giani E, Papa E (2007) Statistical external validation and consensus modeling: A QSPR case study for K_{oc} prediction. J Mol Graph Model 25:755–766

20. Gramatica P, Pilutti P, Papa E (2007) Approaches for externally validated QSAR modelling of nitrated polycyclic aromatic hydrocarbon mutagenicity. SAR QSAR Environ Res 18:169–178

21. Liu H, Papa E, Gramatica P (2006) QSAR prediction of estrogen activity for a large set of diverse chemicals under the guidance of OECD principles. Chem Res Toxicol 19:1540–1548

22. Liu H, Papa E, Gramatica P (2008) Evaluation and QSAR modeling on multiple endpoints of estrogen activity based on different bioassays. Chemosphere 70:1889–1897

23. Papa E, Gramatica P (2008) Externally validated QSPR modelling of VOC tropospheric oxidation by NO3 radicals. SAR QSAR Environ Res 19:655–668

24. Papa E, Gramatica P (2009) QSPR as a support to the EU REACH legislation: PBTs identification by molecular structure. Environ Sci Technol (in press)

25. Papa E, Battaini F, Gramatica P (2005) Ranking of aquatic toxicity of esters modelled by QSAR. Chemosphere 58:559–570

26. Papa E, Villa F, Gramatica P (2005) Statistically validated QSARs, based on theoretical descriptors, for modeling aquatic toxicity of organic chemicals in *Pimephales promelas* (Fathead Minnow). J Chem Inf Model 45:1256–1266

27. Papa E, Dearden JC, Gramatica P (2007) Linear QSAR regression models for the prediction of bioconcentration factors by physicochemical properties and structural theoretical molecular descriptors. Chemosphere 67:351–358

28. Papa E, Kovarich S, Gramatica P (2009) Development, validation and inspection of the applicability domain of QSPR models for physico-chemical properties of polybrominated diphenyl ethers. QSAR Comb Sci. doi: 10.1002/qsar.200860183

29. Todeschini R, Maiocchi A, Consonni V (1999) The *K* correlation index: Theory development and its application in chemometrics. Chemom Int Lab Syst 46:13–29

30. Hawkins DM (2004) The problem of overfitting. J Chem Inf Comput Sci 44:1–12

31. Gramatica P, Pozzi S, Consonni V et al. (2002) Classification of environmental pollutants for global mobility potential. SAR QSAR Environ Res 13:205–217

32. Gramatica P, Papa E, Battaini F (2004) Ranking and classification of non-ionic organic pesticides for environmental distribution: A QSAR approach. Int J Environ Anal Chem 84:65–74

33. Liu H, Papa E, Walker JD et al. (2007) *In silico* screening of estrogen-like chemicals based on different nonlinear classification models. J Mol Graph Model 26:135–144

34. Papa E, Gramatica P (2008) Screening of persistent organic pollutants by QSPR classification models: A comparative study. J Mol Graph Model 27:59–65

35. Papa E, Pilutti P, Gramatica P (2008) Prediction of PAH mutagenicity in human cells by QSAR classification. SAR QSAR Environ Res 19:115–127

36. Breiman L, Friedman JH, Olshen RA et al. (1998) Classification and regression trees. Chapman & Hall/CRC, Boca Raton, FL

37. Frank JE, Friedman JH (1989) Classification: Oldtimers and newcomers. J Chemom 3:463–475

38. Sharaf MA, Illman DL, Kowalski BR (1986) Chemometrics. Wiley, New York

39. Gasteiger J, Zupan J (1993) Neural networks in chemistry. Angew Chem Int Ed Engl 32:503–527

40. Hecht-Nielsen R (1988) Applications of counter-propagation networks. Neural Netw 1:131–139

41. Zupan J, Novic M, Ruisanchez I (1997) Kohonen and counter-propagation artificial neural networks in analytical chemistry. Chemom Int Lab Syst 38:1–23

42. Livingstone DJ (2000) The characterization of chemical structures using molecular properties. A survey. J Chem Inf Comput Sci 40:195–209

43. Stuper AJ, Jurs PC (1976) ADAPT: A computer system for automated data analysis using pattern recognition techniques. J Chem Inf Comput Sci 16:99–105

44. Mekenyan O, Bonchev D (1986) OASIS method for predicting biological activity of chemical compounds. Acta Pharm Jugosl 36:225–237

45. Katritzky AR, Lobanov VS (1994) CODESSA. Ver. 5.3, University of Florida, Gainesville

46. Todeschini R, Consonni V, Mauri A et al. (2006) DRAGON – software for the calculation of molecular descriptors. Ver. 5.4 for Windows, Talete srl, Milan, Italy

47. MollConnZ (2003) Ver. 4.05. Hall Ass. Consult., Quincy, MA

48. Devillers J, Balaban AT (1999) Topological indices and related descriptors in QSAR and QSPR. Gordon and Breach Science Publishers, Amsterdam

49. Karelson M (2000) Molecular descriptors in QSAR/QSPR. Wiley-InterScience, New York

50. Todeschini R, Consonni V (2000) Handbook of molecular descriptors. Wiley-VCH, Weinheim, Germany

51. Todeschini R, Lasagni M (1994) New molecular descriptors for 2D and 3D structures. J Chemom 8:263–272

52. Todeschini R, Gramatica P (1997) 3D-modelling and prediction by WHIM descriptors.5. Theory development and chemical meaning of WHIM descriptors. Quant Struct Act Relat 16:113–119

53. Todeschini R, Gramatica P (1997) 3D-modelling and prediction by WHIM descriptors.6. Application of WHIM descriptors in QSAR studies. Quant Struct Act Relat 16:120–125

54. Consonni V, Todeschini R, Pavan M et al. (2002) Structure/response correlations and similarity/diversity analysis by GETAWAY descriptors. 1. Theory of the novel 3D molecular descriptors. J Chem Inf Comput Sci 42:682–692.

55. Consonni V, Todeschini R, Pavan M et al. (2002) Structure/response correlations and similarity/diversity analysis by GETAWAY descriptors. 2. Application of the novel 3D molecular descriptors to QSAR/QSPR studies. J Chem Inf Comput Sci 42:693–705
56. Xu L, Zhang WJ (2001) Comparison of different methods for variable selection. Anal Chim Acta 446:477–483
57. Davis L (1991) Handbook of genetic algorithms. Van Nostrand Reinhold, New York
58. Hibbert DB (1993) Genetic algorithms in chemistry. Chemom Int Lab Syst 19:277–293
59. Wehrens R, Buydens LMC (1998) Evolutionary optimisation: A tutorial. TRAC 17:193–203
60. Leardi R, Boggia R, Terrile M (1992) Genetic algorithms as a strategy for feature-selection. J Chemom 6:267–281
61. Rogers D, Hopfinger AJ (1994) Application of genetic function approximation to quantitative structure–activity relationships and quantitative structure–property relationships. J Chem Inf Comput Sci 34:854–866
62. Devillers J (1996) Genetic algorithms in computer-aided molecular design. In: Devillers J (ed) Genetic algorithms in molecular modeling. Academic Press Ltd, London
63. Leardi R (1994) Application of a genetic algorithm to feature-selection under full validation conditions and to outlier detection. J Chemom 8:65–79
64. Kubinyi H (1994) Variable selection in QSAR studies. 1. An evolutionary algorithm. Quant Struct Act Relat 13:285–294
65. Kubinyi H (1994) Variable selection in QSAR studies. 2. A highly efficient combination of systematic search and evolution. Quant Struct Act Relat 13:393–401
66. Todeschini R, Consonni V, Pavan M et al. (2002) MOBY DIGS. Ver. 1.2 for Windows, Talete srl, Milan, Italy
67. Kubinyi H (1996) Evolutionary variable selection in regression and PLS analyses. J Chemom 10:119–133
68. Guha R, Serra JR, Jurs PC (2004) Generation of QSAR sets with a self-organizing map. J Mol Graph Model 23:1–14
69. Netzeva TI, Worth AP, Aldenberg T et al. (2005) Current status of methods for defining the applicability domain of (quantitative) structure–activity relationships – the report and recommendations of ECVAM Workshop 52. ATLA 33:155–173
70. Tunkel J, Mayo K, Austin C et al. (2005) Practical considerations on the use of predictive models for regulatory purposes. Environ Sci Technol 39:2188–2199
71. Atkinson AC (1985) Plots, transformations and regression. Clarendon Press, Oxford
72. Hulzebos EM, Posthumus R (2003) (Q)SARs: Gatekeepers against risk on chemicals? SAR QSAR Environ Res 14:285–316
73. Jouan-Rimbaud D, Massart DL, deNoord OE (1996) Random correlation in variable selection for multivariate calibration with a genetic algorithm. Chemom Int Lab Syst 35:213–220
74. Efron B, Gong G (1983) A leisurely look at the bootstrap, the jackknife, and cross-validation. Am Stat 37:36–48
75. Efron B, Tibshirani RJ (1993) An introduction to the bootstrap. Chapman & Hall, London
76. Shao J (1993) Linear-model selection by cross-validation. J Am Stat Assoc 88:486–494
77. Golbraikh A, Tropsha A (2002) Predictive QSAR modeling based on diversity sampling of experimental datasets for the training and test set selection. J Comput Aid Mol Des 16:357–369
78. Zefirov NS, Palyulin VA (2001) QSAR for boiling points of "small" sulfides. Are the "high-quality structure-property-activity regressions" the real high quality QSAR models? J Chem Inf Comput Sci 41:1022–1027
79. Golbraikh A, Shen M, Xiao ZY et al. (2003) Rational selection of training and test sets for the development of validated QSAR models. J Comput Aid Mol Des 17:241–253
80. Leonard JT, Roy K (2006) On selection of training and test sets for the development of predictive QSAR models. QSAR Comb Sci 25:235–251

81. Sjostrom M, Eriksson L (1995) Chemometric methods in molecular design. van de Waterbeend H (ed) Vol. 2. VCH, New York, p 63
82. Marengo E, Todeschini R (1992) A new algorithm for optimal, distance-based experimental-design. Chemom Int Lab Syst 16:37–44
83. Gramatica P (2004) Evaluation of different statistical approaches to the validation of quantitative structure-activity relationships. http://ecb.jrc.it/DOCUMENTS/QSAR/Report_on_QSAR_validation_methods.pdf Accessed April 2008
84. Gramatica P (2007) Principles of QSAR models validation: Internal and external. QSAR Comb Sci 26:694–701
85. Kahn I, Fara D, Karelson M et al. (2005) QSPR treatment of the soil sorption coefficients of organic pollutants. J Chem Inf Model 45:94–105
86. Todeschini R, Gramatica P, Provenzani R et al. (1995) Weighted holistic invariant molecular descriptors. 2. Theory development and applications on modeling physicochemical properties of polyaromatic hydrocarbons. Chemom Int Lab Syst 27:221–229
87. Chiorboli C, Gramatica P, Piazza R et al. (1997) 3D-modelling and prediction by WHIM descriptors. Part 7. Physico-chemical properties of haloaromatics: Comparison between WHIM and topological descriptors. SAR QSAR Environ Res 7:133–150
88. Gramatica P, Navas N, Todeschini R (1998) 3D-modelling and prediction by WHIM descriptors. Part 9. Chromatographic relative retention time and physico-chemical properties of polychlorinated biphenyls (PCBs). Chemom Int Lab Syst 40:53–63
89. Todeschini R, Vighi M, Finizio A et al. (1997) 3D-modelling and prediction by WHIM descriptors. Part 8. Toxicity and physico-chemical properties of environmental priority chemicals by 2D-TI and 3D-WHIM descriptors. SAR QSAR Environ Res 7:173–193
90. Todeschini R, Gramatica P (1997) The WHIM theory: New 3D-molecular descriptors for QSAR in environmental modelling. SAR QSAR Environ Res 7:89–115
91. Todeschini R, Gramatica P (1998) 3D-QSAR in drug design. Kubiny H, Folkers G, Martin YC (eds) vol. 2. KLUWER/ESCOM, Dordrecht, p 355
92. Gramatica P (2006) WHIM descriptors of shape. QSAR Comb Sci 25:327–332
93. Patel H, Cronin MTD (2001) A novel index for the description of molecular linearity. J Chem Inf Comput Sci 41:1228–1236
94. Nikolova N, Jaworska J (2003) Approaches to measure chemical similarity – a review. QSAR Comb Sci 22:1006–1026
95. Gramatica P, Navas N, Todeschini R (1999) Classification of organic solvents and modelling of their physico-chemical properties by chemometric methods using different sets of molecular descriptors. TRAC 18:461–471
96. Gramatica P, Corradi M, Consonni V (2000) Modelling and prediction of soil sorption coefficients of non-ionic organic pesticides by molecular descriptors. Chemosphere 41:763–777
97. Sabljic A, Gusten H, Verhaar H et al. (1995) QSAR modeling of soil sorption – improvements and systematics of Log K_{oc} vs Log K_{ow} correlations. Chemosphere 31:4489–4514
98. Gawlik BM, Sotiriou N, Feicht EA et al. (1997) Alternatives for the determination of the soil adsorption coefficient, K_{oc}, of non-ionicorganic compounds – a review. Chemosphere 34:2525–2551
99. Doucette WJ (2003) Quantitative structure–activity relationships for predicting soil-sediment sorption coefficients for organic chemicals. Environ Toxicol Chem 22:1771–1788
100. Tao S, Piao HS, Dawson R et al. (1999) Estimation of organic carbon normalized sorption coefficient (K_{oc}) for soils using the fragment constant method. Environ Sci Technol 33:2719–2725
101. Huuskonen J (2003) Prediction of soil sorption coefficient of a diverse set of organic chemicals from molecular structure. J Chem Inf Comput Sci 43:1457–1462
102. Huuskonen J (2003) Prediction of soil sorption coefficient of organic pesticides from the atom-type electrotopological state indices. Environ Toxicol Chem 22:816–820

103. Andersson PL, Maran U, Fara D et al. (2002) General and class specific models for prediction of soil sorption using various physicochemical descriptors. J Chem Inf Comput Sci 42:1450–1459

104. Delgrado EJ, Alderete JB, Gonzalo AJ (2003) A simple QSPR model for predicting soil sorption coefficients of polar and nonpolar organic compounds from molecular formula. J Chem Inf Comput Sci 43:1928–1932

105. EPI Suite. Ver. 3.12 (2000) Environmental Protection Agency, USA http://www.epa.gov/opptintr/exposure/docs/EPISuitedl.htm. Accessed 9 February 2007

106. Gramatica P, Consonni V, Todeschini R (1999) QSAR study on the tropospheric degradation of organic compounds. Chemosphere 38:1371–1378

107. Atkinson R (1987) A structure-activity relationship for the estimation of rate constants for the gas-phase reactions of OH radicals with organic compounds. Int J Chem Kinet 19:799–828

108. Sabljic A, Gusten H (1990) Predicting the nighttime NO_3 radical reactivity in the troposphere. Atmos Environ A-General Topics 24:73–78

109. Müller M, Klein W (1991) Estimating atmospheric degradation processes by SARS. Sci Total Environ 109:261–273

110. Medven Z, Gusten H, Sabljic A (1996) Comparative QSAR study on hydroxyl radical reactivity with unsaturated hydrocarbons: PLS versus MLR. J Chemom 10:135–147

111. Klamt A (1996) Estimation of gas-phase hydroxyl radical rate constants of oxygenated compounds based on molecular orbital calculations. Chemosphere 32:717–726

112. Bakken G, Jurs PC (1999) Prediction of hydroxyl radical rate constants from molecular structure. J Chem Inf Comput Sci 39:1064–1075

113. Güsten H (1999) Predicting the abiotic degradability of organic pollutants in the troposphere. Chemosphere 38:1361–1370

114. Pompe M, Veber M (2001) Prediction of rate constants for the reaction of O-3 with different organic compounds. Atmos Environ 35:3781–3788

115. Meylan WM, Howard PH (2003) A review of quantitative structure–activity relationship methods for the prediction of atmospheric oxidation of organic chemicals. Environ Toxicol Chem 22:1724–1732

116. Pompe M, Veber M, Randic M et al. (2004) Using variable and fixed topological indices for the prediction of reaction rate constants of volatile unsaturated hydrocarbons with OH radicals. Molecules 9:1160–1176

117. Oberg T (2005) A QSAR for the hydroxyl radical reaction rate constant: Validation, domain of application, and prediction. Atmos Environ 39:2189–2200

118. AOPWIN. Ver. 1.90 (2000) Environmental Protection Agency, USA

119. Devillers J, Bintein S, Domine D (1996) Comparison of BCF models based on log P. Chemosphere 33:1047–1065

120. Meylan WM, Howard PH, Boethling RS et al. (1999) Improved method for estimating bioconcentration/bioaccumulation factor from octanol/water partition coefficient. Environ Toxicol Chem 18:664–672

121. Lu XX, Tao S, Hu HY et al. (2000) Estimation of bioconcentration factors of nonionic organic compounds in fish by molecular connectivity indices and polarity correction factors. Chemosphere 41:1675–1688

122. Dearden JC, Shinnawei NM (2004) Improved prediction of fish bioconcentration factor of hydrophobic chemicals. SAR QSAR Environ Res 15:449–455

123. Dimitrov S, Dimitrova N, Parkerton T et al. (2005) Base-line model for identifying the bioaccumulation potential of chemicals. SAR QSAR Environ Res 16:531–554

124. Zhao CY, Boriani E, Chana A et al. (2008) A new hybrid system of QSAR models for predicting bioconcentration factors (BCF). Chemosphere 73:1701–1707

125. Todeschini R, Vighi M, Provenzani R et al. (1996) Modeling and prediction by using WHIM descriptors in QSAR studies: Toxicity of heterogeneous chemicals on Daphnia magna. Chemosphere 32:1527–1545

126. Gramatica P, Vighi M, Consolaro F et al. (2001) QSAR approach for the selection of congeneric compounds with a similar toxicological mode of action. Chemosphere 42:873–883

127. Benigni R, Giuliani A, Franke R et al. (2000) Quantitative structure–activity relationships of mutagenic and carcinogenic aromatic amines. Chem Rev 100:3697–3714

128. Gramatica P, Papa E, Marrocchi A et al. (2007) Quantitative structure–activity relationship modeling of polycyclic aromatic hydrocarbon mutagenicity by classification methods based on holistic theoretical molecular descriptors. Ecotoxicol Environ Saf 66:353–361

129. Shi LM, Fang H, Tong W et al. (2001) QSAR models using a large diverse set of estrogens. J Chem Inf Comput Sci 41:186–195

130. Hong H, Tong W, Fang H et al. (2002) Prediction of estrogen receptor binding for 58,000 chemicals using an integrated system of a tree-based model with structural alerts. Environ Health Perspect 110:29–36

131. Tong W, Fang H, Hong H et al. (2003) Regulatory application of SAR/QSAR for priority setting of endocrine disruptors: A perspective. Pure Appl Chem 75:2375–2388

132. Tong W, Welsh WJ, Shi LM et al. (2003) Structure–activity relationship approaches and applications. Environ Toxicol Chem 22:1680–1695

133. Fang H, Tong W, Sheehan DM (2003) QSAR models in receptor-mediated effects: the nuclear receptor superfamily. J Mol Struct (THEOCHEM) 622:113–125

134. Saliner AG, Amat L, Carbo-Dorca R et al. (2003) Molecular quantum similarity analysis of estrogenic activity. J Chem Inf Comput Sci 43:1166–1176

135. Saliner AG, Netzeva TI, Worth AP (2006) Prediction of estrogenicity: Validation of a classification model. SAR QSAR Environ Res 17:195–223

136. Coleman KP, Toscano WA, Wiese TE (2003) QSAR models of the in vitro estrogen activity of bisphenol A analogs. QSAR Comb Sci 22:78–88

137. Roncaglioni A, Novic M, Vracko M et al. (2004) Classification of potential endocrine disrupters on the basis of molecular structure using a nonlinear modeling method. J Chem Inf Comput Sci 44:300–309

138. Asikainen A, Ruuskanen J, Tuppurainen K (2003) Spectroscopic QSAR methods and self-organizing molecular field analysis for relating molecular structure and estrogenic activity. J Chem Inf Comput Sci 43:1974–1981

139. Asikainen A, Kolehmainen M, Ruuskanen J et al. (2006) Structure-based classification of active and inactive estrogenic compounds by decision tree, LVQ and kNN methods. Chemosphere 62:658–673

140. Devillers D, Marchand-Geneste N, Carpy A et al. (2006) SAR and QSAR modeling of endocrine disruptors. SAR QSAR Environ Res 17:393–412

141. Roncaglioni A, Benfenati E (2008) In silico-aided prediction of biological properties of chemicals: Oestrogen receptor-mediated effects. Chem Soc Rev 37:441–450

142. Roncaglioni A, Piclin N, Pintore M et al. (2008) Binary classification models for endocrine disrupter effects mediated through the estrogen receptor. SAR QSAR Environ Res 19:697–733

143. Kuiper GG, Lemmen JG, Carlsson B et al. (1998) Interaction of estrogenic chemicals and phytoestrogens with estrogen receptor ß. Endocrinology 139:4252–4263

144. Liu H, Yao X, Gramatica P (2009) The applications of machine learning algorithms in the modeling of estrogen-like chemicals. Comb Chem High Throughput Screen (special issue on "Machine learning for virtual screening) 12(5) (in press)

145. Joliffe IT (1986) Principal component analysis. Springer-Verlag, New York 490–496

146. Jackson JE (1991) A user's guide to principal components. John Wiley & Sons, New York

147. Gramatica P, Consolaro F, Pozzi S (2001) QSAR approach to POPs screening for atmospheric persistence. Chemosphere 43:655–664

148. Papa E, Gramatica P (2005) PBTs screening by multivariate analysis and QSAR modeling platform presented at 10th EuCheMS-DLE Intern. Conf., Rimini, Italy

149. Papa E, Gramatica P (2006) Structurally-based PBT profiler: The PBT index from molecular structure. Presented at 16th Annual Meeting SETAC-Europe, The Hague, Holland.

150. Vighi M, Gramatica P, Consolaro F et al. (2001) QSAR and chemometric approaches for setting water quality objectives for dangerous chemicals. Ecotoxicol Environ Saf 49:206–220

151. Papa E, Castiglioni S, Gramatica P et al. (2004) Screening the leaching tendency of pesticides applied in the Amu Darya Basin (Uzbekistan). Water Res 38:3485–3494

152. UNEP (2001) Stockholm convention on persistent organic pollutants. United Nations Environmental Program, Geneva, Switzerland. http://www.pops.int. Accessed 9 February 2007

CHAPTER 13

THE ROLE OF QSAR METHODOLOGY IN THE REGULATORY ASSESSMENT OF CHEMICALS

ANDREW PAUL WORTH

Institute for Health & Consumer Protection, European Commission - Joint Research Centre, Via Enrico Fermi 2749, 21027 Ispra (VA), Italy,   e-mail: Andrew.WORTH@ec.europa.eu

Abstract: The aim of this chapter is to outline the different ways in which quantitative structure–activity relationship (QSAR) methods can be used in the regulatory assessment of chemicals. The chapter draws on experience gained in the European Union in the assessment of industrial chemicals, as well as recently developed guidance for the use of QSARs within specific legislative frameworks such as REACH and the Water Framework Directive. This chapter reviews the concepts of QSAR validity, applicability, and acceptability and emphasises that the use of individual QSAR estimates is highly context-dependent, which has implications in terms of the confidence needed in the model validity. In addition to the potential use of QSAR models as stand-alone estimation methods, it is expected that QSARs will be used within the context of broader weight-of-evidence approaches, such as chemical categories and integrated testing strategies; therefore, the role of (Q)SARs within these approaches is explained. This chapter also refers to a range of freely available software tools being developed to facilitate the use of QSARs for regulatory purposes. Finally, some conclusions are drawn concerning current needs for the further development and uptake of QSARs

Keywords: REACH, Regulatory assessment, Validity, Applicability, and adequacy of QSAR

13.1.   INTRODUCTION

Regulatory programmes aimed at assessing and managing the risks of chemicals used in substances or consumer products require information on a wide range of chemical properties and effects, including physicochemical and environmental fate properties, as well as effects on human health and environmental species. In the European Union (EU), for example, information on the properties of chemicals is required under multiple pieces of legislation, including REACH [1, 2], the Biocides Directive [3], the Plant Protection Products Directive [4], the Water Framework Directive [5], and the Cosmetics Directive [6]. While these programmes differ in terms of the types and amount of information required, the risk assessments often need to be performed in the fact of numerous data gaps in hazard and exposure

T. Puzyn et al. (eds.), Recent Advances in QSAR Studies, 367–382.
DOI 10.1007/978-1-4020-9783-6_13, © Springer Science+Business Media B.V. 2010

information. For reasons of cost-effectiveness and animal welfare, these data gaps cannot be completely filled by reliance on a traditional testing approach.

As a means of overcoming this "information deficit", it is now widely recognised that a more intelligent approach to chemical safety assessment is needed. The aims of this approach are to make the assessment process more efficient, more cost-effective, more animal-friendly, and targeted to chemicals of greater concern. Such an approach needs to be based, as far as scientifically possible, on the use of alternative (e.g. in vitro) and so-called non-testing methods (in which data are derived from chemical structures alone). At present, there is no single or harmonised approach for the "intelligent" assessment of chemicals, and regulatory frameworks differ in terms of the extent to which they allow the replacement (or avoidance) of experimental (and especially animal) testing. Indeed, the question of how to optimise the integration of data obtained from multiple sources and generated by different methods is the subject of considerable debate and research [7–9]. Furthermore, the use of integrated assessment approaches will continue to generate considerable debate at the regulatory level. Thus, at present, the design and use of integrated testing and assessment approaches is still in its infancy. Nevertheless, it is clear that a range of non-testing methods will need to be included. In this chapter, the combination of different non-testing methods that generate predictions from chemical structure is deliberately referred "QSAR methodology". The aim of this chapter is to explain the extent to which QSAR methodology is currently accepted within the context of regulatory assessments, either as stand-alone methods, or within the context of chemical category assessments or integrated testing strategies. The chapter will conclude with some reflections about future steps needed for the further uptake of QSAR methodology

13.2.   BASIC CONCEPTS

A non-testing method refers to any method or approach that can be used to provide data for the assessment of chemicals without the need to perform new experimental work (although all such methods are based on the use of previously generated experimental data). The different types of formalised non-testing methods, which can be collectively referred to as "QSAR methodology", include qualitative and quantitative structure–activity relationship (i.e. SAR and QSAR) models; activity–activity relationships (AARs) and quantitative structure–activity–activity relationships (QSAARs) [10]; and expert systems [11]. SARs and QSARs, collectively referred to as (Q)SARs, are theoretical models that can be used to predict in a qualitative or quantitative manner the physicochemical, biological (e.g. toxicological), and (environmental) fate properties of compounds from a knowledge of their chemical structure. These terms are defined in Table13-1. Further explanation and illustration of QSAR concepts is given in Chapter 1.

Non-testing data can also be generated by less-formalised chemical grouping approaches, referred to as the analogue and chemical category approaches [12]. These approaches are further discussed in Chapter 7. All of these non-testing methods are based on the premise that the properties (including biological activities) of

Table 13-1. Definitions of key terms used in this chapter

Term	Definition
Structure–activity relationship (SAR)	A SAR is a qualitative relationships that relates a (sub)structure to the presence or absence of a property or activity of interest. The substructure may consist of adjacently bonded atoms or an arrangement of non-bonded atoms that are collectively associated with the property or activity
Quantitative structure–activity relationship (QSAR)	A QSAR is a mathematical model (often a statistical correlation) relating one or more quantitative parameters derived from chemical structure to a quantitative measure of a property or activity (e.g. a (eco)toxicological endpoint). The term quantitative in QSAR refers to the nature of the parameter(s) used to make the prediction, and hence the nature of the model. The presence of a quantitative parameter enables the development of a quantitative model. Thus, QSARs are quantitative models yielding a continuous or categorical result
Descriptor	A parameter used in a QSAR model

the chemical depend on its intrinsic nature and can be directly predicted from its molecular structure and inferred from the properties of similar compounds whose activities are known.

Formalised approaches, such as (Q)SARs, are best regarded as collections of data packaged in the form of models, whereas non-formalised chemical approaches are best thought of as weight-of-evidence assessments, in which QSAR methodology generally plays a role.

13.3. THE REGULATORY USE OF (Q)SAR METHODS

The regulatory assessment of chemicals involves one or more of the following procedures: (a) hazard assessment (which includes hazard identification and dose-response characterisation), possibly leading to classification and labelling; (b) exposure assessment; (c) risk assessment based on hazard and exposure assessments; and (d) the identification of persistent, bioaccumulative, and toxic (PBT) and (in the EU) very persistent and very bioaccumulative (vPvB) chemicals according to formal criteria.

To address one or more of these different regulatory goals, (Q)SAR methods can in principle be used in various ways, namely (a) to support priority setting procedures (i.e. provide the basis for further assessment work, especially involving testing); (b) to supplement the use of experimental data in weight-of-evidence approaches (i.e. to strengthen the weight-of-evidence and reduce the magnitude of standard assessment factors, e.g. by filling in some of the data gaps, by providing mechanistic information, or by supporting the evaluation of existing test data); and (c) to substitute or otherwise replace the need for experimental (especially animal) data. In practice, the ways in which (Q)SAR methods are used depend on the possibilities foreseen by the regulatory framework (some being more conservative than others) and the specific context (which includes the availability of other information and the possible consequences of relying on an incorrect prediction). Clearly, in the first two applications (a and b), the use of (Q)SAR data is more indirect and most likely not decisive in the final assessment. In the third application (c), the (Q)SAR data is used to directly replace experimental data, and is therefore more influential on the final outcome of the assessment. Accordingly, irrespective of the regulatory framework, it can be expected that the burden-of-proof needed for the third application will be greater than that needed in the first two. This is reflected in the technical guidance for REACH [13], which explicitly allows and encourages the use of (Q)SARs as a means of identifying the presence or absence of hazardous properties of the substance while at the same time minimising the costs of experimental and the use of vertebrate animals. The information that constitutes this burden-of-proof is described below (Section 13.4).

Examples of the use of (Q)SAR methodology under different regulatory programmes are provided elsewhere [14, 15]. These surveys show that (Q)SARs (and especially grouping approaches) have been used quite widely in different regulatory programmes. However, little documentation is available that captures the reasoning why a particular non-testing approach was eventually accepted or not. In addition to these examples of actual use, a number of case studies have been published, which explore possible applications of QSAR methodology [16–19].

At present, the data generated by (Q)SAR methods is most often be used to supplement experimental test data within weight-of-evidence assessments, including chemical categories and endpoint-specific integrated testing strategies (ITS). However, it is expected that (Q)SARs will be used increasingly for the direct replacement of test data, as relevant and reliable models become increasingly available, and as experience in their use becomes more widespread.

The use of (Q)SAR methods implies the need for computational tools and a structured workflow to facilitate their application [20, 21]. This is discussed further below (Section 13.5.1).

13.4.   THE VALIDITY, APPLICABILITY, AND ADEQUACY OF (Q)SARS

REACH provides a flexible framework for the use of alternative (in vitro) and non-testing methods. In principle, it is possible to use data from (Q)SAR models instead of experimental data if each of four main conditions is fulfilled: (i) the model used is shown to be scientifically valid; (ii) the model used is applicable to the chemical of interest; (iii) the prediction (result) is relevant for the regulatory purpose; and (iv) appropriate documentation on the method and result is given. Thus, multiple, overlapping conditions must be fulfilled to use a (Q)SAR prediction instead of data generated by a standard experimental test, as illustrated in Figure 13-1. Compared with the indirect and supporting use of (Q)SAR data, this can be considered a relatively high burden-of-proof. The extent to which these conditions can be relaxed for indirect uses remains to be established on the basis of experience. The following sections will explain the considerations necessary for demonstrating model validity, applicability, and adequacy and will include references to what is considered "appropriate documentation".

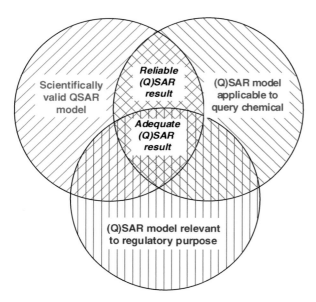

Figure 13-1. The overlapping considerations of validity, applicability, and relevance needed to demonstrate (Q)SAR adequacy

13.4.1. Demonstrating Validity

The first condition for using (Q)SARs is the demonstration of model validity. There is widespread agreement that models should be *scientifically valid* or *validated* if they are to be used in the regulatory assessment of chemicals. Since the concept of validation is incorporated into legal texts and regulatory guidelines, it is important to clearly define what it means and to describe what the practical validation process might entail.

According to the OECD Guidance Document on the Validation and International Acceptance of New or Updated Test Methods for Hazard Assessment [22], the term validation is defined as follows:

> ... the process by which the reliability and relevance of a particular approach, method, process or assessment is established for a defined purpose

This wide-ranging definition is intended to cover all kinds of traditional and alternative testing methods. In the context of (Q)SARs, this definition is rather abstract and difficult to apply. However, in the case of (Q)SARs, a set of five validation principles has been established by the OECD [23]. The OECD principles for (Q)SAR validation state that in order:

> to facilitate the consideration of a (Q)SAR model for regulatory purposes, it should be associated with the following information:
>
> 1. a defined endpoint;
> 2. an unambiguous algorithm;
> 3. a defined domain of applicability;

4. appropriate measures of goodness-of-fit, robustness and predictivity;
5. a mechanistic interpretation, if possible.

These principles were adopted by the OECD member countries and the European Commission in November 2004, following an extensive evaluation exercise [23]. The principles are based on a set of six "criteria" proposed during an international workshop on the "Regulatory Acceptance of QSARs for Human Health and Environment Endpoints", organised by the International Council of Chemical Associations (ICCA) and the European Chemical Industry Council (CEFIC) held in 2002 [24–27].

The OECD principles identify the types of information that are considered useful for the assessment of (Q)SARs for regulatory purposes. They constitute the basis of a conceptual framework, but they do not in themselves provide criteria for the regulatory acceptance of (Q)SARs. Fixed criteria are difficult, if not impossible, to define in a pragmatic way, given the highly context-dependent framework in which non-testing data may be used. The intent of each principle is explained in Table 13-2.

For the purposes of REACH, the assessment of (Q)SAR model validity should be performed by reference to the OECD principles for the validation of (Q)SARs. It

Table 13-2. Explanation of the OECD (Q)SAR validation principles

A (Q)SAR model should be. . .	Explanation
(1) Associated with a defined endpoint	Aims to ensure transparency in the endpoint being predicted by a given model, since a given endpoint could be determined by different experimental protocols and under different experimental conditions. The endpoint refers to any physicochemical property, biological effect, (human health or ecological) and environmental fate parameter that can be measured and therefore modelled
(2) Expressed in the form of an unambiguous algorithm	Aims to ensure transparency in the description of the model algorithm
(3) Associated with a defined domain of applicability	Recognises that (Q)SARs are reductionist models which are inevitably associated with limitations in terms of the types of chemical structures, physicochemical properties, and mechanisms of action for which the models can generate reliable predictions. The principle expresses the need to justify that a given model is being used within the boundary of its limitations when making a given prediction
(4) Appropriate measures of goodness-of-fit, robustness, and predictivity	Expresses the need to provide two types of information: (a) the internal performance of a model (as represented by goodness-of-fit and robustness), determined by using a training set and (b) the predictivity of a model, determined by using an appropriate test set
(5) Associated with a mechanistic interpretation, wherever possible	Aims to ensure that there is an assessment of the mechanistic associations between the descriptors used in a model and the endpoint being predicted, and that any association is documented. Where a mechanistic interpretation is possible, it can add strength to the confidence in the model already established on the basis of principles 1–4. The wording of this principle, while seemingly redundant in its use of "where possible" emphasises that is not always possible to provide a mechanistic interpretation of a given (Q)SAR

is reasonable to assume that the same principles will be used in the context of other legislation (e.g. the Water Framework Directive, Cosmetics Directive). The validation exercise itself may be carried out by any person or organisation, but it will be the industry registrant (i.e. manufacturer or importer) of the chemical who needs to argue the case for using (Q)SAR data in the context of a registration dossier. This is consistent with a key principle of REACH that the responsibility for demonstrating the safe use of chemicals lies with industry. The need to demonstrate the validity of a (Q)SAR does not necessarily imply that the model would have been validated by means of a formal validation process, such as the process that has been applied to some in vitro tests. Practical guidance has been published by the OECD on how to carry out the validation of (Q)SAR models [22]. This guidance is also summarised in the REACH guidance on Information Requirements and Chemical Safety Assessment [13].

The "appropriate documentation" for demonstrating model validity is the QSAR Model Reporting Format (QMRF), which is structured according to the OECD principles for (Q)SAR validation. Information on (Q)SAR model validity, including peer-reviewed documentation, is available from various sources, including the JRC QSAR Model Database [28].

13.4.2. Demonstrating Applicability

Assessment of model validity is a necessary but not sufficient step in assessing the acceptability of a QSAR prediction. Assuming that the model of choice is considered valid, a second essential step is to demonstrate the applicability of the model to the chemical of interest. The evaluation of model applicability is related to the evaluation of the reliability of prediction for the chemical of interest, since a valid (Q)SAR is associated with at least one defined applicability domain in which the model makes estimations with a defined level of accuracy (reliability). When applied to chemicals within its applicability domain, the model is expected to give reliable results. Conversely, if a model is applied to a chemical outside its applicability domain, it is likely that the estimated result is not sufficiently reliable for the purpose.

The applicability domain of a model [29] is a multi-faceted concept and can be broken down into (a) a descriptor domain; (b) a structural fragment domain; (c) a mechanistic domain; and (d) a metabolic domain. In other words, the reliability of a prediction is constrained by whether the chemical of interest has (a) descriptor values within predefined ranges; (b) structural fragments that are not "known" to the model; (c) its mode and mechanism of action; and (d) the likelihood that it may undergo transformation or metabolism, and the characteristics of any products.

There is no unique measure of model reliability, and no criteria for (Q)SAR reliability have been established in regulatory guidance. Model reliability should be regarded as a relative concept, depending on the context in which the model is applied. In other words, a greater or lesser degree of reliability may be sufficient for a given regulatory application. This implies that the applicability domain can be defined to suit the regulatory context.

The assessment of whether a given model is applicable to a given chemical can be broken down into the following specific questions:

1. Is the chemical of interest within the scope of the model, according to the defined applicability domain of the model?
2. Is the defined applicability domain suitable for the regulatory purpose?
3. How well does the model predict chemicals that are similar to the chemical of interest?
4. Is the model estimate reasonable, taking into account other information?

The importance of having an explicit definition of the model domain becomes apparent when addressing question 1. Unfortunately, in practice, there is often limited information concerning the descriptor, structural fragment, mechanistic domain, and metabolic domains.

The second question arises because most currently available models were not tailor-made for current regulatory needs and inevitably incorporate biases which may or may not be useful, depending on the context of prediction. A model can be biased towards certain types of chemicals (e.g. a model optimised to calculate values for those training substances that most closely matched measured ones) or towards a certain type of prediction (e.g. a model optimised to correctly identify positives at the expense of correctly identifying negatives). Such biases do not affect the validity of the model, but they do affect its applicability for specific purposes. Information on these biases can therefore help the user determine whether or how the model is suitable. For example, many QSARs for predicting biodegradation are biased towards predictions of non-ready biodegradability [30]. The predictions generated by such models may be used in a conservative manner to predict non-ready biodegradability, but predictions of biodegradability might not be reliable.

The third question provides a simple way of checking whether a model is appropriate by checking its predictive capability for one or more analogues that are similar to the one of interest and for which measured values exist. This is effectively using a read-across argument to support the reliability of the (Q)SAR prediction.

A more generic check, expressed by question 4, is whether the predicted value seems "reasonable", based on any other information available. This is thus an appeal to an expert judgement, supported with argumentation.

The judicious application of these questions to assess the applicability of a (Q)SAR model is by no means trivial, and needs specialised expertise. Software applications that generate (Q)SAR estimates vary in the extent and manner to which they express the applicability domains of the models they use. The development of methods and tools for assessing model applicability domains is a field of active research.

13.4.3.   Demonstrating Adequacy

The preceding sections explain that in order for a (Q)SAR result to be adequate for a given regulatory purpose, the estimate should be generated by a valid model and the model should be applicable to the chemical of interest with the necessary level of reliability. Fulfilment of these two conditions is necessary but not sufficient for demonstrating adequacy. Exactly what constitutes an adequate (Q)SAR result has been the subject of considerable debate, for example, in the EU Working Group on (Q)SARs, which had a mandate to advise the European Commission on QSAR-related issues between January 2005 and May 2008, prior to the entry into operation of REACH. At the time of writing, there is no detailed and firm guidance on how to demonstrate adequacy, but some general considerations are offered in the technical guidance for REACH [13]. This is perhaps an indication that more experience is needed at the regulatory level to expand on existing guidance or perhaps that the concept of adequacy, by its very nature, means that only general considerations will be possible. In any case, to demonstrate the adequacy of a QSAR estimate generated by a valid and applicable model some additional argumentation is required.

One piece of argumentation is that the model endpoint should be relevant for the regulatory purpose. For some models, in which the model predicts directly the regulatory endpoint (e.g. acute aquatic toxicity), the relevance is self-evident. However, in many QSAR models, and especially a new generation of QSAR models that are focusing on predicting lower-level mechanistic endpoints, an additional extrapolation is needed to relate the modelled endpoint (e.g. nucleophilic reactivity towards proteins) to the endpoint of regulatory interest (e.g. skin sensitisation [31]).

The relevance and reliability of a given prediction needs to be assessed in relation to a particular regulatory purpose, taking into account the availability of other information in the context of a weight-of-evidence assessment. In other words, the question being asked is whether the totality of information is sufficient to reach a regulatory conclusion, and if not, what additional information (possibly including new test data) is needed to reduce the uncertainty and increase confidence in the conclusion. This also needs to take account of the "severity" of the decision (the "principle of proportionality") as well as the possible consequences of reaching a "wrong" conclusion (principle of caution or conservativeness). Thus, the amount and quality of information that is required depends on the uncertainty in the data, the severity of the regulatory decision, and the consequence of being wrong.

It can be seen that the determination of adequacy is not only based on a scientific argument, but also on a policy decision. For this reason, it will be difficult, perhaps impossible, to develop a structured framework for assessing adequacy that would be acceptable within different regulatory frameworks.

Under REACH, there is no formal adoption process for (Q)SARs (or other non-testing methods) and there is no official list of accepted, legally binding models [32]. The (Q)SAR data is used, alongside other information reported to the authorities in the registration dossier to decide whether the information on the substance, taken as a whole, is adequate for the regulatory purpose. This process involves an initial acceptance of the data by the industry registrant and the subsequent evaluation, on a case-by-case basis, by the authorities. In accordance with the legal text, and as explained in the accompanying guidance documentation, it is necessary to demonstrate the validity of the model and its applicability to the chemical of interest, as well as the relevance and adequacy of the model estimate. The QSAR Model Reporting Format (QMRF) is the appropriate format for documenting the characteristics and validity of the model, whereas the QSAR Prediction Reporting Format (QPRF) should be used to justify the adequacy of the QSAR prediction. Further information on these reporting formats can be found elsewhere [13, 33].

13.5.   THE INTEGRATED USE OF (Q)SARS

Traditionally, the use of (Q)SARs for regulatory purposes has been conservative and generally within the context of weight-of-evidence assessments in which the (Q)SAR data is combined with other information (especially test data). This seems likely to continue, although one can imagine the use of (Q)SARs as stand-alone methods to increase as experience with, and confidence in, the methods develops. One of the achievements in recent years has been to develop structured frameworks for assessing and documenting models and their predictions, so in the future (Q)SARs should be used in a more consistent way. Another achievement has been the increasing availability of freely available software that can be used by all stakeholders in the risk assessment process. Some of this software is proprietary, such as the OECD QSAR Application Toolbox [34]. Other tools are open-source applications, for example the AMBIT software [35] developed by CEFIC and the tools developed by the European Commission's Joint Research Centre [33, 36], available to the entire scientific community for further development. An unfortunate consequence of the increasing availability of (Q)SAR models and tools could be an increasing confusion on how best to combine use of the totality of information resulting from the application of these methods, as well as data generated by test methods. The solution is to develop structured frameworks for integrating the data and for performing what is often referred to as a weight-of-evidence (or totality-of-evidence) assessment. This section will explain three approaches for integrating (Q)SAR data with other types of information: a stepwise approach for using (Q)SAR methods, chemical categories, and integrated testing strategies. These should be considered as different ways of combining and thinking about information on chemical approaches, rather than mutually exclusive approaches; most likely all three will be useful when performing a hazard or risk assessment.

13.5.1.   Stepwise Approach for Using (Q)SAR Methods

As mentioned above, the title of this chapter deliberately refers to QSAR methodology, since different (Q)SAR methods, while having different names (SARs, QSARs, QSAARs, grouping approaches, etc.) are all based on the similarity principle, and in practice, different methods are used in combination with each other to produce an overall assessment of the properties of a chemical based on its structure.

A stepwise approach for using QSAR methodology in the context of a hazard assessment has been developed by Bassan and Worth [20, 21] and incorporated into the technical guidance for REACH [13]. The different steps provide a logical approach to the compilation of a datasheet which forms the basis for an assessment of the adequacy of the non-testing data as a whole. In earlier steps, information is gained that can be used to guide the search for information in later steps. For

Table 13-3. A stepwise approach for using QSAR methods

Step	Explanation
0	Collection of existing information on substance of interest
1	Preliminary assessment of reactivity, uptake, and fate
2	Application of classification schemes for modes of action (e.g. Verhaar scheme), levels of concern (e.g. Cramer scheme), and toxicity (e.g. BfR rulebases for skin and eye irritation)
3	Application of structural alerts for toxicological effects of interest
4	Preliminary assessment of reactivity, uptake, fate, and toxicity
5	Apply read-across within chemical groups (analogue and category approaches)
6	Apply QSARs
7	Overall assessment of adequacy of non-testing data and need for additional information

example, structure-based approaches for the mechanistic [37] and toxicological [38, 39] profiling of chemicals can be used in the subsequent selection of QSARs. The approach is summarised in Table 13-3.

13.5.2.   Use of (Q)SARs in Chemical Categories

As explained in Chapter 7, a chemical category is a group of related chemicals with common properties or trends in properties. The members of category are generally related by chemical structure but may also (or additionally) be related by mode or mechanism of action. The commonalities allow for interpolations and extrapolations to be made between chemicals and endpoints, thereby enabling the filling of data gaps, by read-across, trend analysis, QSARs, and AARs. The category approach should increase the efficiency of the hazard assessment process, because it represents a departure from the traditional substance-by-substance, endpoint-by-endpoint approach.

The category approach has many synergies with the SAR and QSAR approaches, which is not surprising since all of these approaches are based on the similarity principle, namely, chemicals with similar structures exhibit similar properties [40]. For example, the identification of a suitable analogue for read-across is essentially the same as identifying a chemical with a common structural alert (SAR) or finding a nearest neighbour in a QSAR training set. Furthermore, the use of a series of category members to perform a trend analysis is essentially the same as developing a mini-QSAR. Indeed, the category approach distinguishes between "internal" QSARs, which are based on the inter-endpoint correlations within the category, and "external" QSARs, which are based on data relating to additional or different chemicals. Also not surprisingly, the same scientific methods and computational tools can be used in the development of (Q)SARs and chemical categories, as explained in [16]. Various software tools can be used to support the formation of categories and the application of read-across, including freely available tools such as the OECD QSAR toolbox [34], AMBIT [35] and Toxmatch [17, 18, 33].

13.5.3. Use of (Q)SARs in Integrated Testing Strategies

There is a growing literature on what is sometimes called "intelligent" or "integrated" testing, and in particular there are numerous proposals for integrated testing (and assessment) strategies (ITS) based on this concept. In this chapter, the term "integrated" is used in line with the technical guidance for REACH. An ITS is essentially an information-gathering and generating strategy, which does not in itself have to provide a means of using the information to address a specific regulatory question. However, it is generally assumed that some decision criteria will be applied to the information obtained in order to reach a regulatory conclusion. Normally, the totality of information would be used in a weight-of-evidence approach. An introduction to integrated testing is given by van Leeuwen et al. [7].

Many ITS have been proposed in the scientific literature (e.g. [41–47]), although relatively few have been agreed at the regulatory level and published in regulatory guidance [13]. While the details of published ITS differ, they are all based on the idea of combining the use of different testing and non-testing methods in an optimal manner, in order to (a) increase the efficiency and effectiveness of hazard and risk assessments; (b) minimise costs; and (c) reduce, replace, and refine animal testing to the extent possible, while at the same time ensuring a sufficient protection of human and environmental health. A general principle is that the need for additional testing should be limited to obtaining only essential information, rather than testing "unintelligently" and to cover all data gaps according to an indiscriminate checklist approach. Thus, testing should be focused on chemicals and properties of concern and properties that are expected to influence the regulatory decision.

An ITS is often presented in the form of flow charts with multiple decision points, to help the assessor determine when there is adequate information for a particular regulatory purpose, such as classification and labelling or risk assessment. The implementation of the ITS principles in terms of endpoint-specific ITS for a particular regulatory framework represents a significant challenge, because the ITS need to be constrained in order to reflect the information requirements of the legislation, which are often tonnage-dependent, and also subject to other adaptations (triggering or waiving of information needs) in a context-specific manner. It is also worth noting that there is rarely, if ever, a single optimised ITS – flexibility is needed to allow for different starting positions (in terms of the availability of existing information) and different options within a cost–benefit analysis framework (the cost of additional information), and thus certainty in the outcome of the assessment needs to be weighed against the additional costs of testing. For these reasons, there is a limit to how detailed any ITS can be, while at the same time being applicable across a broad range of chemicals. Thus, the application of ITS will require specialised expertise.

QSAR methodology plays a central role within ITS, since structure-based approaches form the basis of three of the six common components of ITS: chemical categories and read-across assessments; formalised models such as SAR and QSARs; and thresholds of toxicological concern (TTC) [48]. The other three common components are: exposure-based waiving (EBW) and exposure-based triggering (EBT) of tests; in vitro methods; and optimised in vivo tests.

The development and assessment of ITS for regulatory purposes is a subject of ongoing research (e.g. [9]), including research in the framework of EU-funded projects such as OSIRIS [49]. However, experience in the real-world application of ITS is needed to further develop the regulatory guidance.

13.6.   CONCLUSIONS

In this chapter, the term "QSAR methodology" is used to refer to the variety and totality of so-called "non-testing" methods, which allow predictions of chemical properties and effects to be made on the basis of chemical structure alone. Some of these methods (e.g. QSAR models) are more formalised than others (e.g. read-across), but they are all based on the principle of chemical similarity and thus rely on an appropriate grouping of chemicals according to structure and/or mechanism of action. Increasingly, the differences among SARs, QSARs, and read-across approaches are of little practical importance because current software tools apply multiple methods, which make it easy to obtain multiple predictions.

A great deal has been achieved in recent years in terms of developing and harmonising the formats which report the results of QSAR methods. This has been an essential step towards ensuring the reproducibility of predictions and transparency in their interpretation. Furthermore, an increasing and arguably overwhelming array of different computational tools are being developed to implement QSAR methods. An increasing number of these tools are being made freely available, and in some cases they are also open to the scientific community for further development. However, additional efforts are still required to extend the applicability domains and the accuracies of the underlying models and to create user-guided workflows that facilitate their integrated use.

Another important challenge remains in developing a common understanding of how best to integrate multiple predictions and existing experimental data in weight-of-evidence approaches. The way forward will be to develop a general framework that encourages transparency as well as carefully documented case studies that show how the framework has been applied to specific chemicals for specific regulatory purposes. Any attempt to develop a rigid set of acceptance criteria is unlikely to be productive because this ignores the context-dependent nature of the regulatory decision-making process. Already, it is possible to generate a huge amount of information by simply pressing a button, but this does not replace the need for expert interpretation and consensus in the regulatory use of QSAR methodology.

REFERENCES

1. European Commission (2006a) Regulation (EC) No 1907/2006 of the European Parliament and of the Council of 18 December 2006 concerning the Registration, Evaluation, Authorisation and Restriction of Chemicals (REACH), establishing a European Chemicals Agency, amending Directive 1999/45/EC and repealing Council Regulation (EEC) No 793/93 and Commission Regulation (EC) No 1488/94 as well as Council Directive 76/769/EEC and Commission Directives 91/155/EEC, 93/67/EEC, 93/105/EC and 2000/21/EC. Official Journal of the European Union,

L 396/1 of 30.12.2006. Office for Official Publications of the European Communities (OPOCE), Luxembourg

2. European Commission (2006b) Directive 2006/121/EC of the European Parliament and of the Council of 18 December 2006 amending Council Directive 67/548/EEC on the approximation of laws, regulations and administrative provisions relating to the classification, packaging and labelling of dangerous substances in order to adapt it to Regulation (EC) No 1907/2006 concerning the Registration, Evaluation, Authorisation and Restriction of Chemicals (REACH) and establishing a European Chemicals Agency. Official Journal of the European Union, L 396/850 of 30.12.2006. Office for Official Publications of the European Communities (OPOCE), Luxembourg

3. European Commission (1998) Directive 98/8/EC of the European Parliament and of the Council of 16 February 1998 concerning the placing of biocidal products on the market. Official Journal of the European Union, L 132/1 of 24.04.1998. Office for Official Publications of the European Communities (OPOCE), Luxembourg

4. European Commission (1991) Council Directive 91/414/EEC of 15 July 1991 concerning the placing of plant protection products on the market. Official Journal of the European Union, L 230/1 of 19.08.1991. Office for Official Publications of the European Communities (OPOCE), Luxembourg

5. European Commission (2000) Directive 2000/60/EC of the European Parliament and of the Council of 23 October 2000 establishing a framework for the Community action in the field of water policy. Official Journal of the European Union, L 327/1 of 22.12.2000. Office for Official Publications of the European Communities (OPOCE), Luxembourg

6. European Commission (1976) Council Directive 76/768 of 27 July 1976 on the approximation of the laws of the Member States relating to cosmetic products. Official Journal of the European Union, L 262/169 of 27.09.1976. Office for Official Publications of the European Communities (OPOCE), Luxembourg

7. Van Leeuwen CJ, Patlewicz GY, Worth AP (2007) Intelligent testing strategies. In: Van Leeuwen CJ, Vermeire TG (eds) Risk assessment of chemicals. An introduction, 2nd edn. Springer Publishers, Dordrecht, The Netherlands, pp 46–509

8. Grindon C, Combes R, Cronin MT et al. (2006) Integrated testing strategies for use in the EU REACH system. ATLA 34:407–427

9. Hoffmann S, Gallegos Saliner A, Patlewicz G et al. (2008) A feasibility study developing an integrated testing strategy assessing skin irritation potential of chemicals. Toxicol Lett 180:9–20

10. Lessigiarska IV, Worth AP, Netzeva TI et al. (2006) Quantitative structure–activity–activity and quantitative structure–activity investigations of human and rodent toxicity. Chemosphere 65: 1878–1887

11. Dearden JC, Barratt MD, Benigni R et al. (1997) The development and validation of expert systems for predicting toxicity. The report and recommendations of an ECVAM/ECB workshop (ECVAM workshop 24). ATLA 25:223–252

12. OECD (2007) Guidance on grouping of chemicals. ENV/JM/MONO(2007)28 Series on Testing and Assessment Number 80, Organisation for Economic Co-operation and Development, Paris, France. Available at: http://www.oecd.org/

13. ECHA (2008) Guidance on information requirements and chemical safety assessment. European Chemicals Agency, Helsinki, Finland. Available at:http://reach.jrc.it/docs/guidance_document/information_requirements_en.htm/

14. OECD (2007) Report on the Regulatory Uses and Applications in OECD Member Countries of (Quantitative) Structure–Activity Relationship [(Q)SAR] Models in the Assessment of New and Existing Chemicals. ENV/JM/MONO(2006)25, Organisation for Economic Co-operation and Development, Paris, France. Available at: http://www.oecd.org/

15. Worth A, Patlewicz G (eds) (2007) A Compendium of Case Studies that helped to shape the REACH guidance on chemical categories and read across. European Commission report EUR

22481 EN. Office for Official Publications of the European Communities, Luxembourg. Available at: http://ecb.jrc.ec.europa.eu/qsar/publications/

16. Worth A, Bassan A, Fabjan E et al.(2007) The use of computational methods in the grouping and assessment of chemicals – preliminary investigations. European Commission report EUR 22941 EN, Office for Official Publications of the European Communities, Luxembourg, 2007. Available at: http://ecb.jrc.ec.europa.eu/qsar/publications/

17. Patlewicz G, Jeliazkova N, Gallegos Saliner A et al. (2008) Toxmatch – a new software tool to aid in the development and evaluation of chemically similar groups. SAR QSAR Environ Res 19:397–412

18. Gallegos Saliner A, Poater A, Jeliazkova N et al. (2008) Toxmatch – A chemical classification and activity prediction tool based on similarity measures. Regul Toxicol Pharmacol 52:77–84

19. Pavan M, Worth A (2008) A set of case studies to illustrate the applicability of DART (Decision Analysis by Ranking Techniques) in the ranking of chemicals. European Commission report EUR 23481 EN, Office for Official Publications of the European Communities, Luxembourg, 2008. Available at: http://ecb.jrc.ec.europa.eu/qsar/publications/

20. Bassan A, Worth AP (2007) Computational tools for regulatory needs. In: Ekins S (ed) Computational toxicology: Risk assessment for pharmaceutical and environmental chemicals. John Wiley & Sons, Hoboken, NJ, pp 751–775

21. Bassan A, Worth AP (2008) The integrated use of models for the properties and effects of chemicals by means of a structured workflow. QSAR Comb Sci 27:6–20

22. OECD (2007) Guidance document on the validation of (quantitative) structure activity relationship [(Q)SAR] models. OECD Series on Testing and Assessment No. 69. ENV/JM/MONO(2007)2. Organisation for Economic Cooperation and Development, Paris, France. Available at: http://www.oecd.org/

23. OECD (2004) The report from the expert group on (quantitative) structure activity relationship ([Q]SARs) on the principles for the validation of (Q)SARs, Series on Testing and Assessment No. 49 (ENV/JM/MONO(2004)24). Organisation for Economic Cooperation and Development, Paris, France. Available at: http://www.oecd.org/

24. Jaworska JS, Comber M, Auer C et al. (2003) Summary of a workshop on regulatory acceptance of (Q)SARs for human health and environmental endpoints. Environ Health Perspect 111:1358–1360

25. Eriksson L, Jaworska JS, Worth AP et al. (2003) Methods for reliability, uncertainty assessment, and applicability evaluations of classification and regression based QSARs. Environ Health Perspect 111:1361–1375

26. Cronin MTD, Walker JD, Jaworska JS et al. (2003) Use of quantitative structure-activity relationships in international decision-making frameworks to predict ecologic effects and environmental fate of chemical substances. Environ Health Perspect 111:1376–1390

27. Cronin MTD, Jaworska JS, Walker J et al. (2003) Use of quantitative structure-activity relationships in international decision-making frameworks to predict health effects of chemical substances. Environ Health Perspect 111:1391–1401

28. JRC QSAR Model Database. http://qsardb.jrc.it

29. Netzeva TI, Worth AP Aldenberg T et al. (2005) Current status of methods for defining the applicability domain of (quantitative) structure-activity relationships. The report and recommendations of ECVAM workshop 52. ATLA 33:155–173

30. Pavan M, Worth AP (2008) Review of estimation models for biodegradation. QSAR Comb Sci 27:32–40

31. Patlewicz G, Aptula AO, Roberts DW, Uriarte E (2008) A minireview of available skin sensitization (Q)SARs/expert systems. QSAR Comb Sci 27:60–76

32. Worth AP, Bassan A, de Bruijn J et al. (2007) The role of the European Chemicals Bureau in promoting the regulatory use of (Q)SAR methods. SAR QSAR Environ Res 18:111–125

33. JRC QSAR Tools. http://ecb.jrc.ec.europa.eu/qsar/qsar-tools/

34. OECD QSAR Application Toolbox. http://www.oecd.org/env/existingchemicals/qsar/
35. AMBIT Software. http://sourceforge.net/projects/ambit/
36. Pavan M, Worth A (2008) Publicly-accessible QSAR software tools developed by the Joint Research Centre. SAR QSAR Environ Res 19:785–799
37. Enoch SJ, Hewitt M, Cronin MT et al. (2008) Classification of chemicals according to mechanism of aquatic toxicity: An evaluation of the implementation of the Verhaar scheme in Toxtree. Chemosphere 73:243–248
38. Benigni R, Bossa C, Jeliazkova N et al. (2008) The Benigni/Bossa rulebase for mutagenicity and carcinogenicity – a module of Toxtree, European Commission report EUR 23241 EN, Office for Official Publications of the European Communities, Luxembourg
39. Patlewicz G, Jeliazkova N, Safford RJ et al. (2008) An evaluation of the implementation of the Cramer classification scheme in the Toxtree software. SAR QSAR Environ Res 19:397–412
40. Jaworska J, Nikolova-Jeliazkova N (2007) How can structural similarity analysis help in category formation. SAR QSAR Environ Res 18:195–207
41. Combes R, Grindon C, Cronin MT et al. (2007) Proposed integrated decision-tree testing strategies for mutagenicity and carcinogenicity in relation to the EU REACH legislation. ATLA 35:267–287
42. Grindon C, Combes R, Cronin MT et al. (2007a) Integrated decision-tree testing strategies for skin corrosion and irritation with respect to the requirements of the EU REACH legislation. ATLA 35:673–682
43. Grindon C, Combes R, Cronin MT et al. (2007b) An integrated decision-tree testing strategy for skin sensitisation with respect to the requirements of the EU REACH legislation. ATLA 35:683–697
44. Combes R, Grindon C, Cronin MT et al. (2008) Integrated decision-tree testing strategies for acute systemic toxicity and toxicokinetics with respect to the requirements of the EU REACH legislation. ATLA 36:45–63
45. Grindon C, Combes R, Cronin MT et al. (2008a) Integrated decision-tree testing strategies for developmental and reproductive toxicity with respect to the requirements of the EU REACH legislation. ATLA 36:65–80
46. Grindon C, Combes R, Cronin MT et al. (2008b) An integrated decision-tree testing strategy for eye irritation with respect to the requirements of the EU REACH legislation. ATLA 36:81–92
47. Grindon C, Combes R, Cronin MT et al. (2008c) An integrated decision-tree testing strategy for repeat dose toxicity with respect to the requirements of the EU REACH legislation. ATLA 36: 93–101
48. Barlow S (2005) Threshold of toxicological concern (TTC) – A tool for assessing substances of unknown toxicity present at low levels in the diet. ILSI Europe Concise Monographs Series, International Life Sciences Institute, Brussels, 2005. Available at: http://europe.ilsi.org/publications/Monographs/
49. OSIRIS Project http://www.osiris-reach.eu/

CHAPTER 14

NANOMATERIALS – THE NEXT GREAT CHALLENGE FOR QSAR MODELERS

TOMASZ PUZYN[1], AGNIESZKA GAJEWICZ[1], DANUTA LESZCZYNSKA[2], AND JERZY LESZCZYNSKI[3]

[1]*Laboratory of Environmental Chemometrics, Faculty of Chemistry, University of Gdańsk, ul. Sobieskiego 18/19, 80-952 Gdańsk, Poland, e-mail: puzi@qsar.eu.org*
[2]*Department of Civil and Environmental Engineering, Interdisciplinary Nanotoxicity Center, Jackson State University, Jackson MS 39217-0510, USA*
[3]*Department of Chemistry, Interdisciplinary Nanotoxicity Center, Jackson State University, Jackson MS 39217-0510, USA*

Abstract: In this final chapter a new perspective for the application of QSAR in the nanosciences is discussed. The role of nanomaterials is rapidly increasing in many aspects of everyday life. This is promoting a wide range of research needs related to both the design of new materials with required properties and performing a comprehensive risk assessment of the manufactured nanoparticles. The development of nanoscience also opens new areas for QSAR modelers. We have begun this contribution with a detailed discussion on the remarkable physical–chemical properties of nanomaterials and their specific toxicities. Both these factors should be considered as potential endpoints for further nano-QSAR studies. Then, we have highlighted the status and research needs in the area of molecular descriptors applicable to nanomaterials. Finally, we have put together currently available nano-QSAR models related to the physico-chemical endpoints of nanoparticles and their activity. Although we have observed many problems (i.e., a lack of experimental data, insufficient and inadequate descriptors), we do believe that application of QSAR methodology will significantly support nanoscience in the near future. Development of reliable nano-QSARs can be considered as the next challenging task for the QSAR community.

Keywords: Nanomaterials, Nanotoxicity, Nano-QSAR

14.1.   INCREASING ROLE OF NANOMATERIALS

The history of the "nanoworld" begun on December 29, 1959, being initiated by the classic talk given at the Annual Meeting of the American Physical Society by Richard P. Feyman [1]. "There's plenty of room at the bottom" – he summarized his visionary ideas about libraries as small as a pin head and miniature machines able to penetrate human body via the blood vessel and act as microscopic surgeons. The "nano" prefix, in a chemical context, describes particles characterized by at least

T. Puzyn et al. (eds.), Recent Advances in QSAR Studies, 383–409.
DOI 10.1007/978-1-4020-9783-6_14, © Springer Science+Business Media B.V. 2010

one diameter of 100 nm or less. When nanoparticles are intentionally synthesized to be used in consumer goods, they are called "nanomaterials" [2].

Nowadays, 50 years after Feyman's lecture, nanotechnology has emerged at the forefront of science and technology developments and nanomaterials have found a wide range of applications in different aspects of human life. For example, nanoparticles of such inorganic compounds as TiO_2 and ZnO oxides are used in cosmetics [3], sunscreens [3], solar-driven self-cleaning coatings [4], and textiles [5]. Nano-sized CuO has replaced noble metals in newer catalytic converters for the car industry [6]. Nanopowders of metals can be used as antibacterial substrates (e.g., the combination of the pure nanosilver ion with fiber to create antiodor socks) [7]. Finally, metal salts (i.e., CdSe quantum dots) have found many applications in electronics and biomedical imaging techniques [8, 9].

The discoveries of fullerene (C_{60}) in 1985 by Kroto et al. [10] and carbon nanotubes in 1991 by Iijima [11] opened a new area of the tailored design of carbon-based nanomaterials. Carbon-based nanomaterials are currently used, among other applications, for synthesis of polymers characterized by enhanced solubility and processability [12] and for manufacturing of biosensors [13]. They also contribute to a broad range of environmental technologies including sorbents, high-flux membranes, depth filters, antimicrobial agents, and renewable energy supplies [14].

According to current analysis [15], about 500 different products containing nano-materials were officially on the market in 2007. Most of them (247) have been manufactured in the USA, 123 in East Asia (China, Taiwan, Korea, Japan), 76 in Europe, and only 27 in other countries. It is interesting that the number (500) is two times higher than the number of nanoproducts in the previous year. Investments in nanotechnology industry grew from $13 billion in 2004 to $50 billion in 2006 and – if one can believe the forecast – will reach $2.6 trillion in 2014.

Without doubt, nothing is able to stop such a rapidly developing branch of technology and we should be prepared for (better or worse) living day by day in symbiosis with nanomaterials.

14.2. THEIR INCREDIBLE PHYSICAL AND CHEMICAL PROPERTIES

The astonishing physical and chemical properties of engineered nanoparticles are attributable to their small size. In the nanometer-scale, finite size effects such as surface area and size distribution can cause nanoparticles to have significantly different properties as compared to the bulk material [16]. For instance, by decreasing the size of gold samples one induces color changes from bright yellow through reddish–purple up to blue.

However, from the physico-chemical viewpoint, the novel properties of nanoparticles can also be determined by their chemical composition, surface structure, solubility, shape, ratio of particles in relation to agglomerates, and surface area to volume ratio. All these factors may give rise to unique electronic, magnetic, optical, and structural properties and, therefore, lead to opportunities for using nanomaterials in novel applications and devices [16].

New, characteristic properties of nanomaterials include greater hardness, rigidity, high thermal stability, higher yield strength, flexibility, ductility, and high refractive index. The band gap of nanometer-scale semiconductor structures decreases as the size of the nanostructure decreases, raising expectations for many possible optical and photonic applications [17].

With respect to the size of the grains, it has been suggested that nanomaterials would exhibit increased (typically 3–5 times) strength and hardness as compared to their microcrystalline counterparts. For example, the strength of nanocrystalline nickel is five orders of magnitude higher than that of the corresponding microcrystalline nickel [18]. Interestingly, the observed strength of crystalline nanomaterials is accompanied by a loss of ductility, which can result in a limitation of their utility [19]. However, some of the nanocrystalline materials have the ability to undergo considerable elongation and plastic deformation without failing (even up to 100–300%). Such machinability and superplasticity properties have been observed for ceramics (including monoliths and composites), metals (including aluminum, magnesium, iron, titanium), intermetallic elements (including iron, nickel, and titanium base), and laminates [20]. Although the atomic weight of carbon nanotubes is about one-sixth of the weight of steel, their Young's modulus and tensile strength are, respectively, five and 100 times higher than those of steel [21]. In addition, nanoparticles, because of their very small sizes and surface/interface effects such as the fundamental change in coordination, symmetry, and confinement, they may exhibit high magnetic susceptibility. A variety of nanoparticles reveal anomalous magnetic properties such as superparamagnetism. This opens new areas of potential application for them, such as data storage and ferrofluid technology [22].

According to recent studies, nanoparticles may have also great potential in medical application, mostly due to their good biocompatibility that allows them to promote electron transfer between electrodes and biological molecules. For instance, the high biocompatibility of magnetite nanocrystals (Fe_3O_4) makes them potentially useful as the magnetic resonance imaging contrast agents [23]. One of the unique aspects of nanoparticles is their high wettability, termed by Fujishima [24] as superhydrophilicity. Depending upon the chemical composition, the surface can exhibit superhydrophilic characteristics. For example, titanium dioxide (TiO_2), at sizes below a few nm, can decrease the water contact angle to $0\pm1°$ [24]. Nano-sized composites, due to the chemical composition and viscosity of the intercrystalline phase, may provide a significant increase in creep resistance. It has been demonstrated that alumina/silicon carbide composites are characterized by a minimum creep rate, three times lower than the corresponding monolith [25].

14.3. NANOMATERIALS CAN BE TOXIC

As mentioned in Section 14.1, different types of nanomaterials are increasingly being developed and used by industry. However, little is known about their toxicity, including possible mutagenic and/or carcinogenic effects [26]. Some recent contributions report evident toxicity and/or ecotoxicity of selected nanoparticles and highlight the potential risk related to the development of nanoengineering. Evidently, there is insufficient knowledge regarding the harmful interactions of nanoparticles with biological systems as well as with the environment.

14.3.1.   Specific Properties Cause Specific Toxicity

It is well known that the most important parameters with respect to the induction of adverse effects by a xenobiotic compound are its dose, dimension, and durability. Conversely, it is well established that nano-sized particles, due to their unique physical and chemical properties discussed above, behave differently from their larger counterparts of the same chemical composition [26–31]. Because of the difference between nanoparticles and bulk chemicals, the risk characterization of bulk materials cannot be directly extrapolated to nanomaterials.

The biological activity of nanoparticles and their unique properties causing harmful effects are highly dependent on their size. Nanoparticles, because of their small size, may pass organ barriers such as skin, olfactory mucosa, and the blood–brain barrier [32–34], readily travel within the circulatory system of a host, and deposit in target organs. This is not possible with the same material in a larger form [35]. Indeed, reduction of the particle's size to the nanoscale level results in a steady increase of the surface to volume ratio. As a consequence, a larger number of potentially active groups per mass unit is "available" on the surface and might interact with biological systems [35]. This is one possible explanation why nano-sized particles of a given compound are generally more toxic than the same compound in its larger form [36].

However, Oberdörster et al. [37] suggested that the particle size is not the only possible factor influencing toxicity of nanomaterials. The following features should be also considered:

- size distribution,
- agglomeration state,
- shape,
- porosity,
- surface area,
- chemical composition,
- structure-dependent electronic configuration,
- surface chemistry,
- surface charge, and
- crystal structure.

Natural and anthropogenic nanoparticles gain access into the human body through the main ports of entry including the lungs, the skin, or the gastrointestinal tract. The unique properties of nanoparticles allow them not only to pnetrate physiological barriers but also to travel throughout the body and interact with subcellular structures. Toxicological studies show that nanoparticles can be found in various cells such as mitochondria [38, 39], lipid vesicles [40], fibroblasts [41], nuclei [42], and macrophages [43].

14.3.2.   Oxidative Stress

Depending on their localization inside the cell, nanoparticles can induce formation of reactive oxygen species (ROS), for instance, superoxide radicals, hydroxyl radicals reactive nitrogen [44], sulfur [45], and other species stressing the body in a similar manner to the effect of ROS [46]. This results in oxidative stress and inflammation, leading to the impacts on lung and cardiovascular health [16].

It is worth noting that normally, due to the presence of antioxidant molecules (i.e., vitamin C and glutathione), the body's cells are able to defend themselves against ROS and free radicals damage. However, when a large dose of strongly electrophilic nanoparticles enter the body, the balance between reduced glutathione (GSH) and its oxidized form (GSSG) is destroyed [47] and the unscavenged oxidants cause cell injuries by attacking DNA, proteins, and membranes [48]. At the cellular level, oxidative stress is currently the best developed paradigm depicting the harmful effects of nano-sized particles [31, 49, 50].

14.3.3.   Cytotoxicity and Genotoxicity

The mechanism of oxidative stress occurring at the molecular level is mainly responsible for observed cytotoxic and genotoxic effects induced by nanoparticles. Cytotoxicity of selected nanospecies has been confirmed by many researchers. For example, fullerene (C_{60}) particles suspended in water are characterized by antibacterial activity against *Escherichia coli* and *Bacillus subtilis* [51] and by cytotoxicity to human cell lines [52]. Single multiwalled carbon nanotubes (CWCNTs and MWCNTs) are also toxic to human cells [41, 53]. Nano-sized silicon oxide (SiO_2), anatase (TiO_2), and zinc oxide (ZnO) can induce pulmonary inflammation in rodents and humans [54–56].

Epidemiological studies have shown that nanoparticles might be genotoxic to humans [57]. Irreversible DNA modifications resulting from the activity of ROS may lead to heritable mutations, involving a single gene, a block of genes, or even whole chromosomes. DNA damage may also disrupt various normal intracellular processes, such as DNA replication and modulate gene transcription, causing abnormal function or cell death [16, 44, 58]. Until now, more than 100 different oxidative DNA lesions have been found. The most investigated OH-related DNA lesions is 8-hydroxydeoxyguanosine (8-OHdG) [59], which may be induced by several particles such as asbestos, crystalline silica, coal fly ashes. Oxygen free radicals may overwhelm the antioxidant defense system by mediating formation of base adducts, such as 8-hydroxydeoxyguanosine, and therefore play a key role in initiation of carcinogenesis [60].

14.3.4.   Neurotoxicity

Data on neurotoxic effects of engineered nanoparticles are very limited, but it has been reported that inhaled nanoparticles, depending on their size, may be distributed to organs and surrounding tissues, including the olfactory mucosa or bronchial epithelium and then can be translocated via the olfactory nerves to the central nervous system [61]. There is also some evidence that nano-sized particles can penetrate and pass along nerve axons and dendrites of neurons into the brain [33]. Recent studies confirm the translocation of nanoparticles from the respiratory tract into the central nervous system; for example, inhalation with 30 nm magnesium oxide in rats showed that manganese can be taken up into olfactory neurons and accumulated in the olfactory bulb [34].

The particles at the nanoscale may also gain access to the brain across the blood–brain barrier [2]. There is experimental evidence that oxidative stress also plays an important role in neurodegenerative diseases and brain pathology, for instance, Hallervorden-Spatz Syndrome, Pick's disease, Alzheimer's disease, or Parkinson's disease [62].

14.3.5. Immunotoxicity

The effects of nanoparticles on the immune system are still unclear. Although the reticuloendothelial system (RES) is able to eliminate nanoparticles, several toxicological studies have suggested that nanoscale particles' interaction with the defense activities of immune cells can change their antigenicity and stimulate and/or suppress immune responses. Direct experiments showed that dendritic cells and macrophages uptake of nanoparticle–protein complexes may change the formation of the antigen and initiate an autoimmune response [16]. Several studies have also reported that nanoparticles may induce damage to red blood cells (erythrocytes). Bosi et al. [63] have studied the hemolytic effect of different water-soluble C_{60} fullerenes. Preliminary results indicate that hemolytic activity depends on the number and position of the cationic surface groups. However, no clinically relevant toxicity has yet been demonstrated [64].

14.3.6. Ecotoxicity

Nano-sized particles such as volcanic ash, dust storms, or smoke from natural fires have always been present in the environment. However, the recent progress of industry has increased engineered nanoparticle pollution. The unique size-specific behavior and specific physical–chemical properties, in combination with toxicity to particular living organisms, may also result in harmful effects on the level of whole environmental ecosystems [65].

In the pioneering report on the non-human toxicity of fullerene, Eva Oberdörster [66] observed that manufactured nanomaterials can have negative impacts on aquatic organisms. Water-soluble C_{60} fullerenes cause oxidative damage (lipid peroxydation in the brain) and depletion of glutathione in the gill of juvenile largemouth bass (*Micropterus salmoides*) at a concentration of 0.5 ppm. However, these results might be disputable, because the authors used the organic solvent tetrahydrofuran (THF) to disaggregate C_{60} fullerenes, THF is classified as a neurotoxin [67].

Subsequently, Lover and Klaper [68] observed the toxicological impact of nanoparticles of fullerenes (C_{60}) and titanium dioxide (TiO_2) to *Daphnia magna*: C_{60} and TiO_2 caused mortality with a LC_{50} value of 5.5 ppm for TiO_2 and a

LC_{50} value of 460 ppb for the fullerene. In this case the authors also used THF for solubilization of hydrophobic C_{60}, thus the results are also of lower credibility. Interestingly, in similar experiments by Andrievsky et al. [69] with "fullerene water solutions" (hydrated fullerenes, $C_{60} \cdot nH_2O$), no mortality was observed.

In a later study, Adams et al. [70] confirmed the acute toxicity of selected nano-sized metal oxides against *D. magna*. He observed that SiO_2 particles were the least toxic and that toxicity increased from SiO_2 to TiO_2 to ZnO. A further study by the authors [71] showed that these three photosensitive nanoscale metal oxides in water suspensions have similar antibacterial activity to Gram-positive (*B. subtilis*) and Gram-negative (*E. coli*) bacteria (SiO_2 < TiO_2 < ZnO). All the metal oxides nanoparticles tested inhibited the growth of both Gram-positive and Gram-negative bacteria; however, *B. subtilis* was more sensitive than *E. coli*.

Similar results have been observed for a bath of ZnO, TiO_2, and CuO against bacterium *Vibrio fischeri* and crustaceans *D. magna* and *Thamnocephalus platyurus* [72]. The antibacterial effects of nano-sized metal oxides to *V. fischeri* were similar to the rank of toxicity to *D. magna* and *T. platyurus*; they increased from TiO_2 to CuO and ZnO. It is also very important to recognize that titanium dioxide was not toxic even at the 20 g/l level, which means that not all nanoparticles of metal oxides induce toxicity.

Smith et al. [73] investigated the ecotoxicological potential of single-walled carbon nanotubes (SWCNT) to rainbow trout (*Oncorhynchus mykiss*) showing that the exposure to dispersed SWCNT causes respiratory toxicity – an increase of the ventilation rate, gill pathologies, and mucus secretion. Additionally, the authors observed histological changes in the liver, brain pathology, and cellular pathologies, such as individual necrotic or apoptotic bodies, in rainbow trout exposed to 0.5 mg/l SWCNT.

Mouchet et al. [74] analyzed the acute toxicity and genotoxicity of double-walled carbon nanotubes (DWNTs) to amphibian larvae (*Xenopus laevis*). The authors did not observe any effects at concentrations between 10 and 500 mg/l. However, at the highest concentrations (500 mg/l) 85% of mortality was measured, while at the lowest concentrations (10 mg/l) reduced size and/or a cessation of growth of the larvae were observed.

Summarizing this section, there is strong evidence that chemicals, when synthesized at the nanoscale, can induce a wide range of specific toxic and ecotoxic effects. Moreover, even similar compounds from the same class can differ in toxicity. The available data on toxicity are still lacking; thus, more comprehensive and systematic studies in this area are necessary and very important.

14.4. "NANO-QSAR" – ADVANCES AND CHALLENGES

As demonstrated in this book, quantitative structure–activity relationship (QSAR) methods can play an important role in both designing new products and predicting their risk to human health and the environment. However, taking into account the specific properties of nanomaterials and their still unknown modes of toxic action, this class of compounds seems to be much more problematic for QSAR modelers than the "classic" (small, drug-like) chemicals.

14.4.1.   Description of Structure

Until now, more than 5000 different descriptors have been developed and used for the characterization of molecular structure (Chapter 3). In general, the descriptors can be classified according to their dimensionality. Constitutional descriptors, so-called "zero-dimensional," are derived directly from the formula (e.g., the number of oxygen atoms). Descriptors of bulk properties, such as n-octanol/water parti-tion coefficient or water solubility, are classified as "one-dimensional" descriptors. Topological descriptors based on the molecular graph theory are called "two-dimensional" descriptors and characterize connections between individual atoms in the molecule. "Three-dimensional" descriptors reflect properties derived from the three-dimensional structure of a molecule optimized at the appropriate level of quantum-mechanical theory. "Four-dimensional" descriptors are defined by molec-ular properties arising from interactions of the molecule with probes characterizing the surrounding space or by stereodynamic representation of a molecule, includ-ing flexibility of bonds, conformational behavior, etc. [75–79]. Only a little is known about applicability of those "traditional" descriptors for the characterization of nanostructures. Some authors [80–82] postulate that the existing descriptors are insufficient to express the specific physical and chemical properties of nanoparticles. Thus, novel and more appropriate types of the descriptors must be developed.

A group of nanoparticles is structurally diversified. In fact, this group has been defined arbitrarily in some way, taking into account size as the only criterion of the particles' membership. Therefore, structures as various as nanotubes, fullerenes, crystals, and atom clusters as well as chemical species of such different proper-ties as metals, non-metals, organic compounds, inorganic compounds, conductors, semi-conductors, and isolators were put together into one single group. Since nanoparticles are not structurally homogenous, a common mechanism of toxicity cannot be expected for all of them. In consequence, toxicity and other properties should be studied within the most appropriately chosen sub-classes of structural and physico-chemical similarity.

What is the best way to define the sub-classes? The answer might be given based on a stepwise procedure recommended by the OECD guidance document on the grouping of chemicals [83] (see also Chapter 7). Along with the guidelines, the following eight steps should be performed:

1. Development of the category hypothesis, definition, and identification of the category members. The category can be defined based on chemical similar-ity, physico-chemical properties, toxicological endpoint, and/or mechanism of action, as well as in terms of a metabolic pathway.
2. Gathering of data for each category members. All existing data should be collected for each member of the category.
3. Evaluation of available data for adequacy. The data should be carefully evaluated at this stage according to the commonly accepted protocols (i.e., according to the appropriate OECD guidance).
4. Construction of a matrix of data availability (category endpoints vs. members). The matrix is to indicate whether data are available or not.

5. Performing of a preliminary evaluation of the category and filling data gaps. The preliminary evaluation should indicate if (i) the category rationale is supported and (ii) the category is sufficiently robust for the assessment purpose (contains sufficient, relevant and reliable information).
6. Performing of additional testing (experiments). Based on the preliminary evaluation (especially evaluation of the robustness), additional experiments and group members for testing can be proposed.
7. Performing of a further assessment of the category. If new data from the additional testing are generated, the category should be revised according to the criteria from step 5.
8. Documenting of the finalized category. Finally, the category should be documented in the form of a suitable reporting format proposed by the guidance.

The currently proposed [82] working classification scheme for nanostructured particles includes nine categories:

1. spherical or compact particles;
2. high aspect ratio particles;
3. complex non-spherical particles;
4. compositionally heterogeneous particles – core surface variation;
5. compositionally heterogeneous particles – distributed variation;
6. homogeneous agglomerates;
7. heterogeneous agglomerates;
8. active particles;
9. multifunctional particles.

This classification has been adapted from the original work of Maynard and Aitken [84].

What types of structural properties should be described within the groups? As previously discussed in Section 14.3, the diameter of a nanoparticle is important, but it is not the only one possible factor influencing the mode of toxic action. The additional structural characteristics which must also be appropriately expressed are size distribution, agglomeration, shape, porosity, surface area, chemical composition, electronic configuration, surface chemistry, surface charge, and crystal structure. In contrast to the classic QSAR scheme, an entire characterization of a nanostructure may be impossible only when computational methods are employed. Novel descriptors reflecting not only molecular structure, but also supra-molecular pattern (size, shape of the nanoparticles, etc.) should be derived from both computational and experimental techniques.

The fastest and relatively easy step of characterizing the structure is the calculation of constitutional and topological descriptors. An interesting and very practical idea in this field is to replace a series of simple descriptors by one, so-called "technological attributes code" or "SMILES-like code" [85–88]. For instance, a nanoparticle of ceramic zirconium oxide, existing in bulk form and synthesized at a temperature of 800°C can be expressed by the code "Zr,O,O,CER,%E" [80]. Similar to the simplified molecular input line entry system (SMILES), the international chemical identifier (InChI) might also be used directly as a descriptor of chemical composition [89]. Another possibility is to apply descriptors derived from either molecular graph (MG) or the graphs of atomic orbitals (GAO) theory [90–92]. In the first case, vertexes in the graph represent atoms, while edges represent covalent bonds. In the second method, vertexes refer to particular atomic orbitals (1s, 2s, 2p, etc.), while edges connect the orbitals belonging to different atoms (Figure 14-1). Based on the molecular graphs, Faulon and coworkers [93–96] have developed the signature molecular descriptor approach for the characterization of fullerenes and nanotubes. The signature is a vector including extended valences of atoms derived from a set of subgraphs, following the five-step algorithm:

1. constructing of a subgraph containing all atoms and bonds that are at a distance no greater than the given signature height;
2. labeling the vertices in a canonical order;
3. constructing a tree spanning all the edges;
4. removing of all canonical labels that appear only one time;
5. writing the signature by reading the tree in a depth-first order.

The signature descriptor can be utilized not only for direct QSAR modeling, but also for calculating a range of topological indices (i.e., the Wiener index).

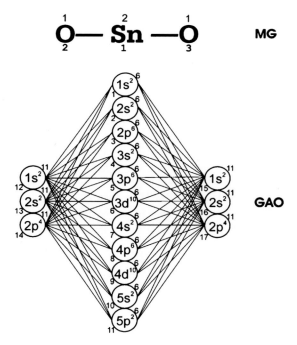

Figure 14-1. Molecular graph (MG) and graph of atomic orbitals (GAO) for SnO_2 (vertex numbering and vertex degrees). [90–92, 132]

Without doubt, simplicity of calculation is the most significant advantage of the topological descriptors. However, in many cases these two-dimensional characteristics are insufficient to investigate more complex phenomena. In such a situation, a more sophisticated approach must be employed to describe the structure appropriately. As mentioned previously, quantum-mechanical calculations can deliver useful information on the three-dimensional features (see Chapter 2). Among others, they include: molecular geometry (bond lengths, valence, and torsion angles), electron distribution, ionization potential, electron affinity, surface reactivity, and band gap. When performing quantum-mechanical calculations, there are always two important assumptions to be introduced. First one is an appropriate molecular model; the second one is the appropriate level of the theory. Both assumptions are closely related: when the model (system) is too large, the calculations at the highest levels of the theory are impossible, because of large computational time and other technical resources to be required [97].

Small fullerenes and carbon nanotubes can be treated as whole systems and modelled directly with quantum-mechanical methods. Among the theory levels, the density functional theory (DFT) recently seems to have been accepted as the most appropriate and practical choice for such calculations. Indeed, DFT methods can serve as a good alternative for conventional ab initio calculations, when a step beyond the means field approximation is crucial and the information on the electron correlation significantly improves the results (e.g., Hartree–Fock – HF method in conjunction with Møller-Pleset the second-order correction – MP2). Unfortunately, even "small" fullerenes and carbon nanotubes (containing between 40 and 70 carbon atoms) are, in fact, large from quantum-mechanical point of view. Therefore, the "classic" ab initio calculations might be impractical because of the reasons mentioned in the previous paragraph, whereas DFT can be performed in reasonable time.

The functional commonly utilized for DFT is abbreviated with the B3LYP symbol. In B3LYP calculations (Eq. 14-1) the exchange-correlation energy E_{XC} is expressed as a combination (a_0, a_X, and a_C are the parameters) of four elements: (i) the exchange-correlation energy from the local spin density approximation (LSDA, E_{XC}^{LSDA}), (ii) the difference between the exchange energy from Hartree–Fock (E_X^{HF}) and LSDA (E_X^{LSDA}), (iii) Becke's exchange energy with gradient correction (E_X^{B88}), and (iv) the correlation energy with Lee-Yang-Parr correction (E_c^{LYP}) [98, 99]:

$$E_{XC} = E_{XC}^{LSDA} + a_0(E_X^{HF} - E_X^{LSDA}) + a_X E_X^{B88} + a_C E_C^{LYP} \qquad (14\text{-}1)$$

Sometimes, when a system is too large from the quantum-mechanical point of view, the calculations are practically impossible. The situation is very common for larger crystalline nanoparticles (i.e., nanoparticles of metal oxides: TiO_2, Al_2O_3, SnO_2, ZnO, etc.) and, in such cases, a simplified model of the whole structure must first be appropriately selected. In general, there are two strategies for modeling of crystalline solids: (i) an application of the periodic boundary conditions (PBSs) and (ii) calculations based on the molecular clusters. In the first approach, calculations for a single unit cell are expanded in the three dimensions with respect to the translational symmetry by employing appropriate boundary conditions (i.e., the unit cell should be neutral and should have no dipole moment). In doing so, the model includes information on the long-range forces occurring in the crystal. However, the cell size should be large enough to also be able to model defects in the surface and to eliminate the spurious interactions between periodically repeated fragments of the lattice [100–102].

In the second approach, a small fragment or so-called "cluster," is cut off from the crystal structure and then used as a simplified model for calculations. The only problem is how to choose the diameter of the cluster correctly? This must be performed by reaching a compromise between the number of atoms (and thus the required time of computations) and the expected accuracy (and hence level of the theory to be employed). It is worth mentioning that the molecular properties can be divided into two groups depending on how they change with increasing size of the cluster (going from molecular clusters to the bulk form). They are (i) scalable properties, varying smoothly until reaching the bulk limit and (ii) non-scalable properties, when the variation related to increasing size of the cluster is not monotonic. Although the cluster models usually avoid the long-range forces, they have found many applications in modeling of local phenomena and interactions on the crystal surface [103].

As previously mentioned, in addition to calculated properties, experimentally derived properties may also serve as descriptors for developing nano-QSARs (Table 14-1). The experimental descriptors seem to be especially useful for expressing size distribution, agglomeration state, shape, porosity, and irregularity of the surface area. Interestingly, the experimental results can be combined with numerical methods to define new descriptors. For example, images taken by scanning electron microscopy (SEM), transmission electron microscopy (TEM), or atomic force

Table 14-1. Experimental properties for possible use as descriptors in nano-QSAR studies [105]

Properties	Instruments and methods*
Diameter	EM, AFM, Flow-FFF, DLS
Volume	Sed-FFF
Area	EM, AFM
Surface charge	z-Potential, electrophoretic mobility
Crystal structure	XRD, TEM-XRD
Elemental composition	Bulk: ICP-MS, ICP-OES Singe nanoparticle: TEM-EDX Particle population: FFF-ICP-MS
Aggregation state	DLS, AFM, ESEM
Hydrophobicity	Liquid–liquid extraction chromatography
Hydrodynamic diameter	Flow-FFF, DLS
Equivalent pore size diameter	Particle filtration

*Abbreviations: EM – electronic microscopy, AFM – atomic force microscopy, FFF – field flow filtration, DLS – dynamic light scattering, LC – liquid chromatography, XRD – X-ray diffraction, TEM – transmission electron microscopy, ICP-MS – inductively coupled plasma mass spectrometry, ICP-OES – inductively coupled plasma emission spectroscopy, EDX – energy dispersive X-ray spectrometry, ESEM – environmental scanning electron microscopy.

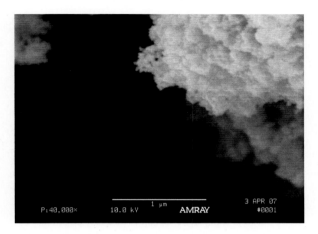

Figure 14-2. Nanopowder – SEM image of nano-sized SnO_2

microscopy (AFM) (Figure 14-2) might be processed with use of novel chemometric techniques of image analysis. Namely, a series of images for different particles of a given nanostructure should first be taken. Then, the pictures must be numerically averaged and converted into a matrix containing numerical values that correspond to intensity of each pixel in the gray scale or color value in the RGB scale. New descriptors can be defined based on the matrix (i.e., a shape descriptor can be calculated as a sum of non-zero elements in the matrix; porosity – as a sum of relative differences between each pixel and its "neighbors," etc.) [104].

Without doubt, an appropriate characterization of the nanoparticles' structure is currently one of the most challenging tasks in nano-QSAR. Although more than 5000 QSAR descriptors have been defined so far, they may be inadequate to express the supramolecular phenomena governing the unusual activity and properties of nanomaterials. As a result, much more effort in this area is required.

14.4.2. Nanostructure – Electronic Properties Relationships

An important step related to the numerical description of chemical structure and QSAR modeling involves establishing a qualitative relationship between the structure of a nanoparticle and its various electronic properties.

The B3LYP functional and the standard 6-31G(d) Polple's style basis set were applied by Shukla and Leszczynski [106] to investigate the relationships between the shape, size, and electronic properties of small carbon fullerenes, nanodisks, nanocapsules, and nanobowls. They found out that the ionization potentials decrease, while the electron affinities increase in going from the C_{60} fullerenes to the closed nanodisks, capsules, and open bowl-shaped nanocarbon clusters. In similar studies performed for capped and uncapped carbon nanotubes at the B3LYP/6-31G(d) level of theory by Yumura et al. [107, 108], the authors demonstrated that the tube lengths, edge structures, and end caps play an important role in determining the band gap expressed as a difference between the energies of the highest occupied and lowest unoccupied molecular orbitals (HOMO–LUMO) and vibrational frequencies. Wang and Mezey [109] characterized electronic structures of open-ended and capped carbon nanoneedles (CNNs) at the same theory level (B3LYP/6-311G(d)) concluding that conductivity of the studied species is strictly correlated to their size. Only very long CNNs structures have band gaps sufficiently narrow to be semiconductors, while the band gaps of very short and thin structures are too large to conduct electrons. Similarly, Poater et al. [110, 111] observed that the Parr electrophilicity and electronic movement described by the chemical potential increase with increasing length of the carbon nanoneedles and very "short" structures (containing four layers and less) have a HOMO–LUMO gap too large to allow conductivity. Moreover, Simeon et al. [112], by performing B3LYP calculations, demonstrated that a replacement of the fullerene carbon atom with a heteroatom results in a significant change of electronic and catalytic properties of the fullerene molecule.

Similar studies have been performed for crystalline metal semi-conductors with the use of the cluster calculations. As mentioned in Section 14.4.1, some electronic properties are scalable. They change with the changing size of the cluster until the bulk limit is reached. Known examples of such properties are the HOMO–LUMO gap (band gap) and the adiabatic electron detachment energy. For instance, the band gap of ZnO nanoparticles decreases with increasing diameter of the particle up to the bulk value observed for about 4 nm [113]. Similarly, the bulk limits of the HOMO–LUMO gap and the detachment energy for titanium oxide anion clusters of increasing size (increasing n) were reached already for n=7 [114, 115].

In the classic formalization of QSARs, electronic properties (e.g., HOMO, LUMO, ionization potential) have been utilized as "ordinary" molecular descriptors. As discussed above, this approach should be revised for nanoparticles, for which the properties vary with size of a particle and this variation cannot be simply described by a linear function. It is not out of the question that similar phenomena might be observed also for other types of the "traditional" descriptors and further studies in this area are required and strongly justified.

14.4.3.   Nano-QSAR Models

Regarding the five OECD principles for the validation of a (Q)SAR as discussed in Chapters 12 and 13, an ideal QSAR model, applicable for regulatory purpose, should be associated with (i) a well-defined endpoint; (ii) an unambiguous algorithm; (iii) a defined domain of applicability; (iv) appropriate measures of goodness-of-fit, robustness, and predictivity; and (v) a mechanistic interpretation, if possible. Unfortunately, it is extremely difficult to fulfill all of these principles for (Q)SARs applicable to nanomaterials. There are two main difficulties related to the development of nano-QSARs. The first one is lack of sufficiently numerous and systematic experimental data, while the second one is very limited knowledge on mechanisms of toxic action.

As we mentioned many times, regarding their structure, the class of nanomaterials is not homogenous, combining a range of physico-chemical properties, as well as possible mechanisms of metabolism and toxicity. Thus, it is impossible to assume one common applicability domain for all nanomaterials. Each mode of toxicity and each class of nanomaterials should be studied separately. Analyzing the literature data (Section 14.3) it must be concluded that even if a class of structurally similar nanoparticles is tested with the same laboratory protocol, the number of tested compounds is often insufficient to perform comprehensive internal and external validation of a model and to calculate the appropriate measures of robustness and predictivity in QSAR. For instance, Limbach et al. [116] have proposed two rankings of cytotoxicity of seven oxide nanoparticles based on the in vitro study of human and rodent cells. The rankings were as follows: (i) $Fe_2O_3 \approx$ asbestos $>$ ZnO $>$ $CeO_2 \approx ZrO_2 \approx TiO_2 \approx Ca_3(PO_4)_2$ and (ii) ZnO $>$ asbestos $\approx ZrO_2$ $>$ $Ca_3(PO_4)_2 \approx Fe_2O_3 \approx CeO_2 \approx TiO_2$, respectively, for human (mesothelioma) and rodent cells. In another paper by the same research group, the authors have found that for four metal nanoparticles – namely, TiO_2, Fe_2O_3, Mn_3O_4, and Co_3O_4 – the chemical composition was the main factor determining the formation of reactive oxygen responsible for toxicity toward human lung epithelial cells [117]. Obviously, the results cannot be combined together and a data set containing five or six compounds is too small to build an appropriately validated QSAR model.

Do these restrictions and problems mean QSAR modelers are not able to provide useful and reliable information for nanoparticles? We do not believe this to be true. The amount of data will increase along with increasing number of nanotoxicological studies. However, no one can expect the accumulation in the next few years of such extensive data for nanomaterials, as it is now available for some environmental pollutants, pharmaceuticals, and "classical" industrial chemicals [118, 119]. Despite the limitations, there are some very promising results of preliminary nano-QSAR studies which are reviewed below.

Toropov et al. [81] have developed two models defining the relationships between basic physico-chemical properties (namely, water solubility, log *S*, and n-octanol/water partition coefficient, log *P*) of carbon nanotubes and their chiral vectors (as structural descriptors). The two-element chiral vector (*n*, *m*) contains information about the process of rolling up the graphite layer when a nanotube is formed. It had been previously known [120] that the elements of the chiral vector are related to conductivity. At this point, Toropov et al. confirmed, using the QSPR-based research, that the vector is also strictly related to other properties. The models developed were defined by the following two equations (Eqs. 14-2 and 14-3):

$$\log S = -5.10 - 3.51n - 3.59m$$
$$R^2 = 0.99, \ s = 0.053, \ F = 126 \tag{14-2}$$

$$\log P = -3.92 + 3.77n - 3.60m$$
$$R^2 = 0.99, \ s = 0.37, \ F = 2.93 \tag{14-3}$$

The study was based on experimental data being available for only 16 types of carbon nanotube. To perform an external validation, the authors divided the compounds into a training set ($n=8$) and a test set ($n_{test}=8$). Statistics of the validation were $R^2_{test} = 0.99$, $s_{test}=0.093$, and $F_{test}=67.5$ and $R^2_{test}=0.99$, $s_{test}=0.29$, and $F_{test}=5.93$, respectively, for the models for water solubility and n-octanol/water partition coefficient. Without doubt, these were the first such QSPR models developed for nanoparticles. However, the ratio of descriptors to compounds (the Topliss ratio) was low, thus the model might be unstable (see discussion in Chapter 12 for more detail).

Another contribution by Toropov and Leszczynski [80] presents a model predicting Young's modulus (*YM*) for a set of inorganic nanostructures (Eq. 14-4).

$$YM = -3720.0(\pm 39.9) + 3950.0(\pm 39.2)DCW$$
$$R^2 = 0.98, \; s = 18.3, \; F = 761, \quad\quad\quad (14\text{-}4)$$
$$R^2_{test} = 0.90, \; s_{test} = 34.7, \; F_{test} = 51$$

The model was calibrated with a training set of 21 compounds and validated with eight compounds, thus the Topliss ratio in this case was satisfactory. The values of *DCW* descriptor were calculated from the Smiles-like code, according to the following equation (Eq. 14-5):

$$DCW = \prod_{k=1}^{N} CW(I_k) \quad\quad\quad (14\text{-}5)$$

where I_k is the component information on the nanostructure (e.g., Al, N, BULK, refer to Section 14.4.1), $CW(I_k)$ is the correlation weight of the component I_k, and N is the total number of these components in a given nanostructure. The values of $CW(I_k)$ were calculated by the Monte Carlo method with the software developed by the authors. The model was correctly validated and the authors demonstrated the possibility of the prediction the Young's modulus for external compounds with QSAR.

Martin et al. [121] have proposed two QSAR models predicting the solubility of buckminsterfullerene (C_{60}), respectively, in n-heptane (log $S_{heptane}$) and n-octanol (log $S_{octanol}$) (Eqs. 14-6 and 14-7):

$$\log S_{heptane} = 3.49(\pm 3.46) + 76.98(\pm 8.11)RNCG - 9.56(\pm 2.25)^2 ASIC$$
$$-1.18(\pm 0.45)E_{ee}^{min}(CC)$$
$$n = 15, \; R^2 = 0.90, \; s^2 = 0.18, \; F = 34.8,$$
$$n_{test} = 3, \; Q^2 = 0.84, \; R^2_{50} = 0.82, \; s^2_{50} = 0.35$$
$$(14\text{-}6)$$

$$\log S_{octanol} = 10.5(\pm 1.30) - 8.40 \times 10^{-2}(\pm 7.71 \times 10^{-3})^1 IC - 1.57(\pm 0.16)$$
$$E_{ee}^{min}(CC) + 0.88(\pm 0.15)RPCS$$
$$R^2 = 0.96, \; s^2 = 0.078, \; F = 97.3,$$
$$Q^2 = 0.93, \; R^2_{50} = 0.96, \; s^2_{50} = 0.10$$
$$(14\text{-}7)$$

The symbols R_{50}^2 and s_{50}^2 refer to leave-50%-out cross-validation. The authors applied CODESSA descriptors, namely, *RNCG* – relative negative charge (Zefirov's PC); ^2ASIC – average structural information content of the second order; E_{ee}^{min} (CC) – minimum exchange energy for a C–C bond; ^1IC – first-order information content; and *RPCS* – relative positive charged surface area. Interestingly, the models were calibrated on 15 compounds including 14 polycyclic aromatic hydrocarbons (PAHs) containing between two and six aromatic rings and the fullerene. Although values of solubility predicted for the fullerene seem to be reasonable, the authors did not validate the applicability domain of the models. Indeed, the structural difference between 14 hydrocarbons and the fullerene is probably too large to make reliable predictions for C_{60} (the polycyclic hydrocarbons are planar, but the fullerene is spherical). In addition, the experimental values of log S for 14 PAHs ranged from −3.80 to 0.22 in heptane and from −3.03 to −0.02 in octanol, while the experimental values for the fullerene were −4.09 and 4.18 in heptane and octanol, respectively.

An interesting area of nano-QSAR applications is estimating solubility of a given nanoparticle in a set of various solvents. In that case, the main purpose of molecular descriptors is to correctly characterize the variation in interactions between the particle and the molecules of different solvents [122]. In fact, it means that the descriptors are related to the structure of solvents rather than to the nanoparticle structure.

Murray et al. [123] have developed a model characterizing the solubility of C_{60} in 22 organic solvents by employing three following descriptors: two quantities, σ_{tot}^2 and υ reflecting variability and degree of balance of electrostatic potential on the solvent surface and the surface area, *SA* (Eq. 14-8).

$$\log(S \times 10^4) = -29.0 \left[\sigma_{\text{tot}}^2/(SA)^{3/2}\right] + 1.28 \left(\upsilon\sigma_{\text{tot}}^2\right)^{1/2} + 1.53 \times 10^{-9}(SA)^4 - 2.72$$
$$(14\text{-}8)$$

Although the model is well fitted (R=0.95, s=0.48), nothing is known about its predictive ability, because the model has not been validated.

A set of linear models built separately for individual structural domains, namely alkanes (n=6), alkyl halides (n=32), alcohols (n=6), cycloalkanes (n=6), alkylbenzenes (n=16), and aryl halides (n=9), was published by Sivaraman et al. [124]. The models were based on connectivity indices, numbers of atoms, polarizability, and variables indicating the substitution pattern as molecular descriptors for the solvents. The values of R^2 for particular models ranged between 0.93 (alkyl halides) and 0.99 (cycloalkanes) with the corresponding values of s from 0.22 (alkyl halides) to 0.04 (cycloalkanes). The authors concluded that it was impossible to obtain a unified model that included all solvents. However, when the first three classes of solvents (i.e., alkanes, alkyl halides, and alcohols) were combined together into one model, the results of an external validation performed were satisfactory.

As well as linear approaches, non-linear models have been constructed. For instance, Kiss et al. [125] applied an artificial neural network utilizing molar volume, polarizability parameter, LUMO, saturated surface, and average polarizability as structural descriptors of solvents. They observed that for most of the solvents studied (n=126) solubility decreases with increasing molar volume and increases with polarizability and the saturated surface areas of the solvents. The reported value of s in that case was 0.45 of log units. The values of R^2 and F were 0.84 and 633, respectively.

In another study [126] the authors proposed modeling with both multiple linear regression with heuristic selection of variables (HM-MLR) and a least-squares support vector machine (SVM). Then they compared both models with each other. Both models were developed with CODESSA descriptors [127]. Interestingly, the results were very similar (the model using SVM had slightly better characteristics). The values of R^2 for the linear and non-linear model were, respectively, 0.89 and 0.90, while the values of F were 968 and 1095. The reported root mean square errors were 0.126 for the linear model (HM-MLR) and 0.116 for the model employing SVM. When analyzing all the results it might be concluded that the main factor responsible for differences in the model error is related to the type of the descriptors rather than to the mathematical method of modeling.

Recently, Toropov et al. [89] developed an externally validated one-variable model for C_{60} solubility using additive optimal descriptors calculated from the International Chemical Identifier (InChI) code (Eq. 14-9):

$$\log S = -7.98(\pm 0.14) + 0.325(\pm 0.0010)\ DCW(\text{InChI})$$
$$n = 92,\ R^2 = 0.94,\ Q^2 = 0.94,\ s = 0.25,\ F = 1540, \tag{14-9}$$
$$n_{\text{test}} = 30,\ R^2_{\text{test}} = 0.94,\ s_{\text{test}} = 0.35,\ F_{\text{test}} = 437$$

The descriptor $DCW(\text{InChI})$ is defined as the sum of the correlation weights $CW(I_k)$ for individual IChI attributes I_k characterizing the solvent molecules. The example of the $DCW(\text{InChI})$ calculation is presented in Table 14-2. The values of $CW(I_k)$ were optimized by the Monte Carlo method.

Table 14-2. Illustration of the DCW calculation using pentane as an example (InChI: 1/C5H12/c1-3-5-4-2/h3-5H2,1-2H3). The value of DCW(InChI)=6.9256652 [89]

I_k	$CW(I_k)$
C5	2.0516145
H12	−0.1385480
/	−0.5043203
c1	0.9127424
-3	0.0975796
-5	0.7976968
-4	0.7174808
-2	0.6093029
/	−0.5043203
h3	0.4292022
-5	0.7976968
H2	−0.4992814
-1	0.4421542
-2	0.6093029
H3	1.1073621

All of the above models refer to physico-chemical properties as the endpoints, thus they are also termed quantitative structure–property relationships (QSPRs). Currently, there are only a small number of QSARs related directly to nano-materials' activity. In 2007 Tsakovska [128] proposed the application of QSAR methodology to predict protein–nanoparticle interactions. In 2008 Durdagi et al. published two papers [129, 130] presenting QSAR-based design of novel inhibitors of human immunodeficiency virus type 1 aspartic protease (HIV-1 PR). In the first work [130] the authors developed a three-dimensional QSAR model with compar-ative molecular similarity indices analysis (CoMSIA) method for 49 derivatives of fullerene C_{60}. The values of R^2 and Q^2 for the training set ($n=43$) were 0.99 and 0.74, respectively. The absolute values of residuals in the validation set ($n=6$) ranged from 0.25 to 0.99 logarithmic units of EC_{50} (μM). The second model [129] were characterized by lower values of the statistics ($n=17$, $R^2=0.99$ and $Q^2=0.56$). However, in that case the predictions for an external set of compounds ($n_{\text{test}}=3$) were possible with an acceptable level of error. In addition, the authors proposed nine novel structures indicating possible inhibitor activity based on the model obtained. They concluded that steric effects play the most important role in the inhibition mechanism as well as electrostatic and H-donor/acceptor proper-ties. However, the last two types of interactions are of lower importance. Similarly, SMILES-based optimal descriptors have been successfully applied for modeling HIV-1 PR fullerene-based inhibitors [131]. The model reported by Toropov et al. [131] was described by the following equation and parameters:

$$pEC50 = -31.6 + 0.125\ DCW$$
$$n = 8\ R^2 = 0.90\ Q^2 = 0.85\ s = 0.35\ F = 58\ \text{(subtraining set)}$$
$$n = 7\ R^2 = 0.52\ Rm^2 = 0.13\ s = 1.27\ F = 5\ \text{(calibration set)} \quad (14\text{-}10)$$
$$n = 5\ R^2 = 0.99\ Rm^2 = 0.96\ s = 0.18\ F = 367\ \text{(test set)}$$

Rasulev et al. [132] developed a QSAR model for the cytotoxicity to the bac-terium *E. coli* of nano-sized metal oxides. They successfully predicted the toxicity of seven compounds (namely, SnO_2, CuO, La_2O_3, Al_2O_3, Bi_2O_3, SiO_2, and V_2O_3) from the model trained on the other seven oxides (ZnO, TiO_2, Fe_2O_3, Y_2O_3, ZrO_2, In_2O_3, and Sb_2O_3). The model employing the SMILES-based descriptor DCW is given by Eq. (14-11):

$$-pLD50 = 1.32(\pm 0.031) + 0.27(\pm 0.0080)\ DCW$$
$$n = 7,\ R^2 = 0.99,\ s = 0.053,\ F = 539; \quad (14\text{-}11)$$
$$n_{\text{test}} = 7,\ R^2_{\text{test}} = 0.82,\ s_{\text{test}} = 0.241,\ F = 23$$

The DCW descriptor in this case is defined as the following (Eq. 14-12):

$$DCW = \sum_{i=1}^{N} CW(SA_k) \quad (14\text{-}12)$$

where the SA_k is a SMILES attribute, i.e., one symbol (e.g., "O," "=," "V") or two symbols (e.g., "Al," "Bi," "Cu") in the SMILES notation. Numbers of double bonds have been used as global SMILES attributes. They are denoted as "=001" and "=002." "=001" is the indicator of one double bond and "=002" is the indicator of two double bonds.

Although we strongly believe in the usefulness and appropriateness of QSAR methodology for nanomaterial studies, the number of available models related to activity and toxicity is still very limited. When analyzing the situation, it seems that the main limitation is insufficient amount of existing experimental data. In many cases, lack of data precludes an appropriate implementation of statistical methods, including necessary external validation of the model. The problem of the paucity of data will be solved only when a strict collaboration between the experimentalists and QSAR modelers is established. The role of the modelers in such studies should not be restricted only to rationalization of the data after completing the experimental part, but also they must be involved in the planning of the experimentation. Since the experiments on nanomaterials are usually expensive, a kind of compromise between the highest possible number of compounds for testing and the lowest number of compounds necessary for developing a reliable QSAR model should be reached. Regarding the limited amount of data and high costs of the experiments, the idea of applying novel read-across techniques enabling preliminary estimation of data (Chapter 7) [82, 133] is very promising. However, no one has yet tried to implement this technique to nanomaterials.

14.5. SUMMARY

Without doubt, a large and increasing aspect of the near future of chemistry and technology will be related to the development of nanomaterials. On one hand, due to their extraordinary properties, nanomaterials are becoming a chance for medicine and industry. But, on the other hand, the same properties might result in new pathways and mechanisms of toxic action. In effect, the work with nanomaterials is challenging for both "types" of chemists: those who are searching for and synthesizing new chemicals and those who are working on risk assessment and protection of humans from the effects of these chemicals.

When analyzing the current status of nano-QSAR, the four noteworthy suggestions for further work can be made:

1. There is a strong need to supplement the existing set of molecular descriptors by novel "nanodescriptors" that can represent size-dependent properties of nanomaterials.
2. A stronger than usual collaboration between the experimentalists and nano-QSAR modelers seems to be crucial. On one hand, it is necessary to produce data of higher usefulness for QSAR modelers (more compounds, more systematic experimental studies within groups of structural similarity, etc.). On the other hand, a proper characterization of the nanomaterials structure is not possible only at the theoretical (computational) level. In such situation, experiment-based structural descriptors for nano-QSAR might be required.

3. It is possible that the current criteria of the models'quality (the five OECD rules) will have to be re-evaluated and adapted to nanomaterials. This is due to the specific properties of chemicals occurring at the "nano" level (i.e., electronic properties change with changing size) and the very limited number of data (problems with the "classic" method of validation which is biased to small, low molecular weight molecules).
4. Greater effort is required in the areas of grouping nanomaterials and nano-read-across. This technique might be useful especially at the initial stage of nano-QSAR studies, when the experimental data are scarce.

In summary, the development of reliable nano-QSAR is a serious challenge that offers an exciting new direction for QSAR modelers. This task will have to be completed before the massive production of nanomaterials in order to prevent potentially hazardous molecules from being released into the environment. In the long term, prevention is always more efficient and cheaper than clean-up.

ACKNOWLEDGEMENT

Tomasz Puzyn thanks the Foundation for Polish Science for granting him with a fellowship and a research grant in frame of the HOMING Program supported by Norwegian Financial Mechanism and EEA Financial Mechanism in Poland. The authors would like to thank for support the NSF CREST Interdisciplinary Nanotoxicity Center NSF-CREST Grant No. HRD-0833178, High Performance Computational Design of Novel Materials (HPCDNM) – Contract #W912HZ-06-C-0057 and the Development of Predictive Techniques for Modeling Properties of NanoMaterials Using New OSPR/ASAR Approach Based on Optimal NanoDescriptors – Contract #W912HZ-06-C-0061 Projects funded by the Department of Defense through the U.S. Army Engineer Research and Development Center, Vicksburg, MS. This work was supported by the Polish Ministry of Science and Higher Education Grant No. DS/8430-4-0171-9.

REFERENCES

1. Feynman RP (1959) There's plenty of room at the bottom. An invitation to enter a new field of physics. The Annual Meeting of the American Physical Society, California Institute of Technology (Caltech). http://www.zyvex.com/nanotech/feynman.html. Accessed 6 January 2009
2. Borm PJ, Robbins D, Haubold S et al. (2006) The potential risks of nanomaterials: A review carried out for ECETOC. Part Fibre Toxicol 3:11
3. Serpone N, Dondi D, Albini A (2007) Inorganic and organic UV filters: Their role and efficiency in sunscreens and suncare products. Inorg Chim Acta 360:794–802
4. Cai R, Van GM, Aw PK et al. (2006) Solar-driven self-cleaning coating for a painted surface. CR Chimie 9:829–835
5. Yuranova T, Laub D, Kiwi J (2007) Synthesis activity and characterization of textiles showing self-cleaning activity under daylight irradiation. Catal Today 122:109–117
6. Zhou K, Wang R, Xu B et al. (2006) Synthesis, characterization and catalytic properties of CuO nanocrystals with various shapes. Nanonechnology 17:3939–3943

7. Consumer Products Inventory of Nanotechnology Products (2009) The Project on Emerging Nanotechnologies. http://www.nanotechproject.org/inventories/consumer/. Accessed 15 January 2009

8. Alivisatos P (2004) The use of nanocrystals in biological detection. Nat Biotechnol 22:47–52

9. Chen FQ, Gerion D (2004) Fluorescent CdSe/ZnS nanocrystal-peptide conjugates for long-term, nontoxic imaging and nuclear targeting in living cells. Nano Lett 4:1827–1832

10. Kroto HW, Heath JR, O'Brien SC et al. (1985) C60: Buckminsterfullerene. Nature 318:162–163

11. Iijima S (1991) Helical microtubules of graphitic carbon. Nature 354:56–58

12. Ravi P, Dai S, Wang C et al. (2007) Fullerene containing polymers: A review on their synthesis and supramolecular behavior in solution. J Nanosci Nanotechnol 7:1176–1196

13. Agui L, Yanez-Sedeno P, Pingarron JM (2008) Role of carbon nanotubes in electroanalytical chemistry: A review. Anal Chim Acta 622:11–47

14. Mauter MS, Elimelech M (2008) Environmental applications of carbon-based nanomaterials. Environ Sci Technol 42:5843–5859

15. Chatterjee R (2008) The challenge of regulating nanomaterials. Environ Sci Technol 42:339–343

16. Nel A, Xia T, Madler L et al. (2006) Toxic potential of materials at the nanolevel. Science 311:622–627

17. Li M, Li JC (2006) Size effects on the band-gap of semiconductor compounds. Mater Lett 60:2526–2529

18. Wang N, Wang Z, Aust KT et al. (1997) Room temperature creep behavior of nanocrystalline nickel produced by an electrodeposition technique. Mater Sci Eng A237:150–158

19. Wang Y, Chen M, Zhou F et al. (2002) High tensile ductility in a nanostructured metal. Nature 419:912–915

20. Xing HL, Wang CW, Zhang KF et al. (2004) Recent development in the mechanics of superplasticity and its applications. J Mater Process Technol 151:196–202

21. Mamalis AG (2007) Recent advances in nanotechnology. J Mater Process Technol 181:52–58

22. Sun X (2004) Magnetic properties of nanoparticle assemblies. In: Schwarz JA et al. (eds) Dekker encyclopedia of nanoscience and nanotechnology. Taylor & Francis, London

23. Li Z, Wei L, Gao M et al. (2005) One-pot reaction to synthesize biocompatible magnetite nanoparticles. Adv Mater 17:1001–1005

24. Fujishima A, Rao TN, Tryk DA (2000) Titanium dioxide photocatalysis. J Photochem Photobiol C: Photochem Rev 1:1–21

25. Ohji T, Nakahira A, Hirano T et al. (2005) Tensile creep behavior of alumina/silicon carbide nanocomposite. J Am Ceram Soc 77:3259–3262

26. Dreher KL (2004) Health and environmental impact of nanotechnology: Toxicological assessment of manufactured nanoparticles. Toxicol Sci 77:3–5

27. Oberdorster G (1996) Significance of particle parameters in the evaluation of exposure-dose-response relationships of inhaled particles. Inhal Toxicol 8(Suppl):73–89

28. Borm PJ (2002) Particle toxicology: From coal mining to nanotechnology. Inhal Toxicol 14:311–324

29. Donaldson K, Stone V (2003) Current hypotheses on the mechanisms of toxicity of ultrafine particles. Ann Ist Super Sanita 39:405–410

30. Kreyling WG, Semmler M, Moller W (2004) Dosimetry and toxicology of ultrafine particles. J Aerosol Med 17:140–152

31. Oberdorster G, Oberdorster E, Oberdorster J (2005) Nanotoxicology: An emerging discipline evolving from studies of ultrafine particles. Environ Health Perspect 113:823–839

32. Semmler M, Seitz J, Erbe F et al. (2004) Long-term clearance kinetics of inhaled ultrafine insoluble iridium particles from the rat lung, including transient translocation into secondary organs. Inhal Toxicol 16:453–459

33. Oberdorster G, Sharp Z, Atudorei V et al. (2004) Translocation of inhaled ultrafine particles to the brain. Inhal Toxicol 16:437–445

34. Elder A, Gelein R, Silva V et al. (2006) Translocation of inhaled ultrafine manganese oxide particles to the central nervous system. Environ Health Perspect 114:1172–1178

35. Donaldson K, Stone V, Tran CL et al. (2004) Nanotoxicology. Occup Environ Med 61:727–728

36. SCENIHR (2006) Modified opinion on the appropriateness of the risk assessment methodology in accordance with the technical guidance documents for new and existing substances for assessing the risks of nanomaterials. Scientific Committee on Emerging and Newly Identified Health Risks. Accessed 15 April 2009

37. Oberdorster G, Maynard A, Donaldson K et al. (2005) Principles for characterizing the potential human health effects from exposure to nanomaterials: Elements of a screening strategy. Part Fibre Toxicol 2:8

38. Li N, Sioutas C, Cho A et al. (2003) Ultrafine particulate pollutants induce oxidative stress and mitochondrial damage. Environ Health Perspect 111:455–460

39. Xia T, Kovochich M, Brant J et al. (2006) Comparison of the abilities of ambient and manufactured nanoparticles to induce cellular toxicity according to an oxidative stress paradigm. Nano Lett 6:1794–1807

40. Penn A, Murphy G, Barker S et al. (2005) Combustion-derived ultrafine particles transport organic toxicants to target respiratory cells. Environ Health Perspect 113:956–963

41. Tian F, Cui D, Schwarz H et al. (2006) Cytotoxicity of single-wall carbon nanotubes on human fibroblasts. Toxicol In Vitro 20:1202–1212

42. Chen M, von Mikecz A (2005) Formation of nucleoplasmic protein aggregates impairs nuclear function in response to SiO_2 nanoparticles. Exp Cell Res 305:51–62

43. Yokoyama A, Sato Y, Nodasaka Y et al. (2005) Biological behavior of hat-stacked carbon nanofibers in the subcutaneous tissue in rats. Nano Lett 5:157–161

44. Risom L, Moller P, Loft S (2005) Oxidative stress-induced DNA damage by particulate air pollution. Mutat Res 592:119–137

45. Giles GI, Jacob C (2002) Reactive sulfur species: An emerging concept in oxidative stress. Biol Chem 383:375–388

46. Theresa F (2007) The environmental implications of nanomaterials. In: Monteiro-Riviere N, Tran C (ed) Nanonotoxicology: Characterization, dosing and health effects. Informa Healthcare, New York

47. Sies H (1997) Oxidative stress: Oxidants and antioxidants. Exp Physiol 82:291–295

48. Brown DM, Wilson MR, MacNee W et al. (2001) Size-dependent proinflammatory effects of ultrafine polystyrene particles: A role for surface area and oxidative stress in the enhanced activity of ultrafines. Toxicol Appl Pharmacol 175:191–199

49. Shvedova AA, Kisin ER, Mercer R et al. (2005) Unusual inflammatory and fibrogenic pulmonary responses to single-walled carbon nanotubes in mice. Am J Physiol Lung Cell Mol Physiol 289:L698–L708

50. Donaldson K, Stone V, Clouter A et al. (2001) Ultrafine particles. Occup Environ Med 58:211–216, 199

51. Lyon DY, Fortner JD, Sayes CM et al. (2005) Bacterial cell association and antimicrobial activity of a C60 water suspension. Environ Toxicol Chem 24:2757–2762

52. Sayes CM, Fortner JD, Guo W et al. (2004) The differential cytotoxicity of water-soluble fullerenes. Nano Lett 4:1881–1887

53. Magrez A, Kasas S, Salicio V et al. (2006) Cellular toxicity of carbon-based nanomaterials. Nano Lett 6:1121–1125

54. Gordon T, Chen LC, Fine JM et al. (1992) Pulmonary effects of inhaled zinc oxide in human subjects, guinea pigs, rats, and rabbits. Am Ind Hyg Assoc J 53:503–509

55. Rehn B, Seiler F, Rehn S et al. (2003) Investigations on the inflammatory and genotoxic lung effects of two types of titanium dioxide: Untreated and surface treated. Toxicol Appl Pharmacol 189:84–95

56. Chen Y, Chen J, Dong J et al. (2004) Comparing study of the effect of nanosized silicon dioxide and microsized silicon dioxide on fibrogenesis in rats. Toxicol Ind Health 20:21–27

57. Donaldson K, Tran L, Jimenez LA et al. (2005) Combustion-derived nanoparticles: A review of their toxicology following inhalation exposure. Part Fibre Toxicol 2:10

58. Wiseman H, Halliwell B (1996) Damage to DNA by reactive oxygen and nitrogen species: Role in inflammatory disease and progression to cancer. Biochem J 313:17–29

59. Marnett LJ (2000) Oxyradicals and DNA damage. Carcinogenesis 21:361–370

60. Floyd RA (1990) Role of oxygen free radicals in carcinogenesis and brain ischemia. FASEB J 4:2587–2597

61. Wang B, Feng W, Zhu M et al. (2009) Neurotoxicity of low-dose repeatedly intranasal instillation of nano- and submicron-sized ferric oxide particles in mice. J Nanopart Res 11:41–53

62. Swaiman KF (1991) Hallervorden-spatz syndrome and brain iron metabolism. Arch Neurol 48:1285–1293

63. Bosi S, Feruglio L, Da Ros T et al. (2004) Hemolytic effects of water-soluble fullerene derivatives. J Med Chem 47:6711–6715

64. Dobrovolskaia MA, Aggarwal P, Hall JB et al. (2008) Preclinical studies to understand nanoparticle interaction with the immune system and its potential effects on nanoparticle biodistribution. Mol Pharm 5:487–495

65. Zuin S (2007) Effect-oriented physicochemical characterization of nanomaterials. In: Monteiro-Riviere NA, Tran CL (ed) Nanonotoxicology: Characterization, dosing and health effects. Informa Healthcare, New York

66. Oberdorster E (2004) Manufactured nanomaterials (fullerenes, C60) induce oxidative stress in the brain of juvenile largemouth bass. Environ Health Perspect 112:1058–1062

67. Andrievsky G, Derevyanchenko L, Klochkov V (2007) The myth about toxicity of pure fullerenes is irreversibly destroyed. Eighth Biennial Workshop "Fullerenes and Atomic Clusters" IWFAC'2007, St. Petersburg, Russia. http://www.fullwater.com.ua/Veb%20picture/End%20of%20Myth.htm. Accessed 1 April 2009

68. Lovern SB, Klaper R (2006) Daphnia magna mortality when exposed to titanium dioxide and fullerene (C60) nanoparticles. Environ Toxicol Chem 25:1132–1137

69. Andrievsky G, Klochkov V, Derevyanchenko L (2005) Is the C60 fullerene molecule toxic?! Fuller Nanotub Carbon Nanostruct 13:363–376

70. Adams LK, Lyon DY, McIntosh A et al. (2006a) Comparative toxicity of nano-scale TiO_2, SiO_2 and ZnO water suspensions. Water Sci Technol 54:327–334

71. Adams LK, Lyon DY, Alvarez PJ (2006b) Comparative eco-toxicity of nanoscale TiO_2, SiO_2, and ZnO water suspensions. Water Res 40:3527–3532

72. Heinlaan M, Ivask A, Blinova I et al. (2008) Toxicity of nanosized and bulk ZnO, CuO and TiO_2 to bacteria Vibrio fischeri and crustaceans Daphnia magna and thamnocephalus platyurus. Chemosphere 71:1308–1316

73. Smith CJ, Shaw BJ, Handy RD (2007) Toxicity of single walled carbon nanotubes to rainbow trout, (Oncorhynchus mykiss): Respiratory toxicity, organ pathologies, and other physiological effects. Aquat Toxicol 82:94–109

74. Mouchet F, Landois P, Sarremejean E et al. (2008) Characterisation and in vivo ecotoxicity evaluation of double-wall carbon nanotubes in larvae of the amphibian *Xenopus laevis*. Aquat Toxicol 87:127–137

75. Todeschini R, Consonni V (2000) Handbook of molecular descriptors. Wiley – VCH Verlag, Weinheim

76. Opera TI, Waller CL, Marshall GR (1994) 3D-QSAR of human immunodeficiency virus (i) protease inhibitors. III. Interpretation of COMFA results. Drug Des Discov 12:29–51

77. Ravi M, Hopfinger AJ, Hormann RE et al. (2001) 4D-QSAR analysis of a set of ecdysteroids and a comparison to COMFA modeling. J Chem Inf Comput Sci 41:1587–1604

78. Mekenyan O, Dimitrov S, Schmieder P et al. (2003) In silico modelling of hazard endpoints: Current problems and perspectives. SAR QSAR Environ Res 14:361–371

79. Kuz'min VE, Artemenko AG, Polischuk PG et al. (2005) Hierarchic system of QSAR models (1d-4d) on the base of simplex representation of molecular structure. J Mol Model 11:457–467

80. Toropov AA, Leszczynski J (2007) A new approach to the characterization of nanomaterials: Predicting young's modulus by correlation weighting of nanomaterials codes. Chem Phys Lett 433:125–129

81. Toropov AA, Leszczynska D, Leszczynski J (2007) Predicting water solubility and octanol water partition coefficient for carbon nanotubes based on the chiral vector. Comput Biol Chem 31:127–128

82. Worth A (2007) Computational nanotoxicology – towards a structure-activity based paradigm for investigation of the activity of nanoparticles. Icon Workshop. Towards Predicting Nano-Bio Interactions, Zurich, Switzerland. Available online at: http://ecb.jrc.it/qsar/information-sources/. Accessed on 05-2008

83. OECD (2007) Guidance document on the grouping of chemicals. Organisation of Economic Cooperation and Development. http://www.oecd.org. Accessed 15 May 2008

84. Maynard AD, Aitken R (2007) Assessing exposure to airborne nanomaterials: Current abilities and future requirements. Nanotoxicology 1:26–41

85. Toropov AA, Benfenati E (2007) Smiles as an alternative to the graph in QSAR modelling of bee toxicity. Comput Biol Chem 31:57–60

86. Toropov AA, Benfenati E (2008) Additive smiles-based optimal descriptors in QSAR modelling bee toxicity: Using rare smiles attributes to define the applicability domain. Bioorg Med Chem 16:4801–4809

87. Toropov AA, Benfenati E (2007a) Optimisation of correlation weights of smiles invariants for modelling oral quail toxicity. Eur J Med Chem 42:606–613

88. Toropov AA, Benfenati E (2007b) Smiles in QSPR/QSAR modeling: Results and perspectives. Curr Drug Discov Technol 4:77–116

89. Toropov AA, Toropova AP, Benfenati E et al. (2009) Additive in chi-based optimal descriptors: QSPR modeling of fullerene C60 solubility in organic solvents. J Math Chem. Published on line: doi: 10.1007/s10910-008-9514-0

90. Toropova AP, Toropov AA, Maksudov SK (2006) QSPR modeling mineral crystal lattice energy by optimal descriptors of the graph of atomic orbitals. Chem Phys Lett 428:183–186

91. Gutman I, Toropov AA, Toropova AP (2005) The graph of atomic orbitals and its basic properties. 1. Wiener index. MATCH Commun Math Comput Chem 53:215–224

92. Gutman I, Furtula B, Toropov AA et al. (2005) The graph of atomic orbitals and its basic properties. 2. Zagreb indices. MATCH Commun Math Comput Chem 53:225–230

93. Churchwell CJ, Rintoul MD, Martin S et al. (2004) The signature molecular descriptor. 3. Inverse-quantitative structure-activity relationship of icam-1 inhibitory peptides. J Mol Graphics Modell 22:263–273

94. Faulon JL, Churchwell CJ, Visco DP Jr (2003) The signature molecular descriptor. 2. Enumerating molecules from their extended valence sequences. J Chem Inf Comput Sci 43:721–734

95. Faulon JL, Collins MJ, Carr RD (2004) The signature molecular descriptor. 4. Canonizing molecules using extended valence sequences. J Chem Inf Comput Sci 44:427–436

96. Faulon JL, Visco DP Jr, Pophale RS (2003) The signature molecular descriptor. 1. Using extended valence sequences in QSAR and QSPR studies. J Chem Inf Comput Sci 43:707–720

97. Jensen F (1999) Introduction to computational chemistry. John Wiley & Sons, Chichester

98. Lee CW, Yang W, Parr RG (1988) Development of the colle-salvetti correlation energy formula into a functional of the electron density. Phys Rev B 37:785–789

99. Becke AD (1993) Density-functional thermochemistry. III The role of exact exchange. J Chem Phys 98:5648–5652

100. Makov G, Payne MC (1995) Periodic boundary conditions in ab initio calculations. Phys Rev B 51:4014–4022

101. Marana NL, Longo VM, Longo E et al. (2008) Electronic and structural properties of the (1010) and (1120) ZnO surfaces. J Phys Chem A 112:8958–8963

102. Beltran A, Andres J, Sambrano JR et al. (2008) Density functional theory study on the structural and electronic properties of low index rutile surfaces for $TiO_2/SnO_2/TiO_2$ and $SnO_2/TiO_2/SnO_2$ composite systems. J Phys Chem A 112:8943–8952

103. Jena P, Castleman AW (2006) Clusters: A bridge across the disciplines of physics and chemistry. Proc Natl Acad Sci USA 103:10560–10569

104. Puzyn T, Michalkova A, Gorb L et al. (2007) A new concept of molecular nanodescriptors for QSAR/QSPR studies. Seventh Southern School on Computational Chemistry and Material Science, Jackson, MS, USA. http://ccmsi.us/sscc_archive/ Accessed 21 August 2009

105. Hassellov M, Readman JW, Ranville JF et al. (2008) Nanoparticle analysis and characterization methodologies in environmental risk assessment of engineered nanoparticles. Ecotoxicology 17:344–361

106. Shukla MK, Leszczynski J (2006) A density functional theory study on the effect of shape and size on the ionization potential and electron affinity of different carbon nanostructures. Chem Phys Lett 428:317–320

107. Yumura T, Nozaki D, Hirahara K et al. (2006) Quantum-size effects in capped and uncapped carbon nanotubes. Annu Rep Prog Chem Sect C: Phys Chem 102:71–79

108. Yumura T, Nozaki D, Bandow S et al. (2005) End-cap effects on vibrational structures of finite-length carbon nanotubes. J Am Chem Soc 127:11769–11776

109. Wang JL, Mezey PG (2006) The electronic structures and properties of open-ended and capped carbon nanoneedles. J Chem Inf Model 46:801–807

110. Poater A, Gallegos Saliner A, Worth A (2007) Modelling nanoneedles: A journey towards nanomedicine. Second Nanotoxicology Conference, Venice, Italy. Available online at: http://ecb.jrc.it/qsar/information-sources/. Accessed 15 May 2008

111. Poater A, Saliner AG, Carbo-Dorca R et al. (2009) Modeling the structure-property relationships of nanoneedles: A journey toward nanomedicine. J Comput Chem 30:275–284

112. Simeon TM, Yanov I, Leszczynski J (2005) Ab initio quantum chemical studies of fullerene molecules with substituents $C_{59}X$ X=Si, Ge, Sn., $C_{59}X^-$ X=B, Al, Ga, In., and $C_{59}X^+$N, P, As, Sb. Int J Quantum Chem 105:429–436

113. Kukreja LM, Barik S, Misra P (2004) Variable band gap ZnO nanostructures grown by pulsed laser deposition. J Cryst Growth 268:531–535

114. Qu ZW, Kroes GJ (2006) Theoretical study of the electronic structure and stability of titanium dioxide clusters $(TiO_2)_n$ with n=1-9. J Phys Chem B 110:8998–9007

115. Zhai HJ, Wang LS (2007) Probing the electronic structure and band gap evolution of titanium oxide clusters $(TiO_2)_n^-$ (n=1-10) using photoelectron spectroscopy. J Am Chem Soc 129:3022–3026

116. Limbach LK, Li Y, Grass RN et al. (2005) Oxide nanoparticle uptake in human lung fibroblasts: Effects of particle size, agglomeration, and diffusion at low concentrations. Environ Sci Technol 39:9370–9376

117. Brunner TJ, Wick P, Manser P et al. (2006) In vitro cytotoxicity of oxide nanoparticles: Comparison to asbestos, silica, and the effect of particle solubility. Environ Sci Technol 40:4374–4381

118. Mackay D, Shiu WY, Ma K-C et al. (2007) Handbook of physical-chemical properties and environmental fate for organic chemicals. Taylor & Francis, Boca Raton, London, New York

119. Toxnet toxicology data network (2009) United States National Library of Medicine. http://toxnet. nlm.nih.gov/. Accessed 1 April 2009

120. Ormsby JL, King BT (2004) Clar valence bond representation of pi-bonding in carbon nanotubes. J Org Chem 69:4287–4291

121. Martin D, Maran U, Sild S et al. (2007) QSPR modeling of solubility of polyaromatic hydrocarbons and fullerene in 1-octanol and n-heptane. J Phys Chem B 111:9853–9857

122. Marcus Y, Smith AL, Korobov MV et al. (2001) Solubility of fullerene. J Phys Chem B 105:2499–2506

123. Murray JS, Gagarin SG, Politzer P (1995) Representation of C_{60} solubilities in terms of computed molecular surface electrostatic potentials and areas. J Phys Chem 99:12081–12083

124. Sivaraman N, Srinivasan TG, Vasudeva Rao PR et al. (2001) QSPR modeling for solubility of fullerene (C_{60}) in organic solvents. J Chem Inf Comput Sci 41:1067–1074

125. Kiss IZ, Mandi G, Beck MT (2000) Artificial neural network approach to predict the solubility of C60 in various solvents. J Phys Chem A 104:8081–8088

126. Liu H, Yao X, Zhang R et al. (2005) Accurate quantitative structure–property relationship model to predict the solubility of C60 in various solvents based on a novel approach using a least-squares support vector machine. J Phys Chem B 109:10565–20571

127. Katritzky AR, Karelson M, Petrukhin R (2001–2005) CODESSA PRO. Comprehensive descriptors for structural and statistical analysis. http://www.codessa-pro.com. Accessed 15 October 2008

128. Tsakovska I, Gallegos Saliner A, Bassan A et al. (2007) Computational modelling of nanoparticles. Second Nanotoxicology Conference, Venice, Italy. Available online at: http://ecb.jrc.it/qsar/ information-sources/. Accessed 05-2008

129. Durdagi S, Mavromoustakos T, Papadopoulos MG (2008) 3D QSAR COMFA/COMSIA, molecular docking and molecular dynamics studies of fullerene-based HIV-1 pr inhibitors. Bioorg Med Chem Lett 18:6283–6289

130. Durdagi S, Mavromoustakos T, Chronakis N et al. (2008) Computational design of novel fullerene analogues as potential HIV-1 pr inhibitors: Analysis of the binding interactions between fullerene inhibitors and HIV-1 pr residues using 3D QSAR, molecular docking and molecular dynamics simulations. Bioorg Med Chem 16:9957–9974

131. Toropov AA, Toropova AP, Benfenati E et al. (2009) SMILES-based optimal descriptors: QSAR analysis of fullerene-based HIV-1 PR inhibitors by means of balance of correlations. J Comput Chem (accepted)

132. Rasulev BF, Toropov AA, Puzyn T et al. (2007) An application of Graphs of Atomic Orbitals for QSAR modeling of toxicity of metal oxides. 34th Annual Federation of Analytical Chemistry and Spectroscopy Societies, Memphis, TN, USA

133. Saliner AG, Poater A, Worth AP (2008) Toward in silico approaches for investigating the activity of nanoparticles in therapeutic development. IDrugs 11:728–732

APPENDIX A

Table A-1. Types of information included in in silico modelling approaches and reference to chapters for further reading

- Data to be modelled
    - o Pharmacological effects (Chapter 9)
    - o Toxicological effects (Chapters 7, 11, 12 and 14)
    - o Physico-chemical properties (Chapters 12 and 14)
    - o Pharmacokinetic properties governing bioavailability (Chapters 9 and 10)
    - o Environmental fate (Chapter 12)
- Chemistry
    - o Physico-chemical properties (Chapters 12 and 14)
    - o Structural properties – 2-D and 3-D (Chapters 4, 5, 8 and 14)
    - o Presence, absence and counts of atoms, fragments, sub-structures (Chapters 3 and7)
    - o Quantum and computational chemistry (Chapters 2 and 14)
- Modelling
    - o Formation of categories of "similar molecules" (Chapters 7, 13 and 14)
    - o Statistical (Chapters 5, 6 and 12)
    - o 3D/4D QSAR (Chapters 2, 4, 5, 9 and 14)
- Other issues
    - o Data quality and reliability (Chapter 11)
    - o Model and prediction reporting formats (Chapter 13)
    - o Applicability domain (Chapters 12 and 13)
    - o Robustness of model and validity of a prediction (Chapters 6 and 12)

T. Puzyn et al. (eds.), Recent Advances in QSAR Studies, 411–413.
DOI 10.1007/9978-1-4020-9783-6, © Springer Science+Business Media B.V. 2010

Table A-2. Summary of the main modelling approaches for the development of (Q)SARs and in silico techniques and where further details are available in this volume

(Q)SAR method	Chapters
Hansch analysis	9
Free-Wilson	9
Structural fragments and alerts	7, 12
Category formation and read-across	7
Linear regression analysis	5, 6, 12
Partial least squares	5, 6
Pattern recognition	6
Robust methods, outliers	6
Pharmacophores	4, 5, 9
3-D models	2, 4, 14
CoMFA	4

Table A-3. Invaluable resources for QSAR

Internet

There are obviously many Internet sites, wikis and blogs devoted to (Q)SAR, molecular modelling, drug design and predictive ADMET. Two of the most well established are

- The homepage of the International Chemoinformatics and QSAR Society: www.qsar.org – this is a good starting place for those in the field of QSAR; it also contains excellent listings of upcoming meetings and resources.
- The homepage of the Computational Chemistry List: www.ccl.net – this also contains excellent listings resources and freely downloadable software.

Journals

Papers relating to (Q)SAR are published in a very wide variety of journals from those in pure and applied chemistry to pharmacology, toxicology and risk assessment and as far as chemoinformatics and statistics. The following is a small number that is commonly used by the author; whilst the reader will hopefully find these suggestions useful, they are, by no means, an exhaustive list (see the resources section of www.qsar.org which lists over 250 journal titles).

- *Chemical Research in Toxicology*
- *Chemical Reviews*
- *Journal of Chemical Information and Modeling*
- *Journal of Enzyme Inhibition and Medicinal Chemistry*
- *Journal of Medicinal Chemistry*
- *Journal of Molecular Modelling*
- *"Molecular Informatics (formerly QSAR and Combinatorial Science)"*
- *SAR and QSAR in Environmental Research*

Books

There are many hundreds of books available in areas related to (Q)SAR. Again, the reader is referred to the resource section of www.qsar.org. A very short list is given below, clearly biased by the author's own interests and experience. Apologies are given for omission of other "favourite" or "essential" books that have not been listed.

- Cronin MTD, Livingstone DJ (eds) (2004) *Predicting Chemical Toxicity and Fate*, CRC Press, Boca Raton, FL.
- Helma C (ed) (2005) *Predictive Toxicology*, CRC Press, Boca Raton, FL.

Table A-3. (continued)

- Livingstone DJ (1995) *Data Analysis for Chemists – Application to QSAR and Chemical Product Design*, Oxford University Press, Oxford.
- Todeschini R, Consonni V (2001) *Handbook of Molecular Descriptor*. Wiley, New York.
- Triggle DJ, Taylor JB (series eds) (2006) *Comprehensive Medicinal Chemistry II – Volumes 1–8*. Elsevier, Oxford.

Software

It is well beyond the scope or possibility of this section to note individual software for use in (Q)SAR. Experienced QSAR practitioners will no doubt be familiar with many of the freely available and commercial packages available. For the novice, in addition to the resources listed on www.qsar.org and www.ccl.net, there is information in the following chapters of this book in the three key areas to formulate a (Q)SAR:

- Activity to be modelled: Pharmacology (Chapters 4, 5, 9 and 10), ADMET (Chapters 4, 7, 10, 11, 12 and 14), physico-chemical properties (Chapters 8, 12 and 14)
- Descriptor calculation (Chapters 2, 3, 4, 5 and 14)
- Statistical analysis (Chapters 5, 6 and 12)

INDEX

Printed by Books on Demand, Germany